D0753402

Textbook of Pollen Analysis

Textbook of Pollen Analysis

by

Knut Fægri
Botanical Institute, University of Bergen, Norway

and

Johs. Iversen

IV Edition

by

Knut Fægri
Peter Emil Kaland

and

Knut Krzywinski

THE BLACKBURN PRESS

Textbook of Pollen Analysis 4th Edition

Copyright © Knut Fægri, Peter Emil Kaland and Knut Krzywinski

Previous editions published in Denmark
1950, 1964, 1975 and 1989

ISBN 1-9306650-016
Library of Congress Card Number: 00-107295

All rights reserved.

No part of this book may be reproduced by any means, or
transmitted, or translated into machine language without the
written permission of the publisher.

The Blackburn Press
P.O. Box 287
Caldwell, New Jersey 07006
973-228-7077
www.BlackburnPress.com

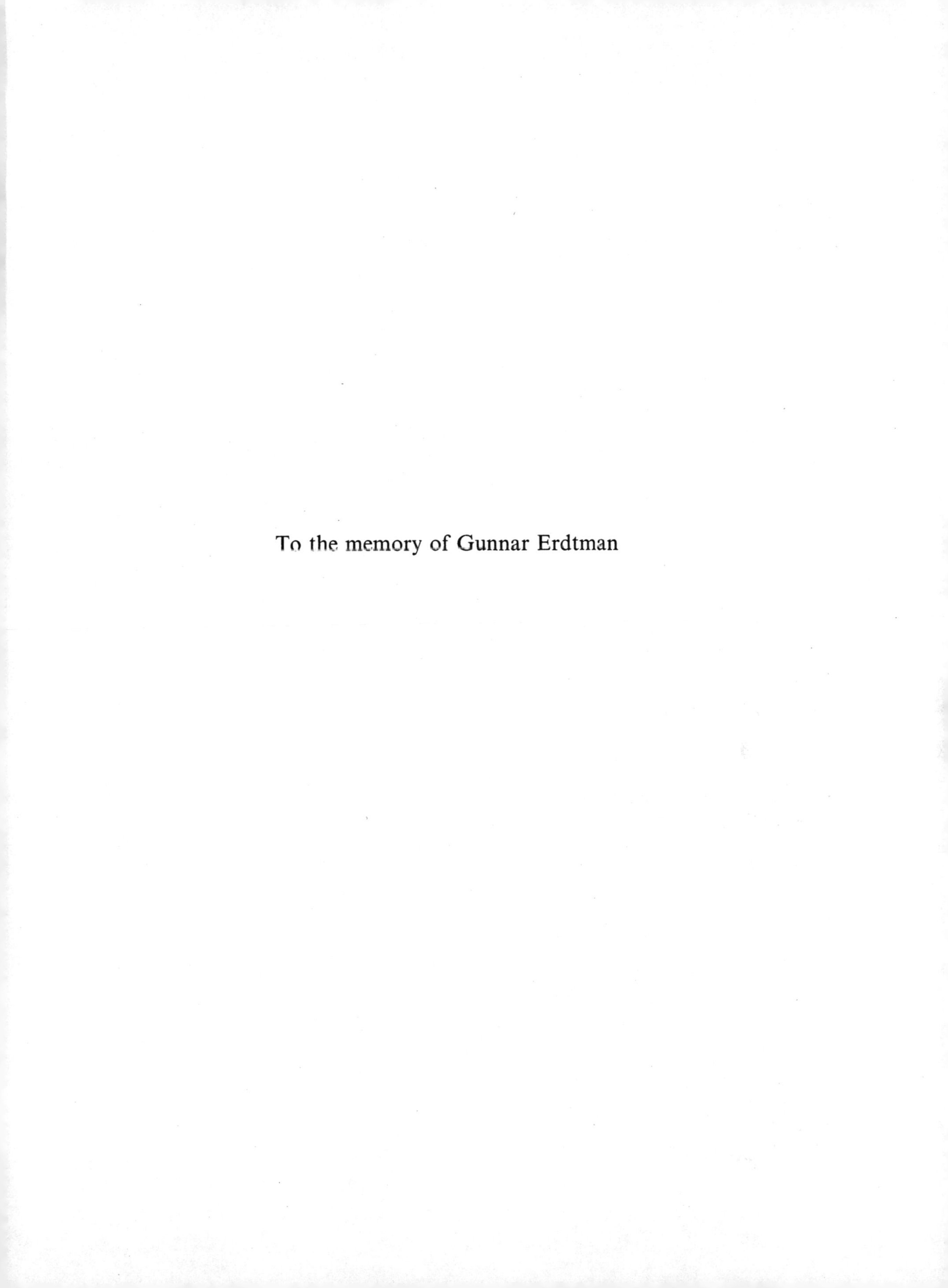

To the memory of Gunnar Erdtman

Contents

Foreword to this Printing

Apart from some minor changes and corrections, especially in the pollen identification keys and the index to same, this reprint of the Faegri-Iversen- Textbook of Pollen Analysis 4th edition by Knut Faegri, Peter Emil Kaland and Knut Krzywinski (1989) is unchanged. When the Blackburn Press first proposed that a reprint edition should be published the authors were initially skeptical of the need for reprint but eventually acquiesced, for the following reasons: (1) the fourth edition has been out of print for some years. In some respects it has been outdated by recent developments but we are persuaded that it continues to have a market and an audience. (2) None of us will have an opportunity to produce an up-to-date revision of the book in the near future. (3) We think that the core of the book is valid as it stands even if some techniques in pollen analysis have advanced beyond (if not necessarily far beyond) what was valid in 1989. A solid knowledge of the pollen grain and a sound ecological background are still and must be the fundamentals of pollen analytical research.

The most important changes in pollen analytical research have, perhaps, come in the field of statistics, a subject which has, on the whole, been ignored in this text. Sound statistics has always been the (not always realized) basis of pollen analytical research. However, there is not much in principle which differs between the application of these methods in pollen analysis and more general applications. Further, even if modern statistics can above all falsify erroneous deductions, ecological understanding is still the basis for drawing biologically correct deductions from the results of the analysis.

The very great advances in laboratory (and field) instrumentation is an area in which this book is not quite up-to-date. However, this is a matter chiefly for the larger laboratories. Smaller ones are still well served, and in many cases better served, with the methods discussed here than with more complicated, and expensive, instrumentation, the added effect of which may be marginal. New advances need specialized technical staff both in the field and in the laboratory, resources generally not available to the users of this book.

Our thanks are due to Frances Reed of the Blackburn Press for pleasant communication during the preparation of this reprint and to Anne Birgit R. Hage at the Botanical Institute who patiently has suffered through the tedious rewriting of the index to the pollen identification keys.

The authors.

Preface

The dearth of textbooks in pollen analysis was the main reason why the first edition of the Fægri–Iversen textbook was written. That situation is now a thing of the past, and one might ask if there is not a sufficient number of such texts available. Two considerations prompted the decision to publish yet another edition.

During the present authors' travels to pollen-analytic laboratories in and also outside Europe they had opportunities to study the methods employed. Methods are a function of the problem to be solved, but also of tradition, economy and the general context. During the past 40 years our own laboratory has changed a great deal. Former methods, even those of Fægri–Iversen 1974, are in many ways obsolete today.

However, we found that they—or even more traditional methods—are still in use in many laboratories, so we think that some of our present methods might be of help if introduced in other laboratories which have not yet adopted them. We shall therefore dwell at some length on them, at the same time bypassing those which we find of little value, whether developed in our laboratory or elsewhere.

The other, more pertinent, reason is that some of the existing textbooks, while often covering practical aspects well, do not penetrate into the theoretical background. Knowing how to do things is important, but even more important is to know why a thing is done, and why it is done in that particular way. Like any scientific method pollen analysis changes with time. A textbook which gives nothing more than the practices of 1988 will be obsolete in 1998. We hope that a book giving also the philosophical background will be as pertinent twenty years hence as it is today, and as it was when the first edition was published in 1950.

This does not mean that the theoretical part of the book has been static, unchanged. The rewriting which, because of Iversen's premature death, could not be properly realized in 1974, has now been carried out, thanks to my two co-authors who have been more active in the field and the laboratory during these years than I have been myself. However, it should be kept in mind that behind them, again, there is a great number of collaborators, students and technicians, who have contributed to the work and the advances made in our laboratory. Some of them are now well established elsewhere; some of them are still with us. Some of them have got their names in print; the work of others is hidden in theses or technical reports. All of them have contributed to the life and work of our laboratory and to the successes we might have achieved. Only one name should be mentioned: John Birks, who recently joined our staff, and whose advice we have sought on many occasions, especially during the later part of the preparation.

We have written this book as biologists. Pollen analysis is essentially a biological technique—it cannot be properly understood nor employed except on a biological background. The utilization of pollen-analytic data for elucidating problems in geology, archaeology, etc., presumes that biological implications are understood. Methods, however sophisticated, that do not yield results which can be translated into concrete biological terms are therefore on the whole neglected or treated very summarily. Modern numerical methods are already of importance in pollen analysis, and will no doubt be even more so in the future. The almost complete absence of numerical registrations of important calibration data is at present one of the main obstacles to a full use of such methods in practice. Also, the extreme complexity of the phenomena necessitates a multivariate approach without which we fear that the theoretical results will be less rewarding. Our aim has been to present methods and principles of immediate utility, well realizing that beyond this there is a large field that may in the (near?) future become fruitful also in practical analysis and interpretation, but which is still so much in the embryonic stage as to be outside our objective.

Some colleagues have very freely placed unpublished material at our disposal. This has been fully utilized in the preparation of the book, even if it may not be expressly quoted in the text. This help is gratefully acknowledged.

A number of our staff and students have collaborated in a series of colloquia, the object of which was to go through the pollen identification keys in the light of recent experiences.

Personally, I shall have to thank my coauthors. Without their persuasion and initial enthusiasm I should certainly not have started the work of revising this book for what is actually the fifth edition (including a Polish one), knowing how much had to be done. The whole text has been thoroughly discussed in innumerable colloquia. Kaland wrote the first drafts to the revision of chapters on laboratory technique and on the use of pollen analysis in archaeology. Krzywinski also contributed to the latter and produced the first draft of the Field technique chapter. The first drafts, of the other revisions—but certainly not the final form—were mine.

Many illustrations have been taken over from earlier editions, some of them more or less modified. Most of the illustrations made for this edition have been executed by the Institute artist, Siri Herland, whom we thank for a difficult job well done. We are grateful to many colleagues and their publishers around the world for permission to use illustrations from their books and papers, illustrations which we have in many cases redrawn more or less completely for our purpose. We thus take the responsibility for their present form. Acknowledgements for the permission to reproduce these illustrations are given in the proper place.

The first editions were dedicated to some of the pioneers of pollen analysis, Lennart von Post, Knud Jessen and Johs. Iversen (after his death). Considering the disagreements there have been—and still are—between Gunnar Erdtman and myself in many crucial matters, especially terminological, the dedication of this edition to him may cause some astonishment. For those who are astonished I refer to my obituary of Gunnar Erdtman (Fægri, 1973).

Our thanks are also due to the Institute clerical staff, especially Annechen Ree and Kari Eeg, for patiently typing and retyping no end of drafts and final manuscripts, helped also by Hilary Birks who, in addition, picked up some linguistic peculiarities here and there. Jan Berge took a great number of photographs (light and SEM microscopy) which were used for producing the illustrations, especially those accompanying the identification key.

Bergen, 1988 Knut Fægri

1. Introduction

1.1. What is pollen analysis? The principle

Basically, pollen analysis is a technique for reconstructing former vegetation by means of the pollen grains it produced. These grains have been preserved in various geologic deposits. They can be retrieved, identified and interpreted through various techniques in the field, in the laboratory and in the later stages also by statistical techniques.

Pollen analysis was born in 1916. In the beginning, the use of the technique was limited to the study of Quaternary lake and bog deposits with the aim of reconstructing Late-Quaternary changes of vegetation. Today pollen is also analysed in pre-Quaternary beds, and a wide range of deposits (marine, lacustrine, terrestric sediments, loose and consolidated) are studied. In addition, the pollen content of the air and many other substances and products is studied more or less regularly.

The identification of pollen grains to the lowest possible taxonomic level is a crucial point in pollen analysis. The size of most pollen grains is in the order of 10 to 100 μm, and the critical morphological features are often at or beyond the limit of resolution of the light microscope. Modern pollen analysis presumes the availability of first-class optical microscopes with phase contrast (and knowledge of how to use it). In advanced studies of pollen morphology (not pollen analysis) the scanning electron microscope has become an indispensable instrument.

The term *palynology* was introduced (by Hyde and Williams 1944 in a privately circulated newsletter) to cover a wider aspect, taking into account not only pollen, but other resistant microfossils as well: diatoms, cryptogam spores, animal remains, etc. Microfossils found in other contexts have also been included: air pollen, pollen found in food, etc. The term 'pollen analysis' (unqualified) is reserved for the original concept of analysis of fossil pollen in a geologic context.

Today, pollen analysis is by far the most important method for the reconstruction of past flora, vegetation and environment. Pollen is such a useful tool because: (1) pollen grains are extremely resilient and can be found in deposits in which other types of fossils have been diagenetically destroyed; (2) pollen grains are produced in enormous numbers; (3) pollen grains are more widely and more evenly spread than larger (sub)fossils and are therefore less dependent on their mother plants having been members of the community forming the deposit; (4) pollen grains can be retrieved in great quantities, limited only by the work one is willing to put into it; they can be treated statistically and quantitative variations can be adequately controlled. All these statements are subject to important qualifications and modifications, and will be discussed further in the following.

In this book the primary approach is still that of the classical pollen analysis, i.e. reconstructing the Late-Quaternary vegetation. However, properly adapted, the same methods are also applicable in a wide variety of other fields which will be dealt with in less detail, as will additional evidence obtainable from non-pollen in the same deposits.

In *vegetation research* pollen analysis adds the time dimension. Plant communities are not only an expression of the ecologic factors active today: they are also functions of a secular succession, the development of ecologic factors and of plant assemblages. It is one of the objectives of pollen analysis to define these data.

A pollen-analytic investigation is carried out to elucidate a specific problem, answer a question. Everything, from the selection of a site to the final presentation of results, depends on the character of the problem, which should therefore be clearly defined at the outset. It may vary from the total vegetation history of an extensive region to the history of a single plant species in one station, or to the vegetation contemporaneous with a single archaeological find. In the following, we shall use as the main case the classical investigation of the vegetation history of a region, an area measurable in hundreds of square kilometres. Modifications of the procedure caused by other problems will then be dealt with separately.

In this context, it is important to stress the word *vegetation* history. Pollen analysis can give information on vegetation only (cf. Fig. 1.1). Conclusions about climate, human disturbance, etc., are secondary deductions from the vegetational record and depend on the closeness of relations between the vegetation and the features studied. Thus, the interpretation of pollen-analytic data implies two different aspects: reconstructing the vegetation and, secondarily, explaining causal conditions.

The flowchart (Fig. 1.1) shows the principle of pollen analysis. The central object of the study is the vegetation, the totality of plant species and their relative quantity at the site. If we possess this information and know the ecologic demands of the individual constituents of the plant community, it is possible to draw conclusions about the climatic and ecologic conditions prevailing at the time of deposition. Changes with time can be detected, and the changes in floristic composition be related to climatic, geologic and human influences.

Figure 1.1 summarizes the events influencing a pollen-analytic registration, starting with the pollen produced by the vegetation of the area. These influences fall into three groups:

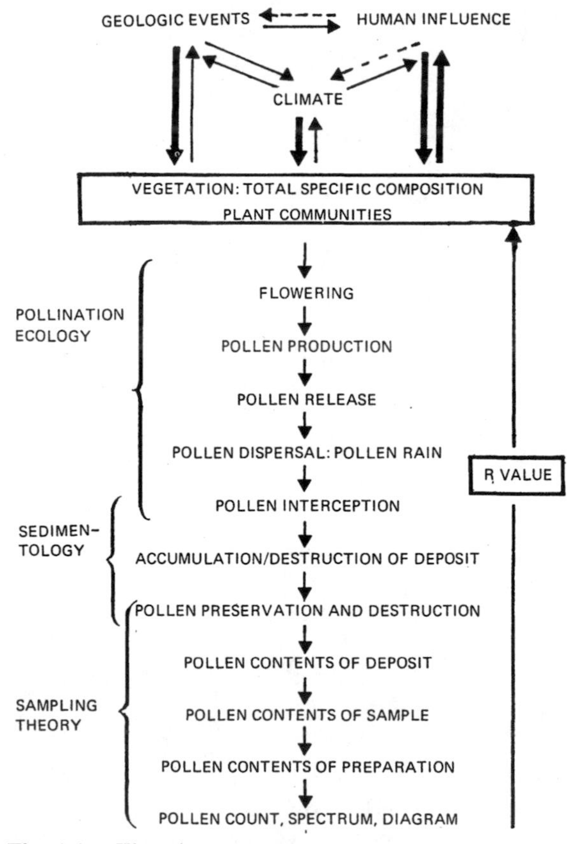

Fig. 1.1. Flowchart of pollen analysis.

1. The pollination ecology of the members of the vegetation which, together with the quantity of the species, determines the composition of the pollen flora deposited on the ground or in material under investigation.
2. The changes occurring in the deposit from the moment of deposition to that of sampling, which influence the quantitative relations between pollen types.
3. The effects of sampling procedures and analysis technique, which influence the relation between the pollen flora of the deposit and the registrations of it by analysis.

There is nothing like a one-to-one relation anywhere, and conclusions from the figures of the pollen count back to the composition of the vegetation can only be indirect, by mentally running all these processes backwards. The relation between the representation of a taxon in the pollen count and in the actual vegetation is frequently referred to as the R (for representativity?) factor, introduced by M. Davis (1963). It is a purely theoretical concept: conditions are far too complicated and variable for general R values to be established. Due to the percentage calculations, the R values, in addition to the processes included in Fig. 1.1, also depend on the pollen production of the other taxa occurring in the vegetation and competing for space not only in the field but also inside the 100% universe. Obviously, R values cannot be transferred from one spectrum to another.

It is also seen from Fig. 1.1 that if the ecologic parameters of the vegetation are not known, conclusions about climatic and other conditions cannot be drawn from the results of a pollen analysis. Nevertheless, such analyses, e.g. in pre-Quaternary deposits, can be extremely valuable in a stratigraphic context, not only as stratigraphic markers, but also as indicators of climatic similarity and potential synchroneity of deposition, even if the climatic data themselves are unknown or known *ex analogia* only, with the consequent danger of circular reasoning.

1.1.1 The basic assumption

As compared with older, macrofossil techniques, pollen analysis is based upon the assumption of the great quantity and high degree of uniformity in the dispersal of pollen grains. In contrast, the occurrence of macrofossils is always more or less accidental: most of the macrofossils found in sediments or peat are remains of the local vegetation, which is frequently climatically indifferent. Species of dry ground are heavily underrepresented or usually not represented at all. The find of a macrofossil is a fairly certain indicator that the species in question occurred in or near the locality at the time of deposition of the sample, but *nothing can be concluded from its absence*. This is the basic rule in all palaeontological work, from which pollen analysis makes an implicit exception. If a wind-pollinated species flowers in the neighbourhood, its pollen grains will be sifted over the whole region, including bogs and lakes: they will be embedded and can be recovered. If no pollen is found, there is only a very small chance that the species grew locally at this time. The absence thus is meaningful.

Although there is, in principle, no difference between identifying, on one hand, pollen grains and, on the other hand, seeds, cone-scales, etc., there is one important practical distinction: the fossils dealt with in pollen analysis are all comparable; consequently, their *relative frequencies* are comparable. Macrofossils are not. The numbers found of various macrofossils: leaf impressions, seeds, twigs, may be comparable between themselves: many seeds of taxon *x* in sample A, fewer in sample B. But the relation between the number of leaf imprints and cone-scales is not a meaningful parameter because of different genesis, dispersal, preservation, etc.

On the other hand, pollen grains constitute a uniform group: origin, dispersal, embedding, preservation are—within limits—the same for all taxa. It is therefore meaningful to treat pollen quantitatively as a uniform group: (1) the total pollen incidence, and (2) as individual constituents of the group, the representation of that taxon in the total. At the time pollen analysis was created there were practically no possibilities for establishing a meaningful absolute measure of the total pollen incidence (1). Even pollen concentration (q.v.) could not be properly defined, and influx (q.v.) even less. The remedy was to express the representation of the individual taxon (2) as a percentage of the total pollen incidence (1), *tacitly assuming that the latter was approximately constant.*

Relative (percentage) pollen analysis was the only technique available during the first 50 years of pollen analysis, and became conceptually identical with that. It still is dominant and probably will always be. However, with the advent of proper dating methods and new counting principles, it became possible to define quantity (1) as well, and to establish the *absolute pollen incidence* as *pollen influx*, which simply means the number of pollen grains deposited per year. Conceptually, pollen influx data are much simpler and more direct representations of the actual vegetation, and therefore permit a more penetrating analysis, leading to more meaningful results. It opens the possibility of a (semi-)quantitative evaluation of former vegetation. It is fair to speak about a revolution in the conceptual framework of pollen analysis, a revolution that is still in progress.

This does not mean that the old methods have become obsolete or redundant. Percentage presentations are inherent parts of all quantitative numerical pollen analyses and are automatically obtained. To transform them to concentration data costs very little, but to add the datings necessary for influx data may be impossible in some sediments, expensive and cumbersome in others. Since in many—most—cases the assumption of a constant total pollen influx holds approximately true, percentage data—with their immanent limitations—give adequate information about the composition of the vegetation and its changes in each locality, provided climatic changes have not been too drastic.

Percentage calculation actually implies that we draw inferences from negative evidence, which is strictly against the basic rule in paleontology. The excuse for this practice is the assumption that the distribution of pollen grains is sufficiently wide so that their absence from a surface can be considered to indicate that the species in question did not occur at that timee in that locality. In absolute pollen frequency dates this is not built in, like in percentage data, but it is obvious that, external conditions being the same, what is permissible in the one case is also in the other.

This further implies that species, the pollen of which is *not* evenly sifted over the region, i.e. strictly entomophilous or autogamous plants, should not be treated on equal terms with the others. They may be important as indicator species, but theoretically, should not be included in the pollen sum (cf. below), since no conclusions can be drawn from their absence. Usually they occur in such small numbers that in practice it does not matter very much whether they are included in a pollen percentage calculation or not. The theoretically most correct procedure would be only to indicate presence or absence both in percentage and absolute data. If one wants a quantitative measure also in percentages, those should be calculated on the basic sum plus the taxon in question. But both in that case and for absolute data the non-relevance of absence should be kept in mind.

1.1.2. The pollen sum

Absolute pollen data are meaningful in themselves. The number of grains of taxon A counted immediately indicates the quantity of A present, as the figure is related to deposition area and time. If these parameters are not available, the counts must be related to some other quantity; they must be expressed as percentages of a statistical universe —a basic sum including all relevant variables (pollen taxa) on which their percentages can be calculated. Since percentage calculations are and will probably remain the most frequent and most versatile way of dealing with the pollen contents of a deposit, this sum is a very important parameter. The establishment of the pollen sum is a very important step in any percentage analysis, as it defines the question asked in the investigation and consequently the framework for the possible answers. The pollen sum defines the plant communities one wants to draw the data from, i.e. the plant communities presumed to occur within the area under investigation. There is nothing like one 'correct' pollen sum; different questions presume different pollen sums even within the same region.

Thus, one may want to study the total vegetational development and include everything, but it is equally justified to study the development of forest only, and omit pollen produced in non-forested areas. Similarly, it is justifiable to try to find out what happened in the open (man-made?) areas and for that purpose to exclude the forest-tree pollen, even in an area which is generally forested. The only prerequisite is that in each case all pollen (and spores) that have a bearing on the actual problem be included, also and even autogamous species if their pollen can, under the given circumstances, be considered meaningfully distributed. Thus, it would be fully justifiable to include pollen of *Lysichitum americanum* or *Ilex aquifolium* in a local forest study, but hardly in a regional analysis. This is not so much related to the (very sparse) positive finds in the regional pollen rain as to the danger of conclusions being drawn from their absence in other samples of the same series.

Von Post's old rule was that only the constituents of the topmost layers of vegetation should be included in the sum. The reason for this was the idea that the curves of the pollen diagram should represent separate areas, and the variations of the curves demonstrate the relative changes of areas occupied by various vegetational types. Consequently only one layer can be represented in each place, viz. the topmost layer of the vegetation that covers that particular spot. However, since the flowering of wind-pollinated plants, e.g. *Corylus*, in the lower layers is normally much reduced and their pollen is usually not transported outside the forest itself, they are badly represented in the regional pollen rain. This fact makes the theoretical question of their inclusion or exclusion unimportant in general practice.

It is essential always to bear in mind the aim of the investigation and, if possible, to exclude the pollen of disturbing species. If a pollen diagram, for example, is to express the regional forest development it may be necessary to exclude the pollen of plants that have grown on the actual site, because local 'macroscopical' pollen supply may otherwise completely distort the general trend of the curves. If, however, the aim is to study the local vegetational succession, we have, of course, to include the pollen from the local vegetation, and even that from lower strata of multi-layered vegetation.

As all manipulation of the pollen sum must be based on certain assumptions, and represents a certain personal judgement, the material should be presented in such a way that the original counts are documented. This is especially important in areas which are not too well known pollen-analytically, whereas in areas that have been analysed earlier, there are usually previous discussions of the pollen sum and its implication, as will be shown in a later chapter.

The considerations outlined above are valid for the application of pollen analysis to problems of Quaternary geology. However, the same principles are valid in whichever field pollen analysis in a wider sense, i.e. the use of pollen grains as indicators, is carried out. The emphasis may shift, and other fields have other basic assumptions. These will be discussed in later chapters dealing with use of some form of pollen analysis in various fields, more or less scientific, more or less utilitarian: ecology, climatology, stratigraphy, general geology, archaeology and history, nutritional and forensic sciences, etc. Pollen grains being present well-nigh everywhere and being, practically speaking, indestructible, there is no reason why they should not be even more utilized in studies of the provenance of vegetable matter—for scientific or practical, criminal or civil purposes.

1.2. The history

1.2.1. The peat diggers

In regions where peat is widespread and plentiful it has been customary to use it for many practical purposes: fuel, house-building, soil improvement, etc. Also ditches had to be cut through bogs to drain them and make them arable or passable. Peat utilization is known from pre-Christian times (Borchgrevink, 1974: 521) in Denmark. The first person mentioned in Nordic literature as user of peat for fuel is Torv-Einar, AD 875.

The observation that remains of former vegetation are preserved in peat bogs is certainly as old as the practice of peat cutting itself. Nobody cutting peat in a deforested region could help noticing trunks and roots or pine cones preserved in the bog; reflections on former vegetation would come automatically. Thus, a local scholar of the seventeenth century, M. Edvardsen (cf. Brattegard, 1951), reports the find, from the Bergen City area, of subfossil pines and oaks and draws the adequate conclusions from the find.

A closer inspection may reveal that remains differ in different depths of the deposit, and whereas some of these differences may be due to the natural development of the deposit itself, e.g. the filling-in of a lake, others demand a more radical explanation, e.g. climatic change.

1.2.2. The first pollen grains

The specific identification of seeds and other small fossils found during the examination of peat required the use of some magnification. Stronger magnification quite naturally led to the discovery of even smaller fossils, which would, in the end, also include pollen grains. Oddly enough fossil pollen grains were apparently first observed in pre-Quaternary deposits, by Göppert (1836 and later) and Ehrenberg (1838 and later, both quoted in Kirchheimer, 1940). The first to utilize the occurrence of pollen grains in Post-Glacial deposits were, as far as we know, Geinitz (1887, analyses by Früh) and C. A. Weber and his school (1893 and later).

These early registrations were qualitative (quantitative in Steusloff, 1905), and usually took into account a very small number of pollen taxa. Registration of the actual quantity of pollen in a deposit presumes a fairly sophisticated approach, but a major step forward was represented by percentage calculations, in which the occurrence of one pollen taxon was expressed as a percentage of the pollen totality. In this way relative variations, caused by immigration, succession and recession of species, could be expressed quasi-quantitatively.

The first percentage calculations seem to have been carried out by Lagerheim (in Witte, 1905), and later by C. A. Weber himself (1910, quoted in H. A. Weber, 1918, although it is not quite clear if the actual calculations were carried out in 1910, cf. Weber, 1918: 259). As pollen analyses these early works are of historical interest only, though Holst (1909: 30) evidently realized the great importance of Lagerheim's calculations.

1.2.3. The birth of pollen analysis: the pioneers

The real potentialities of the method were, in fact, not realized until Lennart von Post, then state geologist, took it up with Lagerheim as his micromorphological mentor. Von Post presented the first modern percentage pollen analyses in a lecture to the Scandinavian scientists' meeting at Kristiania (now Oslo) in 1916, repeated later that same year in Stockholm. A short abstract of the latter was published in 1916, whereas the report of the Kristiania meeting was not published until 1918, and a translation of the Swedish original not until 50 years later (cf. references).

The primary use of pollen analysis as an instrument for the investigation of Quaternary changes of vegetation and climate is a historical accident. If palynology had developed on the basis of Göppert's and Ehrenberg's observations on Tertiary and older material, palynology might have evolved as a purely stratigraphic tool, as it actually did when pollen analysis was extended beyond the Quaternary.

In Scandinavia of the late nineteenth and early twentieth centuries the problem of the effects of climatic change on vegetational history during the Late-Quaternary was a burning one. Instigated by Axel Blytt in 1876, a violent discussion raged over these matters, reaching a high point at the International Geological Congress in Stockholm in 1910. At that time, von Post had already become a student and collaborator of one of the two main protagonists of that discussion, Rutger Sernander, and it was thus natural for him to develop his new method to elucidate the old problem. Because of this historical situation 'Pollen analysis' for a long period was identical with Late-Quaternary vegetation research.

Von Post's students and collaborators took up his methods and continued his work; Sandegren (1916), Halden (1917) and Sundelin (1917) published bog monographs including pollen-analytic data. As these papers were all published in Swedish, partly also because of war-time disruption of communications, pollen analysis passed rather unnoticed outside the Scandinavian language region (quoted by C. A. Weber in H. A. Weber, 1918), and the first pollen-analytical investigations outside Sweden were published in Norway (Holmsen) and Denmark (Jessen) in 1919 and 1920. The first major work based on von Post's methods, published in a congress language, came in 1921 (Erdtman).

1.2.4. Technical development in pollen analysis

After that, pollen analysis rapidly supplemented and supplanted older methods in Quaternary research everywhere. Quaternary pollen analysis extended in two directions: within the 'core area', i.e. northwestern and middle Europe, conceptual and technical advances permitted refinement of methods and conclusions far beyond the simple classical analysis. Second, through expansion into regions with a different vegetation (richer floras, different pollination strategies) new problems arose that had to be solved. Some of them still await their final solution.

Two technical papers were important in shaping pollen analysis. In 1924 Assarson and Granlund introduced HF treatment to remove siliceous matter from the samples to be analysed, thereby making it practically possible to analyse also deposits wholly or partly minerogeneous. In 1934 Erdtman published the acetylation method, which permitted the removal of a major part of the organic matter constituting a sample, leaving a high concentrations of pollen grains. This permitted the inclusion of many pollen taxa which had previously been too scarce for analysis, and also permitted the counting of much greater numbers of pollen, which meant removal of a great deal of statistical noise.

The apochromatic lens did exist in 1916, but it was a very expensive piece of equipment rarely seen in a pollen-analytic laboratory. Also, the early pollen analysts were not very experienced microscopists. The two major advances were the introduction of lenses comparable in quality to the old apochromates, but cheaper and easier to handle, and the introduction, in 1942, of phase contrast microscopy, which permitted a more discriminating understanding of pollen morphology during analysis. The introduction of electron microscopy (TEM and SEM) has been of fundamental importance for pollen morphology, but for practical reasons these techniques have had less direct impact on regular analysis.

The development of field equipment was for a long time rather stagnant. The samplers existing in 1916 (with the Hiller as dominant) were adequate for a long time. Other, end-filling, types did exist and were used in limnology, but pollen analysts were, on the whole, not interested. What was later to be known as the Livingstone sampler was developed after the Second World War, and its general acceptance came because its large-diameter modification could also satisfy the demands of other techniques, especially radiocarbon dating, better than the previous types, including the original narrow-bore Livingstone type.

1.2.5. Conceptual advances

Von Post was a geologist, and even if he had a fine botanical acumen his basic problems were those of geology. After trained botanists had taken up pollen analysis in the late 1930s the botanical dimension of the problems was realized and came into the foreground, as nobody realized more clearly than von Post himself in his review (1950) of the first edition of this book. This botanical dimension permits a more in-depth penetrating analysis of pollen-analytic data.

The first pollen analyses came from botanically very simple areas: they were presumed to be, and to have been forested, throughout, and the history of vegetation was essentially a forest history. A major conceptual advance came in 1934 with Firbas' demonstration of the relatively greater role played by non-tree pollen (NAP) in relation to tree pollen (AP) in non-forested areas. What was to become known as the AP/NAP ratio was established as the instrument for study of the degree of forest cover and the vegetation of non-forest areas.

A continuation of this was Iversen's demonstration, in 1941, of the pollen-analytic registration of man-made deforestation during the Neolithic: the change of Europe from a primaeval forest to the cultivation, pasture and arable fields landscape of agricultural man.

In 1940, Aario, by analysing recent pollen incidence across the arctic forest limit, could demonstrate that the pine dominance in the AP diagram from subarctic pine forests was succeeded by birch dominance in the birch belt, but then pine pollen took over again in the tundra because of greater number and better dispersion. Misleading pollen representation due to local over- or under-representation of a taxon was already known. Aario's observation was important, because it demonstrated that records could be systematically 'false' even under quite regular conditions. M. B. Florin (1945) identified a parallel case connected with the regression of the sea.

Late-Glacial clayey deposits used to contain a puzzling pollen flora with a number of thermophilous taxa, the occurrence of which under Late-Glacial conditions seemed very unlikely. Iversen solved that problem in 1936 by demonstration of the 'secondary' pollen, i.e. pollen redeposited from pre-Glacial deposits and recuperable because of the small contemporaneous pollen production contained in those deposits. The wider aspect and importance of redeposition have only been generally recognized during the last decennia.

A very important, though widely overlooked, event in the history of pollen analysis was the meeting which took place under von Post's aegis in Stockholm in the autumn of 1945 (cf. Geologklubben, 1947), patching together the ties that had been broken during the war. At that occasion the old geological/dating pollen analysis was matched against a more modern vegetational approach, and although it was not realized then, it marked the turning point in the philosophy of pollen analysis. A similar meeting was held a year later in Cambridge under the aegis of Harry Godwin.

Since one of the major uses of pollen analysis during the first decennia was for dating purposes two developments should be mentioned which, though not directly concerned with pollen analysis, influenced it profoundly. The first was Granlund's (1932) demonstration of the recurrence surfaces in peat bogs, i.e. assumed synchronous surfaces at which the bog resumed its growth. Although these ideas have undergone strong modifications they still are of importance in peat-bog geology, if not so much for dating any longer. That task has been taken over by the second development: radiocarbon dating, which has obviated the use of the slow and expensive pollen analysis for most dating purposes and now, instead, furnishes pollen analysts with the absolute dates they need for calibration of their relative chronologies.

Pollen analysis is essentially a statistical technique, and formal statistics were brought in fairly early (Fægri and Ottestad, 1948). The enormous potentiality of statistical control presented by modern electronic data techniques has gradually been realized also in pollen analysis. Methods and techniques are discussed in Birks and Gordon (1985), to which reference is made, as a comprehensive discussion would exceed the frame of this book.

Most of the advances mentioned in this short historical sketch (cf. also Fægri, 1981) will be detailed in the following chapters. The pollen-analytical literature is exceptionally well documented. The early literature till 1955 was registered by Erdtman in *Geologiska föreningens i Stockholm förhandlingar*, from 1927 onwards. After 1955 the literature has been registered by van Campo as yearly supplements to *Pollen et spores* (1957 seq.) now running into 26,340 items. There are two major journals exclusively dedicated to palynological problems, viz. *Pollen et spores* (1959–) and *Grana palynologica*, later *Grana* (1954–), cf. also *Palynology* (1977–) and *Journal of palynology* (Lucknow, 1965–). In *Review of palaeobotany and palynology* there is a great proportion of palynology, but much of the literature is scattered in journals of general micropaleontology and Quaternary geology, and also those of archaeology and wider coverage.

Obviously, it is impossible in a book like this —in any book—adequately to cover the pollen-analytic literature. For each field we have quoted (some of) the papers that have influenced our own thinking. We have also tried, on one hand, to quote some of the grand, classical papers, tracing the origin of ideas and techniques. On the other hand, we have included some of the more recent publications in the field, but we have seen no reason to replace a 1965 publication with one about the same subject from 1985 unless the latter represents a distinct advance. If one of our own papers illustrates a point well, we have used that as an example. It does not mean that we do not realize that many other papers deal with the same subject equally well.

2. Where does the pollen go? Production and dispersal of pollen grains

Motto: Gone with the wind

The first part of the processes and phenomena that constitute problems and possibilities of pollen analysis (cf. Fig. 1.1) belong to the area of pollination ecology. Various pollination strategies influence and decide the number of pollen grains ultimately at disposal for analysis, and their distribution.

The essence of pollination is transport, the transport of pollen from anther to stigma. The modes of transport and the adaptation of plants to them are fundamental for the evaluation of pollen-analytical data. This transport takes place in different ways, and can for the community be summarized in Fig. 2.1, which shows the pollen budget. It is maintained by the local production plus the pollen entering the community from outside. It is

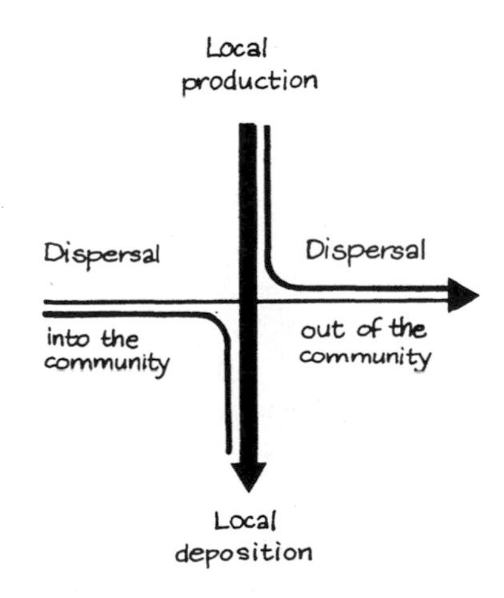

Fig. 2.1. Pollen budget, model.

depleted by pollen going out of the community and by deposition (including pollination) inside the community. A greater part of the latter component is lost by decay, etc. Some of it is preserved and available for later analysis. The understanding of the transport processes in pollination ecology is one of the main keys to the understanding of pollen analysis and its results.

2.0.1. *Self-pollinating plants: pollination by water*

A few aquatic plants are pollinated under water (hyp-hydrogamous: e.g. *Zostera, Ceratophyllum, Zannichellia*). *Per se*, there should be excellent opportunities for pollen of these species to be embedded in sediments and fossilized. Usually, such pollen grains have extremely thin exines and remains are not recognizable; the small group of hyp-hydrogamous species cannot be traced in pollen analysis.

Another heavily underrepresented group is that of obligate autogamous and apogamous plants. In extreme cases the flower is also cleistogamous; it does not open, and the pollen germinates on the stigma within the closed perianth. No pollen is exposed. In other cases the flower opens, but only after pollen grains have germinated and the pollen mass is woven through and glued to the stigma by the pollen tubes. As a consequence of the very great effectivity of the pollination process, few pollen grains are produced and they are not liberated from the mother plant. Even in such autogamous plants where some pollen grains may eventually be exposed, e.g. in ordinary wheat, the number is extremely small compared with that of related allogamous species, e.g. rye (cf. I. Müller, 1947). In 15 chasmogamous (mostly wind-pollinated) taxa of North American *Plantago*, Basset and Crompton (1967) found ca. 600 pollen grains per anther; in 13 cleistogamous taxa they found an average of 30. The liberation of pollen of autogamous plants into the air varies with external conditions.

2.0.2. *Pollination by animals*

In the very comprehensive group of zoophilous blossoms pollen is carried from the anther of one flower to the stigma of another by some animal vector: insect, bird, bat. In extreme cases these blossoms are highly specialized and pollen is released only when the right animal visits the blossom and behaves correctly. The pollen is then (firmly) deposited on the vector, and is removed from it by the stigma only. Unless the pollen-bearing animal itself perishes in a lake or bog, or whole flowers (or anthers) drop accidentally into the water, pollen grains from this type of blossom have extremely small chances of being preserved as fossils in ordinary deposits. When they are occasionally found in peat or sediments, they are always infrequent; their occurrence is accidental and unreliable, even if their number in an exceptional sample may be high. Consequently they should not be treated on a par with grains of anemophilous species. Their occurrence may furnish indications of the greatest value, but nothing can be concluded from their absence.

Whereas zoophilous pollen grains are rare in the type of deposits normally used for pollen analysis, they may be found in great numbers if the soil beneath zoophilous plants is analysed. Plants like *Ilex aquifolium*, the pollen of which is extremely rare in ordinary deposits, may be amply represented in forest soil, on to which the male flowers drop after anthesis. Pollen of such plants may be found in great numbers both in forest bottom deposits and in excavations from former farms and cultivated soil, e.g. a mediaeval field with substantial quantities of *Vicia faba* pollen (Krzywinski and Fægri, 1974).

The more specialized and the more effective the zoophilous pollination, the smaller the production of pollen dispersal units and the fewer units are liberated into the air. On the other hand, some zoophilous species produce such great quantities of pollen that their production is fully comparable to that of the wind-pollinated species, e.g. *Tilia* (Hyde and Williams, 1945a: 457) or *Calluna* (Pohl, 1937: 440). They may be facultatively anemophilous, e.g. *Salix* (Argus, 1974) or the bees' pollen-collecting technique may disperse part of the pollen in the air (cf. Parker, 1926), especially by buzz pollination (cf. Buchman *et al.*, 1985). Part of the output is regularly 'lost' in the air and behaves like that of wind-pollinated species. Some of these pollen grains possess a coating of oil, e.g. *Tilia* (Zander, 1935: 218), whereas others are dry, e.g. *Calluna* (Zander, 1935: 241). Such high pollen production may occur in typical nectar blossoms like those of the above-mentioned plants. It is self-evident that the production is also high in pollen blossoms, e.g. *Filipendula ulmaria*, where surplus pollen is the lure that attracts animals to the flowers.

Accordingly the group of zoophilous flowers comprises all types, from species which are never to be expected in pollen analysis to those that are regularly encountered, form constituents of the pollen rain, and are calculated in ordinary analysis.

2.1. Wind pollination, anemophily

The last, and for our purposes most important, group is that of wind-pollinated plants, which produce very great quantities of pollen that is liberated into the air and scattered all over the surroundings as the 'pollen rain'. A negligible fraction of this settles on the stigmata; the bulk of it is lost, some on bogs and in lakes. The typical pollen of wind-pollinated species is dry and the individual grains separate, thus producing the maximum number of *pollen dispersal units* and securing a regular distribution of the grains over a wide area. Wodehouse (1935: 351) has pointed out that pollination by wind tends to bring about a reduction of exine 'with an attendant loss or reduction of its structure such as furrows, pores, and sculpturing' (Fig. 2.2; cf. Knoll, 1930: 629). Most of the pollen types of northern European wind-pollinated species are smooth or almost so, the strong sculpturing being characteristic of zoophilous species.

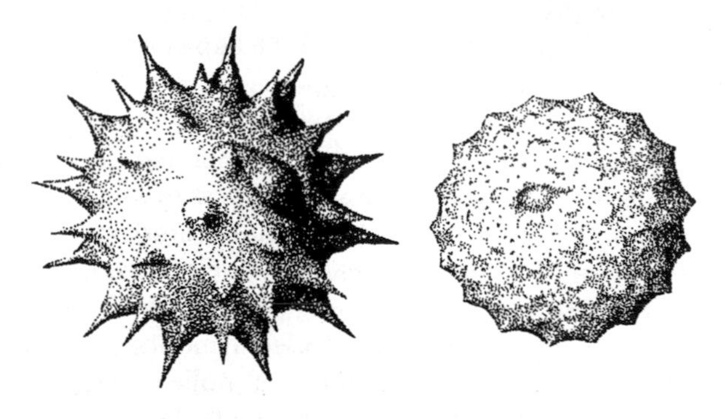

Fig. 2.2. Pollen grains in Compositae Asteroideae. Left: Insect-pollinated (*Helianthus annuus*). Right: wind-pollinated (*Xanthium catharticum*). From Wodehouse, 1935. Reproduced by permission of the copyright-holder: McGraw-Hill, New York.

Whereas the anthers of zoophilous species are more or less concealed, the more so the more intricate the pollination mechanism, those of anemophilous species are generally freely exposed, projecting beyond any perianth. Flowering frequently takes place before the development of the foliage. A number of mechanisms tend to scatter the pollen grains as effectively as possible, and anthers, e.g. in grasses, are often empty very soon after dehiscence (cf. Hyde, 1945). On the other hand, Hyde and Williams (1946: 274) observed that pollen of *Plantago lanceolata* remains in the very exposed anthers during exceptionally calm mornings.

Pollen grains of zoophilous species frequently possess a heavily armoured surface. Layers of sticky oil on the surface of the grains cause them to stick better to each other and to the body of the animal. Extreme cases are the massulae of orchids and Asclepiadaceae or the viscin threads binding together pollen masses, e.g. in Ericaceae and Onagraceae, cf. Hesse (1980 and earlier). As such masses are dispersed as one unit, their existence tends to reduce heavily the number of pollen dispersal units. Also, units consisting of several grains stuck together are perforce heavier than individual grains, which restricts their possibility for wind dispersal. In some cases at any rate pollen-analytical experience indicates that such masses may break up in fossilization. According to Nakagava and Katsuda (1975) the pollen concentration of the typically entomophilous *Chrysanthemum cinerariaefoloum* in the air was great enough to cause pollinosis a kilometre from a massive cultivation, which seems as exceptional dispersion for such a pollen. Compare, however, the quantities of pollen raised by people walking through the vegetation: 1.5 m above ground there were 54 *Taraxacum* and 235 *Chrysanthemum* pollen grains per cubic metre (Kupias *et al.*, 1981).

The tendency of pollen grains to stick together has been investigated by Rempe (1937: 103; cf. also Knoll, 1930, 1936). In most of the typical wind-pollinated genera, *Picea, Fagus, Corylus, Alnus, Betula, Carex* (*montana*), *Secale*, in *Erica carnea* and *Calluna* tetrads and, rather surprisingly, in *Acer platanoides* and *A. pseudoplatanus*, there is practically no clumping of the pollen grains. In the rest of those forest trees that are included in routine analysis: *Quercus, Tilia* (2 spp.), *Ulmus* (2 spp.), *Acer* (2 spp.), *Larix, Salix* (2 spp.), *Castanea*, and most unexpectedly, *Pinus*, 25–40% of the grains stick together to form relatively small clumps. In typical insect-pollinated species: *Salix* (3 spp.), *Malus, Pyrus, Galanthus* (cf. Troll, 1928; Knoll, 1930: 662), and *Lilium*, clumps are both more frequent and bigger. The same pertains to a few wind-pollinated species, e.g. *Mercurialis*.

2.1.1. *Pollen production*

The starting point in the processes detailed in Fig. 1.1 is the quantity of pollen produced by individual plants and by the plant community. This is not a constant magnitude; it varies both specifically and individually and in relation to ecological parameters, especially climatic changes.

The most characteristic feature of wind-pollinated plants is the great number of pollen grains produced per ovule (cf. Fægri and van der Pijl, 1979: 36). However, in pollen analysis the chief interest concerns the number of grains produced per area unit.

One single anther of hemp may contain 70,000 pollen grains. In north European forest trees the number is considerably lower, e.g. *ca.* 10,000 in birch. Species that are habitually insect-pollinated tend to produce smaller numbers, *ca.* 1000 in *Acer* (all seq. Pohl, 1937); in *Linum catharticum* (frequently autogamous) the number is *ca.* 100, and in autogamous species it may be even smaller. Ikuse (1965) quotes figures between 64 (*Malva*) and 44,500 (*Hydrangea*). In anemophiles Reddi and Reddi (1986) found between 32 (*Bothriochloa*) and 89,000 (*Phoenix dactylifera*) pollen grains per anther with very great individual variation, e.g. 32–1990 in *Bothriochloa*.

As each flower usually possesses a number of anthers, and each shoot many flowers, the figures for total production reach enormous dimensions. One shoot of hemp produces more than 500 million pollen grains, whereas a male plant of *Rumex acetosa* produces 400 million and a large panicle of a *Sorghum* cultivar over 100 million (Stephen and Quinsby, 1934), but even a large specimen of *Linum catharticum* hardly reaches 20,000. Forest trees produce great quantities: a male strobilus of *Pinus contorta* contains almost 600,000 pollen grains, and a shoot 8–9 million (Ho and Owens, 1973), a 10-year-old branch system of beech (*F. sylvatica*) produces more than 28 million; birch, spruce and oak a little more than 100 million, and *Pinus sylvestris ca.* 350 million (nearly as much as a male plant of *Rumex acetosa*). The production per hectare of forest runs into billions (all seq. Pohl, 1937). Hesselman (1919a: 41) concludes that the spruce forests of southern and middle Sweden produce *ca.* 75,000 tonnes of pollen annually when flowering freely. According to Smirnov (1964) the Rybinsk reservoir (N. lat. 58–59°) receives 6 kg of spores and pollen per hectare per year. Similar figures are quoted by other investigators, e.g. Koski (1970), who calculates that Finnish pine (*P. sylvestris*) forests produce between 10 and 80 kg of pollen per hectare per year, corresponding to 30,000 to 280,000 grains per cm^2 per season. Such figures are perforce based upon extensive extrapolations.

Flowering and pollen production of and in a forest are highly variable. A tree that is freely exposed, e.g. in a field, produces many times as much pollen as one that grows in a dense stand, even if the latter reaches the upper stratum of the canopy. Individuals that do not reach this upper stratum, because they are too young or are suppressed, produce very little pollen. Many forest practices tend to reduce drastically the flowering of forest trees (Niederwald, palina, etc.). Species that habitually belong to lower strata—shrubs and forest bottom herbs—flower also inside the forest, but flowering is usually not very profuse, and many of them are insect-pollinated. Even if they are wind-pollinated, the wind velocity is low. Inside a beech forest without under-storey Kiese (1972) found a wind velocity of 24% of that above the canopy; Andersen (1974) registered similar velocities in a mixed forest before leafing. In more complicated forests, the figures are considerably lower; Andersen (l.c.) found that after leafing wind velocities had halved. The usually dense forest margins prevent most of that pollen from getting out of the forest. This effect is especially marked for species

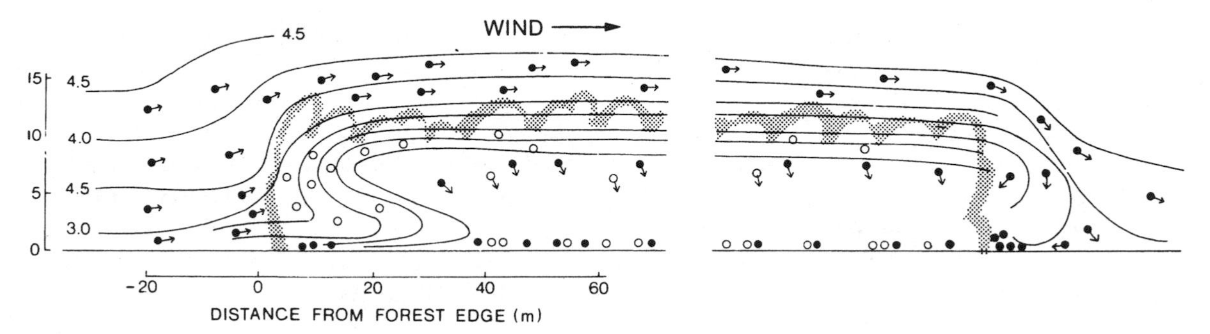

Fig. 2.3. Isopleths for wind passing over and through a forest (indicated by the irregular contour). Dots: regional pollen; rings: local pollen. From Raynor, 1971, modified. Reproduced by permission of the copyright-holder: Society of American Foresters; Bethesda.

that flower during summertime when the forest canopy is closed by foliage.

Figure 2.3 illustrates the changes of wind velocities when wind enters and leaves a forest. At any rate with narrow hedgerows there is a distinct concentration of pollen in the eddies at the lee side. Probably something similar is the case at the lee side of a forest (cf. Fig. 2.4).

In a forest region the undergrowth therefore plays a secondary part in the pollen rain, but pollen productivity of vegetation on open, non-forested ground, measured in number of pollen grains produced per area unit, is in many cases equal to that of a forest. Consequently, if the forest cover disappears, conditions change: the increased amount of light favours flowering of the lower vegetation, and wind can disperse the pollen. According to Pohl (1937: 440) the pollen production of a rye field per m^2 is higher than that of *Carpinus* (720 millions); *Arrhenatherum elatius* produces more pollen per unit area than oak or beech, and *Calluna* twice as many units (eight times as many pollen grains) as *Pinus sylvestris*. *Calluna* is zoophilous, but at the end of the flowering season great quantities of pollen are liberated in the air. In contrast, a field of maize is said to produce only 70 million per m^2 (Goss, 1968). Within the same climatic region the pollen production of an area is roughly of the same order of magnitude whether the area is forest-clad or covered with lower vegetation, and the relation between the pollen produced by forest trees and that produced by ground vegetation is an indication of the density of the forest. There is, however, one important difference: most of the pollen produced by the ground cover is almost immediately deposited on the ground: in the lowermost layers of vegetation air turbulence is insufficient to bring pollen grains up into higher air strata where horizontal transport takes place.

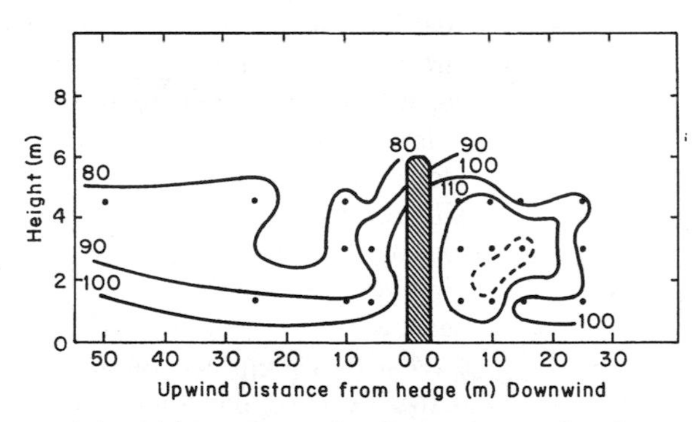

Fig. 2.4. Enhancement of pollen concentration downwind from a hedgerow. Figures give percentages of a reference value. Dots indicate measuring points. From Raynor *et al.*, 1974. Reproduced by permission of the copyright-holder: Elsevier Scientific Publishing Co., Amsterdam.

Potter (1967) has shown that water tanks, the rims of which are situated 2–3 m above ground, receive substantially smaller percentages of ground-cover plants in relation to trees than does the ground itself. In Finland, Hicks and Hyvärinen (1986: 225) caught *ca.* 50% more pollen at ground level than 1.1 m above (Tauber traps). Similar results were obtained by Hyde and Williams (1945b), whose observations from the roof-top of the Cardiff hospital gave values much smaller than those from ground level. As roof-top stations have frequently been used for pollen deposit registrations this factor must be taken into account.

Because of scavenging by standing vegetation the incidence of pollen on the vegetation itself may be very high. Fokkema (1971) reports 3000 grains per cm^2 leaf surface, i.e. a cover of 11%. Greig (1982: 58) registered 9500 pollen grains per 10 cm straw of wheat and 1900 on barley straw. Most of such deposits are usually washed off with the first rain and redeposited on the ground. Even in dry weather the life-time of such pollen on the vegetation is short. Krzywinski (1976) found a reduction of *Pinus* pollen on *Betula* leaves from 3350 to 20, two weeks after flowering had ceased (cf. Fig. 2.5). The remainder seemed to adhere for a long time.

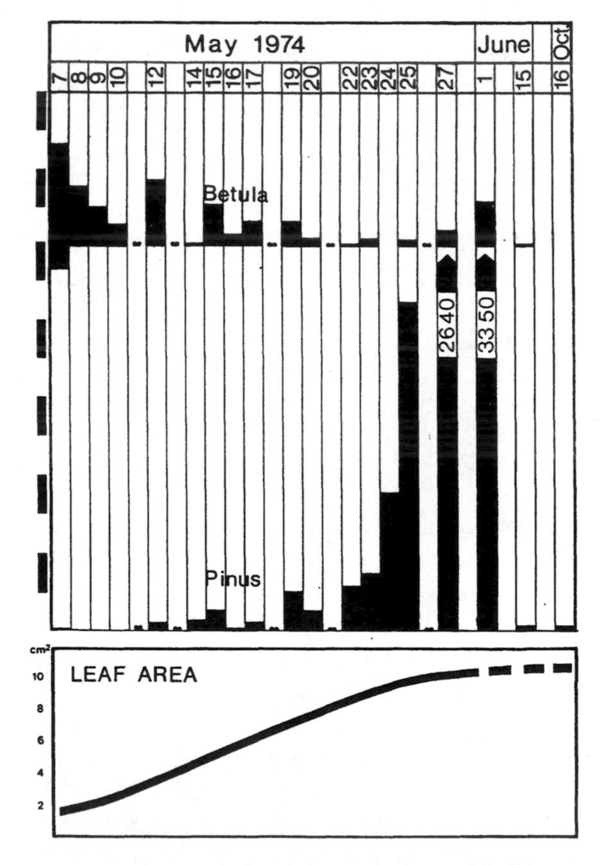

Fig. 2.5. Birch and pine pollen adhering to birch leaves. The small signatures indicate days when no counts were made. Scale units left side: 100 pollen grains/cm^2 leaf surface. Bottom curve: Average total surface of birch leaves (upper and lower-side). From Krzywinski, 1976.

Enormous quantities of pollen produced by anemophilous plants are liberated and float in the air for a shorter or longer period before being deposited on the ground or in water. These floating pollen masses are usually referred to as the *pollen rain*, and are over time supposed to be more or less evenly sifted over the ground. Knowledge of the composition of the pollen rain is the key to understanding pollen deposition.

2.1.2. *Trapping pollen grains*

For understanding of the following steps in Fig. 1.1, and for the final interpretation of the pollen record, it is important to possess a detailed knowledge of the travels of pollen in the air between release and deposition. Pollen sampling from the air is not easy (cf. May, 1967). A multitude of apparatus has been described (cf. Ogden *et al.*, 1974) with the objective of obtaining a quantitative measure of the composition of the pollen rain, but very few of them give information directly applicable to pollen analysis problems, and few investigators clearly define what they are doing. Theoretically, one may sample three different quantities: (1) the pollen contents of the air, i.e. the number of grains in a static volume of air; (2) the pollen transport in the air, i.e. the number of grains passing through a cross-section per time unit; and (3) the pollen deposition from the air, i.e. the number of grains settling on a horizontal surface in a given time interval. As far as sampling goes, this means pollen which is (1) immobile, (2) in horizontal, and (3) in an integrated downward movement. The three quantities are obviously related, but there is no constant, linear regression between them. Wind velocity and turbulence, precipitation and surface character change the relations all the time.

For registration of pollen contents the logical method would be to isolate a known volume of air and count the grains in it. This presents great practical difficulties and no workable method has been found (cf. the 'Automatic grab sampler' discussed by Ogden *et al.* (1971) and earlier reports, cf. also Ogden *et al.*, 1974: 47).

The alternative method is to use a vacuum

cleaner or similar instrument, an idea introduced by Erdtman (1937). A filter stops the pollen grains. Afterwards, the filter can be dissolved and the grains counted. As all palynological laboratories have the equipment to dissolve cellulose, there is no reason to use anything more difficult than an ordinary paper filter. Glass filter (HF treatment) is just another complication. The main problem is to define the volume of air passing through the apparatus. Strictly speaking, this method does not give the static picture, even in calm air, as the suction itself causes an air stream. However, in this case the trajectories of pollen grains will hardly deviate very much from that of the ambient air itself. A more frequent, but less reliable, method is to let a known volume of air impinge upon a sticky surface (coated microscope slide) and count directly, but this introduces problems of the efficiency of capture (cf. May, 1967). The Burkard trap, which works on this principle, is today the standard instrument in aerobiological research. It is mechanically reliable, easy to use, and has the advantage of high time resolution, giving an hour-to-hour picture of the pollen content of the ambient air.

Originally constructed for the trapping of fungus spores, the Burkard trap suffers from several shortcomings as a pollen trap. The different capture effect of heavy vs. light grains is one of these, cf. Fig. 2.7. The effect of wind speed is another. The often-quoted Fig. 5-4 in Ogden *et al.* (1974) is misleading inasmuch as it shows the same number of pollen entering the trap at all speeds in spite of maintaining the opposite. The effect of wind speed on pollen concentration in the air is largely unknown, especially as regards gustiness vs. constant wind (cf. Burrows, 1975 a, b), infinite vs. finite pollen sources, continous vs. interrupted pollen release. Another complicating factor is the slow windwane effect in turning the orifice of the trap against a turbulent wind. Possibly the various irregularities annihilate each other, giving, in spite of everything, a fairly consistent, if disorderly, picture. With increasing wind speed the pollen concentrations registered by the Burkhard trap increasingly represent the dynamic concept, the pollen transport, as compared to the static. For a further evaluation of the efficiency of this counter, cf. Käpylä and Penttionen (1981).

For registration of the second quantity, *transport*, the method is to place a surface perpendicularly to the path of the air current and count the number of grains passively trapped on the surface. One method is to place a sufficiently dense, greased grid in the path of the air, again using a wind vane to keep it at right angles to the air current (cf. Cour, 1974).

Another instrument for this purpose is the rotorod/rotoslide used by Ogden and collaborators (1974). The principle is whirling a vertical, greased rod or edge of a slide at fairly high speed. Owing to the great speed the instrument is very effective; overloading is a problem.

The result of measurements with such traps depends not only on the number of particles in the air, but on air speed, and physical characteristics of both the particles and the trapping surface in a very complicated way (May, 1974). Whereas results are consistent in themselves, they are very difficult to translate unless the efficiency of the trap has been calibrated by extensive wind-tunnel tests. One of the major points is that the trapping surface must be very narrow to avoid eddying. The narrower the surface, the more effective the trap (cf. Fig. 2.6). Other parameters of importance are wind speed and weight (inertia) of the particle (cf. Fig. 2.7). Since pollen transport is so much a function of air speed, transport values are very low at canopy level, due to the calmness of the air in the canopy (Ammann, unpublished).

These two quantities, content and transport, are of major importance in allergy and pollination studies. For discussions of methods and problems, cf. Hyde (1972: 163 and Ogden *et al.* (1974). Grosse-Brauckmann and Stix (1979) have discussed the relation between the pollen count in the air and the deposition on the ground in a number of stations in Germany, and found that there are so many parameters involved that no practical rule can be set up. Each station follows its own pattern, probably chiefly due to wind and precipitation condition, and to which of the components of the pollen rain (cf. below) is locally dominating. The only general correlation seems to be between size of

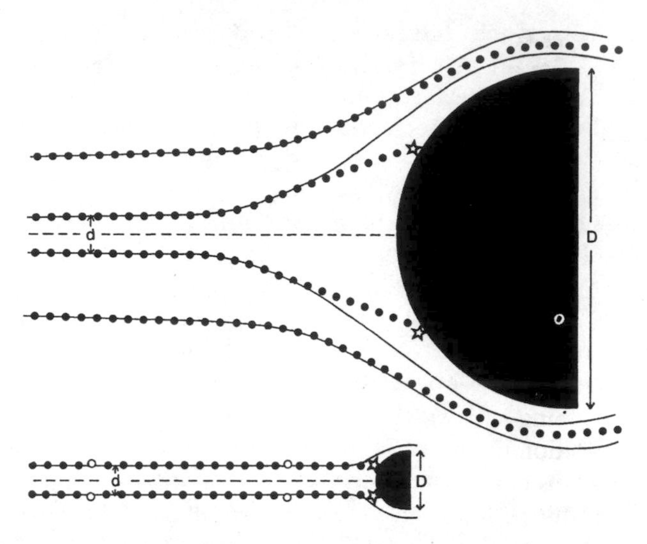

Fig. 2.6. Capture of particles on a broad and a narrow obstacle. At the broad obstacle the airflow is diverted from relatively far off, and particles (pollen grains), following the diversion, are led round the obstacle. The width of the surface from which pollen is captured is less than half the diameter of the obstacle. At the thinner obstacle the air stream is not diverted until very close, and the heavier particles, following their inertia, impinge on the obstacle: the effective width is more than half the diameter of the obstacle. Wind speed and particle size are assumed constant. Cf. also Fig. 2.7.

the pollen grain and deposition: the smaller the grain, the fewer are deposited. The authors ascribe this to the rate of fall; more probably it is a function of both the smaller capture diameter and lower impact inertia in relation to ground obstacles; cf. the discussion of scavenging below.

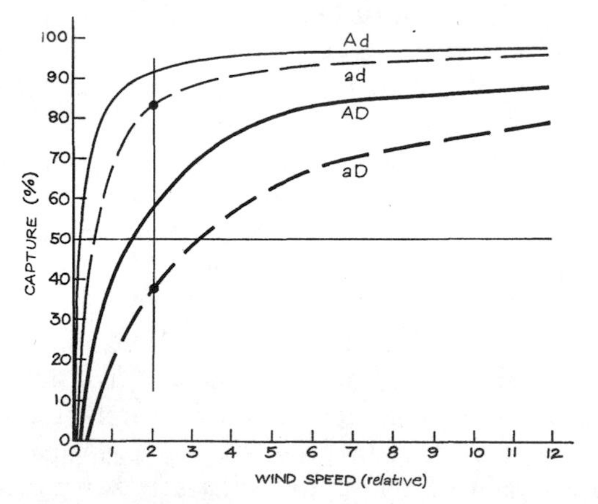

Fig. 2.7. Impaction of a small (*a ca.* 20 μm) and a larger (*A ca.* 30 μm) particle on a thin (*d ca.* 1 mm) and a thicker (*D ca.* 6 mm) cylinder at varying wind speed. The vertical line represents the case of Fig. 2.6. Data from Ogden *et al.*, 1974.

2.1.2.1. Pollen deposition traps

The third quantity, *deposition*, is more relevant in pollen analysis. There are three main ways of studying pollen deposition: by direct experiment, by short-term trapping in artificial traps, or by study of the results of long-term trapping in natural (or sometimes man-made) pollen traps. All three approaches should be explored in order to give a proper understanding of the deposition process—which still is sorely missing.

Various traps have been constructed and used, none of which can be considered to give 'correct' data. Their results must be integrated to give a more complete picture of pollen deposition. Pollen deposition is the capture and immobilization of pollen by ground obstacles (including open water). Under special circumstances pollen can be liberated again and redeposited. This regularly happens when pollen is primarily caught in vegetation, etc., but also pollen captured on the ground may be redeposited after *reflotation*.

Very few of the *experiments* on pollen transport and deposition that have been carried out so far are of direct practical value for pollen analysis problems. Mostly, the distances covered are much too short. For such experiments a marked pollen type must be released under circumstances approaching the natural ones. Because of the strong attenuation of the pollen cloud, such experiments would demand excessive quantities of pollen to give meaningful results. Pollen can be marked either by being released at a time when the corresponding species does not flower, by using exotic pollen species, including one coming naturally from a small, isolated stand of an exotic tree, or pollen that is permanently stained and thus recognizable (cf. Raynor *et al.*, 1966). Staining with fluorescent stains would ease the recognition of the pollen in question, and so would perhaps also some type of radioactivity marking. Inorganic pollen imitation is another possibility.

For *settling* pollen the logical artificial trap is a horizontal surface exposed as near the ground as possible. As early as 1919, Hesselman (1919b) carried out registrations that way, and later many such samplings have been made with apparatus that are, on the whole, variations of the same type. The main problems are:

1. A sampler at ground level is subject to splash from the surroundings, trampling animals, etc. Most samplings have therefore taken place above ground level, which introduces some bias (cf. above).
2. A sampler exposed above ground level creates aerodynamic problems (common to all precipitation gauges), cf. below.
3. The sampler must be sufficiently deep to accommodate also the rain falling during the registration period; values from covered samplers may be biased.

Hesselman trapped his pollen grains in Petri dishes, the bottoms of which were covered with a glycerol-drenched filter paper. To prevent overflowing, the dishes had to be closed when rain began. The values are therefore minimum values. Commonly used (Lüdi and Vareschi (1936), and many others) are cylindrical glass jars with a little glycerol; after having been exposed, the jars are washed with distilled water and the pollen contents concentrated on the centrifuge. Fægri used partly this method, and partly cylindrical zinc vessels with conical bottoms and a glycerol-drenched filter paper, excess water being drained off through a cock in the bottom of the vessel. Both methods suffer from the fact that pollen-bearing insects sometimes perish in the vessels; in such cases total pollen counts are too high. The quantity of pollen of anemophilous species should not, however, be greatly influenced by this. Ritchie and Lichti-Federovitch used covered collectors, the efficiency of which is hardly the same as that of open ones. The greased grids used by the Montpellier school (Cour, 1974) can also be used horizontally, at ground level or higher up.

Some of the early data from this kind of trapping are presented in Table 2.1. Modern investigations have, on the whole, corroborated them.

One of the problems met with in most artificial traps is increased eddying at the mouth of the trap —a well-known problem in registering meteorology. However, the greater rate of fall of raindrops simplifies that problem, and fairly crude measures are satisfactory. For the very slowly settling pollen grains Tauber (1974) has constructed a more sophisticated variant with an aerodynamic collar that does not interfere with the existing turbulence at the orifice, not creating any eddies (cf. tests by Bonny and Allen, 1983). The Tauber trap is widely used, and has become a standard instrument. In experiments Krzywinski (1976) found the Tauber trap less efficient than the simple cylindrical jar (cf. modification by Cundill, 1986).

The data obtained by such traps have contributed greatly to the understanding of pollen transport and deposition. However, the deposition is far from being natural: the basic philosophy of traps (the Tauber model and others) is that of the *settling* pollen grain (*ex analogia* with the settling raindrop). There is every reason to believe that settling is effective only in very calm weather, especially at night, whereas deposition in moving air is a *scavenging* process: pollen adhering to rough surfaces across and through which it is blown by a turbulent wind; cf. Tauber's registration (1977: 102 *et seq.*) of scavenging by a *Salix* shrub; scavenging by the bottom layer of vegetation should be even

Table 2.1. Number of pollen grains per cm² per season.

	Weather ship M 1949 (Fægri, unpublished)	Davos high valley 1934, 1935 (Lüdi and Vareschi, 1936; Lüdi, 1937)		Lightship in the Bothnian Gulf, 35 and 50 km offshore (Hesselman, 1919a)		Transect across N. Canada. (recalculated from Ritchie and Lichti-Federovitch, 1967)					
						Rock desert	Sedge-moss tundra	Dwarf shrub tundra	Forest tundra	Conifer forest	Winnipeg: deciduous forest landscape (the same, 1963)
Picea	0	4.7	4.6	700	400	0	0.8	7.4	157	613	213
Pinus	5.5	5.0	49.1	200	100	0.5	5.6	15.2	107	748	441
Betula	4.7	—	2.5	200	300	2.1	5.7	11.7	319	1032	290
Alnus	—	—	8.3	—	—	0	2.2	7.2	95	681	96
Corylus	—	—	2.7	—	—	—	—	—	—	—	29
Quercus	0.8	—	—	—	—	—	—	—	—	—	568
Populus	—	—	—	—	—	0	0	1.6	7	166	667
Salix	—	—	—	—	—	0	4.7	22.5	24	209	145
Gramineae	5.5	16.7	150.3	—	—	0	4.1	32.8	88	602	429
Total	—	—	—	—	—	5	43.7	301	1087	4955	5250

more effective. A realistic trap should therefore be a scavenger.* The effectivity of scavenging is probably enhanced by electrostatic forces establishing themselves between the passing air and its particle contents on the one side and the fixed object on the other. At any rate, pollen grains are also found on the under-side of sheet collectors.

2.1.2.2. *Natural traps*

Whereas the momentary pictures given by sampling in artificial traps are of great interest, especially because it is possible to relate them to actual meteorological situations, they suffer from extreme local and temporal variability due to variation in pollen output, changing meteorological conditions (wind direction), etc. Of greater direct impact on the evaluation of pollen-analytic data are registrations over a period which has integrated short-term fluctuations. The classical natural traps are moss and lichen cushions, but also other surficial deposits have been used, including bottom deposits of (artificial) water tanks (Potter, 1967), ingrown bark (Adam *et al.*, 1967) and *Sarracenia* leaves (Terasmäe and Mott, 1964). Ritchie (1974) found that recent samples from lake mud reflect accurately the surrounding vegetation, whereas samples from moss cushions are subject to large fluctuations, which are ascribed to the influence of local vegetation (cf. also Cundill, 1984).

* Some experiments have been carried out at Bergen (Ammann, unpublished), using a glycerol jelly-drenched nylon 'scour-puss' as a trap. The experiments were promising, but were not carried through. Currier and Kapp (1974) have used a scour-puss for sampling airborne (not depositing) pollen. Benninghoff (1965) also carried out some tentative—successful —experiments with this type of trap, coated with glycerol jelly or silicone. He suggests placing a filter below the trap to intercept any grains washed off after capture.

Heim (1971) has shown that the use of natural pollen traps may introduce additional complications. Pollen spectra at different levels of a moss cushion vary, and so do registrations from mosses on the ground and arboreal mosses, as well as cushions in different exposures. It is not possible to decide whether the statistically significant differences between tops and bases of moss cushions are due to (short-term) changes in flowering or to selective destruction*). At any rate the observations cast some doubt on the usefulness of moss cushions as pollen traps. Only the tips of the moss plants should be used for comparison with actual conditions in order to avoid selective destruction, but then again the integration effect over time may be lost.

The accumulation zones of glaciers represent enormous natural pollen traps, which so far have not been much studied. Even in the High Alps, Bortenschlager (1969) registered a total of more than 200 pollen grains per cm^2 per year. This may be of importance for the interpretation of Late-Glacial deposits with very little autochthonous pollen (cf. Larsen *et al.*, 1984: 150).

When results of such registrations are used for comparison with fossil spectra, one must also take into account the rates of destruction of different pollen types during the fossilization process, a parameter which is usually unknown. Such data therefore have obvious shortcomings, especially for quantitative comparisons, but they may be used with some caution. They may furnish a qualitative, non-numerical understanding of relevant processes. They are not useful for numerical modelling.

Profiles across a vegetation border give a consistent picture (e.g. Anderson, 1955; Fedorova, 1959a; Eisenhut 1961): at the borderline between two vegetation types, the composition of the recent pollen rain changes abruptly. At the border between forested and non-forested area the incidence of forest tree pollen declines very rapidly in the course of 100–200–300 m to some (3–)10 (–30)% of its value at the forest edge. Forest edge trees generally flower well. On the other hand, Hicks (1986) found very small differences in Tauber trap registrations across *orographic* vegetation borders. The complications of up-winds may explain this.

Further away, values are remarkably constant up to a distance of at least 3–4 km. It is assumed that these lower, constant values represent the composite pollen rain of the whole region, not only of the stand under investigation (cf. Fig. 2.12).

Heim's registrations from recent moss cushions (1970) indicate that the AP/NAP ratio reacts very sensitively to the forest cover. Even quite small clearings (± 200 m) depress the AP percentage from *ca.* 70 to *ca.* 20, and on the other hand a small birch copse in heathland raises it from *ca.* 30 to almost 90 (cf. Fig. 2.8).

Very sharp decreases in pollen incidence are also registered at the transition between land and sea deposits (J. Muller, 1959). Especially when pollen incidence can be compared to that of remains of marine organisms, precise indication of the location of former beaches—a problem of great importance in oil geology, among others—may be obtained by this registration.

2.2 Aerial travels of pollen grains

Each pollen dispersal unit travels its own way, and only by statistical inference can the average result be predicted. However, there is always the chance that grains may defect from this because of unusual external conditions, e.g. wind directions during flowering time.

The uniformity of the pollen rain should not be over-rated. In a mixed vegetation plants flower at different periods, and wind and turbulence may vary. Thus, the pollen rain from a particular patch of vegetation may be deposited in different places in different years, as shown in Lüdi (1947). The further away from pollen-producing vegetation, the less pronounced this effect, and in the course of a few years the variations will have equalized over the surrounding surfaces. As pollen analysis samples usually represent the deposit of many years, the effect is not noticeable in ordinary diagrams, but its possible presence should not be forgotten if there are inconsistencies in the record. In sediments,

* This statement is based upon the original counts, copies of which have kindly been placed at our disposal by Dr Heim; cf. also Cundill (1984).

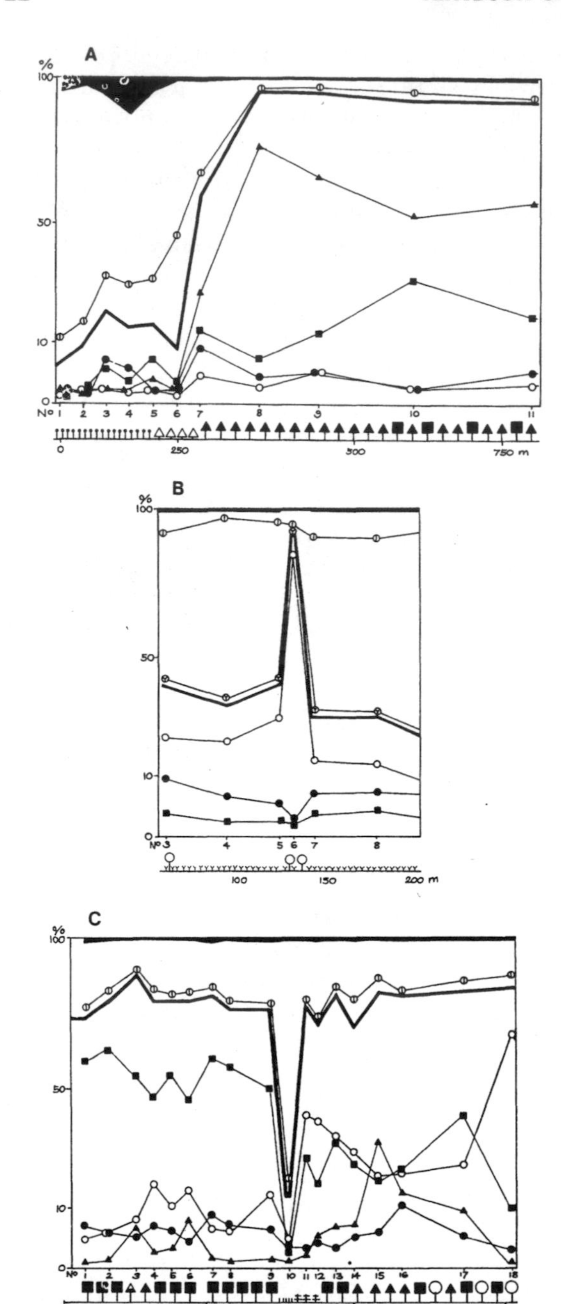

Fig. 2.8. Percentage pollen records from moss cushions. A: Transition between cultivated field and beech forest; B: birch copse in heath; C: clearing in oak and beech forest. Black: Cerealia. Other signatures as in Fig. 6.3. Subordinate constituents omitted. Data from Heim, 1970. Reproduced by permission of the copyright-holder: J. Heim.

curves seem, on the whole, to be remarkably smooth, which may be due to an integrating effect of the deposition mechanism in water: bioturbation (cf. Swain 1980). Possibly, this would be different in a varved deposit, but the problem has not been taken up in those investigated until now. Trap sampling shows large variations.

Sometimes remarkable concentrations of unexpected pollen types are met with in diagrams. The registrations are not reproducible in other sections from the same deposit, and if contaminations can be ruled out it is tempting to ascribe such short, local maxima to an effect corresponding to Rempe's experience with the *Picea* pollen rain over Göttingen (cf. below). Even without such contributions registrations may differ by 100% or more from one year to another.

The concept of the pollen rain has been criticized by Tauber (1965, cf. 1967*), who maintains that a major part of the pollen transport takes place not above the canopy, but via the trunk space. Apart from the fact that there is no trunk space in the case of NAP, his conclusions are hardly tenable even for high-stemmed forests.

1. His own measurements show that the quantity of pollen in the trunk space is much lower than one would expect if that were the main transport route for pollen grains (1967, Table III).
2. Even more decisive: S. T. Andersen (1967) has shown conclusively that in a forest of varying composition there is an almost perfect correlation between the composition of the pollen deposited on the forest floor and that produced in the canopy immediately overhead. Such a good correlation, which is corroborated by other investigations (cf. Heide and Bradshaw, 1982), would be unthinkable if there were an appreciable horizontal pollen transport inside the forest.

However, Tauber's division of the pollen rain into various components with a different dispersal

* The idea that the concept of pollen rain should implicate 'a more or less vertical fall of pollen grains through the air' (Tauber, 1977: 78, orig. Danish) hardly exists anywhere but in the author's imagination.

history is in itself of merit, even if the components are not entirely distinct. The first consists of pollen falling straight down, the *gravity component*,* which is transported (horizontally) a few dekametres at most. It corresponds to Tauber's trunk space component, apart from the fact that it is not transported horizontally. Semerikov and Glotov (1971) consider 80 m the maximum dispersal distance inside a forest, and S. T. Andersen (1967) even considers 20–30 m. Obviously, this distance very much depends on the (arbitrary) limit for detection, as well as on the character and the scavenging capacity of the undergrowth. A dense undergrowth both impedes horizontal air movement and increases scavenging. Edges of natural forests (Tauber's were plantations) are usually dense and more scavenging than the interior.

Probably, a major part of the gravity component is formed by pollen which for various reasons failed to separate into single grains, coming down with remains of flowers after anthesis, with leaves in the autumn, or even coming down with the seeds when fruits break up (cf. also Tauber's data (1977) for 'refloatation . . . for *Betula*').

A large part of the gravity component is redeposited, including wash-down by rain of pollen grains originally adhering to leaves and branches in the canopy or in the upper levels of the under-vegetation. It is the only component in which redeposition and rain-out is part of the regular deposition mode.

The second component, the *local pollen*, comprises that part of the pollen output which went into a diffusion cloud the centreline of which remained more or less parallel to the ground, increasing in diameter and being scavenged by the ground cover, (cf. Fig. 2.9). The ground cover in this case is the upper surface of the vegetation, which may be anything from a *Calluna–Sphagnum* mat to the top of the tree crowns in a mature forest. A small distance is frequently registered between the source and the maximum pollen concentration. It may be due to ineffective scavenging during the first metres of flight, and also to thermic updraughts at the sunny side of trees.

The third component, the *regional component*, is truly airborne, comprising the pollen caught by updraughts and transported to and via greater altitudes than the tree-tops and the air-current conditions created by them, (cf. Fig. 2.10).

Little is known about the effect of the mode and timing of pollen relase, both on local and regional pollen rain components. Release of pollen from an anemophilous blossom is probably always discontinuous, and may be so fast and complete as to approach puff release. Such releases are transported in narrow and unpredictable diffusion clouds (Fig. 2.11). Release at a period of active thermic turbulence and updraft brings a large proportion of the pollen output into the regional component, but also in this case quantitative data are missing. As a general, qualitative rule it may be said that the longer the release, the more diffuse the dispersion.

For the evaluation of the two components it is of interest to know a few actual parameters, even if the theoretical calculations based upon them have often given figures that are very far from being

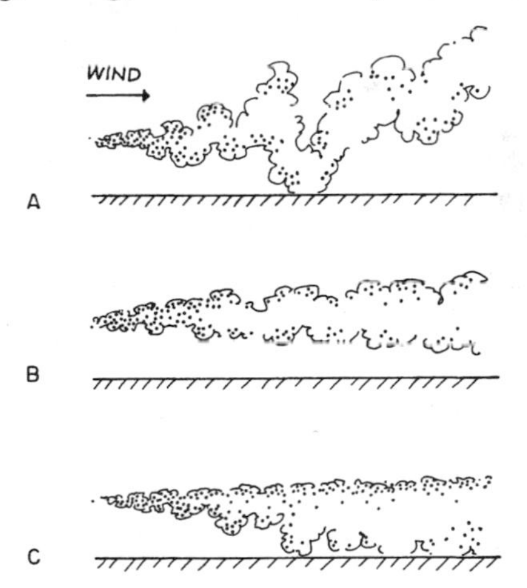

Fig. 2.9. Characteristic plume patterns after release from a point source under different atmospheric conditions. A: Unstable; large eddies; B: stable; small eddies; C: inversion layer above; eddying downwards. Dimensions vary with height of the release point. From Bierly and Hewson, 1962; redrawn. Reproduced by permission of the copyright-holder: American Meteorological Society, Boston.

* Terminology is not consistent. As used by Janssen (1966) 'local' corresponds to our gravity and 'extra-local' to our local. 'Regional' and 'extra-regional' both correspond to our regional.

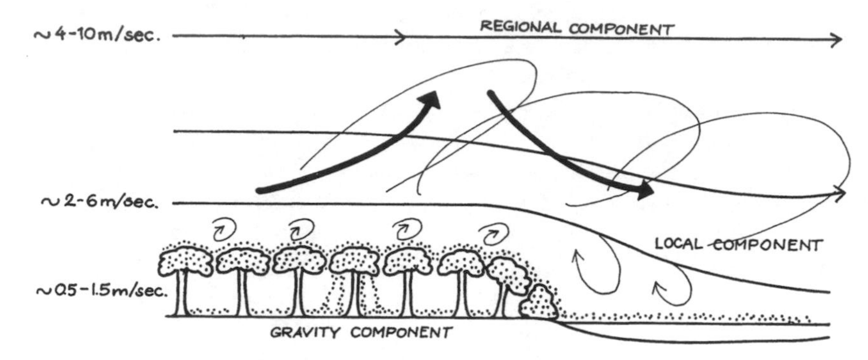

Fig. 2.10. Normal wind conditions (lines), pollen transport and deposition (dots) in a forest and a lake. Thin lines: horizontal wind and turbulence; heavy lines: resultant updraft. From Tauber (1965); modified. Reproduced by permission of the copyright-holder: H. Tauber.

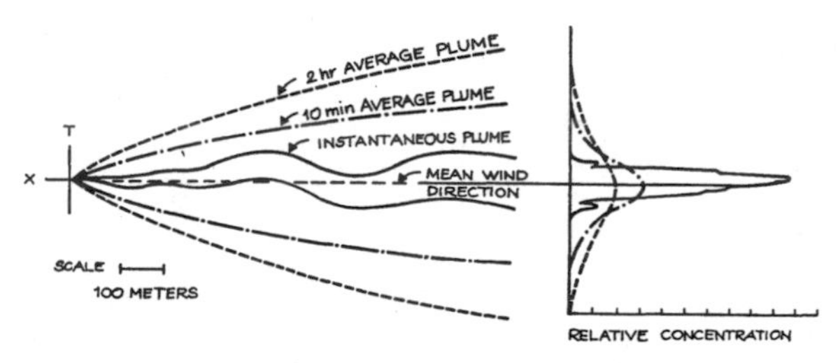

Fig. 2.11. Plume shape in dependence of length of observation time. Data from Slade, 1968. Reproduced by permission of the copyright-holder: US Department of Energy.

realistic (Fig. 2.12). This is also a reason why we do not enter upon the various aerodynamic theories and formulae. They are discussed in the literature referred to.

Di-Giovanni *et al.* (1988) have subjected the local (i.e. gravity plus) pollen component to a stringent numerical treatment, one of the main results of which is 'the problem of the multiplicity of undistinguishable sources that occurs in natural woodland' and also an almost complete absence of the numerical data necessary for calibration and calculation, which therefore have to be based upon a multiplicity of assumptions instead. The paper demonstrates very succinctly how complex are the conditions in the forest and how impossible to encompass everything into one mathematical reasoning. One of the 'undistinguishable sources' in this case is the through-fall, the redeposition with the first rainstorm of the great quantity of pollen that never came out into the air, but was trapped in the canopy of the mother tree. Another is the unspent pollen falling down with the flowers. These two factors certainly change the pollen incidence at the first 5 m off the stem (Di-Giovanni *et al.*, 1988: Fig. 7) drastically in relation to the calculated attenuation curve.

The rate of fall of pollen grains in calm air varies from *ca.* 2 cm/sec for small and very buoyant ones to *ca.* 50 cm/sec for the heaviest wind-borne pollen types (Knoll, 1932; Dyakowska, 1937; Ichikura and Iwanami 1981). Even if such values suffer from great experimental errors (cf. Durham, 1946), they give an indication of the magnitude.

The horizontal air speed in a medium wind is 5–10 m/sec, thus almost two magnitudes greater. However, wind is never laminar; it is strongly turbulent with vertical velocities similar to the horizontal ones: 1–2, up to 5 m/sec. It is easily seen that, possibly apart from the heaviest pollen grains, the rate of fall is a rather unimportant quantity compared to the velocities of the air masses, and pollen grains in the open can be considered part of the air mass transporting them. The movements of air masses have been extensively studied, both from a theoretical and a practical point of view, but especially the former literature is of little importance for the evaluation of pollen trasfer data.

One of the most advanced models, which is often used as a basis for calculation of pollen transport, is Sutton's formula. Even if it represents an elegant mathematical approximation to conditions in turbulent air it is still far from describing

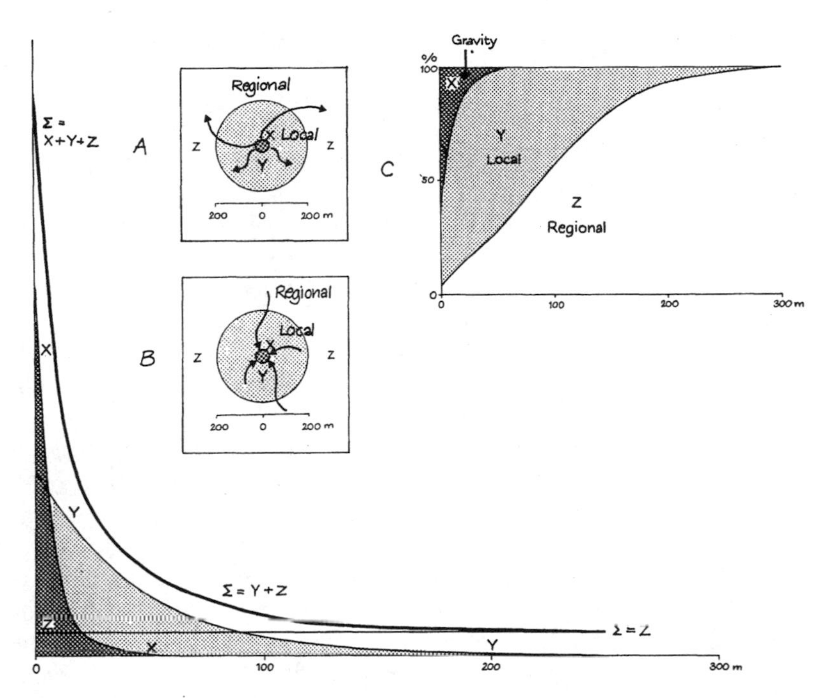

Fig. 2.12. Dispersal types in relation to distance from source. Assumptions: 64% enter the gravity component (X), attenuation 50% per 5 m; 32% enter the local component (Y), attenuation 50% per 30 m; 4% enter the regional component (Z), no attenuation. Abscissa: distance from source; ordinate: pollen content of the air (arbitrary scale). Heavy shading: gravity deposition area. Light shading: area of local deposition. *Inset A:* dispersal of components from a single source. *Inset B:* contribution of the three components to the pollen deposition in one point. *Inset C:* percentage composition of pollen emission from an individual source in relation to distance from the source.

what does actually take place. There are three main reasons:

1. The source of pollen is usually very diffuse, the point source of models is replaced by an undefinable source area. Transformation of the model to account for this process presumes so many approximations that the validity of results is questionable.
2. The formula deals only with one fraction of the pollen rain, viz. plumes with an approximately horizontal centreline, i.e. the local component.
3. The formula does not account for the scavenging function, different scavenging efficiency of different surfaces and of the same surface under different conditions (dry/wet, etc.). These points are elaborated below.

Individually, pollen grains are emitted from a point source, but the usual comparison with a point source emitter at some level above ground (a smoke stack) is invalid in one important respect: apart from the not-too-important case of isolated trees, pollen grains are emitted from what is in reality ground level, namely the surface of the vegetation, even if that surface happens to be a canopy 20–30 m up in the air. Thus, the starting point of the diffusion cloud is in itself situated at altitude zero, in a position to give a much higher initial rate of scavenging of solid particles than usualy theoretical considerations assume. This may explain some of the discrepancies between theoretical data and actual observations. The different character of the air layers immediately above ground is an important factor in this connection.

Immediately above the surface, i.e. the top of the vegetation, turbulence is negligible, and air movement in the extreme bottom layer is even laminar. The thickness of this laminar boundary layer varies with wind velocity and with the roughness of the surface. Very few pollen grains are released by explosion—and that would not bring them very far anyhow; a realistic magnitude is 1 cm (cf. May, 1967, Fig. 1), although Cuellar (1967) reports 6 cm in *Celtis laevigata.* This theoretical possibility of bringing grains out into turbulent air can be neglected. However, many pollen grains are released by being shaken out of anthers or off whole branch systems, which creates local turbulences holding the grains in the air for some time and, as the case may be, bringing some of them out into permanently turbulent air strata. But many grains are picked up by the low-turbulent boundary layer and transported a longer or shorter distance without ever becoming what Ludlam (1967) calls 'effectively airborne'. Interestingly enough, he sets the upper diameter for spherical particles to become effectively airborne at 60 μm, which means that most pollen grains at least have a theoretical chance of becoming airborne. On the other hand, unless a particle reaches *ca.* 10 cm above the surface its fate is very uncertain, and there is a great probability that it will never become airborne. Here is a point where the differential rate of fall may be of consequence.

According to Scott (1970) the pollen concentration 46 cm above the flowers in a sugar beet field was about 80% of that at flower level. This would indicate that 20% of the pollen originally released into moving air never becomes effectively airborne. Presumably, the latter figure is higher in a tree canopy.

Comparisons between dispersion and deposition patterns give an idea of the quantity of pollen that is scavenged close to the source. Studies of *Zea* pollen released from a limited plot, diameter 18.3 m, show attenuation 50:1 of airborne grains between the source and 60 m off (Raynor *et al.*, 1972a) whereas attenuation is 2500:1 for deposition. The much heavier deposition near the source corresponds to the grains that never became effectively airborne. For *Phleum pratense* the equivalent figures are 20:1 and 100:2 (Raynor *et al.*, 1972b). The effect of size is very distinct (*Zea* $\approx$ 80 μm, *Phleum* $\approx$ 30 μm).

Turbulence in itself is a circular motion that would not bring air masses or pollen grains either up or down. With increasing downward diffusion the air collides with the ground and may deposit its load of particles. In this way the air is gradually 'cleaned' of its pollen. However, most anemophilous pollen grains are liberated during day-time and in dry weather. There then usually exist thermals, columns of air heated from below and rising turbulently. As long as heating goes on, more air will rise than goes down, so turbulence has a positive effect in bringing pollen grains up into higher strata.

The upper limit for the convective ascent of air is given by the thermal inversion over which cumulus clouds indicate the position of actual updraughts. According to Ludlam (1967) the individual updraughts do not last 'much longer than is needed to move air from its bottom to top: thus the duration of even the biggest, as indicated by the life of the individual cumulus, seldom exceeds about 20 min.' During this period the whole bubble will also move downwind.

The existence and structure of such air cells can be demonstrated directly by radar observation. The elementary cell is some 1–3 km in diameter, reaches some 1–2000 m in height, and has an individual lifetime of 20–30 min. Such cells form composite convection cells, diameter 5–10 km, persisting for several hours. Upward velocity of cell tops reaches 0.5–1.5 m/sec, and horizontal expansion 0.5–1.0 (Hardy and Ottersten, 1968).

Thus, some pollen grains will have reached the inversion layer when the bubble collapses, whereas others will be somewhere in between, which explains the relative constancy of the pollen content of turbulent air in a vertical profile up to a certain level, well known from Rempe's pioneer investigations in the 1930s (Rempe, 1937; cf. also Hirst and Hurst, 1967).

Sometimes this under-the-cumulus transport of pollen has interesting effects. Rempe (1937: 134) has shown that an upward air current under a cumulus cloud took a *Picea* pollen rain from the Harz mountains 34 km southwest to Göttingen, where the pollen dropped in a beechwood district when the cloud dissolved. A total of 39 610 of

44 625 pollen grains collected by Hesselman in the Bothnian Gulf (cf. Section 2.1.2.1) were caught in the course of 2 of the 40 days of observation: it is probable that this concentration may be due to similar transports with cumulus clouds.

After the collapse of a thermal air bubble pollen grains move downwards again, but as long as there is a positive convective air movement the net result will be to keep pollen in the atmosphere. Only when cooling starts, i.e. during evening and night, is there a negative resultant movement which will bring pollen down. Low-level inversion due to radiant cooling can interfere with this pattern. Late-night pollen concentration maxima in the lower strata of air are well known from literature (cf. Spieksma and den Tonkelaar, 1986).

In the very calm night air pollen grains will have a tendency to sort themselves out according to their rate of fall, helped by downward air movements due to cooling (cf. Rempe, 1937, Table 5).

This return of pollen grains to the lower air strata at night-time is shown by some of the hourly ragweed pollen trapping data (1.5 m above ground) published by Ogden and collaborators (1971). In addition to the morning maximum of pollen being released, there is another one at about 22–24 hours, sometimes reaching as high values as the former one (cf. also Fig. 2.13).

Even if nightly temperature inversions near the ground may stabilize the air and prevent particles from being transported further down, it is to be supposed that a major portion of the pollen grains brought up into the air at day-time, settles out at night. According to Stix and Grosse-Brauckmann (1970) there may be a complete fall-out at night, even of small pollen grains (*Betula*). This has the important repercussion that the normal maximum 'life-time' of a pollen grain in the air is 1 day, and the corresponding maximum transport distance will be that of 1 day, which seems to be equivalent to some 50–100 km, but it is well known that much longer transports do occasionally take place.

In connexion with the division of the pollen rain into three components it is of interest to quote the results of pollen trapping in the Bialowieza forest (Borowik, 1963, 1966):

1. The greatest catches of pine and oak pollen were not made in the forest, but in the open, partly on an open bog (with some pine) inside the forest area, partly on grassland outside.
2. Whereas the AP count in the grassland station was of the same magnitude as in the forest, practically no NAP came into the forest from the grasslands. The results of the counts in 1961 are summarized in Table 2.2. (For various reasons the figures do not add up exactly.) The counts in 1963 gave similar, but lower figures.

Borowik's figures seem to indicate that the flow of pollen immediately above the surface is more important than the large-scale turbulence, at any rate for the NAP. It is more difficult to evaluate which factor is more important for AP, especially as there must be a strong negative convective air movement over the open areas during the night, which would tend to concentrate pollen deposition there anyhow, operating as a kind of pollen trap. The nightly fall-out will be selective inasmuch as heavy grains will be more completely eliminated from the air.

It should be added that our knowledge of the details of pollen transport in the air in the Tropics is very meagre and inconclusive. Existing records from tropical areas give unexpectedly small pollen deposition values (Schnetter, 1972: 1163 grains per cm^2 per year in Colombia. Liu and Colinvaux (1988) arrive at some 5000 grains in the Amazonian rain forest). The predominance of insect-pollinated plants may be one of the explanations for low figures.

2.2.1. *Pollen transport distances*

The problem of pollen transport distance is of importance in the breeding of anemophilous plants, and a great deal of material has been presented by forest researchers (cf. Scamoni, 1955). The net result of the various means of pollen transport usually gives a characteristic curve of pollen deposition from individual trees, sinking from very high values at the source toward an asymptotic background value, which is usually reached at

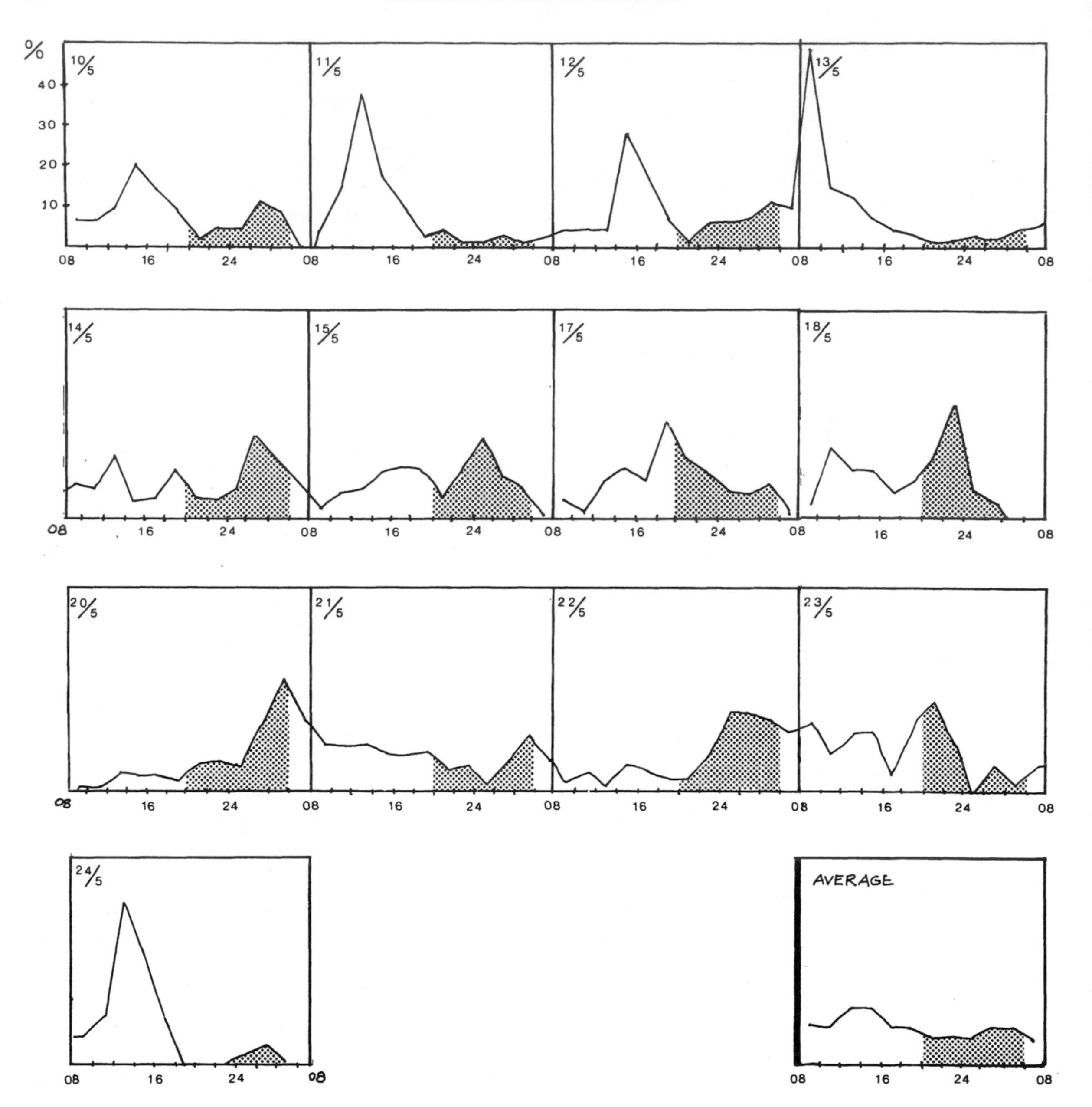

Fig. 2.13. Birch pollen in the air, 2.5 m above the ground (Burkhard trap): 2-hourly variations as percentage of the day's total. Steady meteorological conditions, no precipitation. Night observations shaded. From unpublished report by Svenn Sivertsen (Bot. Inst., Bergen). Reproduced by permission of the author.

some 300 m distance (Fig. 2.12; cf. Lanner, 1966). Beyond this distance the contribution of the individual tree cannot be traced in the general pollen rain.

The ascending air at a large tree may sometimes disturb this pattern of pollen deposition as a function of the distance from the source. Murkaite (1970) quotes figures from *Picea abies* (grains per cm^2 per 14 days) in dry, warm weather (see Table 2.3).

Table 2.2. Pollen data from Bialowieza (after Borowik, 1963)

	1961					1963 Pollen transport
	Pine	Oak	AP	NAP	Total	
Grassland	4180	428	4709	7932	12,642	2379
Open bog	9470	338	9827	299	10,036	3015
Pine forests (average)	3504	87	3606	52	3660	1156
Oak forests (average)	904	123	1039	54	1090	451
Total	33,771	11,590	35,640	8727	44,331	

Kozumplik and Christie (1972) have shown that the pollen deposition curve of ground-cover plants (*Dactylis glomerata*) declines toward the background value much more abruptly than that of forest canopy species. At 25 m distance an approximately asymptotic value was reached. That would mean that the pollen influence of the individual ground-cover plant would be concentrated within an area approximately 1:100 of that of a tree. Considering that the area production of pollen may be of the same magnitude in trees and non-trees, this means that the local influence of the ground cover must be much higher than that of the forest canopy. This, again, explains the usually very violent reactions of the curves of the NAP diagram when locally produced pollen is included in the registration (cf. Fig. 7.19).

Hirst and Hurst (1967) quote a case in which a pollen cloud generated over Britain could later be found as a pollen concentration over the North Sea. The following day it was 300–400 km off the coast. Between the pollen cloud and the coast there was another cloud of fungus spores belonging to species that liberate their spores during the night. This observation is in apparent contradiction to the night-and-day pattern discussed above, but it should be taken into consideration that the transport took place over the sea, where convection conditions are different, and fungus spores are usually more buoyant than pollen grains.

Table 2.3. Pollen dispersal from forest edge, after Murkaite, 1970

Distance from forest edge, m	5	20	200	320
Pollen deposition	720	2085	6564	6484

As indicated above, the distance 50–100 km forms a natural limit of aerial pollen dispersal. It is self-evident that the greatest quantities are deposited long before this limit has been reached; on the other hand, there are other grains which remain in the air for more than 1 day, and which can therefore be transported over great distances. Naturally, there is in this respect a great difference between buoyant pollen types that are not completely eliminated during the night, and the others. The long-distance transport problem for species like *Fagus* and *Picea* is quite different from that of *Pinus* or *Betula*. Under equal circumstances McAndrews and Wright (1969) found that pine pollen in the prairie area in North America was transported in quantity 300 km from the nearest pine forest, whereas pollen of *Picea, Abies* and *Pseudotsuga* (and deciduous trees) was only transported some tens of kilometres.

The long-distance transport problem will be taken up from the pollen diagram point of view later (Section 7.6.3). Here, we shall confine ourselves to dealing with the problem as it affects pollen dispersal. Even if maximum transport distances are very large, only a very small fraction of the pollen will have any chance of spreading outside the region of production, i.e. beyond the frequently rather insignificant area covered by the day's horizontal transport. Erdtman's observations show that the atmosphere over the Atlantic Ocean does contain pollen several hundred kilometres from the nearest shore (Erdtman, 1937: 189), but also that the numbers met with are negligible compared with those of a forested area (6 grains per 100 m^3 of air vs. 18,000).

Observing that the pollen catches at sea *ca.*

20 km offshore were equal to those above the forest, Sarvas (1955) concluded that the updraught responsible for keeping pollen afloat in the air over land is ineffective over (cold) water where the convection currents collapse. This would explain the heavy pollen deposition immediately outside a forested shore, and also the marine over-representation of pollen types especially susceptible to this phenomenon. Hyde's observation (1945) that pollen deposition at Bishop Rock, 80 km offshore, was of the same magnitude as on land would indicate that pollen may travel a long distance over the sea, perhaps longer than explained by Sarvas (cf. also the observation by Hirst and Hurst, referred to above).

There are many spectacular examples of long-distance pollen dispersal, such as Bassett and Terasmäe's (1962) finds of *Ambrosia* pollen 600 km north of its northernmost occurrences in Canada, or Maher's (1964) finds of *Ephedra* pollen all over the USA, including interception of grains in the atmosphere over Iowa. Compare also the data from weather ship station M given in Table 2.1. The position is lat 66°N, long 2°E Greenwich, 450 km from the Norwegian coast and some 1700 km from the Middle European oak region, from which the oak pollen must have come, if not across the Atlantic.

Bortenschlager's observation (1969) of *Ephedra* pollen that had travelled 3000 km is a spectacular example of long-distance transport, as are Hafsten's finds of pollen grains of *Ephedra* and also *Nothofagus* (1960a) in Tristan da Cunha and Gough Islands. If these grains came from South America, as Hafsten presumes, the transport distance would be 4500 km. On the desolate arctic Devon Island, McAndrews (1984) registered an almost exclusively long-distance transported pollen flora with minimum travel distances over 1000 km. Apparently, pollen had been unable to cross the polar front: even *Pinus* pollen was scarce (McAndrews, 1984: Fig. 2).

Clearly, long-distance dispersal is not covered by the tranquil mode involved in the day-and-night model discussed so far, but must be connected with the major movements of air masses in the atmosphere (Ludlam, 1967). A frontal storm can lift air masses several kilometres up in the air in a very short time, and thus place pollen grains and other terrestric material far above the day-and-night cycle. Mandrioli *et al.* (1984) found attenuation of the pollen and spore concentration in the air up to 3–5 km altitude, after which fungus spores regularly, and pollen to a much smaller extent, again increased. They suggest that this upper population of particles 'indicate the existence of differing situations in the atmospheric layers involved' (Mandrioli *et al.*, 1984: 51).

By following trajectories backwards it is possible to indicate that air masses with a characteristic component of pollen grains, pathogenic fungus spores, or even dust, may come from, or rather via, rather far-away places and still preserve their identity without too much attenuation of particle contents from diffusion and fall-out. Such situations are rare, but not so rare as to be negligible in the interpretation of pollen diagrams averaging long periods of time. The air masses in question may have been under way for 2 or 3 days (Lundqvist and Bengtsson, 1970; Christie and Ritchie, 1969). Tyldesley (1973) found appreciable quantities of arboreal pollen (up to 30 per m^3) in the air in the treeless Shetlands, 250–380 km away from the nearest forests, in connection with favourable meteorological conditions, i.e. cyclonic storms.

The extreme transport distances are conceivable if grains are incorporated in higher air masses, penetrating 'beyond the weather', beyond any inversion layer. The formation of the usual summer cumulus type of inversion layer is dependent on air humidity. The possibility of particles rising higher is therefore greater in dry areas, and thus it is interesting that pollen of typical steppe/desert plants such as *Ephedra* and Chenopodiaceae are frequently recorded in long-distance transport, e.g. in sediments from the high seas (Stanley, 1967: 199, or in the examples quoted). But Newmark and Salley (1972) caught no pollen in Antarctica during the period October to February. Nor did Fægri (unpublished) observe pollen in moss samples from Bouvet Island. Kappen and Straka (1988) observed quite a number of pollen grains in a moss cushion from the American side of Antarctica, but practically none in samples from the Australian sector.

Cross-wind mountain ranges force the air masses upwards and create 'lee waves' at high level for a longer or shorter distance before again coming down to ground level behind the mountains (cf. Ludlam, 1967). If, as is often the case, the vegetation on the two sides of the mountain range is different, the transport of pollen from the windward side is a factor to be taken into account. Changes in the incidence of pollen from the windward side may then simply reflect changes in wind pattern, and not have any relation to vegetational changes behind the mountains. On the other hand, it is difficult to imagine a general change of wind pattern that would not also cause some vegetational change. During one flowering season Scamoni (1949), in a profile across the Riesengebirge (Sudeten), found fairly uniform deposition (grains/cm^2 day) of *Pinus* and *Picea* between 450 m and the timberline at *ca.* 1300 m. Above the timberline values of *Picea* remained the same, whereas those of pine were almost 6 times as as great. Since these were absolute counts, the usual percentage effect is absent and the possibility may be that this is a lee wave (eddy) effect.

2.2.2. *Precipitation and pollen dispersal*

The ordinary dispersal and deposition pattern of pollen grains in dry air is complicated enough. In addition, other phenomena, especially meteorological ones, disturb any regularity there might have been. The first of these is precipitation. Observations indicate that even moderate amounts of precipitation clear the atmosphere of most of its suspended particles in a short time, and after extensive rains very little remains (McDonald, 1962; Chamberlain, 1967). From incidental observations Tauber (1967) concludes that the rain-out proportion of pollen deposition is *ca.* 15%, but the problem needs more thorough investigation than given it up till now.

According to Scott (1970) the effect of rain on pollen concentration in the air varies with the time of day. In the morning, i.e. at pollen release time, when the maximum concentrations are still in the lower strata, rain washes pollen out very rapidly. In the night (after a dry day) when there is a great concentration of pollen in higher air strata, rain may cause an immediate temporary increase of pollen concentration near ground level.

Some very preliminary counts made in a *Pseudotsuga* stand in Washington (Fahnestock, unpublished) indicate that pollen deposition with precipitation as throughfall is 50 times higher than with precipitation in the open. Most interestingly, very little pollen is transported with the stemflow. Probably, the throughfall pollen was initially deposited within the tree crowns. Since most deciduous trees flower before leafing this scavenging effect is probably smaller with them than with *Pseudotsuga*.

Rain-out of particles from the air is an important factor in long-distance pollen transport: sooner or later, rain is generated in an air mass, and that will be the end of its pollen transport. Pollen grains may form condensation nuclei or at any rate primary inclusions in rain droplets, or they may be carried down by impact. This again would mean that only pollen grains that come up above the rain cloud strata have a chance of remaining in the air for longer periods.

2.2.3 *Deposition*

On the whole, pollen grains behave as parts of the relevant air masses, since their rate of fall is so much smaller than the movement of the ambient air. However, there are two important qualifications: Whereas the air itself is unaffected by touching the ground, or any solid object, pollen grains may stick and thus be taken out. That means that in a diffusion cone touching the ground, solid particles are all the time being taken out of the lower parts. This scavening effect is a function both of the morphology of the surface and of its physical properties, a wet surface being 3 to 10 times as effective as the corresponding dry one (Chamberlain and Chadwick, 1972). In a comparison between pollen deposition in a hummock and an adjacent hollow Dupont (1985) found a distinctly higher scavenging effect of the rougher surface of the hummock as compared with the smoother one of the hollow (Fig. 2.14). The author's explanation

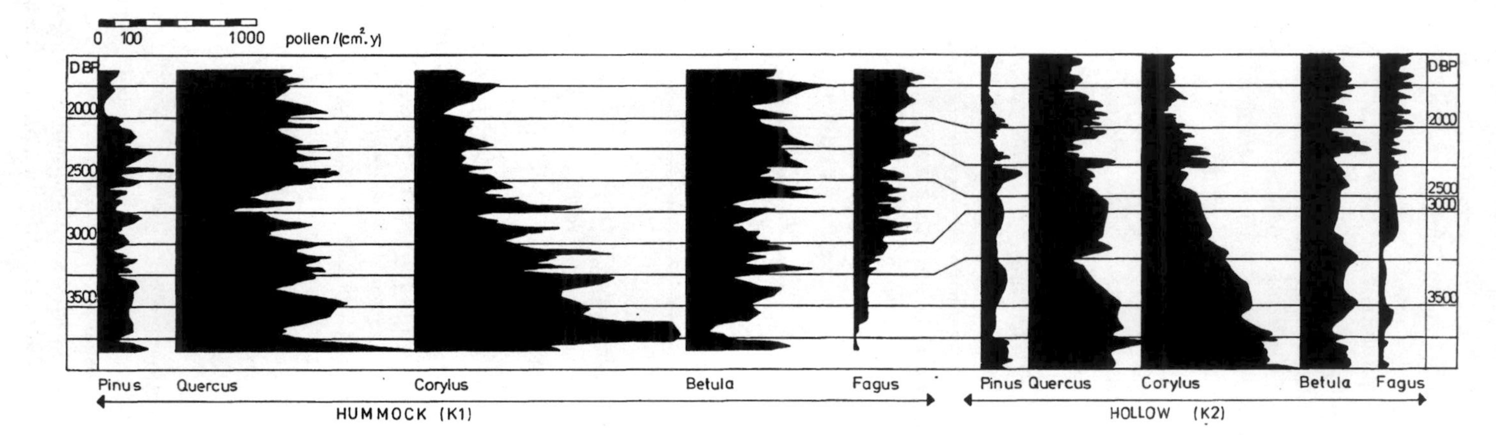

Fig. 2.14. Pollen influx in a hummock and the adjacent hollow. The diagrams encompass the same period, but spectra are not synchronous. From Dupont (1985). Reproduced by permission of the copyright-holder: L. Dupont.

(pollen being floated away from the hollow) is not convincing.

2.2.3.1. Redeposition

Another problem is that of redeposition of pollen grains which have already been taken out of the air. Erdtman described this in 1938 (p. 130), finding *Zea* pollen in the air in Sweden during midwinter. Tauber (1967) registers a very large percentage of 'refloated' pollen, up to 95% of the total catch (of *Fagus*) being refloated with an average percentage of all species of 55 to 34, depending on the size of the basin. Apparently, the shore vegetation contributes most of this pollen, whereas the refloatation percentage for more distant trees (*Pinus*) is considerably lower: 5% compared with 2%. As Tauber's 'flowering period' stretches till 1 August, some redeposition may have taken place also during this period. His own explanation is that the material represents pollen trapped on leaves and twigs, washed off again by the autumn rains (cf. Fig. 2.15. Most of it would then probably be washed straight down, but some drops disintegrate and splash out into the lake with their contents. As this is assumed to be a rainfall effect, no further dispersal would then take place. Tauber's redeposition data have not been generally corroborated. Most investigations show very little or no pollen in the air outside the flowering period. This pertains both to the many analyses of the pollen contents of the air and to actual registrations of pollen deposition. Fægri (unpublished) has found the values given in Table 2.4. Redeposition is not included.

It seems remarkable that pollen should adhere surficially to twigs, etc., for such a long time as presumed by Tauber. A more likely explanation would be that it was trapped in a more secure way, e.g. like the *Betula* pollen exines found by Krzywinski (1976) trapped in the catkins and released at seed dispersal time. Apart from such cases, larger redeposition quantities in natural terrestric habitats should be a more or less freakish event. The reflotation registered by Erdtman (1938) refers to pollen that had blown up from the ground again. Janzon (1981) registered considerable concentrations of pollen (up to 15 per m^3) on a roof-top in Stockholm during the winter. No rain-out is possible. Most remarkably, there is very little, if any, effect of snow-cover. Evidently there are urban effects which we do not yet know (cf. Gifford, 1976).

Pollen trapping data should refer to the *deposition year* starting with the flowering of the individual species and comprising the whole redeposition period (Krzywinski, 1976).

2.2.4. Deposition in water

Deposition of pollen and spores in lakes or in the sea represents special problems with regard to scavenging and redeposition. Over a completely calm water surface, pollen can settle by gravity—like over terrestric sites. Such conditions exist chiefly in the late night and early morning hours when the water surface has cooled off and there is no low-level thermic updraught. This is also a time

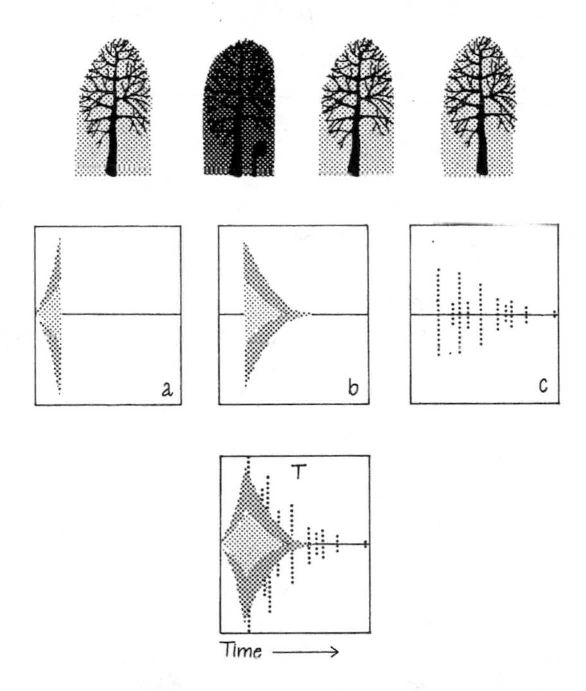

Fig. 2.15. Model of the total gravity component, *T*, measured under a tree in the forest (dark shading) with contribution also from the neighbouring trees (light shading). a: Pollen dispersed and trapped immediately during flowering; b: pollen redeposited by direct wash-off during or shortly after flowering; c: pollen deposited with and in litter. From Krzywinski, 1976; redrawn.

Table 2.4. Pollen deposition data, 1942

Station	Observation period	Pollen type	Flowering season	Pollen deposition per cm^2 per ½ month		
				In flowering season	Post-flowering season	Total non-flowering season
Bergen	II/1–II/9	*Corylus*	I/4–II/4	0.5	0.02	0.02
		Betula	I/5–II/5	13.2	0.8	0.4
		Pinus	I/6–II/6	17.6	0.8	0.4
Voss	I/2–I/9	*Betula*	II/5–II/6	39.0	0.8	1.0
		Pinus	I/6–II/6	39.0	2.0	0.9
		Picea	II/5–I/6	25.1	0.5	0.1
Slirå	I/3–I/9	*Betula*	II/5–II/6	2.4	0.2	0.2
		Pinus	I/6–II/6	5.4	0.1	0.1
		Picea	II/5–I/6	0.6	0.02	0.01

Slirå lies above the timber-line, in regio alpina media. The Roman numerals give, respectively, the first and second half of the month.

of the day when there is little pollen in the air, so the situation is not a very important one. Also, during the main pollination period large bodies of water usually represent a cold surface with concomitant downdrafts. Under windy conditions there is no boundary layer of the same type as over firm ground. Air turbulence goes down to the water surface and meets the waves (cf. Gifford, 1976). The scavenging effect is not well known, but is probably fairly high. One of the effects is capture by water drops blown off the crests of waves.

More important is the fact that in open water scavenging and final deposition are never the same: pollen grains float on the surface (and later are suspended in the water) and are caught by wind-driven and thermic currents after they have left the airstream. Even more than in the air, they become part of the water-masses. The deposition process is therefore immensely more complicated than under telmatic–terrestric conditions, where redeposition is subordinate once pollen has settled. In lakes, there are both (1) internal resedimentation due to processes in the lake itself, most important of which is the spring and autumn turnover in holo- or meromictic lakes; and (2) redeposition of pollen brought in by rivers, which forms a major part of the input to the lake deposition of pollen. On one hand this makes the lake a better integrating instrument for the pollen rain both over time and over area. On the other hand, it makes it difficult to compare limnic deposition with the other modes. A third factor, which especially complicates APF analysis, is the focusing effect of resedimentation (cf. below). This explains the much too large influx registered in the deeper parts of lakes.

Sedimentation processes in lakes have been studied by interception of settling pollen grains in mid-water by submerged traps (open jars held up by floats), but the picture is far from clear. As might be expected, larger lakes have lower incidence of pollen than smaller ones—other conditions being equal. Also the shore vegetation and the weak flyers are more dominant in the small lakes, the upland vegetation and the strong flyers in the larger ones. Thus the larger lakes—like the larger open bogs—integrate the pollen rain from a wider area. Whether that is an advantage or not can only be decided in relation to the actual problem under consideration. According to M.B. Davis (1967a) the results of trapping are unexpectedly consistent within each lake. Probably a great deal of redeposition and relodging by currents takes place before the grains actually settle.

Also, Davis (1968) found that only 20% of the pollen deposited from the water of a lake is new input; the rest is redeposited. Pollen is an important item of food for many small animals living in the water, which will then displace the exines before excreting them again (cf. Smirnov, 1964).

Since water has a much higher and more consistent transport potential than air, water transport is an important factor in the travels of pollen grains over long distances. In the study of pollen sedimentation in lakes in Britain, Bonny (1976: 666) found 'that region pollen, recruited from more than 1 km distant, may form an important factor of the total entering these lakes'. The transport of this pollen need not necessarily have been by wind: watercourses leading into the lakes may carry heavy loads of pollen. Bonny (1976) concluded that resuspension of grains influenced the momentary picture of pollen sedimentation in mictic lakes during the autumnal turnover, while river transport played a great part during the winter season. This pertains to relatively mild and humid winters. In a cold dry winter river transport is probably negligible.

In contrast to that of seeds, water transport of pollen is not necessarily directly downwards. Rowley and Walch (1972) describe an experiment in which pollen grains injected in a turbulent stream were recovered above the point of release. Probably, pollen from bursting bubbles was picked up by wind and transported in any direction. Alternatively, pollen grains may originally have been scavenged by submersed vegetation and blown off that again at low water.

The seasonality in submerged traps does not correspond to that in air traps on the surface of the same lake (cf. also Davis, 1973; Davis and Brubaker, 1973). Experiments by Bonny (1978) indicate that in a (small) basin with relatively great influx of water from rivers, the airborne–waterborne pollen gives a better picture of the composition of the surrounding vegetation than the airborne alone—probably the latter was too much influenced by the local component. In a lake, incoming pollen, especially if it is airborne, has a primary tendency to sediment near the shore. During the spring and autumn turnover in mictic lakes resedimentation brings pollen and other matter down into the hypolimnion and to sedimentation in the deeper parts of the basin. By repeated turnovers this may amount to a focusing of sediment (including pollen) in the deepest parts of the basin.

The slow settling rate of pollen grains in water is a major cause of differential deposition: slowly sinking grains will be driven with wind and currents toward the lee side, and will be deposited in the shallower parts, whereas heavier grains may penetrate into the hypolimnion more or less directly. Davis and Brubaker have studied the problem and find distinct differences between the sedimentation modus of a small grain (*Ambrosia*) and a larger (*Quercus*). A special case is represented by saccate conifer grains, which may remain on the surface for a long time. This might explain observations like the one from 'Michael Sars', quoted below (cf., however, Fig. 2.16). Drifting masses of conifer pollen are often observed close to the shore, both of lakes and the sea. In extreme cases they may be preserved as a special sediment, *fimmenite*, chiefly consisting of pollen. On the other hand, once the

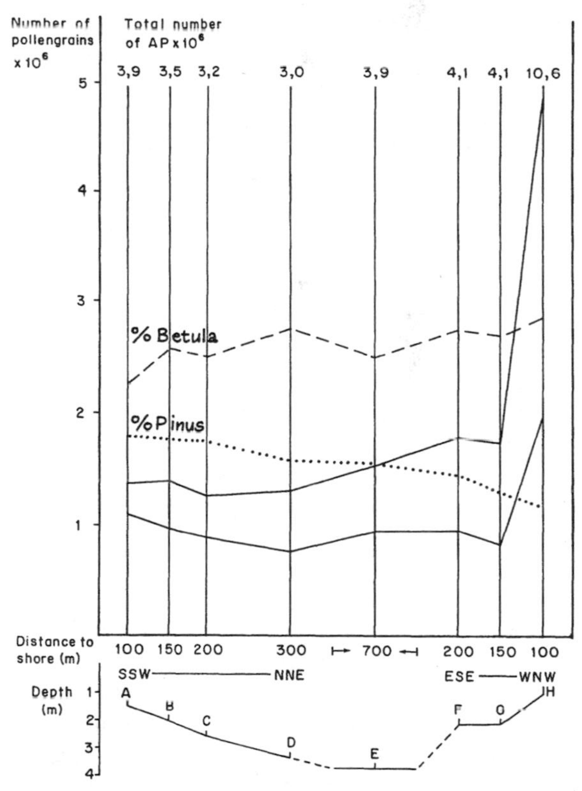

Fig. 2.16. Absolute (full lines) and relative (stippled) pollen registrations in a Southern Swedish lake. After Berglund, 1973; subordinate curves omitted. Reproduced by permission of the copyright-holder: Blackwell Scientific Publications, Oxford.

air bladders of a saccate grain have filled with water it sinks very fast. Littoral accumulations of pollen are not restricted to conifers: in the sagebrush vegetation in Nevada, Fægri (unpublished) registered similar belts of *Artemisia* pollen around small rain-water pools.

In a south Swedish lake Berglund (1973) registered concentration of pollen toward the shore at one side. However, percentages were not affected (cf. Fig. 2.16).

Naturally, the wettability of the pollen plays a great part. If there is a differential rate of sinking this may influence the pollen record, as the more buoyant pollen may be transported further (cf. also R.B. Davis *et al*., 1969).

In larger bodies of water (including the sea), where many rivers come in, a large proportion of the pollen deposited is waterborne, and waterworn as well, and may be redeposited, perhaps from considerably older deposits. Rivers coming from different vegetation regions will bring into the basin pollen loads of different composition, establishing differences in the pollen flora of sediments in different parts of the basin, (cf. McAndrews and Power, 1973). Obviously, sheet-floods may contribute as well as rivers, especially where a ground cover is missing (cultivated soil) or sparse (tundra). Water is a very powerful dispersal agent. Fedorova 1952b: 60–1) quotes observations of *Lycopodium* and *Sphagnum* spores being transported with flowing water 700 and 1300 km, respectively.

2.2.4.1. Marine deposition

The strong currents prevailing in shallow-water marine surroundings greatly influence the deposition of particles, including pollen grains. These are therefore found essentially where sediments of similar dynamic grain size are deposited (Traverse and Ginsburg, 1967). Taking into account the greater density of water, it is obvious that water turbulence is much more efficient in keeping pollen grains afloat than air turbulence. Thus great quantities of pine pollen are reported months after the flowering season and hundreds of kilometres off the nearest forest, e.g. in recent samples from the Arctic (Wille, 1878:15). The journal of the 'Michael Sars' expedition of 1914 records the observation, on 24 June at 17.30 hours, of a *Pinus* pollen rain on the surface of the Arctic Ocean. *ca*. 100 km from the nearest coniferous forest. The number of pollen grains was calculated at 2500 per cm^2, which is certainly too high owing to the method of sampling (skimming). The pollen grains may have been brought together from a wide area by surface currents.

Marine sediments are, on the whole, rather disappointing from the point of view of pollen-analysis. The deep-sea sediments are largely barren (Stanley, 1969), and even the fine-grained coastal sediments, where pollen grains should be expected, are usually rather poor. Estuarine and delta sediments are exceptions.

In the North Pacific Heusser and Balsam (1977) registered (1) concentration of pollen outside the mouths of the two major river systems, the Columbia River (and Strait of Juan de Fuca) and The Golden Gate; (2) that maximum deposition took place at some distance from the shore, presumably when the diameter (4–16 μm) and sinking rate of the mineral grains corresponded to those of the pollen grains concerned. A selective effect was chiefly observed with regard to pine pollen, the relative importance of which increased with increasing distance from the shore.

2.3. Other deposition modes

In addition to the transport and deposition modes discussed so far, there are others, refereable to less common conditions, but certainly of importance where they occur.

Caves. One such deposition system is deposition in caves, which have been important as abodes for man and animals. In caves with free air and/or water passage, and especially in caves that have been inhabited by man or animals, conditions are not *a priori* very different from those in open air, except that the air pollen transport attenuates very rapidly beyond the cave mouth. However, pollen is also found—in small quantities—embedded in concretions in caves which have been hermetically sealed. It may have trickled in with groundwater and been left behind, together with the minerals of the concretion, when the water evaporated. The relation between the composition of such pollen floras and that of the surrounding vegetation is obscure (cf. papers by Bastin (1982) and Renault-Miskovsky and Girard (1978 and earlier), with discussion of technique).

As caves are usually formed by dissolution of sedimentary, calcareous rock, autochthonous cave sediments contain residue from the dissolution process, which may include pollen grains and spores deposited in the rock when it was formed, thus giving a highly metachronous pollen/spore spectrum (cf. Groner, 1985). Depending on local circumstances, cave deposits may give anything from a fairly good representation of the ambient vegetation, to registrations which have no relation to any vegetation picture (cf. O'Rourke and Mead, 1986).

Faecal transport. As usual, human activity brings in a series of complications. A case of faecal transport by farm animals has been described by Moe (1974a): domestic cattle trekking across the mountain have left considerable quantities of lowland pollen, including that of cultivated plants, with their dung at high-alpine altitudes. This dung has later become incorporated in the deposits and actually caused difficulties in the interpretation of pollen diagrams from near a traditional resting place for cattle. Faecal transport usually implies concentration of pollen grains from a larger grazing area.

Faecal transport (or as the case may be, stomach content) has also been registered from subfossil feral animals, including the Siberian mammoths (Kuprianova, 1957). Whereas macrofossils have also been analysed from human bodies, we know of no case that has been investigated palynologically. Pack-rat middens are chiefly renowned for their seed hoards, but pollen grains are also part of the deposit in such middens (cf. Thompson, 1986).

Human-made deposits. In such deposits—archaeological excavations, etc.—one has to take into account, in addition to the usual modes of transport and deposition, that caused by man's growing, collecting and using food and materials from various sources. If the substances involved contained pollen, the exines will be found embedded in the deposits, in fields, in storerooms and, not least, in latrines (Martin and Sharrock, 1964) and other refuse receptors (Krzywinski, 1979). Cereal products contain large quantities of pollen (Greig, 1982).

Under rural conditions most of this pollen is of local origin, but in commerical settlements vegetable matters may have been imported from far away. The pollen flora is *a priori* a very mixed one, and does not relate directly to the local flora (cf. Fig. 9.16). Under such conditions the distinction between primary deposition and redeposition becomes meaningless.

O'Rourke and Lebowitz (1984) find an average of $\frac{1}{2}$ million pollen grains per gramme house dust, and very little pollen in the indoor air, and suggest

Table 2.5. Transport modes of pollen into types of deposits

Site/deposit type	Autochthonous pollen	Allochthonous pollen	Exotic pollen	Redeposition	Preservation
Forest soil	Gravity	0	0	+–++	0–+++
Peat bog	Air (gravity)	0	0	0–+	+–+++
Lake	Air	Water	0	+–+++	++–+++
Cave	0	Air, water, animals			
Archaeological (general)	Air	Food, refuse	Food, refuse	+–+++	0–++
Urban deposits	Air	Food, refuse	Commercial goods	+–+++	0–++

that pollen is 'most likely carried . . . on the feet and bodies of people and pets' (p. 58). This transport mode was certainly no less important in an archaeological or historical context.

Summing up, the transport modes shown in Table 2.5 are of primary importance in the respective deposits. The two last site types are discussed in more detail in Chapter 9.

3. Where pollen is found. Organic deposits, origin and description

Motto: The mire is the historical archive of the landscape

3.1. The deposit

From the moment the pollen grain is finally deposited on a terrestric surface or in water it becomes part of the local sediment and is subject to the same diagenetic processes (decay, pressure, etc.) as the other organic and inorganic constituents of the deposit. The genesis, structure and character of the deposit provides a separate set of data for the final interpretation of the pollen record. The study of the deposit itself is an important part of the investigation, and for the evaluation of the pollen diagram it is important to possess knowledge of the character of the matrix from which the grains were recovered. A careful study of the pollen-bearing sediment in the field yields valuable data, which can be supplemented by, but not replaced by, later laboratory investigations. The time spent on this in the field is amply repaid by time saved later, and by a better understanding of the conditions under which the deposit was formed. Understanding of the stratigraphy of the deposit also helps in selecting representative points* for pollen analysis, cf. Chapter 4.

The information gained by the field study of the deposit is of importance for the field work, for the subsequent analysis and for the final interpretation:

1. The physical and chemical characteristics of the deposit determine the way in which samples should be selected and treated for optimum results by indicating where to obtain the most complete stratigraphic record, and also to be aware of stratigraphic inconsistencies: resedimentation, hiatuses, etc. which may either be avoided or properly evaluated.

* Since terminology is vague and confused we define the following terms: a *deposit* is the material in or by which the work is carried out. A deposit is three-dimensional, but may also refer to the substance alone (non-dimensional). A single vertical sampling, whether taken by a corer or studied directly, is referred to as a *profile point*. The graphic or material representation of a point is the stratigraphic *column*. A *section* is a two-dimensional vertical representation of the stratigraphy of a deposit. It can be based upon a sequence of profile points or studied directly on an *exposed face*, i.e. a wall exposed by excavation. A *horizon* is the (sub)horizontal equivalent of a section. A horizon may be synchronous or metachronous.

2. These characteristics also influence the state of preservation of pollen and spores (sometimes to the extent of making further investigations unprofitable).
3. Plants forming the deposit may have had a pollen production of their own, interfering with whatever pollen rain is the subject of investigation.

Knowledge of peat-forming plant communities themselves is also of great value in interpreting past climatic developments. The great controversy in northern European vegetational historical research at the turn of the last century was precisely between one school that chiefly drew inference from the development of deposit-forming plant communities, the so-called palaeophysiognomic school of Sernander, and on the other hand Gunnar Andersson's palaeofloristic school, which stressed the importance of changes in occurrence of individual species.

The two approaches could at the time have supplemented each other instead of being part of a confrontation. When pollen analysis appeared, it went its own way, and soon took over not only as the main, but to a great extent as the only, way of studying development of vegetation and climate. The two older methods fell into oblivion or were conducted as minor sidelines.

Strictly speaking, pollen analysis is palaeofloristic, dealing with pollen grains instead of macrofossils. The main difference—apart from the obvious one of size—is the omnipresence of pollen, which represents a much wider area than the local macrofossils. However, interpretation of pollen diagrams also includes a strong element of palaeophysiognomy in sediment analysis, peat-bog physiography, etc. The latter field has also benefited from corresponding phytosociologic and ecologic studies. With the increased understanding of the importance of biological evidence in other fields, especially archaeology, macrofossil analysis has also seen a renaissance. All these techniques provide mutually supplementing evidence, and even today, with the much more detailed knowledge of vegetational development afforded by pollen analysis and modern dating techniques, the information provided by the character of the deposit is—or could be—decisive in the interpretation. Within the framework of this text we cannot deal with the subject exhaustively. The intention of this chapter is to stimulate the student to further efforts in description and interpretation of pollen-bearing deposits. We refer to Overbeck (1975) and, for his special methods, to Troels-Smith (1955).

Based upon the origin of the constituting material, deposits may be classified as *allochthonous*, i.e. consisting of material originating elsewhere and deposited or redeposited in the place where it is recovered, and *autochthonous*, i.e. consisting of remains of the vegetation that once lived in the particular place where the sample is recovered. Allochthonous deposits are usually referred to as sediments, autochthonous as peats. Lithologically, peat is of course a sediment, but the traditional terminology neglects this for the sake of conciseness. Many peats contain a great deal of allochthonous material in addition to that locally produced. The concepts are to a certain extent relative, according to how widely one defines the 'place of recovery'.

Organogenic sediments are chiefly formed in open water, below the low-water level, more exceptionally also higher up, e.g. diatom mud in the inundation zone of rivers. Peat, on the other hand, is formed both below low-water level (*limnic* peat), in the periodically submerged zone between low-water and high-water levels (*telmatic* peat), and also above high-water level (*terrestric* peat, cf. Fig. 3.1). Whether a peat belongs to one or the other of these types also to some extent depends on the production rhythm of the plant community in relation to its seasonal submergence or emergence. Thus, a true terrestric peat may form under conditions involving submergence during a non-productive season.

'It is hardly possible to establish a peat type classification that is directly applicable in all circumstances' (von Post and Granlund, 1926: 41), but the distinction between these three primary hydrologic classes is of fundamental importance for the understanding of the development of a deposit anywhere. The ordinary sequence during the filling in of a basin, the hydrosere, passes through the stages: sediment—limnic peat–telmatic peat–terrestric peat. The types of sediment and peat

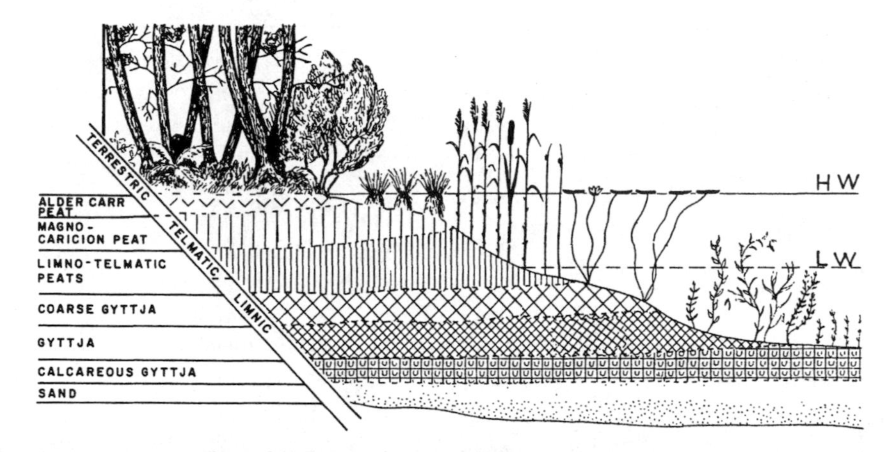

Fig. 3.1. The hydrosere at a eutrophic basin in the European broad-leaf zone: some plant communities and the corresponding deposits. Fægri-Gams notation. After Overbeck, 1975; redrawn. Reproduced by permission of the copyright-holder: Wachholtz, Neumünster.

differ according to the chemical and physical conditions under which they have been formed, but the above sequence (or some truncated type) is always recognizable if the filling in has proceeded normally.

D. Walker (1970) studied the successional pathways in 159 British hydroseres, (cf. Fig. 3.2). Some of Walker's categories are hydrologically ambiguous ('aquatic sphagna', 'marsh'). If these are discounted, only four out of 129 successions go the 'wrong' way between hydrologic classes.

On the other hand a reversion of this sequence, if it occurs, indicates that the regular hydrosere has been interrupted, and further investigations may disclose whether this interruption is due to climatic irregularities or to local causes. The distinction between the main hydrologic classes is therefore of primary importance. Figure 3.3 gives an example of an inverted sequence and its interpretation.

3.1.1. *Allochthonous (sediments)*

Sediment (and peat) types divide themselves between two ecological series, one oligotrophic and one eutrophic,* with intermediate types. The terms may simply be defined as low- and high-productive, which, however, leads to incongruous results unless further qualified. The productivity of a basin depends on several conceptually disparate variables. Favourable climatic conditions under all circumstances increase productivity: tropic basins being high-productive, arctic ones low-productive, but within both regions there is a differentiation of productivity between basins. Favourable topographic conditions do the same: a shallow basin is more productive than a deep one with steep sides. In each case the character of the water defines the relative trophy level (cf. Fig. 3.4). Consequently, we shall use the trophy terms primarily to indicate water quality.

Whereas oligotrophy is relatively easily defined by absence of plant nutrients, the lower limit of eutrophy is diffuse and completely arbitrary. On the other hand a too high nutrient level may produce *saprotrophic* conditions with very characteristic deposits, usually formed anaerobically and stinking from sulphureted hydrogen.

Marine deposits are usually eutrophic, even at coasts where oligotrophy is the rule ashore. The nutrient status of seawater is generally high; the productivity is lively and decomposition rapid, mobilizing nutrients again. In enclosed (brackish water) basins, productivity can reach saprotrophic levels, producing a characteristic zone in the stratigraphic column.

Even if productivity is also climatically controlled the distinction between trophy levels is so

* All these terms are derived from Greek, *trephein* = nourish. *oligo* = few, *eu* = full, *meio* = diminishing, *auxo* = increasing.

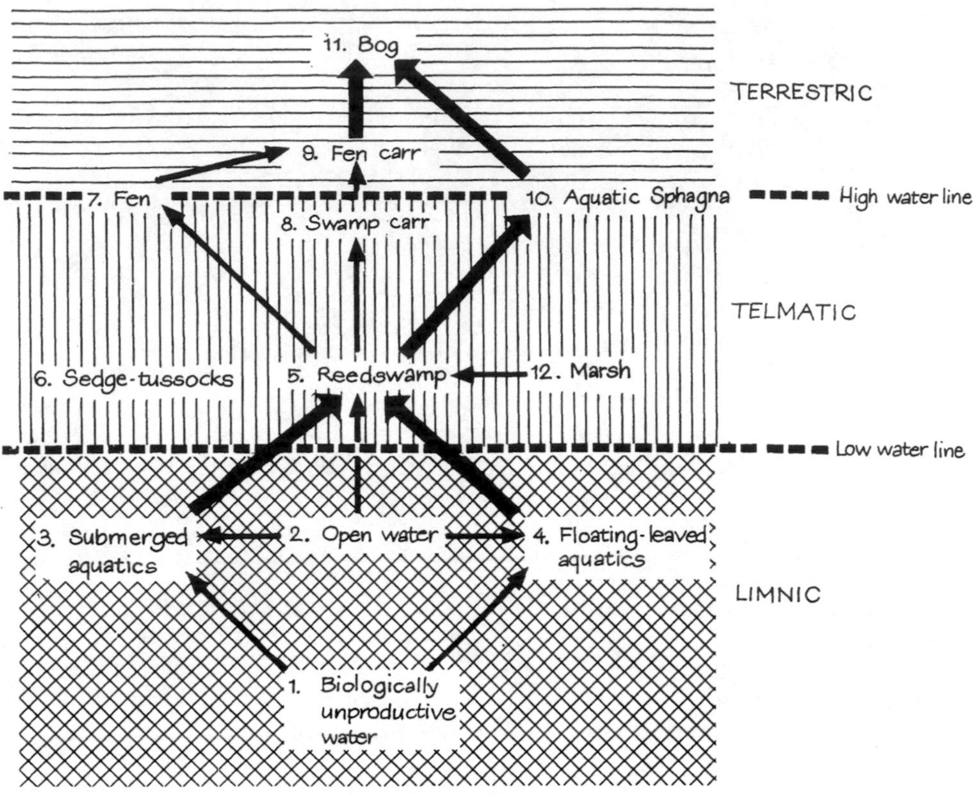

Fig. 3.2. Successional pathways in British hydroseres according to D. Walker (1970). Heavy arrows indicate the main pathways. Redrawn from data in Birks and Birks, 1980.

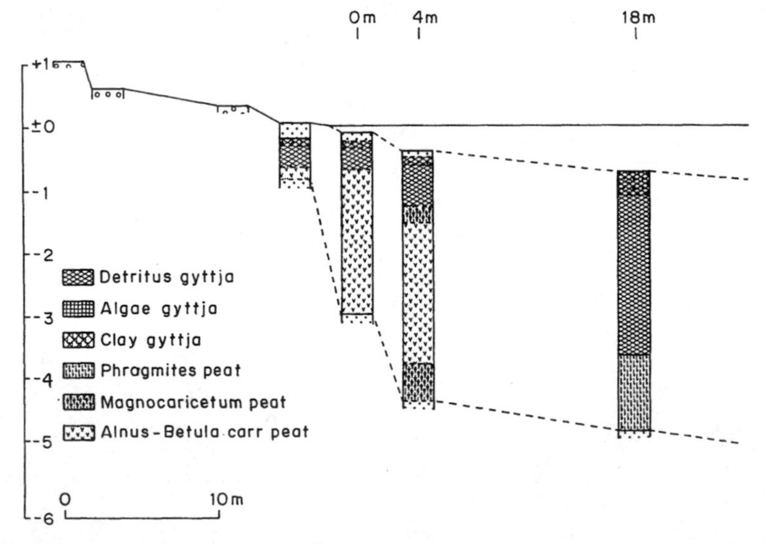

Fig. 3.3. An inverted hydrosere. Bottom layers indicate (telmatic) fen peat at stations 4 and 18, (terrestric) carr peat nearer the shore. The regular hydrosere leads to carr at 4, while fen still covers the area at 18. This telmatic–terrestric series is overlain by (limnic) gyttja, at 4 with a layer of fen peat between. The inverted succession is interpreted as the registration of a rise in water level. On the top there is a layer of sand: indication of erosion frequently resulting from rising of the water level. From Digerfeldt, 1976. Reproduced by permission of the copyright-holder: G. Digerfeldt.

strongly ecological that it is rarely possible to draw any climatic inference from the trophic status and development of a basin. Changes in trophic status are not uncommon. In oceanic areas there is under undisturbed natural conditions a general tendency toward more oligotrophic conditions *(meiotrophication)*, which is the characteristic end-stage of a hydrosere under humic conditions. In non-oceanic areas this tendency is less pronounced or absent. Eutrophication *(auxotrophication)* as the end-stage of the hydrosere is usually due to contamination from human activity.

The three main parameters in trophy (cf. Fig. 3.4) are the quantity of available plant nutrients, hydrogen ion concentration and the concentration of humus colloids. Plant nutrients may be nitrogen, phosphorus, potassium, depending on which substance is at any time in minimum: ample quantities of phosphorus are of little use if there is a dearth of nitrogen. Hydrogen ion concentration depends on the calcium status, but calcium as such is not particularly important and has therefore been omitted: its effect is via the influence on pH. Humus colloids have a double effect: by absorbing light they exclude the growth of plants or the exposure of their leaves under water, except very shallowly. In addition, humus colloids seem to have some unspecified toxic effects. Even if turbidity has a similar effect with regard to light extinction, the two factors cannot be entered on a common (light extinction) axis, and primary turbidity has been omitted as being less important under natural conditions.

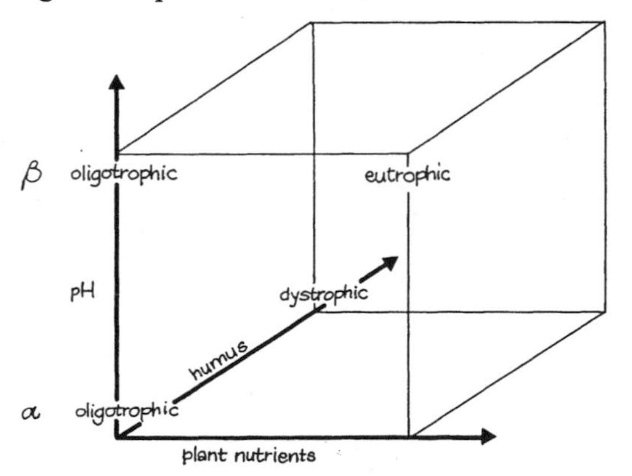

Fig. 3.4. The major trophic types. Theoretically, the three axes should have been drawn to the saturation point, but trophic types establish themselves at much lower concentrations. From Fægri, 1954b; redrawn.

In a lake, trophy is an expression of the productivity of the body of water, and so also is the rate of sedimentation of bottom deposits. These are therefore abundant in eutrophic lakes, whereas in oligotrophic ones organic sedimentation is insignificant and in the (theoretical) extreme dominated by particulate matter coming from outside, including pollen.

Strictly oligotrophic lakes (with clear water) are, however, rather rare; lakes poor in plant nutrients are more frequently dystrophic, with brown water. The sediment is the *dy* (Swedish*) or gel-mud (Godwin, 1938), a blackish-brown, colloidal precipitate that dissolves by treatment with KOH (in extreme cases pollen grains are almost the only solid particles left after such treatment).

Dy is plastic and sticky, and shrinks on drying. It is constituted of acid humus colloids which represent the final product of decomposition in the moist interface between air and—acid—soil. After having dried out, the colloids are very difficult to hydrate again. Humus colloids are not, or to a small extent, produced in the basins themselves, but are brought in by ground or surface water from the surrounding vegetation. In this way also, eutrophic lakes may receive a dy component in their sediments by influx from a surrounding oligotrophic vegetation. The term *mixotrophy* has been used to describe this situation.

Humus colloids are sooner or later precipitated from the brown dystrophic water. If there is an excess of di- or multivalent ions, such precipitation is very rapid and the water remains clear in spite of a high influx and sedimentation of humus colloids/dy.

Other types of oligotrophic sediments are less frequent, at least in northwestern Europe. The pure oligotrophic marls produced by the β-oligotrophic lakes may be mentioned. These lakes are deficient in plant nutrients, and precipitate an almost pure $CaCO_3$. Marl is also produced by eutrophic lakes, but does in such cases contain more gyttja elements.

The sediment of eutrophic basins is the *gyttja*†.

* The vowel is pronounced like the French *u* or German *ü*.

† Grosse-Brauckmann (1961) has given a survey of the confusing older terminology of organic sediments, and especially of the notorious *sapropel* concept, which is so compromised as to be useless.

Its colour varies. Pure gyttja consists of microscopic and submicroscopic remains of the flora and fauna of the basin. It is elastic and not sticky. As the gyttja-producing faunas and floras vary, so do the sediment, but they have in common that they are insoluble in KOH and that the extract is never dark brown, but yellowish, greenish, or almost colourless. Unchanged plant pigments, especially carotenes, are frequently found in gyttjas, and so are also some green pigments, which are most probably chlorophyll derivatives (cf. Baudisch and Euler, 1935; Treibs, 1934). Wieckowski and Szczepanek (1963) have isolated various assimilation pigments, including chlorophyll, from *Abies* needles also in oligotrophic peat. These pigments are often seen after acetic acid extraction in the course of the acetolysis method of preparing samples for analysis (cf. Chapter 5).

The actual sediments observed in nature frequently occupy an intermediate position between the extreme types, and contain both KOH-soluble and KOH-insoluble fractions. They are characterized by their origin from a special community or by having been deposited under special conditions; resultant variants are lake marl, clay gyttja, alga gyttja, diatom gyttja, detritus gyttja, etc.

Whereas it is in some cases sufficient to characterize the sediments as gyttja, or dy, it may in other cases be necessary to distinguish between a long series of finely divided types. It all depends on the problem to be investigated (cf. Lundqvist, 1938; Troels–Smith, 1955).

3.1.2. Autochthonous (peat)

Sediments formed in open water originate in mixed organic production from the whole of the basin, or even also from outside. Peat, on the other hand, mainly consists of substance formed exactly at the place of recovery. In order to remain there after plants have died, the parts must be more or less fibrous. Depending on the original composition of the vegetation, peats are therefore more variable in composition than sediments, each bog community producing its own peat type. The main types from subcontinental northern Europe have been summarized by von Post and Granlund (1926); in other parts of the world other types will be found. It should, however, always be possible, and is desirable, to distinguish between the main hydrologic types although limnic and telmatic peats, which also contain allochthonous elements, are hydrologically rather uninformative. Terrestric peats, on the other hand, react sensitively to changes of hydrology. They are formed in areas of impeded drainage, where waterlogging prevents a total (oxidative) breakdown of organic material formed. Since peats are so narrowly dependent on the mother formation, some knowledge of types of peat-forming vegetation is also necessary for a proper evaluation of pollen diagrams from the corresponding deposits.

As a neutral term peat-forming areas and their vegetation are referred to as *mires* which, according to the quality of the water, may be eutrophic—*fens*, or strictly oligotrophic—*bogs*. The term bog is usually more or less reserved for the extreme-oligotrophic types. The concept of an oligotrophic fen is valid for intermediate cases.

Classifications of such complex concepts as mires are bound to be difficult and never satisfactory for all purposes. One possible starting point is the origin and quality of the water, as used by von Post and Granlund (1926; cf, Fig. 3.5). A mire not rising above a more or less horizontal groundwater level, e.g. in a lake basin, they called *topogenous*. If it built itself up against a ground-water gradient it was termed *soligenous*. The third type, *ombrogenous*, occurred where the mire was unable to transgrede and had to build up its own groundwater level.

Since the term 'raised bog', corresponding to Swedish 'hömosse' and German 'Hochmoor', is ambiguous and compromised through indiscrimi-

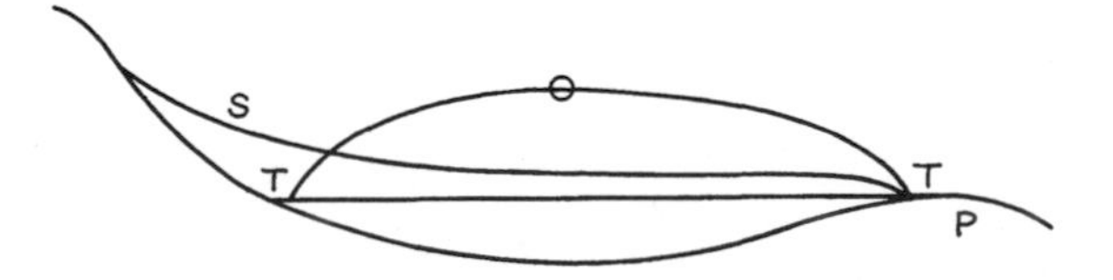

Fig. 3.5. The three major mire types according to von Post and Granlund. P: Pass level of former lakes; T: topogenous mire; O: ombogenous bog; S: soligenous mire.

nating usage, the term 'domed bog' is better used for the ombrogenous type. That term distinguishes the type from other bogs which may rise unilaterally over the general terrain.

A topogenous mire may be anything from extreme-eutrophic to strongly oligotrophic, according to the character of the ground-water. Theoretically, the soligenous type might be the same, but under the circumstances where soligenous mires occur the ground-water is usually oligotrophic. Since the ombrogenous bog cannot receive any water from below, it is dependent on precipitation for its mineral nutrients, which means that it is extreme-oligotrophic. If a soligenous bog grows so high up as to cover the watershed—*blanket bog*—it also becomes *ombrotrophic*, i.e. nourished by precipitation only. The alternative is *minerotrophic*, for water which has passed soil layers and has had the possibility of picking up mineral plant nutrients. Even minute quantities suffice to change the vegetation from the extreme-oligotrophic aspect of the domed bog (cf. du Rietz, 1954).

The character of terrestric peats depends on the composition of the plant community producing the material, and on the conditions for decomposition of the same—this is generally referred to as *humification*. This is a very complex process during which different fractions are broken down more or less easily. The most resistant major constituent is lignin, with cellulose occupying an intermediate position. Dark-coloured acid humus colloids are the final organic product of decomposition under mire conditions. Humification depends on climatic factors: faster at higher temperatures, but depends even more so on the ambient ecological conditions. At high pH and availability of oxygen decomposition goes towards total oxidation, at lower pH towards complete humification. If oxygen is sparse decomposition processes slow down and humification is less complete. This is the case where the watertable is very high and stagnant. In the absence of humus colloids peat is light-coloured.

Holling and Overbeck (1960) have shown that a very great percentage of the living substance is lost even at a very low degree of humification. With great variations the figures centre around 50%. The greater part of this loss is probably due to the production of gaseous substances from easily disintegrating parts of the living plants. According to Reader and Stewart (1972), about 10% of the net primary producton is preserved, but it is obvious that the figure varies enormously with the type of peat. It is also possible that at the same degree of decomposition some plants produce more humus colloids than others. The great initial losses, even in less decomposed peat, may explain why the difference in peat accumulation rate between light and dark peat often turns out to be smaller than expected. Rowell and Turner (1985) found no simple relation between humification and peat accumulation rate (cf. also Aaby and Tauber, 1974).

In extreme cases of peat decomposition there is very little organized material left, and the resulting humus colloid deposit may be more or less indistinguishable from the corresponding limnic oligotrophic deposit. Actually it is the same substance, autochthonous in the one case, allochthonous in the other. Pollen grains are highly resistant to decay by humification.

3.1.2.1. Regenerative peat

Some controversy surrounds a spectial type of terrestric peat, the so-called *regenerative* peat, assumed to be formed by a cylic succession of *Sphagnum* communities on the bog. The classical principle of regeneration was first described by von Post and Sernander (1910); it presumes a cyclic succession between wet and dry communities that form a mosaic on the surface of the bog. Since rapidly growing peat is light and slightly humified, whereas (the smaller quantity of) peat formed by drier communities is dark and highly humified, the resulting section of the bog shows alternation between broad, lens-shaped layers of bright peat and narrow, undulating bands of dark peat ('bacon peat') (Fig. 3.6). It may be questioned whether the connection between dry conditions and dark peat and vice versa is a rule without exceptions.

According to the classical concept the domed bog grows up above the general ground-water level and establishes itself as a separate hydrologic unit, dependent on the excess of precipitation over evapotranspiration. There is, however, no logical

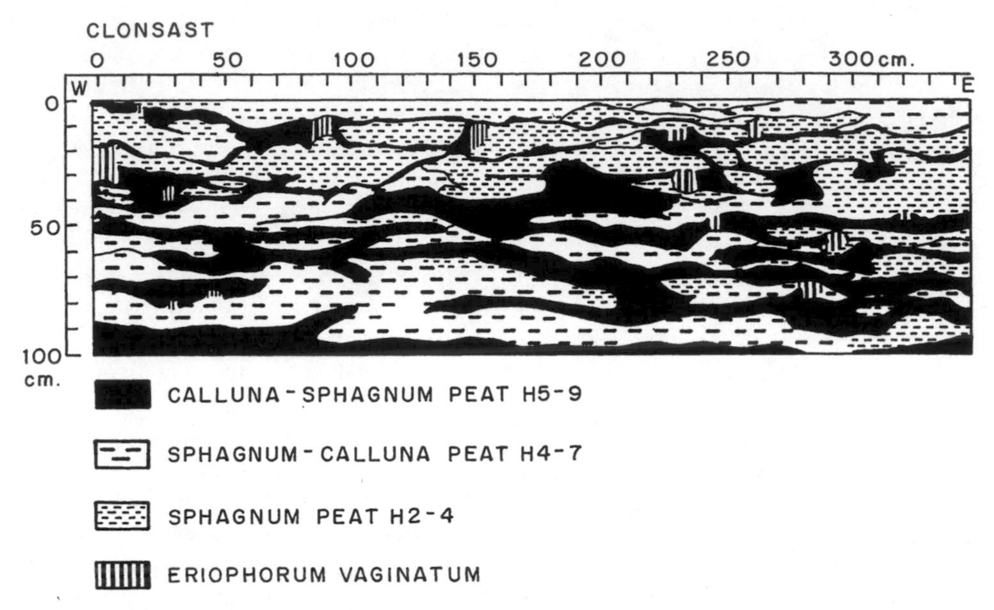

Fig. 3.6. Section of a regenerative peat deposit. From D. Walker and Walker (1961). Reproduced by permission of the copyright-holder: Blackwell Scientific Publications. Oxford.

necessity why ombrogenous bogs should always be regenerative, nor why regenerative peat should always indicate ombrotrophy.

The ground-water deriving from the surroundings is drained away from the bog by an encircling brook or fen, the *lagg*. Into this lagg also drains, centrifugally, the water from the bog, and with increasing elevation of the bog above the surroundings this drainage becomes more effective. According to Granlund (1932) the height above the surroundings to which a domed bog can rise, within the same climatic area, is entirely dependent on precipitation.

The classical concept of regeneration (Fig. 3.7) has been criticized and partly refuted by various recent investigations. D. Walker and Walker (1961) have published data on detailed sections of Irish peat walls with regeneration structures. On the whole, a detailed analysis confirms earlier opinions as to the genesis of such peats, though both hollows and hummocks appear to be more permanent features than originally assumed.

The same conclusion is arrived at by Casparie (1969), who queries the existence of a cyclic regeneration system in Dutch bogs, whereas Dupont (1985, Fig. 3) published a profile from the same area, looking like a textbook illustration of the classical regeneration theory. The most interesting part is that a similar structure is seen also in the Schwarztorf, the highly humified lower deposit, in which such structures are usually not visible.

Aaby (1975) also registers a distinctly cyclic structure in Denmark. In this connection it should be remembered that the effect of winter freezing upon regeneration structures has been suggested many times, but never demonstrated nor refuted. If such an effect exists it may explain some of the discrepancies registered.

Apparently there are places in the bog which are more or less constant hummocks, and others that are equally constant wet hollows. The cyclic development takes place between such nodal areas, rarely across them. The effect, however, is probably the same: under dry conditions the hygrophilous communities withdraw to the permanent hollows and the hummock vegetation spreads into the 'neutral' area, producing dark peat. Under wetter conditions the reverse process takes place. The visual effect will be the 'bacon' peat.

However one wants to explain them, domed

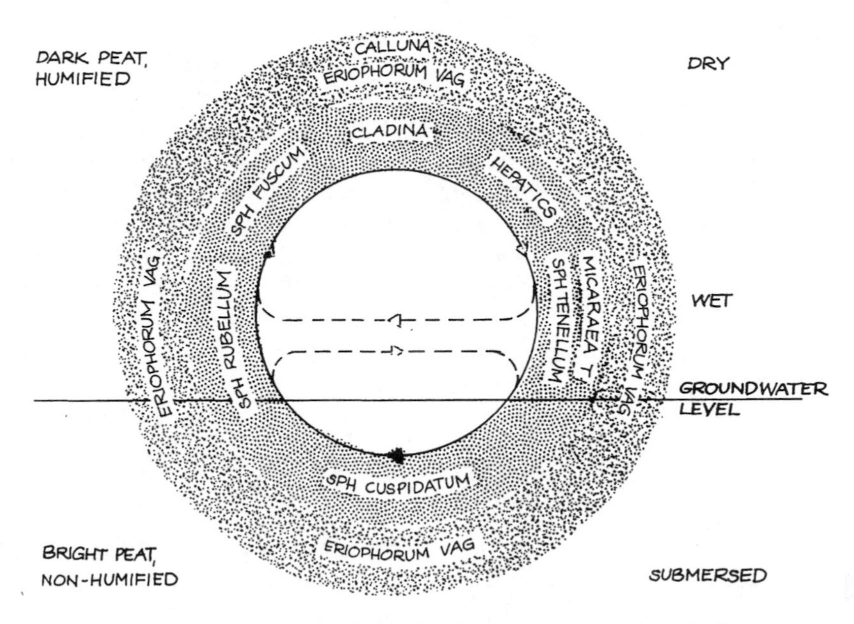

Fig. 3.7. Idealized classical regeneration cycle with indication of some dynamically important species. Succession directions indicated by arrows and related to the water level, which is presumed to rise gradually. On the left-hand side of the diagram the peat grows faster than the rise of the water level. The wet peat-forming communities are replaced by communities forming less and darker peat, then by some not forming peat, and therefore gradually submerged by the rising water table. On the right-hand side are also some plants that usually form no peat, even when growing wet (*Sphagnum tenellum*) or directly destroy peat already formed (lichens, hepatics). Under wet conditions the cycle is truncated at top and only light peat is formed. If the mire dries up the cycle is truncated at the bottom and only dark peat is formed. Some plants such as *Eriophorum vaginatum*—the same tuft—can grow through all phases of a cycle.

bogs exist. Plants growing on them receive practically no nutrients but meteoric water, which is very poor in electrolytes. The extreme sensitivity to chemical stress of the typical flora of ombrotrophic bogs makes these plants easy victims to contamination effects, like many British bog systems which have been destroyed by industrial contamination (Birks and Birks, 1980: 48) killing the flora, especially the sphagna.

Since the typical domed-bog plants are unable to transgrede into the lagg, the water of which is usually richer in electrolytes, the bog eventually reaches a maximum elevation, maximum drainage, and maximum marginal gradient. Owing to the better drainage, whether horizontal or vertical, drier communities become increasingly important in relation to the wet until the bog is completely dominated by the so-called stand-still complex (Osvald, 1923) when very little, or no more, peat is formed. Correspondingly, the peat section gradually becomes dominated by the dark bands, the brighter parts thinning out and disappearing. According to this view the development from dominant bright, slightly humified peat to dominant dark is the normal development of a regenerative peat, just as development from limnic to terrestric is normal within the hydrosere. Reversions, which are considered to be climatically conditioned, are therefore important indicators of climatic change, especially of change in humidity, which is usually not too well registered in pollen diagrams from temperate regions. Such reversions register only if the bog is in a suitable state of development. A young, low bog in rapid growth, far from reaching a level in equilibrium with the existing climate, will not be affected by a minor drying-out, which may completely stop growth of a more mature bog in the same area, close to its maximum level.

Granlund (1932) maintained that southern Swedish domed bogs contained five different reversions of general distribution. Each of these *'recurrence surfaces'* ('rekurrensytor", RY) should represent a climatic change from drier to wetter conditions, permitting the establishment of another regeneration complex instead of the previous stand-still. Other RYs have since been described, and the matter at present seems to be rather unsettled (cf. Olausson, 1957; Aaby, 1975). Some of the discrepancies may be due to careless investigation: it is scarcely possible to demonstrate the existence of an RY except on an exposed peat face, if then.

In the development of a domed bog there are many complications, some of which are difficult to reconcile with the theory, and for which no synoptic explanation can at present be given (cf. Godwin, 1954).

Long ago Weber demonstrated the presence of a similar surface, the *Grenzhorizont*, in northern German bogs. The principal difference was that the humification, especially of the dark peat (Schwarztorf) below, was described as being more uniform than would be expected from Granlund's theory (Overbeck, 1947). On the other hand, conditions on the enormous northwestern German flatland were certainly different from the restricted bogs in the more varied landscape studied by Granlund. They may have resembled those of the North American muskegs today.

Originally, the Grenzhorizont was presumed to be a synchronous phenomenon, but both the existence of more than one such surface in some bogs, and even more so, ^{14}C datings have proved that the Grenzhorizont may have formed at various periods, even if the classical date, corresponding to Granlund's RY_3, viz. *ca.* 500 BC, is the one most frequently observed (Overbeck *et al.*, 1957). Also, a single Grenzhorizont has been proved to be metachronous within one and the same bog. If we follow Granlund this is not unexpected, as various parts of extensive bogs may be in different stages of development in relation to the maximum level, and consequently enter and leave the stand-still stage at various times. Equally, the layer of *Sphagnum cuspidatum* peat sometimes found extensively at the bottom of the younger peat is the metachronous beginning of the new growth. A synchronous so-called Vorlaufstorf is a botanical impossibility.

Reversions of the normal hydrologic development may be due to some non-climatic cause, e.g. a small landslide damming a basin up to a higher level than previously. However, if such causes can be ruled out, the sequence of layers may give very important information about the climatic development: the objective of the old palaeophysiognomic school. By means of radiocarbon and pollen analysis such climatic events may be dated, and their synchroneity and general significance established.

3.2. Classification and description of deposits

The usual pedologic classifications are not well suited for pollen-analytic purposes; separate, more discriminating classifications have been worked out (cf. Grosse-Brauckmann, 1961). In the following we shall restrict the discussion to two main approaches and their notation systems.

The characters which can and should be taken into account by the classification of deposits are the identification of deposit-forming vegetation (and fauna) and the physical status of the components. Many of these may be expressed quantitatively or semi-quantitatively on an arbitrary scale even in the field. Laboratory work gives better discrimination.

For details of methods we refer to the original literature, and also emphasize that field methods can never be learnt properly from literature alone: the main teaching must take place in the field. Most field studies need subsequent verification and checking in the laboratory.

In the field, pieces broken from the fresh sediment can be examined dry at low magnification or dispersed in water. Identified particles can be counted and the results quantified. The ordinary hand-lens should be a standard instrument for peat study in the field.

The degree of humification is a very important parameter for terrestric peats. A simple ten-point field scale was worked out by von Post and Granlund, based upon the colour of the water and/or the consistency and quantity of the substance squeezed out between the fingers when the sample is squeezed by hand. Laboratory methods (Bahnson, 1968; Overbeck, 1975) give approximately the same results, but provide more detail (Fig. 3.8). On the other hand a simpler scale is often sufficient in practical work.

Mineral content is another parameter which can be roughly estimated in the field (by the grittiness of the sample), but a more discriminating evaluation, loss on ignition, must be reserved for the laboratory.

The aim of the study of the units of the more or less diffuse matrix of deposits is to obtain the maximum of information for the knowledge of, and understanding of, the genesis of the deposit, based upon a synthesis of information from the floristic, physical and chemical composition and the stratigraphy of the deposit.

This principle was clearly formulated by von Post and Granlund (1926) at a time when few other methods were available for the interpretation of pollen analyses. Later, two different approaches evolved, the synthetic–interpretative, based on the said principles further developed by Fægri and

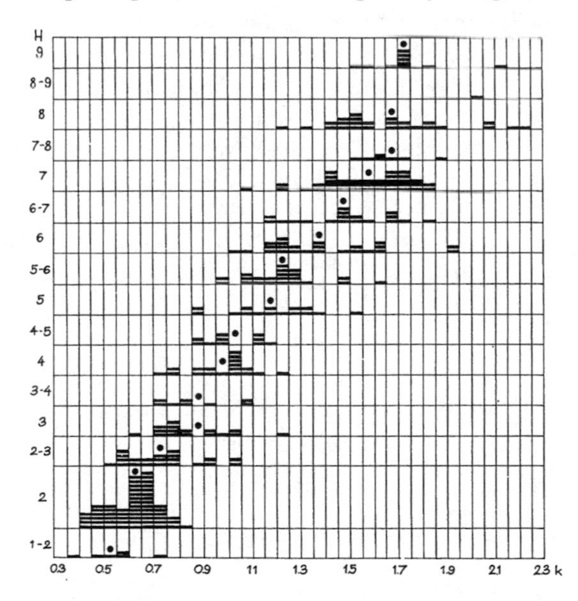

Fig. 3.8. Comparison of the result of huminosity determinations by squeezing the sample after von Post and Granlund (H scale, vertical) and by colorimetry (k scale). After Overbeck (1947). Reproduced by permission of the copyright-holder: Springer Verlag, Heidelberg.

Gams (1937), and the analytic–descriptive of Troels–Smith (1955). The former system implies the description in its genetic terms, the latter makes description the primary object and may—or may not—derive a concept of the genesis of the deposit on that basis.

The ultimate aim of the sediment description should be the elucidation of the origin of the deposit. The difference is that, based on experience of the morphology and genesis of sediment types, the von Post and Granlund approach does not always repeat the detailed morphological analysis of known sediment types, more or less as a trained botanist does not need a formal morphological analysis to recognize a common plant species. However, if the plant (deposit) is unknown, the analysis is necessary for identification and/or recognition. The Troels–Smith approach provides just that. For an inexperienced worker a systematic approach like that of Troels–Smith is certainly a good support in the field, but it may in extreme cases lead to the accumulation of a large amount of dead data, which are not used further. Whether such data should be part of a publication or not is a matter of suitability for the purpose. It is a case of conclusions drawn and presented by a competent worker from immediate contact with the deposit in the field vs. conclusions drawn post-hoc from a system of field notes. The one does not exclude the other; most investigators performing Troels–Smith analysis in the field also include a von Post–Granlund evaluation. In any case the credibility of conclusions is enhanced by simultaneous presentation of the material upon which they are based.

Both systems are flexible: they can be used to extract and present very detailed information, but they can also be truncated if details are unnecessary for the problem at hand. However, taking field notes cannot be postponed. If they are not taken then and there they are lost.

3.2.1. Graphic presentation of stratigraphic information

If we want to transmit the information given by bog development, the stratification of the bog should be presented to the reader in the clearest possible way. Pollen diagrams should be accompanied by a graphic column representing the composition of the deposit where samples were collected. For the evaluation of pollen-analytical data it is indispensable to know something about the deposit; percentages in carr peat and in a sediment mean different things. This holds good even for the arboreal pollen, whilst non-arboreal pollen data from an unknown deposit are of very limited value.

One may enter the name of the peat and sediment types directly on a diagrammatic drawing—this is the easiest way out for the author and the least satisfactory for the reader who, apart from the omnipresent language difficulties, has to form a mental image from a complicated text. When the text—as is sometimes seen—is replaced by some shorthand formula, these complications become almost insuperable. A genetic interpretation of the data cannot be immediately comprehended by a reader and should not be presented as a set of raw data. It is necessary to use a symbol system that gives an immediate impression of essential features of the deposit. Many symbol systems have been devised to cover immediate needs; the consequence is that one and the same symbol has meant different things in different publications. This is, of course, highly inconvenient: symbols are conventional and their effectivity depends upon the universality of their acceptance.

On the other hand the purely analytic data resulting from a Troels–Smith analysis, which do not pretend to represent any synthesis, may be replaced by tables or shorthand formulae, especially in stratigraphic columns (cf. examples below).

Because of the intrinsic importance of the hydrosere for understanding the development of a basin, Fægri and Gams (1937) proposed the following general principles for the presentation of the results of a genetic interpretation: Symbols for (a) sediments, (b) limnic and telmatic peat, and (c) terrestric peat should be immediately recognizable as such in the diagrams. This is obtained by varying the general direction of the symbol lines. Sediments should have symbols consisting of crossing lines, peat symbols of parallel lines: linmic and telmatic peat symbols having vertical lines, and terrestric peat symbols horizontal lines. Because of its importance for the understanding of bog development, the degree of humification should be immediately recognizable. This is achieved by thinner or thicker signature lines (alternatively by spacing of lines of equal thickness—cf. Fig. 3.9).

The notation system is basically very simple, and immediately permits any reader to follow the main features of the development of a basin, even if both the peat and sediment types and their mother formations are unknown.

Sediments are relatively uniform all over the world, and the same symbols ought therefore to be

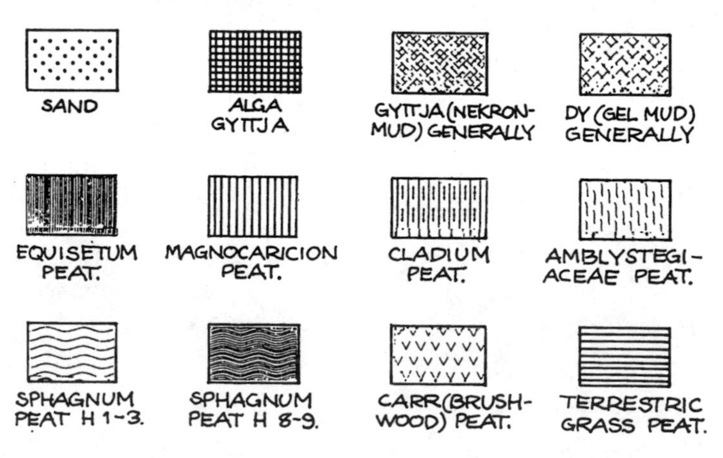

Fig. 3.9. Signatures from the Fægri–Gams notation system. Top row: limnic sediments; middle row: telmatic peats; bottom row: terrestric peats.

used everywhere in the same way. The applicability of peat symbols is less general, owing to the great regional differences between mother formations and consequently between peats. Some peat signatures have become more or less standardized for North and Middle Europe or the Holarctis. It should not be imposible to reserve them for related peat types also in other parts of the world, e.g. the wave-line for *Sphagnum* peat, or the 'v' signature for carr peat. In a genetic system it is necessary to have a signature for hydrologically indeterminable deposits. An oblique line is the simplest. For generalized diagrams it will also be necessary to use symbols meaning peat in general, etc., but such symbols are easily extemporized, and in such simple diagrams consistency is not of the same importance as in more complicated ones.

The non-interpretative notation system corresponding to the Troels-Smith approach is intended to cover the pysical properties of the deposit and its material composition. Once learned, the abbreviated nomenclature is simple to use, also in the field, even if the learning period may cause some agony. A signature in Troels-Smith notation gives an objective description of the material, immediately understandable to the reader who has any experience with the lithology of organic sediments. On the other hand, to draw conclusions with regard to the genesis of the deposit presumes local experience and knowledge. The *physical properties* recognized are colour ('darkness', nigrositas), stratification (stratificatio), elasticity (elasticitas) and water content ('dryness', siccitas). They are graded by a scale of 0 to 4 with 4 as the (arbitrary) maximum of the property in question. A shorthand description of mixed dy-gyttja might then run: nig. 2, strat. 0, elast. 1, sicc. 2.

Composition comprises sixteen main constituents and the mixtures in which they occur. Their relative quantity is expressed in steps of 25%. Less than 12.5% is indicated in text and formula by a +, but not in diagrams. Grade 4 comprises more than 75%. The sum of constituents should total 4. The degree of decomposition/humification of the various elements is indicated by an index figure (scale 0 to 4) added to the name of the element. In short notation a deposit may come out as $Th^2 1$, $Ld^3 2$, Ga1, meaning 25% 'turfa herbacea' (i.e. herbaceous peat substance, unspecified), 50% organic fine fraction consisting of three-quarters acid humus colloids and one-quarter detrital material, and 25% mineral deposit grain size 0.06–0.6 mm.

For graphic display each element is signified by a separate symbol or set of symbols in which distance between symbol elements (lines, signatures) indicates relative quantity on the 1 to 4 scale. The thickness of lines or symbol figures indicates destruction ('humicity') (cf. Fig. 3.10).

For obvious reasons, many symbols in the two notation systems are graphically identical, but they do not, and cannot mean the same thing—the one indicating origin, the other composition.

Signatures for the Troels-Smith system (which cover signatures of the Fægri-Gams system as well) have been worked out and are available from Letraset.

A weakness of the Troels-Smith system is to

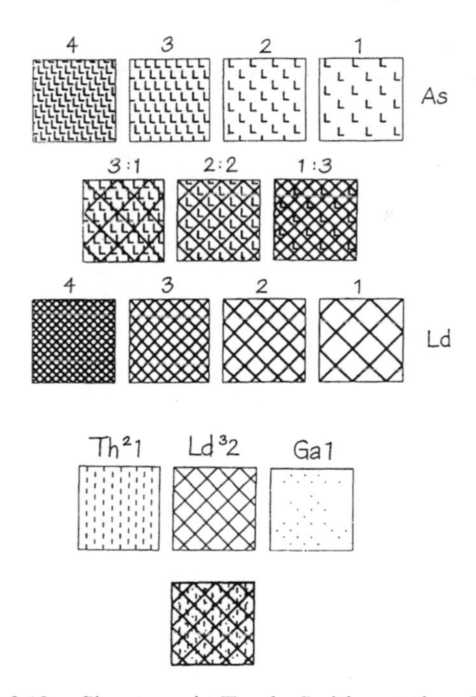

Fig. 3.10. Signatures in Troels-Smith notation. Upper three rows: signatures for anorganic sediments grain size 67 μm (As) and organic sediments 0.1 μm (Ld) in relative quantity 4/4 to 1/4. Between: As/Ld mixtures. Lower two rows: the individual constituents of the formula on this page and the combined signature.

obfuscate the vitally important distinction between the three principal hydrologic classes immediately visible from the older system and the actual basis for all genetic conclusions about the development of a basin under natural conditions.

If used logically the approaches and notation systems of Fægri–Gams and Troels-Smith are incompatible. In spite of assertions to the opposite (1955: 17) the Troels-Smith notation system is not concerned with hydrologic conditions: an unstructured humus (Lh) is a definite lithological specimen and must be indicated by the same signature notwithstanding whether it is a limnic dy or the matrix of an archaeological excavation. The inconvenience of this was realized by Troels-Smith 1955: 30, but his solution—to use two different signatures for what is the same composition—is to break the logic of the system.

The main purpose of the stratigraphic analysis —whether in Troels-Smith or von Post and Granlund style—is to permit conclusions about the genesis of the deposit. If a system cuts the worker away from the genetic conclusion which is the subject of the study, the system takes the character of a straitjacket. That much said, the Troels-Smith approach should be recommended for its attempt to achieve a more perfect objectivity by restricting itself to a purely descriptive approach, indicating what is there and leaving the interpretation to a following step in the work.

4. How pollen is recovered. Field technique (Quaternary pollen analysis)

Motto: Conduct the field work as if you will never be able to return to the site.

The field work provides not only the material basis for the work that is to follow, but also the foundation upon which conclusions are built. Without a solid foundation, structures are unsafe. So also in pollen analysis.

Pollen-analytic field technique has two different aspects: collection of reference material and sampling of (geological and archaeological) deposits.

Exines do not change by drying. Whether pollen for reference is taken from herbaria or from living plants is immaterial from a pollen-analytical point of view. In the latter case voucher material should be placed in a public herbarium.

Pollen may be procured from commercial sources, but such material should not be used for reference: both provenance and taxonomic identification are often unreliable. Some pollen species are very difficult to secure from herbaria, especially grass pollen. As soon as the thecae of grasses open the pollen is quantitatively dispersed, and flowers of herbarium specimens are usually empty, or some completely foreign material may adhere to them. Grass pollen is therefore best collected in the field just before flowers (anthers) burst open. The problem of recent pollen material for reference is discussed in Chapter 11.

4.1. Sampling of geological deposits

In this chapter we shall mainly discuss sampling problems in geological deposits and for geological and biological purposes. The special problems of archaeological deposits will be dealt with in a separate chapter, and so will sampling for special purposes: aerobiology, etc.

Sampling is the first step in the investigation into the composition of former vegetation, which, again, is the basis for further conclusions.

The aim is to obtain for subsequent analysis *uncontaminated* material *representative* for the deposit under investigation. The italicized terms embody the two overruling considerations in this phase of the work.

The samples carried back yield one set of data by analysis in the laboratory—more or less exhaustive according to the nature of the problem and the work and equipment available. Another, independent set of data were furnished by the deposits themselves in their natural setting. These data must be registered there and then—otherwise it is too late. The importance of careful and extensive field work cannot be stressed too strongly. Also, there is no way to compensate for insufficient or sloppy field work or for contamination of samples. In many cases it is impossible to replace unacceptable material by repeating the field work: the site may have been obliterated by roadwork, the archaeological excavation proceeded beyond that point, etc.

No two sites are identical, nor are two problems. It is therefore not possible in a book like this to give information and advice down to the last detail. What is valid for one site may be irrelevant for another. A manual covering all potential cases would be useless in any situation at hand. A very detailed set of prescriptions valid for one problem and one type of deposit would be inapplicable next time. The aim is to develop the student's understanding of the underlying philosophy. Examples are quoted as illustrations more than as precepts. The practical solution of the problem at hand must always be at the investigator's discretion. Our aim is to assist in making this decision, not make it for the investigator. Also, we discuss only cases which are commonly encountered, not the exceptional ones. Sampling from the bottom of very deep lakes or on the high seas are examples of rare events which we do not cover (cf. Wright *et al.*, 1965).

Records of field work should comprise labelling of samples, records of their depth, and data for the construction of stratigraphic columns and sections. They should also contain a description of the present vegetation of the site, a flora list with notes on the plants flowering at sampling time, and a survey of the plant communities of the surrounding area. As in all geologic work, the present is here the key to the past. Further, for the study of sea-level changes, the altitude above sea level should be recorded. The precision of—and time spent on—description of the macroscopic composition of the deposit will vary with varying aims of the investigation. The Troels-Smith approach to sediment description (cf. Section 3.2) has the advantage of a standard formula for the recording of such information. Figure 4.1 shows a sample card for registering physical data of a stratigraphic unit.

In heavy rain field work is difficult. The notebook becomes wet, the pencil tears holes in the paper, rain drops into the samples and splashes the substance all over, and everything becomes hopelessly dirty in a short time. In such circumstances it is almost impossible to work with sufficient precision and to obtain pure samples in the field. Only if some sort of shelter is to hand (*not* a tree: danger of throughfall!) can sampling from exposed faces or opened samplers be attempted, but even then it is very difficult to exclude contamination. Discrimination between deposit constituents is obscured. Under such conditions it is difficult to maintain the enthusiasm necessary for adequate field work. Modern recording equipment—waterproofed dictaphones, etc.—helps a great deal, but not all the way.

Deep frost is another meteorological situation that is counterindicated. Connections and samples freezing and having to be thawed in the field represent contamination hazards. Some of this can be remedied by placing a heated tent over the sampling station and, if on the ice, cores can be prevented from freezing by being placed in water under the ice after having been properly sealed. Freezing of cores inside sample tubes is absolutely ruinous, as the material freezes centripetally and, expanding during freezing, will in the end extrude material from the centre of the core, thus destroying the stratigraphy. If freezing cannot be avoided, tubes must be opened longitudinally, or samples extruded, permitting lateral expansion during freezing.

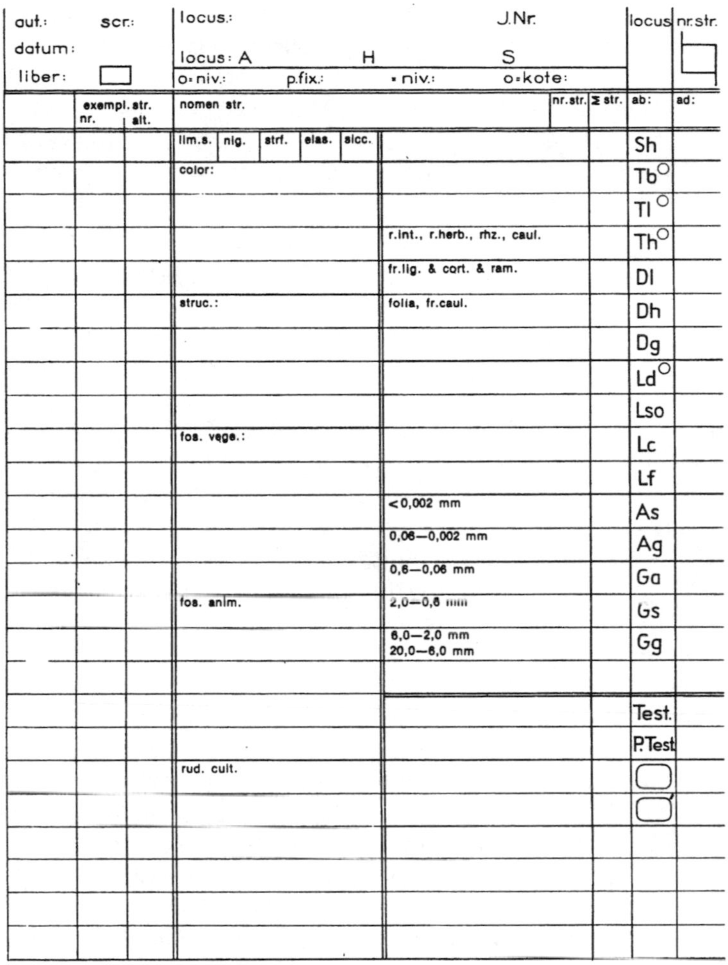

aut.: scr:
datum:
liber: ☐

locus.: J.Nr.
locus: A H S
o-niv.: p.fix.: • niv.: o-kote:

locus nr.str.

exempl. str. nr. | alt. | nomen str. | nr.str. | Σ str. | ab: | ad:

lim.s. | nig. | strf. | elas. | sicc.
color:
r.int., r.herb., rhz., caul.
fr.lig. & cort. & ram.
struc.: folia, fr.caul.
fos. vege.:
<0,002 mm
0,06—0,002 mm
0,6—0,06 mm
fos. anim. 2,0—0,6 mm
6,0—2,0 mm
20,0—6,0 mm
rud. cult.

Sh
Tb°
Tl°
Th°
Dl
Dh
Dg
Ld°
Lso
Lc
Lf
As
Ag
Ga
Gs
Gg
Test.
P.Test

Fig. 4.1. Field note form. Data according to the Troels-Smith notation.

As a general rule, cores for analysis should never be opened in the field if they can be brought back unopened. However, circumstances may enforce the use of samplers, the cores of which cannot be extruded, and discrete subsamples must be taken in the field. If one is compelled to work under conditions where absolute purity of samples cannot be guaranteed, the only resort that remains is to take relatively large samples and to include the entire material of each sample for analysis. Thus one may hope that the contamination, if any, may form an insignificant part of the material studied. But both sample and analysis must be considered very poor substitutes anyhow. No method can compensate for contamination of samples!

Days when pollen production and dispersal are particularly active are, of course, very unfavourable for pollen-analytic field work.

4.2. Selection of sampling site

It is not possible to give hard-and-fast rules for choosing the ideal site: too many variables are involved. An obvious first constraint is represented by the availability of equipment, transport and personnel. Methods and techniques must be adapted to the problem under investigation—and vice-versa. Adequate investigation cannot be carried out with inadequate equipment.

Some general rules always apply. A profitable

investigation presumes a concrete—but not inflexible—plan of action. The first item of the plan is the selection of the site, both of the basin and the particular sampling site inside the chosen basin. In an archaeological context the site is given, but for a regional investigation there is a wider or narrower range of possibilities. Careful preparatory work with detailed maps and air photographs increases the chances of selecting the best sites with a minimum of time lost in the field, also of identifying the surroundings of archaeological sites.

The sites selected must be so situated as to register as clearly as possible the vegetation history of the area under investigation without any masking of the record by allochthonous influx, e.g. from rivers bringing in material from outside, or by erosional phenomena, mixing up the record. The sedimentation rate should not be too low, telescoping the sequence with loss of detail. Apart from a little extra work of sampling, too long-drawn-out records are easily compensated for by greater interval between samples analysed. The character of the deposit is also important: some preserve pollen better than others and/or are easier to work. With some previous knowledge of the area it is possible to evaluate much of this from an air photograph.

The necessary first step in any investigation is usually to give an integrated synthesis of the vegetational history of the area (what might be called the classical pollen analysis). For this purpose the (main) sampling site(s) should be situated as far away from local pollen producers as possible. As the influence of the marginal zone—the ecotone—of a basin is inversely proportional with the size, one should for general studies avoid too small basins (lakes or bogs) and, of course, also avoid marginal zones of larger basins. These are the sites richest in macrofossils and were therefore the favoured sites in pre-pollen analysis days. Today material from such sites is sometimes used to elucidate special problems in spite of distorted pollen records, or sometimes just because of them (cf. also Chapter 9).

In an ideal deposit for classical pollen analysis the mother formation has not influenced the state of preservation of pollen grains, nor the composition of the pollen flora. Such deposits are rather uninteresting from the palaeophysiognomical angle. Older generations of pollen analysts preferred peat bogs but these are, on the whole, not the most suitable for general analysis since the mother formation produces a great deal of pollen, at least of herbs and small shrubs. The central parts of extensive tree-less bogs are relatively free from local influence and may be used with advantage if the upland forest pollen is the sole object of investigation and the pollen produced by the mother formation can be recognized and taken out of the calculations. If the objective of the investigation is to give an integrated picture of the whole vegetational development—not only the forests—pollen types produced on the bog are serious contaminants, both because of their local character and because of the difficulty of distinguishing them from those of other open spaces.

Generally the state of preservation of pollen grains is better in lake deposits than in telmatic or terrestric peats. The pollen produced by the mother formation is more easily recognized in lake deposits and can be more easily discounted there than the corresponding local pollen counts in other deposits. But the marginal vegetation may greatly influence the pollen from small lakes. In comparison with telmatic and especially terrestric deposits, lakes and lake deposits have the disadvantage of being more susceptible to secondary redeposition. This and some other objections which have been brought forth (some of them with good reason) against our preference for lakes are discussed in Chapter 7.

Lakes that have filled in are easily worked, because their sediments can be recovered from the surface when it is firm enough to support investigator and equipment.

In other investigations one is not primarily interested in the integrated picture of what has happened within a large area, but rather in the local development in relation to a small site—archaeological, phytosociological, etc. Obviously, in such cases the criteria for selection of a sampling site are completely different, and the only rule that can be given is to evaluate the possibility for pollen from outside the site to be deposited inside it.

Selection of the best site presumes knowledge of the bottom configuration and sediment stratigraphy of the basin. If not available from existing maps or charts, bathymetric information can be obtained by small battery-operated echo-sounders made for leisure boats, and easily operated from a dinghy or a raft (cf. Fig. 4.10). Investigation of the depth of sediment usually also presumes more sophisticated echo-sounding equipment. In colder regions working from the ice in winter is a very efficient way of sampling open basins, and also for echo-sounding, impregnating the ice with a non-freezing liquid (alcohol) to obtain acoustic contact.

Under ordinary circumstances one is advised to sample where the deposit is deepest, which is usually at or near the centre of the basin. Such deposits have presumably suffered less from erosional loss, but may have become contaminated by redeposited material. One or more sounding sections with records of the stratigraphy are often necessary to identify the most suitable point. Obviously, sites near the influx of major watercourses should be avoided both because erosion is more imminent there and because of the possible influx of allochthonous material if the watercourse has sources outside the area of investigation (cf. Pennington, 1947). The character of the bottom sample often gives an indication of the age—and implicitly the completeness—of the record. In very small basins a single, central sampling point may be sufficient as chances of complications in the horizontal direction are small.

Even an ideal deposit cannot alone give the total picture of the vegetation history of a region. Supplementary evidence must often be obtained from less ideal deposits with more local influence. By integration of all such data the complete picture may be formed.

Pollen-analytically, the size of a basin must be considered a function of travel distances of pollen grains. Near-shore deposits are always strongly influenced by the local, low-ground marginal vegetation, which is usually different from the upland regional climax and its successional stages. The dispersal limit for recognizable pollen output from an individual tree gives the minimum limit for a basin that can be expected to give a regional record. In the model (Fig. 2.12) that would be the point of intersection between curves Y and Z. In practice the limit might be set at 50 m, which means that the middle of a basin *ca.* 100 m across could be expected to give a fairly representative sample of the regional pollen rain, with due consideration of prevailing wind direction. Heide (1984) found very little difference between pollen registrations in a lake of 25 ha and a small pond of 0.4 ha (radius *ca.* 35 m). Smaller basins down to *ca.* 25 m radius register a mixture of local and regional pollen deposition, and in the very small ones the local influence is dominating, i.e. the site lies inside the zone of dominant (gravity) pollen deposition from individual trees, the radius of which is of the magnitude 10 m. Such diagrams are valuable for local problems; they give a distorted picture of the regional development, but if the rules of that distortion are known, conclusions may also be drawn about the regional development. The composition of such local vegetation, especially that around a basin, responds more to ecologic changes of the habitat than to general climatic influences. By analysing several such small basins within a region one may obtain an integrated regional picture.

Bog deposits are subject to the same influences as lakes: regional pollen rain dominating on large, treeless bogs and increasing influence of local marginal vegetation with decreasing size. In the terrestric phase the probability of redeposition is small —especially in domed bogs. On the other hand the admixture of pollen produced by the bog vegetation itself, and indistinguishable from that belonging to the regional pollen rain, represents a difficulty. Inside forests there may be small bogs or simply humid places with a certain deposition. Inasmuch as the forest is often fairly uniform, the edge effect may be negligible, and the record from such basins may give a faithful picture of the history of a limited part of the forest. The possibility of extending this picture to greater parts of the forest depends on its ecologic uniformity.

The establishment of sediment sections both in bogs and lakes presumes some length measure, which may be anything from a tape-measure to optical equipment and, on land, also levelling equipment. The more complex the sequence, the shorter the distance between the individual tests points (cf. Fig. 4.2).

If the sequence is very intricate, and especially in cases when archaeological objects or other traces of human activity are preserved in the deposit, open

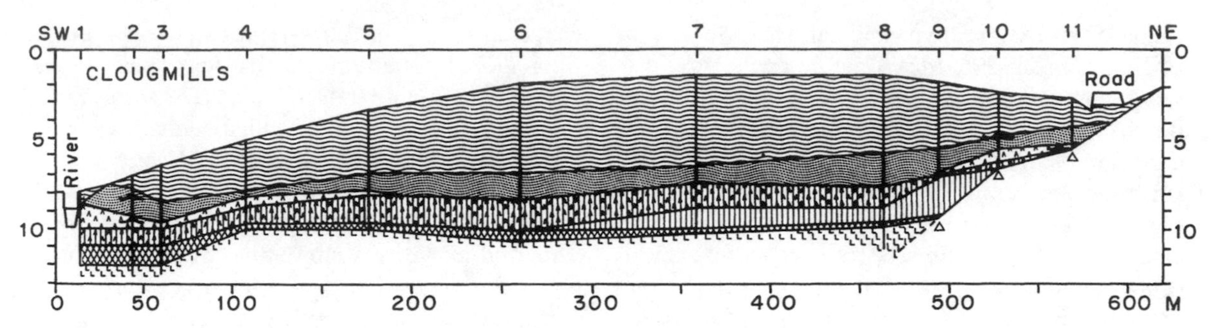

Fig. 4.2. Peat-bog profile (Clougmills, Co. Antrim, Ireland, from Jessen, 1949). Deposits marked according to Fægri–Gams notations. The bog rests on clay and started as a eutrophic lake (gyttja bottom layer), being partly overgrown by *Phragmites* (from the right-hand shore), later by a birch carr with *Phragmites*, drying out to a birch copse near the shore. With subsequent meiotrophication the forest was replaced by *Sphagnum* bog, initially topogenous, producing dark, highly humified peat, later forming an ombrogenous bog producing less humified peat. The angular outlines in this diagram are manifestly artificial. If boundary lines are drawn as more 'natural', sweeping curves unconformities are more difficult to spot.

profiles are necessary. Important points should be marked, e.g. by matches or more elaborate markers. Afterwards the whole profile should be levelled; and the position of all markers either levelled directly or measured from a levelled reference line, e.g. the surface of the mire. This is necessary to secure sufficient detail to identify former bog surfaces, etc.

4.3. Samplers and sampling procedures

The primary distinction is between continuous and discontinuous sampling. In the former case a continuous column through the deposit (alternatively those parts of the deposit that are of interest) is brought back into the laboratory. Continuous sampling is the prerequisite for contiguous analysis, which is the only safe way to catch all changes of pollen curves, albeit statistically smoothed by the time span represented by the formation of each subsample.

In discontinuous sampling discrete subsamples are taken out of the deposit at certain intervals and brought back, while the rest of the material is left behind. The former procedure is technically more cumbersome; the latter demands a much greater insight in the genesis of a deposit and understanding of the evidence presented by the stratigraphy. In the continuous core stratigraphic details may be studied at leisure in the laboratory; with discrete subsamples details of the stratigraphic evidence must be identified in the field in order not to be lost.

The other important difference is between sampling from open sections and from top of the deposit. Both techniques can give both continuous and discrete samples.

4.3.1 *Sampling from open sections*

Open sections are rarely made expressly for pollen-analytic purposes, but may be available, e.g. in excavations for practical or scientific purposes. They have the inestimable advantage of exposing clearly any depositional irregularity present. Sampling can therefore be conducted so as to obtain representation of all layers in a typical facies. Figure 4.2 shows how no single point can represent all layers, not even the most complete ones, points 7 and 8. To include also the dry marginal facies of the birch moor additional samples must be taken from points 4 or 10. To unravel even such a simple sequence from above is much more time-consuming than if an exposed face is available.

Open sections should be as vertical and as smooth as possible. Apart from very shallow ones they must usually be stepped (Fig. 4.4). If samples cannot be taken from one continuous sequence care must be taken to obtain sufficient overlap to have a safe representation of the whole face. Continuous samples from an open section are most easily taken by forcing a three-sided box horizontally into the deposit. Material is removed with a spade from both sides (cf. Fig. 4.3), exposing the box which is then cut loose by means of a string or a sliding lid pushed down from above. Instead of a spade, a 'peat knife', shaped from a scythe blade, can be used with excellent results in *Sphagnum* or other fibrous peat (Vuorela, 1985). The sliding lid is not compatible with the demand for sharp cutting edges of the box. For obvious practical reasons—stability and weight—the individual boxes cannot be too long. The whole depth of deep sections must be covered by two or more boxes placed on top of each other. The more fine-grained the deposit, the easier the sampling goes—within limits. In very dense deposits it may be necessary to punch a hole or two in the bottom of the box to prevent the build-up of an air cushion. Such holes should be sealed as soon as possible, at any rate on return to the laboratory, and measures should be taken against drying out of the material. Cold storage is to be preferred, but is not necessary.

Discrete sampling may be necessary in very coarse or tough material, sediment with rocks or stringy peat. The interval between samples should follow a consistent pattern—every 5 cm, or whatever is considered the suitable interval. In addition extra samples should be intercalated to represent any layer not caught in the regular samples and to give control of distinct sediment boundaries. Also, layers not represented in the main sampling column must be taken up separately by auxiliary columns.

The safest method of taking discrete samples from an open face is to press sample tubes into the deposit at regular intervals (plus possible additional samples) along a predetermined vertical. Again, a perforation prevents the build-up of an air cushion in plastic tubes. Glass tubes have to be inserted at an angle to let as much air escape as possible. In very stringy or coarse deposits it may be necessary to take samples out of the wall by forceps and spatula. As with all discrete sampling it is important to take out a sufficiently large sample for all tests contemplated. It is usually not possible to return for additional samples: open sections rarely last very long.

The continuous vertical column is always to be preferred, but sideway translations may be necessary, especially in coarse deposits (cf. Fig. 4.4). There should be clear overlap between subseries. Sediment boundaries are not always also chronological boundaries!

Taking discrete samples in the field does not differ markedly from taking them in the laboratory, except that contaminations are more difficult to control. Drifting pollen grains may adhere to the

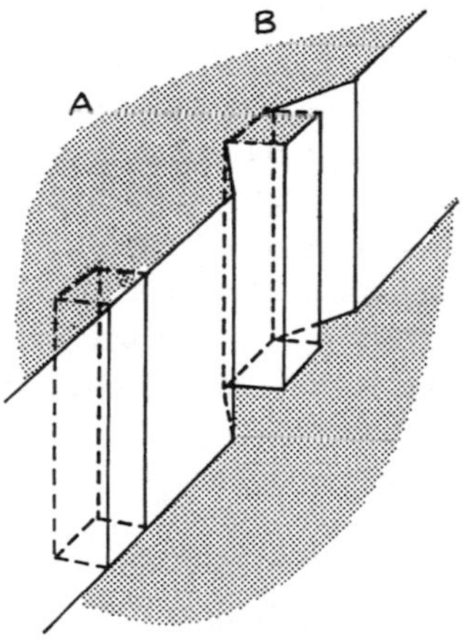

Fig. 4.3. Sampling from open face. A: a peat monolith has been cut out (with a box or something like that); B: the monolith is dug out and ready to be cut free.

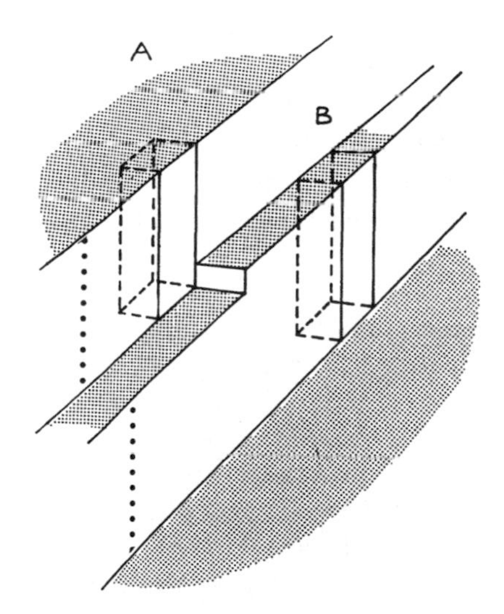

Fig. 4.4. Stepped sampling from an open face. Black dots: discrete sampling in the field (pressing tubes into the material). A and B: box samples. Note overlap.

section face, which should therefore be cleaned by horizontal cuts immediately before sampling. The overriding consideration is to obtain absolutely pure samples. To attain this, great care is necessary when taking samples which should *never* be touched by hand. Use a pair of broad, smooth forceps (the milled type is worse than anything) and/or a smooth spatula. Forceps work better in peat; a spatula, or better two, is preferable in fine-grained sediments. The disposable syringe technique (cf. Chapter 5) is also useful in the field if the deposit is suitable. How to keep instruments clean is a constant problem, the solution of which depends on local circumstances. There is always one way out: to lick them clean. Never use a wet piece of cloth, a towel or similar aid, which very soon becomes so dirty as to present a serious contamination hazard. Expendable absorbent paper towels may be used.

4.3.2. Sampling from the top of the deposit

If no open section is available, samples must be collected from the top of the deposit with some type of sampler. All samplers work on the same principle: a closed chamber is pushed down to desired depth, opened in such a way that it fills with a representative and uncontaminated core of the deposit at that depth, then closed again and pulled up to surface. There are two main types: side-filling and end-filling. Both exist in many variations. With one exception samplers should never be turned on the way down, as all unnecessary turning tends to make a mess of the deposit; they should be pushed straight down. The terms 'borer' or even 'auger', seen in the literature, are dangerously misleading and should be avoided, and so should 'drilling'.

Usually, the topmost sample cannot be collected by means of ordinary samplers, partly due to its unconsolidated nature, partly to the working principle of the implement. The topmost 10–20 cm of a terrestric deposit are therefore usually sampled from a sod cut for the purpose. This should always be the *very first* step in any field work, before trampling has destroyed the vegetation and contaminated the site. Further sampling is then carried out next to the sod hole. Surface samples from a lake bottom demand special precautions.

Depending on the length of the sampler chamber and the depth of the deposit one is usually obliged to extract several chamber lengths of material before reaching the bottom. All samplers have a—stronger or weaker—tendency to mess up the material at the lower end during sampling, and there is always the possibility of material from the sides of the upper part of the borehole being torn loose by repeated sampling in the same hole. Sampling should therefore alternate between two adjacent holes. In ordinary sediments this will obviate the necessity of using a casing in the borehole—a very inconvenient alternative. Theoretically, boreholes should be reamed up before taking the next sample. This is usually not done in actual work. Cores should overlap to safeguard continuity.

No one sampler is useful for all purposes and in all field situations. The character of the site and of the deposit, the maximum depth, the quantity of material needed in each sample and—not least—transport facilities constitute limiting factors. Much field equipment used to be, and still is, produced for and in the individual laboratory, but the more generally used samplers are also commercially available. Under all circumstances it is advisable to plan the equipment in such a way that as much as possible is interchangeable, avoiding an expensive confusion of sampling tubes, extension rods, driving weights, etc.

The selection of equipment also depends on the aim of the investigation. For a reconnaissance, e.g. of the sediments in a sounding profile or section, demands both to quantity and—to a certain extent—purity of the material brought up are fairly relaxed. A simple, light-weight sampler suffices. When it comes to taking samples for microscopical, chemical, etc., studies, larger quantities are necessary and there is no compromise in the demand for purity. More sophisticated samplers are then necessary.

4.3.2.1. Side-filling samplers

The prototype of the side-filling sampler is the Hiller sampler (which is actually much older than Mr Hiller, cf. Fries and Hafsten, 1965), (cf. Figs 4.5 and 4.6). Filling is accomplished by turning the

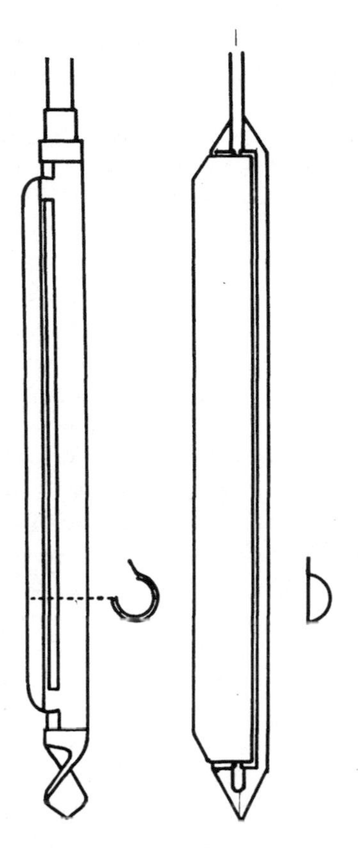

Fig. 4.5. Hiller (left) and Russian sampler: side view and cross-section.

apparatus anti-clockwise. An outer, loose jacket with a projecting lip remains stationary while the inner chamber turns so that longitudinal slits of jacket and chamber coincide. By further anti-clockwise turning the jacket is rotated as well, the chamber fills with sediment and is then closed by being turned the other way before being pulled up. After the sampler has been taken up, the chamber is again opened and the samples taken out.

In the original construction it is difficult to preserve the undisturbed sequence if one tries to remove the sample complete. It is better to take out discrete subsamples with forceps or spatula. Before opening the sampler one must clean the surface with a knife (not a cloth). Parts of the sample that have been in contact with the inner walls of the chamber must be considered contaminated and should be avoided. The exposed surface should be thoroughly cleaned. It is better to remove too much than too little!

A serious error may be caused by the gradual sinking of the sampler (any sampler) during the operation. One should also take care (1) that the edge of the lip, especially the lower edge, is cutting sharp, not blunted; and (2) that the jacket is not too wide. After some use, especially in coarse sediments, the fit of the jacket may become so loose that contamination becomes a danger: such sam-

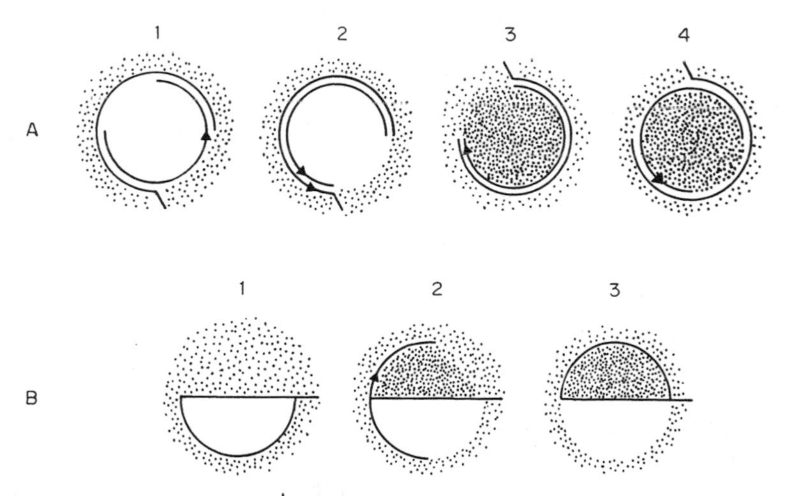

Fig. 4.6. Working principle of Hiller (A) and Russian (B) sampler. Open signature: undisturbed deposit. Dense signature: Sample in chamber. A: (1) Closed sampler pushed down to sampling level; (2) sampler opened by turning inner chamber; (3) both chambers turn, sampler fills; (4) sampler closed by rotating inner chamber, ready for extraction. B: (1) Closed sampler pushed down to sampling level; (2) sediment cut out by turning the half-cylinder jacket; (3) sampler filled by complete turning of jacket, ready for extraction.

plers should not be used. The sampler should be sufficiently leak-proof to hold air when it is pushed down, releasing the air when opened.

Perhaps even more so than other samplers the outer layers (and both ends) of a Hiller core should be regarded as contaminated and be discarded. If these precautions are taken, and the chamber and jacket are thoroughly cleaned between each sampling, one should be able to avoid contamination of the sample. The Hiller sampler is a very reliable instrument. Nejshtadt (1961) reports that he has used it through a 40 m deposit. Especially in fibrous/undecomposed peat, and also in peat rich in woody remains, the Hiller sampler may be the only means of obtaining samples at all, though in such deposits samples are never absolutely safe from contamination. In sandy sediments results are not reliable, and in sticky or very loose muds it may be necessary to use an extra-broad lip on the jacket to make it work. Various other minor modifications have also been introduced by different manufacturers. To our knowledge the Hiller sampler is the most robust side-filling sampler available, and its extension rods are superior to the threaded types.

K. V. Thomas (1964) has proposed some improvements of the Hiller sampler, of which we shall mention two.

1. The screw in the bottom end of the regular sampler is replaced by a cone. This gives a cleaner penetration and obviates the tendency to screw the sampler down.
2. The cone/screw can be taken off, and an inner lining taken out through the bottom of the sampler. This makes it possible to obtain continuous cores by means of a Hiller sampler—which is not possible with the regular type.

Thomas's model has a movable inner jacket (a principle otherwise abandoned years ago), but the idea of the displaceable bottom cone/screw can also be adapted to samplers with an outer loose jacket. The ordinary Hiller-type sampler is commercially available.

The other side-filling sampler in general use is the so-called Russian sampler (cf. Tolonen, 1967). It is not as sturdy as the Hiller sampler, but is simple and reliable, and in most sediments it gives a cleaner result because the sample is cut without any disturbance of the stratigraphy. Even the finest lamina can be seen, which would have been obliterated by the more rough treatment in the Hiller sampler. The Russian sampler (Figs 4.5 and 4.6) consists of a half-circular tube with a centrally hinged door. After being pushed down to sampling depth with the door closed, the tube is turned 180°, cutting out a half-cylinder of sediment. By the end of the turn the sampler is closed against the opposite side of the door, and the sampler is withdrawn without any further turning. The door remains stationary during sampling and is the only part that has been in contact with the sediment on the way down. Deposit close to the door must be suspected of contamination and discarded, but the rest of the sample can be taken for analysis.

Returned to the surface the sampler is opened by the reverse procedure; the sediment lies cleanly exposed and, if desired, subsamples can be taken out or the entire sample be transferred to a split plastic tube and brought back to the laboratory. The sampler can use the same extension rods as the Hiller, and the two complement each other.

A very simple side-filling sampler, chiefly for reconnaissance purposes, consists of an open half-cylinder which is pushed down, rotated 180° and pulled up again. The same hole must be used every time. A sharpened ring at the bottom helps in cutting out the sample. It works best in coherent material. The risk of contamination is not much greater than in other side-filling samplers, but the risk of losing the sample is considerably higher.

4.3.2.2. *End-filling samplers*

In principle, an end-filling sampler is a tube forced down through the deposit, cutting out a monolith. Whereas there are few theoretical difficulties in pushing the sampler down, extracting is more complicated. If the chamber is open at the top the monolith will have a tendency to fall or be dragged out (sucked out by the vacuum) when the sampler is pulled up. If the chamber is closed at the top, an air cushion builds up and prevents material from entering the sampler. The chamber must therefore be open at the top during sampling and have some possibility for closing during the extraction phase. An open sampler tube forced down from the surface can be closed on top with some kind of stopper before being pulled up. Leaks must be prevented by some soft (pollen-free) material at the top. If any air is left in the sampler on top of the core it will expand at extraction and lead to (sometimes total) loss of material.

Numerous end-filling samplers have been constructed and used successfully for various purposes. For geotechnical purposes very heavy equipment is commercially available.

Compared with side-filling samplers, end-filling ones have several advantages. Above all, it is easier to take cores of large diameter. In the former type the diameter can only be a few centimetres without causing technical problems. In the latter the diameter is limited only by the strength of the equipment. A usual wide-corer diameter is *ca.* 10 cm. End-filling samplers also give cleaner cores with less distortion of the stratigraphy. However, it should be kept in mind that the top of the core has travelled the whole length of the chamber and may have left impurities on the walls. If there are strong differences in pollen flora between top and bottom of the core such impurities may introduce errors. The outer parts of the core should therefore always be cleaned away in the laboratory before individual subsamples are taken out for analysis.

On the other hand, end-filling samplers are not effective in coarse sediments and may be useless if there are large, hard inclusions, e.g. rocks or hard wood. On the whole they are heavier than side-filling samplers.

Open-end samplers with sharpened lower edges also work in fibrous peat or wood peat. They are the only samplers which may—and should—be rotated when being pushed down. It is important that the cutting edge be situated at the inner surface of the sampler tube (Fig. 4.7); otherwise material is compressed on entering the tube, and will therefore not proceed properly through the sampler.

Any kind of tubing may, theoretically, be used. Waterpipes have done the job, but cannot be recommended. They are never quite smooth inside (friction problems!) and they are heavy. Plastic (PVC) tubing is better. Since it is so much lighter, large-diameter tubing can be used. It can be cut into suitable lengths and is easily machined with simple tools. It is sufficiently strong, penetrates easily and the adhesion between material and tubing is low, which lowers friction and resultant compression of the core. In the laboratory the tube can easily be split open for inspection and sampling (see Fig. 5.2).

4.3.2.3. *Top sampling*

The simplest end-filling sampler is a tube pressed or hammered down into the sediment from above. This primitive method can be very effective in soft, homogeneous sediments down to a depth of 1–2 m. It is useful in areas where transport of more heavy and sophisticated equipment is impracticable. The tube must be closed at the upper end before being extracted (cf. below). Extracting long tubes, especially from tough terrestric deposits, is heavy work. If mechanical aids (cf. below) are not available, it may be necessary to dig the tube out, wholly or partially, and loosen it by bending, which will also help to break the core off at the bottom and, together with the air being let in, helps to prevent it from being dragged (sucked) out of the tube again.

In theory, one may use an end-filling sampling tube of the same length as the entire depth of the deposit to be penetrated. However, the practical length is limited by: (1) the danger of breakage and (2) the friction, both the external friction which makes penetration difficult and the internal friction may lead to serious compressions and distortions of the core, and also to loss of material, which is under such circumstances pushed aside at the bottom end instead of entering. The force necessary to extract long samplers is also an important point.

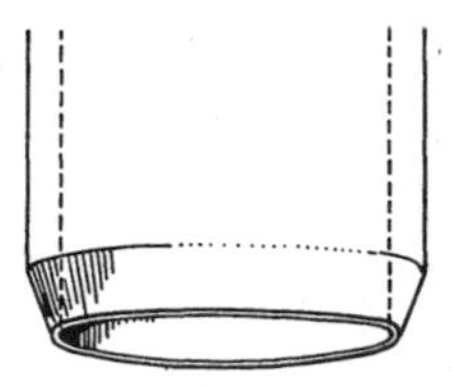

Fig. 4.7. Cutting lower edge of open-ended sampler.

4.3.2.4 Deep sampling

Usually, sampling must carry on beyond the depth that can be worked by the simple tube. In that case the bottom of the chamber must be closed while the sampler is being pushed down.

Modern end-filling samplers are of the stationary piston type, going back to the now obsolete Dachnowsky sampler. The bottom of the sampler chamber is closed by a piston. Sampling level reached, the piston is locked at a stationary level in relation to the surface, while the chamber travels past it on its way further down. The modern version is usually referred to as the Livingstone sampler (Fig. 4.8), although others have also contributed to its development. The piston is formed by rubber stoppers pressed together by nut and screw. The primary coupling between piston and chamber is by friction, and both are pushed down by a (Hiller) rod working on the piston. A separate cord —steel wire—runs from the piston through a hole in the top of the chamber to the surface. When sampling level has been reached the cord is secured at the surface, the rod is retracted to the top of the chamber and locked in the upper position with a bayonet lock. When the chamber is now pushed further down, past the level of the stationary piston, it cuts out a cylindrical core which gradually fills the chamber from below, sticking to the piston in front of it. When the sampler has been pushed down the length of the chamber, the piston is secured in top position with the cord and the sampler withdrawn.

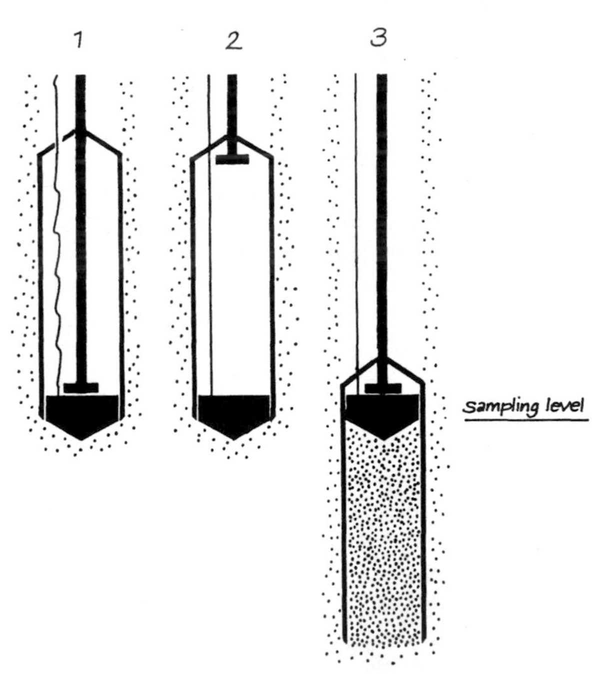

Fig. 4.8. Working principle of Livingstone sampler. (1) Closed sampler pushed down to sampling level; (2) pushing rod pulled up and locked in upper position, wire pulled tight; (3) chamber pushed down its own length, filled and ready for extraction, wire kept tight.

Since the core is in contact with the piston all the time, no vacuum forms at the top of the sample, and the material is prevented from falling out of the chamber. Coarse, incoherent sediments have a tendency to fall out anyhow; the problem of constructing a mechanical lock for the bottom of the sampler has not found a practical solution (we have not tested the valve described by Strøm, 1934). Nor has there been found a practical solution of the problem of relieving the vacuum establishing itself in the borehole when large-diameter samplers are extracted.

With this elegant solution of the two principal problems—the air cushion during filling and the vacuum inside the chamber when extracting—the Livingstone sampler has been developed to such perfection as to make most other constructions obsolete and mainly of historical interest for working normal deposits at moderate depths, i.e. down to 10–20, perhaps 50 m, depending on the character of the deposit. For more difficult work special modifications have been introduced (cf. Wright *et al.*, 1965).

Pieces of PVC tubing with an effective length of 2 m form excellent sampling tubes. The diameter depends on the problem at hand. Narrow tubes are easier to operate; large ones, while more cumbersome, yield more material as may be necessary for comprehensive analyses of micro- and macrofossils, ^{14}C dating, chemical analysis, etc.

After being pulled up to the surface the chamber is disconnected from the rest of the sampler, cleaned outside, properly marked (do not forget to indicate which end is top!), securely wrapped air- and watertight and taken to the laboratory. Monoliths intended for analysis should not be opened in the field. The diameter of the working parts of the sampler must be geared to that of the tubing, whereas the same rods can be used in all normal cases.

For very tough sediments or greater depths this sampler type suffers from two weaknesses: (1) the cord stretches, giving bad depth control; (2) the rods are too weak and may bend or even break. In either case there is, again, no proper depth control and no control with the working angle of the chamber—which may deviate a great deal from the vertical. The remedy is to use a double rod instead of the rod-and-wire system described above. The inner rod (Hiller type) is connected to the piston; an outer, hollow rod to the chamber tube. The working principle is the same. The two rods are initially coupled, locking the piston in bottom position during the pushing down. After sampling level has been reached, rods are uncoupled and the piston fixed at that level by means of the inner rod. Since that rod is not used for any further penetration it remains in constant contact with the piston, and the chamber is pushed down with the outer rod. After the chamber has been filled, the piston is locked in the upper position by locking the two rods again, and the sampler is extracted (cf. Fig. 4.12).

The Reissinger modification of the Dachnowsky sampler (Fig. 4.9) is of some merit. In it, the piston and the rod are permanently locked together (but should be able to rotate against each other). The rod has a thread. When the sampling depth has been reached, the chamber is screwed down its own length by rotating the rod. The outside of the chamber should have two—or more—flanges to prevent it from rotating in the deposit during sampling. It is a slow-working instrument, but lightweight and easily operated and a good substitute under difficult transport conditions. Very little force is needed to drive it down, and it is therefore effective in hard and difficult-to-penetrate deposits. Because of a sponge effect it is not suitable in fibrous peat. The chamber should be solidly constructed in metal. A diameter of 5 cm seems to be the upper limit. Samples must be pushed out by the opposite process, which takes time, or the chamber may be constructed with a detachable lining, in which case it should have a detachable shoe at the bottom, holding the lining in place.

Because of the small manual force needed to drive the Reissinger sampler down there is the danger that in very hard deposits it will rise up unnoticed instead of penetrating downwards. A locking ring placed under a ground platform will prevent unnoticed rising of the sampler. As in all end-filling samplers it is imperative that the piston remains stationary during sampling, and that the adhesion between piston and sample is not broken. If this happens a vacuum may establish itself at the top of the core, with the result that water in the sample starts boiling, ruining the performance of the equipment.

Digerfeldt (1966) has described a simpler instrument for shallow sediment series: a three-sided chamber of quadratic section is pushed down. After the necessary depth is reached the fourth side is closed by pushing down a metal strip, running in a guiding furrow. The latter is bent at the bottom of the instrument, thus closing the chamber. As the chamber is open at the top, neither air cushion nor vacuum builds up.

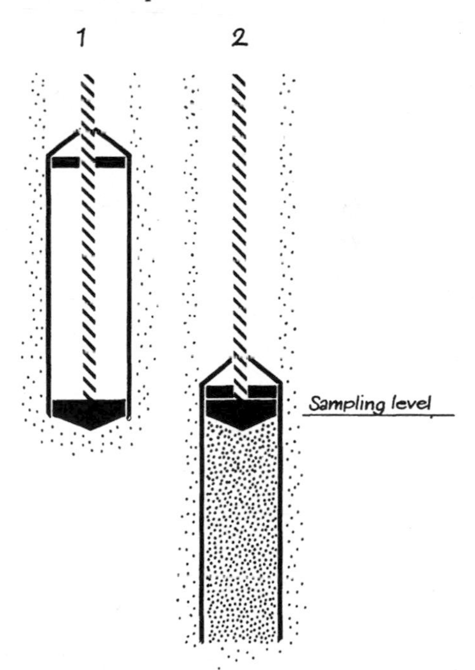

Fig. 4.9. Working principle of Reissinger sampler. (1) Closed chamber pushed down to sampling level; (2) Chamber screwed down its own length, filled and ready for extraction.

A very elegant and powerful sampler is described by Kjellman *et al.* (1950), in which the interior of the instrument lines itself with foil during the pushing down, thus eliminating all friction. It is too complicated and heavy for general pollen-analytic use.

Sampling a lake bottom through open water presents special difficulties and presumes the availability of a raft (cf. Figs 4.10 and 4.11). A single dinghy can only be used for light-weight equipment. The raft must be securely anchored, either by three or four grapnels or by strong wires to the shores (of small lakes). In shallow water the site may be marked by a stake driven into the bottom sediment, and the boat or raft may be moved up to the marker when the next sample is to be collected.

The force needed to drive down a sampler cannot exceed the weight of the equipment plus that of the operator(s). If more weight is needed it can be achieved by (water-)ballast on a ground platform or a raft. The alternative is to anchor the platform by means of soil or ice screws. On ice a simple (and disposable) anchor can be made by making a hole in the ice and putting a suitably long piece of wood (e.g. a stick of firewood) underneath the ice. A loop of rope around the piece of wood anchors the equipment.

On top of lake deposits there is often a loose flocculent layer, which is very difficult to sample. Usually it represents a very short time and is unimportant, but in some cases completeness of the record is imperative. In such cases the ordinary open tube may be used—the tube should then be transparent to allow a visual check of what has been sampled. A modified Livingstone procedure, starting sampling above the lake bottom, also works. Special samplers have been developed for the purpose, such as Züllig's 'Schlammstecher' (1956a:17), the Jenkin sampler (cf. Ohnstad and Jones, 1982—an older, simpler model also exists), and the core freezer. The original model (Shapiro, 1958) freezes the core centripetally with inherent hazard of extrusion and destruction of stratigraphy. A 'cold finger' freezer, which freezes the sediment on the outside of the sampler, obviates this inconvenience, but does not produce a core. A three-sided or semi-cylindrical freezer would be better. By careful diving it is also possible to secure undisturbed samples of the upper, flocculent layers.

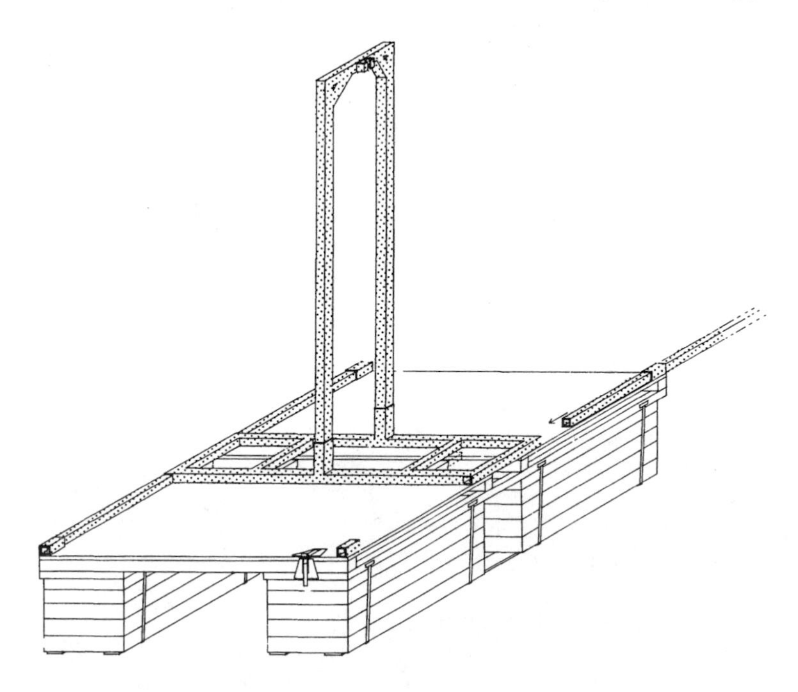

Fig. 4.10. Transportable raft, assembled. Gallows superstructure and locking posts (square tubes) shaded.

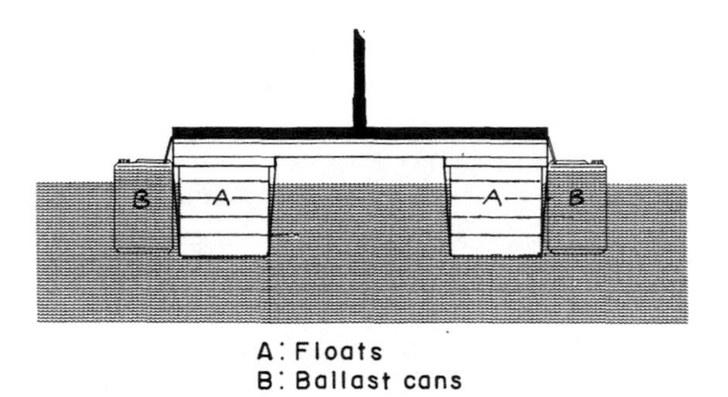

Fig. 4.11. Section of raft with ballast tanks for stability. Superstructure black.

The real problem with this type of material is, however, the extraction and preservation of samples and transport of the core. Unfrozen unconsolidated sediments cannot be transported, and frozen material demands a complicated and cumbersome freeze-chain from the site to the laboratory. Extracting discrete subsamples from this kind of deposit is also extremely difficult if the core is not frozen. Some of the special samplers can be partitioned by insertion of separating walls after sampling. The compartments are then emptied either by scooping or by opening (taped) perforations in each compartment. Such samplers usually presume a bottom valve—cf. the Digerfeldt–Lettevall (1969) sampler.

Even apart from the flocculent top layers, recent lake sediments are usually fairly loose, and a sampler may penetrate far down on its own weight. Pushing rods have no sideway support and may easily bend. If double-rod equipment is used there is less bending, and even less if an outer casing is also used. However, by this time the equipment becomes rather awkward. If water is so deep that pushing rods cannot be used, sampling must be done by instruments lowered on a wire and driven down by their own or added weights.

Three types of weights are in use; driving, pounding and expendable. Driving weights are the simplest and commonest. They are permanently fixed to the sampler or can be detached before the sampler itself is pulled up. Pounding weights have been described by Züllig (1956b) in his 'Ramm-Kolben-Lot'. Expendable weights are used, e.g., in Moore's (1961) 'free cover'. They may be physically detached or they may be removed by dissolution, e.g. of a salt bag. Moore's corer also had a float and surfaced on its own, which means that there is, in principle, no limit to the depth that can be worked. However, expendable weights can also be used with samplers that are lowered and pulled up by wires.

If sampling must go beyond the depth reached by the first core, weights cannot be attached directly to the sampler chamber, but need intermediate rods, which presents a problem. It is therefore tempting to use very long sampling tubes the first time. Such tubes have a tendency to topple over, unless they can be guided (in shallow water) by a diver. Züllig's 'Ramm-Kolben-Lot' (1956b) is in principle a (narrow-bore) Livingstone sampler with a pounding weight and an attached stabilizer-guide. In order to hit the same hole each time (for multiple sampling) Reissinger (1936) used a funnel resting on a piece of sheet iron on the surface of the sediment. We have no experience with this, or with the following.

Table 4.1. Summary of field techniques.

Working from	In	Equipment
Open faces above ground-water level	All types of peat and sediment	Spade, knife and spatula/forceps
Surface of deposit	Fibrous peat or coarse organic sediment	Hiller or Russian sampler
	Highly humified peat or medium to fine-grained sediment	End-filling samplers
	Stiff clays and layers of sand	Heavy core samplers with casing pipes
Surface of lakes (from ice or raft)	Top deposit	'Schlammstecher', core freezer, etc.
	Ordinary deposit	Livingstone sampler, free corer, Mackereth sampler

A very refined, but complicated, device for sampling from deep water is the Mackereth sampler (cf. A. J. Smith, 1959) which is driven down by a combination of vacuum and hydrostatic pressure. We understand that the almost explosive ascent of this sampler has caused some near or full accidents.

Table 4.1 summarizes the most expedient methods to use in the majority of cases, and where conditions are normal. Thus a variety of samplers should be to hand, but extension rods, etc., should be standardized and interchangeable.

When cores are short (and water shallow), samplers can be pushed down and pulled up by hand, but especially the large-diameter samplers soon become too heavy and need some mechanical device for the purpose. The four main parts are a platform, a gallows, ball cone clamps* and a chain-hoist or a jack (cf. Fig. 4.12). For work on open water the platform must be carried on a raft with a working well, floating on styrofoam blocks or inflatable rubber floats. The platform, or the raft, must be ballasted to give counterweight both for driving the equipment down and pulling it up again. Water-filled plastic cans form suitable ballast units for the raft. They can be used for floats as well. All parts of this equipment should be divisible into easily portable units.

Some of the items mentioned above may be commercially available—beware of the quality!—others must be machined for the purpose if not already available at the institute of research. Descriptive and constructional details can be obtained from our institute for a nominal fee, covering copying and postage. The very large and heavy—and expensive—rigs used for extra-long cores of hundreds of metres fall beyond the aims of this book.

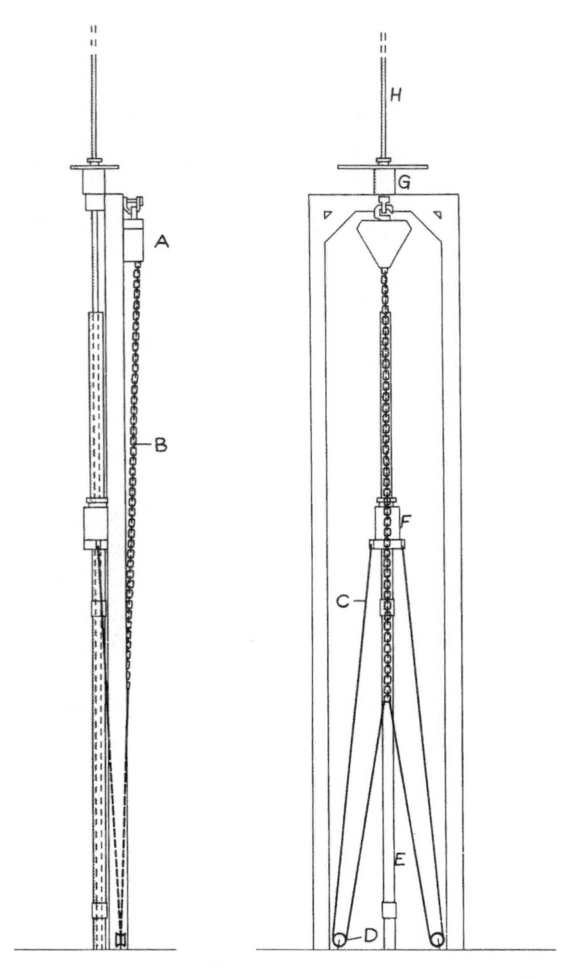

Fig. 4.12. Gallows for working heavy samplers (large-diameter Livingstone, etc.). Profile and face view. A: Differential chain hoist; B: hoist chain; C. wire to (outer) driving rod; D: pulleys for downward movements; E: outer (hollow) driving rod; F: ball clamp for outer rod; G: ball clamp for inner (Hiller-type) rod; H: inner rod. The set is shown rigged for driving the sampler down.

* Geonor manufacture: P. O. Box 99 Røa, N-0701 Oslo, Norway. We understand that this firm has developed a self-moving unit for sampling also from mire surfaces or from aboard a raft. We have as yet no experience with this.

5. Finding the grain. Laboratory technique

Motto: Consider the purpose of the study and how this can most easily be accomplished before deciding on procedure.

Laboratory technique in pollen analysis covers the work from the moment the material reaches the laboratory until pollen counts have been completed.

The tasks comprise storing sediment material, subsample extraction from it, preparation of sediment samples for counting, staining, mounting and microscopy. The laboratory work is usually the most time-consuming part of the total research process.

Laboratory technique also covers work with recent pollen for morphological study and similar problems. Special aspects of this will be treated in Chapter 11.

5.1. Laboratory furnishing and equipment

The laboratory room should be separated from the room in which microscopy takes place. The laboratory atmosphere is not good for microscopes, and the activity in the lab is disturbing for the person(s) at the microscope(s). The laboratory air should be filtered pollen-free. All water used should also be distilled or filtered pollen-free and guarded against secondary contamination.

The laboratory equipment should be arranged in a rational way to avoid crossing traffic lines, especially if more than one person works there at the same time.

The range of processes carried out in a palynological laboratory is limited, and so is the equipment. It is presumed that physical and chemical analyses of samples—if undertaken—are carried out elsewhere. Refrigeration, deep-freeze and freeze-drying equipment are useful, especially for storage of samples, but not absolutely essential.

Most reactions can and should be carried out in the (same) centrifuge tube to avoid loss of material and contamination by transference. If the quantity of suspension is too large for one tube it is more prudent to fill up the same tube repeatedly after decanting instead of transferring between tubes.

For chemical preparation at least one hood is necessary. Some of the fumes produced during the work are extremely poisonous. Centrifuging is one of the most important processes, and there should be at least one centrifuge, speed and size adapted to local circumstances: people working in the lab, number of samples to be worked up, etc. An 8 or 16 × 15 ml centrifuge with speed up to 5000 r.p.m. is ample. Tubes should swing out to a horizontal position.

A mechanical stirrer to shake up precipitates is useful, but generally not sufficient—auxiliary glass rods are a necessity.

In addition to ordinary heating—burners, sandbaths, waterbaths—it is useful to have a heating bench for slow evaporation. Other items of equipment and reagents are partly self-evident, partly indicated in the following under various processes.

The chemial preparation of samples takes time, and even in a one-person lab it pays to prepare batches instead of single samples. In large labs this is necessary. Confusion of samples during preparation is avoided by having a standard set of centrifuge tubes, funnels, etc., for one run of samples —or two to be used alternately—permanently numbered 1–8 or 1–16, depending on the size of batches. For each run one must start by making an identification list of the samples being treated (storage list or some similar identification). Alternatively, one may use unnumbered equipment and number it ad hoc with the identification number of samples (e.g. the last two figures of storage numbers) for each batch. Paper labels are not recommended: they often do not outlast a single run. Non-woven ('Millipore') surgical tape is remarkably resistant and stays several runs.

During laboratory work samples may be contaminated in various ways, and the cleanliness in a pollen-analytical laboratory must be of the same standard as that of a bacteriological one. Contaminations due to dirty laboratories or utensils are unpardonable. Some sources of impurity are especially serious: in spring and summer the air of most laboratories, if not filtered, contains substantial quantities of pollen which may be caught in the preparations. If KOH preparation is used alone, recent pollen grains may be recognized by their cell contents; in other cases it is more difficult. Sometimes it is tempting to ascribe the report of single pollen grains of southern species in Arctic depositions to such laboratory contamination, particularly if there is no information to the contrary in the publication in question. Especially when working with samples that are poor in pollen one must be very careful to avoid contamination from the laboratory air and other sources. The same applies if infrequent pollen types are sought. On the other hand, if the sample is carefully treated from the moment it is taken out of the sampler until the analysis has been completed, contamination is unlikely.

Checks for contamination during handling in the laboratory are obtained by going back to the original material. An empty sample may be run to check on systematic contamination in the laboratory (M. B. Davis, 1961).

Some contaminations are less important than others. A few pine pollen grains, added to a sample containing thousands of them, do not affect results materially, but if oak grains contaminate an Arctic sample the effect is rather unpleasant. Even worse is the contamination of pre-agricultural samples by weed and cereal pollen, which are often very frequent in the surroundings of a laboratory.

5.2. Storage

The first problem presenting itself after return to the institute is storage of material until subsamples are taken out for pollen analysis, macrofossil analysis, radiocarbon dating, and physical and chemical analysis. In many cases material must also be stored for a considerable period both before and possibly after samples have been taken out.

The problem of *conservation* of the sediment varies considerably both with the type of sediment (some peats are self-disinfecting) and with the climatic situation and facilities of the laboratory.

Traditionally, subsamples for pollen analysis are stored in small glass vials, cork-stoppered and sealed with paraffin wax. In most cases these small vials can be stored for years without any further treatment except, perhaps, the addition of a small amount of thymol or phenol to prevent moulding. Mould spores can be very annoying during analysis. Addition of a small quantity of glycerol prevents drying out. Alternatively, the material may be freeze-dried.

Material that has dried out has usually shrunk, and humus colloids have been denatured. It is often time-consuming to disperse such samples, but pollen exines seem to survive excellently providing normal procedures are followed.

If it is desirable to store whole cores or monoliths it is important to keep them well sealed, preferably under refrigeration, and to prevent conditions favourable for microbial activity. Under refrigeration unopened cores in the original plastic tubes (cf. Section 4.3.2.2) can be stored for years without ill-effects. If tubes have been opened it is, on the other hand, very difficult to prevent moulding and drying out even if cores are resealed and refrigerated. Heat-sealed foil-lined plastic wrapping serves best.

If the available refrigeration capacity is limited, or moulding a serious problem, there are alternative ways of storing:

1. to replace the water of the sediment with a suitable substance, e.g. polyethyleneglycol,
2. freeze-drying;
3. drying at 105°C—after these preparations cores can be stored at room temperature for years;
4. deep freeze

The *polyethyleneglycol* method (cf. Løvlie *et al.*, 1979) is laborious and expensive, but has the advantage of great permanency of the results and easy handling of the material. It is indicated if it is important to keep intact the structure or stratigraphy of the material, e.g. for the study of laminations or other physical characteristics. It is practicable only for small quantities of material at a time. A three-walled metal box is pressed into the flattened side of the core and cut loose with a knife or a wire loop (like taking monoliths from an exposed face—cf. Fig. 4.3). The box is put into a bath of low-weight polyethyleneglycol (molecular weight *ca.* 600) for 1–2 days at room temperature. The formation of air bubbles may float the sample, which should be kept in the box by wrapping either in gauze or some similar material. The sample is then, through one or two intermediates, transferred to a high-molecular weight polyethyleneglycol (2000–4000), which is solid at room temperature. The process must be carried out at some 50–60°C in a heating cabinet. After 6–8 days the sample can be taken out. It hardens at room temperature and can be cut with a microtome. Polyethyleneglycol-impregnated sediment 'bars' are also very useful for X-ray photography (Digerfeldt *et al.*, 1975). For subsequent analysis of pollen or other fossils the polyethyleneglycol can be dissolved in hot water and normal procedures followed. Any other water-miscible, non-drying liquid can be used for such infiltration, e.g. glycerol, but samples imbibed with hardening ones are easier to handle afterwards.

Freeze-drying is very gentle towards humus colloids, and simple to use for fibrous peats, etc. The results disperse easily during preparation and can be treated like fresh samples. If carefully handled, non-fibrous sediments freeze to a sponge-like structure which is very brittle, but with great care they can be handled. Unsuccessful freeze-drying of non-fibrous material destroys the structure under mechanical stress and changes it into a fine powder. To prevent stratigraphic mix-up it is therefore necessary to cut such cores into short pieces (which later function as stratigraphic units) before freeze-drying. Such short pieces also dry out much faster than full-length cores; they are less laborious and it is possible to use smaller and less expensive equipment.

Drying at 105°C is the most immediately effective, but brutal, preparation for storage. The material shrinks unequally, the structure is lost and the sediment becomes difficult to disperse for future work. But it may be the only practicable method of conservation before storing. Measuring water content, volume and weight within uniform lengths of the core before freezing permits later calculations of the original volume of the dried material. Ordinary room temperature air drying takes time, but is also possible and has the same result, adding the contamination danger. All drying from liquid phase destroys the micellar structure of sediment colloids in contrast to drying by sublimation from solid phase.

Straight deep-freezing of cores is a simple and effective, but time-limited, way of storing material. Cores expand at freezing, due to their water content. If they are frozen within the tube, expansion will be longitudinal only, which will distort the stratigraphy (the inner parts extruding). Before deep-freezing, tubes must therefore be opened in such a way that cores can expand three-dimensionally. Frozen cores can be wrapped and kept for a fairly long time. Samples must be taken out with a saw or similar equipment. In the long run, frozen cores dry up (theoretically, they could be kept in ice blocks, but that is hardly practicable). The resulting material will then have the same characteristics as freeze-drying material, and the handling of such a core would present insuperable difficulties. If it is decided to keep an open deep-frozen core for a longer period it should be cut up and treated as for freeze-drying.

The length of time over which to store preparations and material is a problem. Obviously, for checking on data until the final writing-up of results, the microscopical preparations actually used are immediately important, as they are also, but to a lesser degree, for later documentation. Every manipulation of the material—extraction of samples, chemical preparation, mounting, etc.—is a potential source of error and contamination. The closer one is to the original, i.e. the material in situ, the safer. If a paper contains a debatable statement, inspection of the preparation actually used can verify or falsify the statement in relation to that preparation, but not in relation to the original material. If material is to be stored for future reference it is therefore important to store the original material as brought in from the field. This is especially important if the original site is, or is in danger of, disappearing. Preparations made from the material are of limited value—the more limited the more processes they have been through. The main exception is the finding and reporting of an occurrence of a single grain of a rarely observed pollen taxon, e.g. of an extreme insect pollinator. Such grains should be stored (as single-grain preparations, cf. below, if possible) as a voucher and documentation *of the identification*. The conclusions drawn from the occurrence are still subject to the other constraints suggested above.

The practical conclusion is therefore that microscopical preparations should be kept until some time after the investigation has been written up and published. By that time most of them have gone bad anyway. An untreated sample should be stored indefinitely for reference. To keep the whole core, while being technically possible, will in the long run create storage volume problems, but should nevertheless be contemplated if the site is destroyed or in other ways unattainable.

5.3. The laboratory work

The flowchart (Fig. 5.1) shows the usual laboratory routine procedures as related to pollen analysis. The framed processes are carried out in practically every case, the other ones only come into the picture if called for, e.g. HCl deflocculation comes in if the material contains appreciable quantities of $CaCO_3$, of if marker chalk pollen tablets have been added. The sequence between HF treatment and acetylation is to a certain degree optional, depending on the character of the material. Also, the need for HF treatment may not be realized until after acetylation. Filtration may come in anywhere after deflocculation. We have preferred to remove as much material as possible before filtration.

5.3.1. Taking subsamples for pollen analysis

Because of the danger of contamination by recent air pollen it is advised, if possible, not to take subsamples for pollen analysis outdoors during field work, but to postpone this for the more controlled atmosphere of the laboratory.

If the core has been taken with a piston sampler the first problem is to gain access to the material. Simple extrusion often compresses the material. Plastic sampling tubes can be opened by being cut longitudinally with a circular saw (Fig. 5.2). Be careful to cut only the plastic, not into the sediment, as the saw-cut contaminates. Some sediment types have a tendency to adhere to the wall of the top side of the split tube. A wire half-loop pulled along the inside (Fig. 5.3), combined with light hammering on the outside, usually suffices to loosen the plastic.

Before subsampling for pollen analysis the core surface should be carefully cleaned with a knife or scalpel. Cleaning movements should go perpendicularly to the axis of the core, i.e. parallel with the strata, to avoid inter-strata contamination. After cleaning, the stratigraphy of the core should be checked against field notes.

Usually subsamples are taken from the core at selected intervals. The distance between samples depends on the problem in hand, but also on other factors such as sedimentation rate. The time within which the investigation must be completed (if a time limit is imposed) also plays a great role for the number of samples. At any rate it is recommended to take subsamples all at one time (not one by one) and at shorter (half) intervals than expected to be necessary for the study. Normally, this extra work does not take much time, and at a later stage it is then easy to make additional pollen counts of samples in between the ones originally counted, if some unexpected change in the pollen assemblages so indicates. Every opening of the core is an invitation to contamination from the air and between parts of the core, not least if the material has been dried and the surface as a consequence forms dust. Techniques should be developed locally to reduce the number of times necessary to expose cores to air and contaminants.

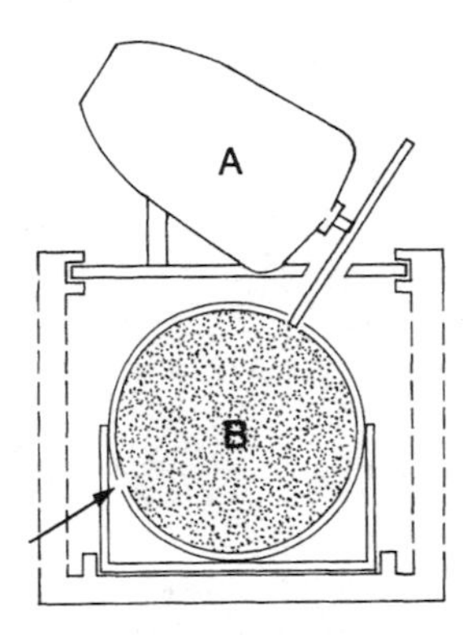

Fig. 5.2. Cutting open a plastic tube for removal of sediment samples. A: Circular saw gliding in a frame; B: Sample tube also gliding in a frame and turned for cutting the other side. The saw cuts through the tube wall and must not enter into the sample. The arrow indicates the other cut.

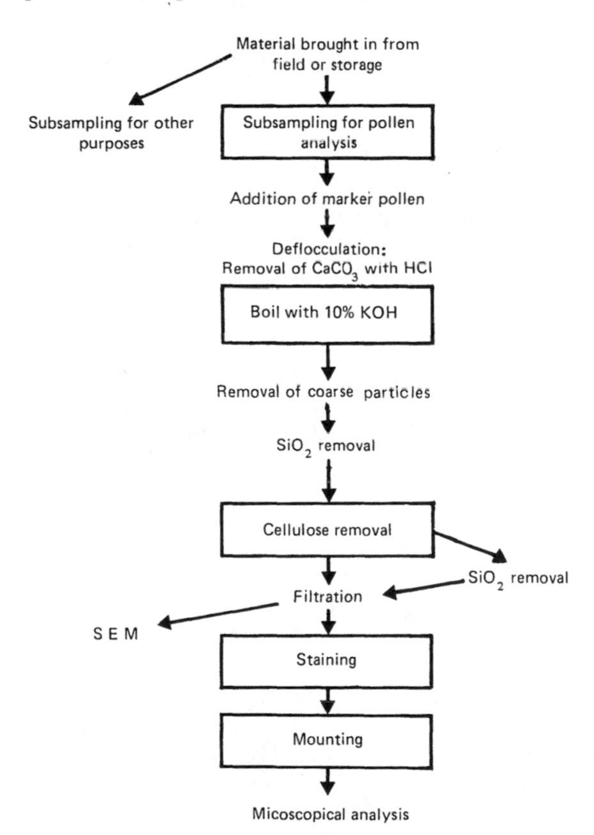

Fig. 5.1. Flowchart for preparation of material for microscopical analysis. Operations indicated in frames are almost always carried out. The others are optional.

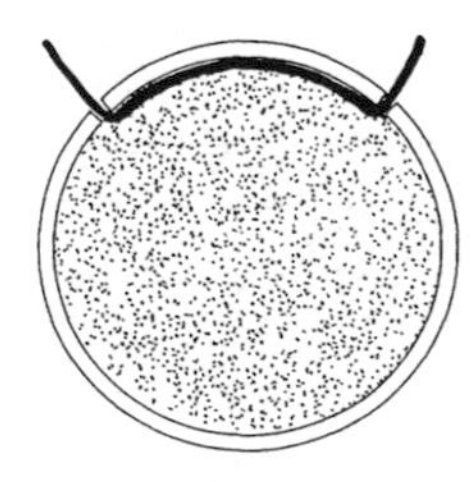

Fig. 5.3. A wire loop or a thin plate is slid along the lid of the tube to cut it loose from the sediment.

The techniques for taking out subsamples for pollen preparations differ according to what type of analysis is intended, and also with the type of material. For *relative* (percentage) analyses an approximately 1 cm^3 sample, taken out with a spatula or scalpel, is sufficient. Such subsamples are conveniently stored in small glass vials with a waxed cork, or in plastic vials with air-tight caps. They are treated and stored like pollen samples taken in the field, mentioned above.

For *absolute* pollen calculations high-precision volumetric determination of subsamples is necessary. For homogeneous (lake) sediments the most common method is to take out a measured length of material by means of a sharp-edged pipe of known cross-section area. In some cases it is difficult to ensure accurate volume data by this method because of difficulties to obtain, even with a razor, a sufficiently sharp end cut without disturbance by harder macrofossils (straw, seeds, etc.). Once taken out, such samples can be transferred to ungraduated vessels. Relative errors are smaller in larger samples than in small ones. In difficult material one may be obliged to take out and measure a comparatively large volume. By homogenizing and taking out a weighed aliquot one obtains a well-defined small volume to work with.

Volume-defined samples can also be taken out in small gelatine capsules of known volume, as used in pharmacy. This has the great advantage that the capsule is dissolved by ordinary preparation procedures, and the sample therefore can be prepared in the capsule. As originally described (pushing the capsule into the core) the method does not give adequate volume control: an air cushion builds up in the capsule (perforate the bottom), and the gelatine softens under the influence of humidity from the sample and warmth from the fingers of the operator, distorting the capsule and changing the volume. Kryzwinski (Fig. 5.4) has modified the method by sucking the material up from the core with a cut-off disposable syringe. After a narrow-bore outlet tube has been added to the syringe, material is again squeezed out into the gelatine capsule, avoiding air bubbles. Volume control is maintained by keeping the capsule in a tight-fitting (wooden or plastic) holder during the operation. Surplus material from the syringe can be used for other purposes.

The volume of fibrous, more or less porous peat is difficult to define and measure, especially as it may change when the sample is brought up to the lower atmospheric pressure at the surface. However, if measurements are consistent such errors will cancel themselves out. Direct measurement as above is not possible, but a similar technique can be used if the sample is deep-frozen in advance or fixed in ethyleneglycol. Samples of known volume can then be cut out, most easily with a cork-borer of known diameter.

The volume samples can also be measured by pyknometer. The pyknometer fluid should be one that is not miscible with water or imbibed by the (wet) peat. Since pyknometer measurements are cumbersome, the density of the sediment as measured may be applied along uniform lengths of the core (test at both ends) and the volume of intermediate samples calculated from their weight.

Taking samples for physical or chemical analyses, ^{14}C dating, etc. does not differ very much from taking samples for pollen analysis except for the larger quantities (still) necessary for some such techniques. Samples for ^{14}C dating should represent the shortest possible period, i.e. be taken as thin slices across the whole core, (cf. Fig. 5.5). Such samples must be taken out before any chemical whatever has been added to the material. For details of methods cf. Berglund (1986).

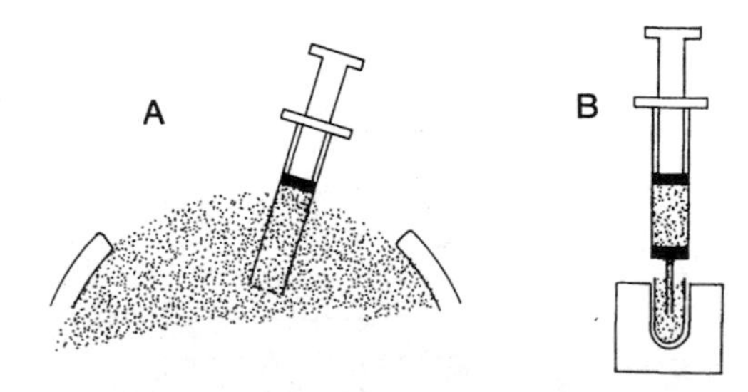

Fig. 5.4. Taking out volume constant samples from the sediment in a tube. A: A cut-off dispensable syringe is filled with sediment; B: a nozzle is put on the syringe and a constant-volume gelatine capsule filled from the bottom up to avoid air bubbles. The gelatine capsule must be placed in a holder to prevent collapse.

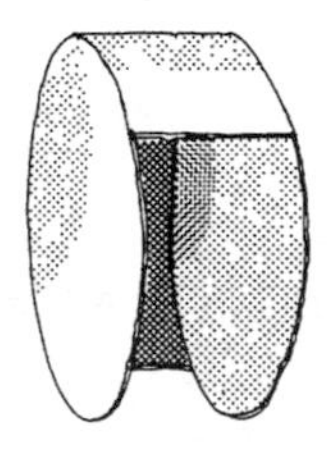

Fig. 5.5. Metal cutter for taking slices of deposit out of a tube, e.g. for ^{14}C analysis. After having been inserted (pry open the tube), the cutter is turned 180° and withdrawn.

5.3.2. Preparing samples for percentage analysis

The objective of preparation is to concentrate pollen, spores and other microfossils in the sediment by removing extraneous matter, and to render them as visible as possible by staining and/or embedding them in a suitable medium.

The fewer the pollen grains are in a given unit of deposit, the more radical will be the measures needed to achieve a reasonable concentration of pollen. Scarcity of pollen grains in a given sample may be due to three causes: (1) rapid growth of the deposit, the pollen rain is highly 'diluted' by matrix substance; (2) subsequent destruction of pollen grains during fossilization in aerated deposits; (3) low pollen production of adjacent vegetation. Deposits of the first type are easily dealt with by acetolysis or HF treatment. Deposits of the second type should be discarded altogether if possible since their pollen flora is probably not representative of the contemporary pollen rain. Besides, the matrix is in this case generally so resistant to chemical treatment that radical measures are necessary, involving danger of further distortion of the pollen flora. Deposits of the third type present real difficulties, especially since they are frequently of great interest, e.g. glacial samples. Concentration by chemical treatment must in these cases be carried out as far as possible without attacking the pollen grains themselves. Even after such treatment the analysis can be extremely trying to one's patience.

An alternative to chemical concentration is staining with fluorochromes after deflocculation, as described by Shellhorn *et al.* (1964): In the fluorescence microscope pollen grains light up against a black, non-stained background and can therefore be detected at low magnifications. Total time for preparation and analysis will not be much shorter than with other methods, but it may be useful when more vigorous treatment is contraindicated.

Too 'successful' concentration of a few pollen grains found in a large sample may be inherently hazardous. The poorer in pollen the original deposit, the greater is the influence of contaminants, and the danger that there has been differential destruction. It is in many cases extremely difficult to find out what is really represented by the few grains extracted from a great lump of matrix. Such 'triumphs of preparation technique' should be viewed with suspicion.

Studies of the technique of preparation of samples have been published repeatedly. We shall here deal only with methods that we have found practicable, and shall make no attempt at completeness. Nor shall we go too far into detail: conditions vary so much in different laboratories that principles must be adapted to local requirements. A simple method of dispensing should be found for the most frequently used reagents. As most of them do not attack ordinary plastics they are efficiently dispensed from plastic squirt-bottles. Sulphuric acid is one of the exceptions. It can be dispensed from an ordinary pipette flask. The water-bottle should permit a strong jet to be formed for washing out residues on sieves, etc.

Flowcharts over the work bench should show the ordinary procedure. It is also a good idea to collect, in a local lab manual, the various methods adapted for regular work with their local modifications. Such a manual may be made more specific than a textbook, and may also contain important, but strictly local, information, e.g. where to obtain materials, etc.

Protective gloves, goggles and aprons should be used when appropriate. In any case, hands should be rinsed under running water *after each single step* in preparation, notwithstanding what the procedure was and which chemical was used. Gloves must be perfect: perforated gloves are more dangerous than nothing, as chemical can penetrate and are then protected from being washed off again, acting on the skin for dangerously long periods. The laboratory staff should receive instruction how to deal with emergency situations due to corrosive chemicals like HF, H_2SO_4 and HNO_3—even acetic acid. It goes without saying that legal demands for employee protection must be complied with. On the other hand hysterical reactions due to ignorance and intimidating propaganda should not be tolerated.

Concentration of pollen can be achieved both by chemical and physical methods. In most procedures both are combined. The large variation of sediment types and pollen preservation makes it necessary to use a variety of procedures. Some procedures for difficult sediments may affect the shape, size and preservation of the pollen, or may lead to differential loss of taxa. These procedures must therefore be applied with great care: the investigator has to balance the possibility of, on the one hand, obtaining an inferior, possibly skewed preparation, and on the other hand, having nothing at all. Details about inherent weaknesses of the processes are given under the individual descriptions.

Reitsma (1985) published a study on size modifications of pollen caused by various treatments. The following points seem evident: dry material increases in size on being wetted, notwithstanding whether is wetted by KOH treatment or by means of a detergent or wetting agent. This is a purely physical process, due to uptake of water. When a sample is subjected to the acetolysis mixture the size of the grains increases suddenly; with further treatment it decreases again and reaches a plateau after some 3 min. These changes must be due to chemical reactions within the exine. According to Reitsma's data the preparation methods described in this book should give consistent results.

Each step in the various processes is usually followed by centrifuging to concentrate the pollen suspension. In the following, a centrifuge speed of approximately 3000 r.p.m. for 2–3 min is presumed if nothing else is described. Pollen grains sink fairly fast in aqueous suspension. Washing the residue with water is a usual step. Tap water may contain pollen grains and other microfossils, and should not be used directly. An empty test run to check on the purity of water may be advisable.

Whereas a few standard methods suffice in nearly all cases, some deposits present difficulties. Such difficulties have to be solved individually, and no textbook can deal with all without assuming undue proportions. Among the apparently insoluble problems are pyrite, pure carbon, and also some cutinous(?) bark substances which seem to have the same chemical properties as exines. Chitin also belongs to the group of substances that cannot be removed from a sample, but generally that is of less importance in fossil material. Beware of methods that produce precipitates which are difficult to remove again.

5.3.2.1. Deflocculation

The general scheme for preparation of pollen-bearing material starts with deflocculation. In most cases this is an automatic result of the removal of some of the substances during the following steps, but in some cases special treatment is necessary.

Pre-treatment with a surface-active substance, e.g. ethanol or a detergent solution, generally serves the purpose, but requires some time: leave the sample for a day or a few days and it will frequently disintegrate by itself. An old-established method of deflocculation is to imbibe the sample with a warm saturated solution which crystallizes when cooled. This method may give results even in very obstinate cases that resist other treatments. Strongly consolidated sediments have to be crushed (hammer or vice) and sieved before any other treatment, but this is rarely necessary in Quaternary sediments.

Bates *et al.* (1978) recommend the use of sodium pyrophosphate as a deflocculant of clay. After a first treatment with (sodium) hydroxide or HCl as usual, the sample is washed till neutral and immersed in hot (water bath) 0.1 M sodium pyrophosphate (NaP_2O_7) for 10–20 min. After centrifugation the clay particles remain in suspension and can be poured off, while pollen grains, silt and organic debris have settled to the bottom. In very clayey samples the treatment may have to be repeated.

Various other surface-active deflocculants have been suggested and tested, but hardly merit special description.

Ultrasonic treatment, which is particularly useful in clayey sediments, may be hazardous, since exines rupture under the influence of ultrasonic vibrations. According to Cerceau *et al.* (1970) *Pinus* exines rupture after only 1 min treatment at 40 kc/c (5.7 W/cm^2). The effect is strongest on heavy exines. Thin exines can stand much longer treatment. Beware of possible health hazards.

5.3.2.2. Removal of extraneous matter

This presents problems varying with the character of the substances in question. Removal may be effected by chemical or physical means. In the following, substances are dealt with by chemical groups. The sequence of treatments in this chapter is the one generally followed in practice—provided the substrate in question occurs in the sample—but reversions are frequently indicated by the material itself.

Calcium carbonate—HCl treatment. Ten per cent cold HCl is in most cases sufficient to remove carbonates. Cold HCl is safe, but prolonged heating with this reagent entails the risk of degrading exines. Froth formed by the reaction is subdued by a drop of 96% alcohol. Calcium should be removed carefully if the sample is later to be treated with HF. If not removed, Ca forms a dark, almost insoluble precipitate, which can destroy the preparation.

Humic acids—KOH Treatment. 'Humic acids', i.e. unsaturated organic soil colloids, are removed by a short boiling with 10% KOH (or NaOH). The treatment is humic acid removal and deflocculation at the same time. The concentration of KOH should not exceed *ca.* 10%: long boiling with stronger solutions attacks the exine of some species. If prolonged boiling is necessary, e.g. to soften and dissolve dried humic substances, concentrations should be kept constant, most easily under a reflex cooler. If the sample contains much humic acid the resulting fluid becomes very dark brown, almost black, and it may be necessary to repeat the treatment to remove all of it. In such highly humified material scarcely anything but pollen is left after treatment, and the following procedures are redundant (pollen size is affected).

The KOH treatment was von Post's original pollen preparation. He boiled a few cubic millimetres of substance on an ordinary microscope slide over a spirit flame (a gas flame is too hot), holding the slide with an ordinary clothes-peg. The method is perfectly safe, but some water must be added to the sample during the process to keep the KOH concentration down. This method is still useful for a rapid survey of the pollen content of a sample both in the laboratory and even more in the field, where one has no access to laboratory equipment.

Coarse particles—sieving. After deflocculation, coarse mineral particles may be removed by repeated decanting between two test tubes or beakers. Another procedure consists of boiling the sample with a fair amount of liquid in a beaker which is afterwards kept in rotation. The heavy particles very soon collect in the vortex at the centre, and the liquid with the organogenic material can then be drained off. But it should not be forgotten that pollen grains also sink very fast. Sieving is safer and also takes care of large organic particles.

For sieving, samples should be shaken vigorously after boiling with KOH (a few drops of alcohol removes froth), and the suspension strained through a metal (copper or nickel), nylon or ceramic strainer, mesh size *ca.* 0.2 mm. The residue on the strainer should be washed with a powerful jet of water from a squirt-bottle. The fine suspension that has passed the strainer—containing, among others, the pollen grains—is concentrated by centrifuging.

Siliceous matter—HF; heavy liquids; flotation. Several methods describe how to remove siliceous matter: sand, clay, etc. The nature of the sediment, equipment of the lab and personal preferences contribute to a decision about which method to use in the individual case.

The dissolution of siliceous matter in hydrofluoric acid was introduced by Assarson and Granlund in 1924, and is the oldest chemical treatment (next to KOH boiling) used in pollen analysis. Some palynologists prefer to use strong (30–40%) boiling acid for a few minutes; others use cold, dilute acid (10%) for days or weeks.

The HF treatment does not appreciably attack organic remains, pollen grains included, but grain size is affected, and expansion at subsequent acetolysis is less pronounced. If size changes must be avoided one of the other methods should be resorted to. Samples with a high proportion of clay are difficult to treat by gravity separation or flotation alone, and should at any rate by pretreated (deflocculated) with HF. A coarse sample (sieving!) may need very long boiling with HF if the larger mineral grains have not been removed. For such samples other methods may be more suitable.

Since HF is a dangerous, corrosive substance, the cold treatment has some psychological advantage. The long period of treatment (up to a month or more) is a negative point. In any case it is important to be careful with HF which is strongly corrosive to skin and eyes both as a liquid and as fumes. Direct contact must be avoided (cf. the safety precautions described above); also take precautions against damage to drainage pipes. Waste HF-containing liquids should be collected in a plastic container (polyethylene or polypropylene) where they are neutralized in a $CaCO_3$ suspension or in dry powdered cement.

(a) Hot, strong HF treatment

1. Wash thoroughly after deflocculation and KOH treatment. Centrifuge.
2. Boil (*ca.* 3 min.) with *ca.* 40% HF in a platinum crucible* or digest in a plastic centrifuge tube in a water bath.
3. Transfer to a plastic centrifuge tube if boiled in a crucible. Centrifuge.
4. Heat with 10% HCl, without boiling, to remove colloidal SiO_2 and silicofluorides. Centrifuge while still hot. This is a dangerous operation, since exines are sensitive to hot HCl. If the original sample contained much SiO_2 the process may have to be repeated until the supernatant is colourless or light greenish.
5. Wash repeatedly with water until neutral.

(b) Cold dilute HF treatment

1. Place the washed residue from KOH treatment in a plastic centrifuge tube and add cold 10% HF. Leave for 24 hours or longer. Stir occasionally. Centrifuge.
2. Follow points 4 to 5 in the description above.

Intermediates between and combinations of methods (a) and (b) are used in various laboratories. Much depends on the size of samples and the quantities of SiO_2 that must be removed. Large quantities are better treated in a beaker initially.

In cases when HF treatment is contraindicated or ineffective *gravity separation* (often, but erroneously, referred to as flotation) can be carried out by means of a heavy liquid, the density of which is such that mineral fragments ($d > 2$) sink, whereas pollen and other organic constituents ($d < 1.7$) float. A common feature of all heavy-liquid processes is the necessity after separation to dilute the pollen-bearing supernatant so much that grains can be centrifuged down.

In cases when it is of interest to save the heavy liquid, and the quantity is not too great, centrifuging can be carried out in a piece of tubing, closed at the bottom end or U-bent, inserted in a water-filled centrifuge case (to prevent collapse of the tubing). After centrifuging the tubing is pinched below the floating sediment and the supernatant with pollen poured off for further treatment.

A large number of heavy liquids has been proposed and used. Among the more common ones are bromoform (immiscible with water, the sample must be dehydrated), stannic chloride (must be kept in acid solution to prevent the formation of oxychloride), Thoulet's solution (KJ and CdJ_2, very expensive) and zinc chloride. Most of the heavy liquids are expensive and poisonous, some of them are highly viscous and it takes a long time for the particles to separate; the fumes are as obnoxious, if not as poisonous, as those from HF treatment.

The two most commonly used heavy-liquid

* The frequently used nickel crucible is attacked by HF and may collapse, with serious consequences. At any rate it should not be used too many times before being replaced.

methods are the bromoform and the zinc chloride methods:

(a) Bromoform/ethanol (acetone) mixture
1. Deydrate the residue twice with concentrated ethanol (96% followed by 100%).
2. Add a bromoform/ethanol mixture with a specific gravity of 2.1 to the residue in a centrifuge tube. Stir.
3. Centrifuge at 200 r.p.m. for 10 min or at 3000 r.p.m. for 3 min to perform the separation.
4. Decant the floating organic material in another centrifuge tube and wash with 96% ethanol.

The specific gravity of the bromoform/ethanol mixture must be checked by hydrometer and adjusted by adding bromoform or alcohol.

(b) Zinc chloride/water solution
1. Disperse residue from deflocculation in a saturated solution of $ZnCl_2$ (specific gravity 1.96). Stir well and centrifuge.
2. Decant supernatant into distilled water with a few drops of HCl to prevent precipitation of $Zn(OH)_2$. Centrifuge and wash with water.

This method was first described by Funkhouser and Evitt (1959), and has the advantage over the bromoform method that it can be carried out in aqueous solution, obviating dehydration. $ZnCl_2$ is also less poisonous and much cheaper than bromoform. The specific gravity of the solution is somewhat lower than that of the bromoform mixture, but in practice this does not seem to matter. $ZnBr_2$ has been proposed to increase the gravity, but has not been commonly used.

One disadvantage with heavy-liquid separation is that some Dinophyceae cysts are so heavy that they follow the siliceous matter to the bottom of the tube and are lost. There are no reports that this is differential between species, but it may well be. Heavy-liquid separation should be used with care, and only if other methods fail or are contra-indicated. Coarse mineral matter can be removed by sieving, and very fine by filtration (see filtration, below).

Flotation (sensu stricto) is based upon the principle that one component is wetted by a froth-forming liquid and thus concentrated in the froth phase, whereas the other is not. For pollen-bearing sediments the method is described by Dumait *et al.* (1963). The medium is a 1‰ polyvinyl alcohol/water mixture. The sample is reduced to a fine powder (possible calcareous substances removed by HCl in the usual way) and placed at the bottom of a glass tube, diameter *ca.* 5 cm, length *ca.* 50 cm. The upper part of the tube is bent over into a beaker. Froth is formed by the introduction of N_2 into the tube from a fine-pored sintered filter in the bottom. The froth gradually brings pollen grains over into the beaker, while mineral particles are left behind. A few small clay particles are easily removed afterwards by gentle HF treatment. The separation is not really quantitative, and nothing is known about differential effects. We understand that other, not published, flotation methods are (have been) used by industrial laboratories.

Cellulose—Acetolysis, oxidation, etc. In organogenic deposits the next type of material that has to be removed, also the bulkiest one, is the cellulose of moss leaves, rootlets, etc. In a water-free medium this can be done with concentrated sulphuric acid. A more practical method is the so-called acetolysis, originally introduced by Erdtman in 1934 (cf. Erdtman, 1969). In our laboratory the method is practised as follows:

1. The residue after KOH deflocculation is washed and dehydrated with glacial acetic acid. Centrifuge.
2. Treat with a fresh mixture of *ca.* nine parts anhydric acetic acid and one part H_2SO_4 conc. Heat gently to the boiling point, e.g. by immersion for some 3 min in a boiling water bath. Centrifuge.
3. Wash with glacial acetic acid. If the sample is transferred directly to KOH or water, cellulose acetate might precipitate again. If so, it can be dissolved in glacial acetic acid. If there was not much cellulose in the original sample this step is unnecessary.
4. Heat very briefly in 10% KOH. This step stops

further action by the acetolysis mixture and prepares the pollen for staining, embedding, etc. Centrifuge.

5. Wash repeatedly with water until neutral. Centrifuge.

The techniques dealt with so far are comparatively safe, as has already been stated, and they will, unless used to excess, not appreciably affect the morphology of exines except, perhaps, the most delicate ones—the preservation of which in deposits is in any case questionable. Exine-less pollen grains (Musaceae, etc.) disappear completely by acetolysis, but such grains are not preserved in deposits.

If a necessary concentration of pollen and spores has not been achieved by these methods, the next step will be to try *oxidation*. Since exines are relatively much less resistant to oxidants than to the treatments so far described, oxidation should not be resorted to unless necessary, and in any case with the greatest care. *Lignin* is one of the substances that can be removed by oxidation, and as it oxidizes more easily than cellulose the latter substance should have been removed in advance.

A great number of oxidants have been proposed. Because exines are more easily broken down in alkaline solution, acid ones should be preferred. Chloric acids have been used, but the usual Schulze reagent ($KClO_3$ and HNO_3) is too strong to be used for Quaternary material. Erdtman introduced $KClO_3$ and HCl instead. The reaction should be carried out in the centrifuge tube.

1. Suspend the sample in glacial acetic acid in a 15 ml tube. Add five or six drops of saturated $NaClO_3$ solution and 1 ml HCl conc. A violent reaction sets in immediately, and the sample is generally bleached very promptly. It is often necessary to stop the reaction instantaneously after a few seconds, which is best done by pouring the contents of the tube into a beaker of distilled water which should always be at hand. Chlorite or hypochlorite may be substituted for chlorate, if preferred.
2. Centrifuge and wash very thoroughly with water.

An alternative treatment is to use nitric acid. Again there is danger of destruction of the exine, and both concentration of the acid, temperature and reaction time must be adapted to the individual sample. Concentrations above 15% cold HNO_3 can ruin the material. The process does not run as fast as chlorate oxidation, and can be stopped by adding KOH to neutralize the acid, followed by washing. Hot 10% HNO_3 has been recommended for the removal of pyrite. Extreme discretion in use is advised.

Oxidation is the least satisfactory process in a pollen-analytic laboratory and should be avoided if at all possible. One of the problems is the great variability in resistance of exines against oxidation. Both original preservation of the material and concentration of chemicals have an influence. In addition one has to take into consideration that even if the pollen safely survives the treatment itself, oxidation combined with other, also normally safe, treatments may be disastrous. Hafsten (1959) describes an experience in which, after original HCl treatment, HF treatment plus acetolysis or oxidation (bleaching) gave consistent results, whereas HF plus acetolysis *and* bleaching resulted in the loss of almost 100% of the pine pollen originally present.

Many investigators have found that the oxidation process continues after the treatment is supposed to be finished, probably because the KOH/water washing has not been sufficiently effective. However, a photo-oxidative effect has also been suggested.

Tatzreiter (1985) described a method of cleaning pollen by the use of a mixture of rhodanide, chloride and sulphate. The method is superb for cleaning recent pollen, but we have not (yet?) found it useful for removing extraneous matter from deposits (cf. p. 212). We may add that neither have we had any success trying to use Schweizer's reagent for solution and removal of cellulose (cf. Strassburger–Koernicke, 1923; 175).

It sometimes happens that after acetolysis (with no preceding HF treatment) it is found that the sample contains disturbing quantities of mineral matter. In such cases HF treatment should be set in at this point.

Fine insoluble particles—filtration. Whereas the objective of sieving as described under point 3, above, is aimed at withholding coarse particles and letting pollen grains through, fine filters have also been produced which withhold pollen grains and let through very fine organic debris, e.g. charcoal dust and other insoluble particles. They may occur in such great quantities that they cover the pollen grains and preclude a meaningful analysis. In such cases filtration with sub-pollen mesh size may be useful, generally in combination with other treatments (cf. Kidson and Williams, 1969; Cwynar *et al.*, 1979).

Filters of appropriate mesh exist in sintered glass, metal and nylon. With so fine a mesh (5–10 μm) clogging can easily take place. The area of the filter should be large enough, and the suspension kept in constant, rapid (shaking) motion, to prevent this. Suction filtering should be avoided —small pollen grains may be sucked through the filter pores.

Filtration techniques have been used more in pre-Quaternary than in Quaternary geology.

5.3.2.3. *Staining*

The morphological criteria used in the identification of pollen taxa are very subtle and often difficult to perceive in the light microscope. Correct staining (Fægri, 1936) is one of the useful ways of increasing the contrast of exine features, facilitating the microscopy work. Pollen exines absorb certain stains more strongly than other substances present in the preparation, and are therefore easier to detect in a stained mount. When there are few grains it is possible to find them at a low magnification. The advantages of counter-staining are usually rather marginal.

Whether staining should be resorted to, or not, is in many cases a matter of personal opinion. A suitable choice of embedding medium (cf. Berglund *et al.*, 1959) and the use of phase contrast may bring out as much detail as even a successful staining. On the other hand staining, besides bringing out structural details of the pollen grain, serves two other important functions in practical analysis: fragmentary or badly crumpled grains are less likely to be overlooked if they are stained, and the staining reaction may in some cases help to differentiate between pollen and other microfossils of similar form, but different chemical composition, e.g. animal remains.

Final preparation for making a mount embraces two processes, (1) washing out all reagents used during the preparation of samples and dehydration, if necessary, and (2) staining. Generally the two processes are carried out simultaneously, the stain being added to the last washing medium in which it is soluble. Thus, there is no sense in staining with basic fuchsine prior to dehydration with alcohol, because the alcohol would remove the stain. On the other hand, it is too late to add the stain, e.g. in xylene or benzene in which it is insoluble. Consequently, in a dehydration series that particular stain should be added in the last alcohol bath.

The choice of stains is to a certain extent a matter of opinion, since a very great number of ordinary laboratory stains of the azine and triphenylmethane groups are more or less equally effective in differentiation, and also in their durability. Azo stains, on the other hand, generally do not stain exines. For theoretical reasons a stain should be used that gives maximum contrast with shortwave light, the use of which, e.g. in microphotography, would correspond to an increase of numerical aperture of the microscope objective. This would mean that red or black stains are to be preferred: neutral red, safranin or basic fuchsin being among the most useful in glycerol or glycerol-jelly preparations. Alcohol-soluble nigrosin, though difficult to handle, gives superb staining for recent preparations, but is less useful for fossil material. For silicone oil preparations safranin may be better than fuchsin.

Staining may also be carried out on the slide, the stain being added on the latter, or it may be incorporated in the embedding medium beforehand. Owing to differential stainability the stain will, after some time, have concentrated in the pollen grains. This method is frequently used in aeropalynology, the grains being caught on slides precoated with stained glycerol jelly. In fossil pollen analysis the (better) alternative is to stain in the centrifuge tube by adding a suitable quantity of staining solution to a bath (one drop of fuchsin solution is sufficient) and then to shake gently. Staining is effected in a few seconds, whereas differentiation on the slide may take some hours.

Staining not only renders pollen more easily recognizable in a preparation; it also contributes to differentiation. Colour differences between various pollen types are partly due to the effect of adding the self colour of the grain to that of the stain, the result often being more easily distinguishable than the original colours, e.g. the clear pink Gramineae contrasted with the saturated red of Betulaceae (fuchsin staining—the slightly more yellowish colour of safranin does not bring this out so well). Also the more bluish colour of semi-tectate grains under low magnification contrasts with the more deep-red one of tectate exines (*Salix* vs. *Artemisia*).

Staining is also dependent on the state of preservation of the exine, and as exines apparently change gradually with time, older exines may stain differently from younger ones, which may help in distinguishing between autochthonous and secondary pollen in a deposit (Stanley, 1965). Owing to different rates of change of exines in different types of deposit, such staining differentiations are probably only diagnostic within the sample.

Staining with fluorochromes is mentioned above. Fluorescence microscopy of the autofluorescence of pollen and spores also differentiates both between species and between stages of preservation with increasing geologic age (van Gijzel, 1967). For distinction between Quaternary and pre-Quaternary pollen, fluorescence is helpful, but the technique is too complicated for routine use. Fluorescence microscopy adds nothing to the optical resolution of the microscope.

The techniques discussed so far are those applicable to fossil deposits which are usually characterized by low pollen concentrations, diluted with great quantities of inert material, and by grains in which only the exine is preserved. However, under other circumstances recent pollen grains are represented, often concentrated, or at any rate mixed with material that is very easily removed, simply by water. Consequently, pollen grains both for comparison and for analysis can be treated in a much simpler manner and observed with cytoplasmatic contents, intine and pollenkitt. There is no doubt that, in some cases, quantity and properties, especially color, of adhering oils, and other features which can be observed only on living pollen grains, contribute to a correct identification. On the other hand, comparative tests have shown that problems may arise due to optical disturbance by the same features. Honey analysts, for example, find difficulties in identifying certain pollen taxa which are easily identified by workers analyzing fossil pollen. Hence, some workers in these fields work with complete grains, others 'fossilize' them (cf. Section 11.3.1). In the end, much depends on personal preferences, even if some techniques, e.g. in aeropalynology, presume direct observation without any pretreatment.

5.3.2.4. Mounting

Before the mounting procedure is decided the aim must be clear: should the mount be permanent or not; is a liquid mount in which grains may be turned desired, or a solid one, slides of which will withstand rough handling and vertical storage.

Solid mounts have the great advantage that grains are easily rechecked for identification. The advantage of a liquid mount is that pollen grains may be pushed about and turned over by slight pressure on the coverslip or by careful pushing at its edges. This is essential for the identification of critical specimens of which it may be necessary to study both equatorial and polar projections, apertures seen both face on and obliquely, etc.

One important factor when choosing an embedding medium is the refractive index, which should differ clearly from that of pollen exines: 1.55 to 1.60 (Christensen, 1954; Berglund *et al.*, 1959). Too strong contrasts may be unpleasant and even obliterate subtle structural details. Embedding media with a refractive index above that of exines present no practical advantage, being more difficult to use. Details of microscopic techniques are discussed in Sections 5.3.4 and 11.3.3.

The simplest medium for a liquid mount is glycerol (refractive index 1.4), a drop of which may be added to the residue in the centrifuge tube after the last washing, or it may be added on the slide after the material has been placed there. The remaining water is left to evaporate. Open glycerol mounts are short-lived and cannot be recommended for permanent use since they become sticky and are easily destroyed. More permanent glycerol preparations can be made with water-free glycerol and sealed with nail polish, or better with a varnish which remains more elastic upon drying. To avoid the varnish creeping under the cover-glass it is advisable to put a small weight on top of the cover-glass while drying. If properly made, such preparations can last for some years. The relatively low viscosity of pure glycerol makes it possible to turn pollen during analysis by pressing the cover-glass with the tip of the pencil, even in sealed preparations.

Silicone oils (Andersen, 1960) are more permanent and have a lower refractive index. Viscosity should be at least 2000 centistokes, but may advantageously be higher, possibly up to 40,000–60,000. The size of pollen grains does not seem to change with storage, and that of fossil and recent grains of the same species seems to be equal after equal treatment. Sizes are smaller than in glycerol jelly preparations—conversion factor *ca.* 0.8. The smaller size is a disadvantage when minute exine details on the resolution limit of the light microscope are essential for identification. Another disadvantage is the necessary dehydration before mounting, which introduces an extra step compared with glycerol-jelly mounting. The permanence of the mount is not unlimited.

The procedure for mounting acetolyzed fossil or modern pollen in silicone oil may be summarized as follows (Andersen, 1960: 16):

1. Wash with water.
2. Wash with a few drops of water and 96% alcohol.
3. Wash with 99% alcohol, stain with fuchsin if desired.
4. Wash with benzene.
5. Add about 1 ml benzene, transfer to small vial, add silicone oil, leave for evaporation for about 24 h. Surplus material may be kept in the vial for future use.
6. Add the amount of silicone oil needed for optimal concentration of pollen. In making slides the smallest possible amount of liquid should be used. The droplet spreads under the cover-slip very slowly. For fossil slides small cover-slips (18 × 18 mm) are recommended. Slides with recent pollen may be sealed with nail polish.

5.3.3. Preparing for absolute frequency counts

Absolute pollen analysis requires data on the density of the individual taxa in the sediment. This presumes knowledge of sediment volume and number of pollen grains per volume. Three different techniques have been proposed for counting absolute pollen frequencies:

1. Volumetric techniques (MB. Davis, 1966).
2. Weighing techniques (Jørgensen, 1967).
3. Exotic marker techniques, introduced by Benninghof (1962).

Since the marker technique found a practical solution (Stockmarr, 1972) it has become prevalent. The others are now mainly of historical interest and will not be further discussed.

The principle of the exotic marker method is to add a known number of marker grains to a known volume of material. Markers may be an exotic pollen taxon, spores, or pollen-sized polystyrene spheres (Cushing, according to Craig, 1972: 50). During analysis the markers are counted with the native pollen of the sample. Because the numerical relation between the native pollen of the sample and the markers is constant throughout the sample, the total number of fossil pollen grains in the sample is given by the formula:

$$\text{Total fossil pollen} = \frac{\text{Fossil pollen counted} \times \text{Total number markers}}{\text{Markers counted}}$$

Concentration of fossil pollen (pollen per cm^3) is thereafter given by division if the volume of the sample differs from unity.

Pollen influx, i.e. number of pollen grains per

cm^2 and year, is obtained by multiplying concentration by rate of sedimentation.

The simplest and most effective technique (Stockmarr, 1971) makes use of the pharmacy technology of producing chalk tablets with a calibrated quantity of *Eucalyptus* pollen or *Lycopodium* spores, both commercially available. The markers are often treated so as to be distinguishable from fossil material, but should not occur naturally in the deposit.

The advantage of the technique is that the tablets, i.e. the known number of spores, come in before any preparation process has begun and thus follow the sample throughout all procedures. Any loss of material during the process, e.g. by filtering or transfer between vessels, has the same numerical effect on marker as on native pollen. As it is in reality almost impossible to get material quantitatively out of centrifuge tubes this advantage can hardly be overrated. An adequate number of marker tablets is added before HCl treatment, which dissolves them, and therefore the markers disperse with the sediment. It is important to homogenize the sediment + chalk suspension by careful stirring. After that, preparation continues as described earlier. Statistical reliability demands that the number of markers should not be less than 20% of the (expected) total of fossil pollen.

The marker should be easily recognizable and not expected to occur naturally in the deposit, nor in the area concerned. Naturally, it must also be easy to obtain in sufficient quantity. With this technique the establishment of the APF in the individual sample is no more complicated than an ordinary analysis. It is possible to use a native pollen as indicator, but the technique is time-consuming and not so accurate. The formula for calculation of PF must be multiplied by $(100/100 - i)$, where i is the original percentage of the marker in the sample. This must be established by a separate count.

A similar technique has been used for absolute diatom analysis performed contemporaneously with pollen preparation (Kaland and Stabell, 1981; Battarbee and Kheen, 1982).

5.3.4. *Light microscopy technique*

In pollen analysis the investigator sits continually at the microscope for long periods. It is important to establish ergonomically favourable working conditions. The days of microscopes without built-in-light should belong to the past, and so should the monocular microscope. On the other hand, no other device gives the observational detail obtained by direct visual observation. Level of eyepieces above the table/chair, and inclination of the ocular tube, are very important factors. So also is the light intensity. Beginners often use too bright fields of vision, with resulting headaches and other discomforts. The ambient light of the lab should not be too bright, and should be lowered in sunny rooms.

The critical magnification range for identification of pollen grains lies between 300 and 1000 ×. In many cases the critical morphological elements are just at, or a little beyond, the resolution limit of an optical microscope. The research pollen analyst therefore needs a good knowledge of the physical basis of optical microscopy to be able to get the most out of the instrument. Electron microscopy techniques should also be mastered even if the lab has not (yet) got an electron (SEM) microscope of its own.

Light microscopy will always remain the main vehicle for routine analysis of fossil pollen, since it is so much faster and more convenient than electron microscopy, even if SEM microscopes have become much more user-friendly, both regarding preparation techniques and general handling. In pollen studies mainly the lower magnification range of the SEM is used, and the new, smaller desk-top models are generally sufficient. The price of these instruments has now come down to the level of that of a good optical reference microscope. SEM microscopy will undoubtedly increase its importance in palynology in the years to come, not only for morphological investigations, but possibly also in analysis.

A modern light research microscope for pollen analysis must change the resolution power from moderate to the attainable maximum quickly and easily. In addition it must give easy access to phase contrast and, conditionally, interference contrast. The light power must be sufficient for adequate and comfortable viewing.

A comfortable magnification for running analysis is approximately 400 ×. There are several objective/ocular combinations giving such magnification. The basic rule in microscopy is to get the highest possible resolution power on the objective keeping the 'empty' magnification on the ocular low. A 63 × dry objective with numerical aperture (n.a.) of 0.95 and an ocular of 6.3 × (total magnification 390 ×) gives a superior picture compared with the same magnification (400 ×) obtained with objective 40 × (n.a. 0.40) and 10 × oculars. 25 × objectives (n.a. 0.22) with 12.5 × oculars (total magnification 312 ×), as still used in some laboratories, give too low a resolution and also magnification for sound pollen identifications.

Most pollen analysts still use medium-resolution (n.a. < 1.0) dry objectives for the running analysis and run into practical difficulties when it is necessary to change to high-resolution immersion objectives (n.a. 1.0–1.4) for the study of details of the pollen surface or struction. Especially if the preparation is not sealed, the easiest way out is to use anisole or some other low-viscosity volatile oil for immersion, which can be wiped off with a piece of blotting paper. Anisole is perfectly safe. The smell may be annoying to some persons after prolonged exposure. Also anisole may, by prolonged immersion, attack the plastics parts of equipment and in some cases the cement of compound lenses. If the preparation is sealed, ordinary immersion fluids can be used. They are inodorous, but have the disadvantage that they must be washed away laboriously with some other liquid (some of them not innocuous!), which is inconvenient in routine work.

Many workers overlook the fact that in order to obtain the full resolution of their objectives they must also have a condensor of n.a. > 1.0, which must be oiled as well. For this purpose anisole is not useful, being so fluid that it runs down from the top of the condensor.

The extra work implied in changing between dry and immersion objectives makes it tempting to skip the closer inspection of grains under higher magnification, which detracts from the quality of pollen identification. The alternative is to use immersion objectives also for the low-magnification running analysis. Main microscope manufacturers produce immersion objectives down to 40 ×, sometimes 23 × magnifications. These low-power immersion lenses have a higher numerical aperture than their dry counterparts and give a superior performance. The changes between immersion objectives do not present special difficulties. In our lab we use a 63 × objective n.a. 1.4 with oculars 6.3 × for analysis, which gives remarkably clear and comfortable pictures. The only disadvantage is the necessity to seal the preparation to prevent the immersion oil from creeping in under the coverslip. However, with a little training it is possible to manipulate individual pollen grains in sealed preparations by careful pressing with the pencil tip.

Phase contrast is essential in the identification of pollen grains. It is possible, and we recommend it, to carry out the whole analysis with phase contrast objectives. They have practically the same optical characteristics as ordinary ones, and can also be used for ordinary light. Phase contrast light can then be used constantly, or should be so constructed that it can easily be switched on and off. The only necessary operation is a change of condensor, which is very simple in some makes. It should be a primary consideration in the purchase of a microscope.

Most microscope manufacturers offer both wide-angle and ordinary oculars. Wide-angle oculars give a very pleasant picture with a feeling of being close to the object. For morphological studies of a single pollen this type of ocular is splendid. For running pollen analysis, during which the preparation is moving almost continuously, wide-angle oculars are tiring and easily give one a headache. One has to make much greater eye movements to observe pollen grains along the edges of the field of vision. For analysis work ordinary oculars should be used.

5.3.4.1. *The analysis*

When the mount has been prepared, analysis proper can begin. Normally a small quantity of a suitable suspension of the sample is transferred to a slide and covered. Care should be taken to spread the suspension over almost as large area of the slide as will be covered by the cover-slip. This counteracts the well-known sorting of the pollen grains (Brookes and Thomas, 1968) when the cover-slip is put on: the smaller grains tend to float towards the edges. If the analysis does not include the whole of a preparation this must be taken into account. It is comforting, though, that, after very few fields of vision in from the edge have been analysed, the percentages have stabilized so much that this effect is no more discernible (Brookes and Thomas, 1968).

Care should also be taken that the suspension fills the entire space under the cover-slip and that nothing is pressed out. This is especially important if the cover-slip is to be sealed with varnish. No useful preparation can be made with bent cover-glasses (not infrequent in low-quality makes).

By means of the mechanical stage the preparation is moved from one side to another under the microscope, and all pollen grains observed are noted. The preparation is then shifted at least $1\frac{1}{2}$ diameters of the field of vision perpendicularly to the first direction of movement, and moved back again. It is practical to note transects on the counting form during work, distinguishing between odd and even numbers. Provided counting always starts at the same corner of the preparation (as it should) this also indicates the moving directions of the transect when counting has been interrupted. As pollen may be unevenly distributed within the preparation, transects should be placed equidistantly so that the whole surface is covered; also in rich preparations where it is unecessary to analyse the whole preparation. The counting work is facilitated if a preprinted form is used where regularly occurring pollen taxa have their permanent place. Forms should contain space necessary for x and y coordinates for special pollen grains which are later to be checked or documented, and for calculations. If the pollen frequencies are later to be loaded into a computer it can be useful to include as many taxa as possible in the counting form, and number each taxon. In this way data punching is reduced to the taxon number followed by its frequency.

The counting of pollen is considerably facilitated by the use of mechanical/electronic counting devices with push-buttons for the most frequent pollen taxa. Less frequent taxa must be entered on the form anyway. Some counters have a transmission arm from the microfocusing screw of the microscope to a similar screw on the push-button desk of the counting device. This makes it possible to keep the hand continuously on the push-button desk, focusing by the thumb and pressing the buttons with the other fingers (as in touch typewriting). Our experience is that well-designed counting devices speed up counting by at least 20%. This development is not without inherent dangers, as extended uninterrupted sessions at the microscope cause fatigue, with blunting of observation power. In a large lab, with constant consultations between workers, this is not too dangerous, but it is so when people work alone or for other reasons without communication with others.

We foresee further development in counting devices, where the signals from the push-buttons are loaded directly into a computer. In this way the total procedure from the pollen count to the finished graphed pollen diagram is computerized and dramatically rationalized in relation to traditional production of pollen diagrams by hand. This perspective also has a negative side: the danger of losing direct contact with the material; mistakes are more likely to slip by unnoticed; the draft diagram produced during the work, however crude, is a necessary checking tool.

5.3.5. Electron microscopy technique

Transmission electron microscopy is important in ontogenetic and morphological studies of pollen grains. It has no place in pollen analysis.

Scanning electron microscopy (SEM) has so far not been used in direct analysis work, and perhaps will not be, except in quite extraordinary cases, due to the meticulous preparation procedures necessary and the limitations of SEM. On the other hand it is a very powerful technique in pollen morphology, due to great resolution and depth of focus. In pollen-morphological research SEM is now routine, but the difficulties of prepara-

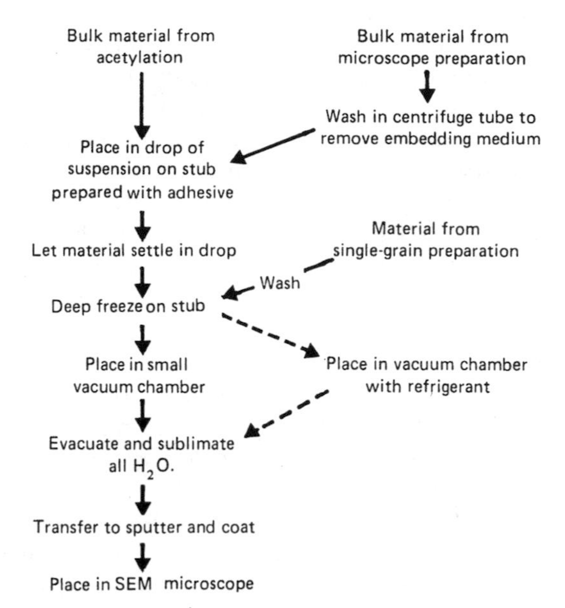

Fig. 5.6. Flowchart for preparation of sample for SEM observation.

tion have prevented its general use to identify pollen grains met with in practical analysis.

Whereas the preparation of recent pollen for SEM does not differ very much from the regular SEM routine, some precautions are necessary when dealing with pollen from a fossil deposit. SEM of recent pollen can be carried out on non-prepared material, but usually pollen material is taken out after acetolysis.

Pollen coming from a pollen analysis preparation has usually been immersed in glycerol or glycerol jelly and must be cleaned. Bulk cleaning is simply carried out by ordinary washing (warm water to dissolve glycerol jelly) in a centrifuge tube until all the medium has been washed off.

Cleaning of single-grain preparations is more meticulous: after having been fished out the grain is transferred to a drop of fuchsin solution on a slide. The staining makes it easier to follow the grain under the following steps. After staining the grain is successively transferred (low-power binocular microscopic control) to two more drops of clean water on the same slide. After the last washing the grain is usually sufficiently clean. We have not been able to clean properly grains that have been immersed in silicone oil.

Stubs for SEM observation of pollen should be coated with a suitable adhesive which should be applied in a very dilute solution and left to evaporate, leaving an ultra-thin adhesive film on the stub. The best (blackest) background is obtained by mounting a small cover-glass on top of the stub and working on that.

The adhesive should be firm enough to prevent the grain from sinking in; on the other hand it should be soft enough to make proper contact. The simplest adhesive we have found is the glue from ordinary adhesive ('Scotch') tape dissolved in diethyl ether. Stubs remain sticky and can be coated in advance.

A drop of a suspension of pollen in water is placed on the top of a stub and left to settle for a moment before freezing. After having been frozen, it is transferred to a vacuum chamber. It is imperative to keep the suspension at freezing temperature until the ambient pressure is sufficiently low to keep the water frozen at room temperature (to avoid boiling). Usually, this goes by itself in a small, pre-frozen vacuum chamber which is rapidly evacuated. In a larger chamber the stub should be placed in a deep-frozen (metal) container that can keep the temperature down for a sufficiently long time.

When the material is completely dry—time depending on the capacity of the equipment—stubs are transferred to some sputtering device. Coating is necessary to prevent building-up of electrostatic charges with consequent discharges and blurring of pictures. Fresh, wet pollen grains can be observed under SEM for a short time, but burn up very soon unless the frozen material is observed on a cryostat. Coating can be by carbon or gold/palladium, or a combination. Too heavy a coating may blur the finest morphological details. SEM stubs can be stored for long periods (protect against contamination).

Treatment of single-grain preparations differs in the more pressing necessity to avoid drying-out before freezing, which makes pollen grains collapse. After the drop has been placed on the stub it must therefore be very quickly transferred to a deep-freeze chamber and protected against thawing again until the necessary vacuum has built up in the freeze dryer. In a small chamber (thick-walled vial 10–15 ml) this goes very fast.

SEM gives a surface picture only. Internal structures must be studied in sectioned or cracked grains.

5.4. Documentation

Important finds should be preserved in one way or another for future reference and checking. Unfortunately, documentation in pollen analysis has often been neglected; in particular the older pollen-analytical literature often contains statements much in need of corroboration, which is now impossible since no attempts at documentation have been made.

The actual preparation used for analysis can, if necessary, be preserved by sealing. If the location of a pollen grain is marked on the cover-slip (with India ink), it is usually possible to recover it for later reference. Ordinary India ink is sufficiently resistant against immersion oil and against careful cleaning. Alternatively, it may be applied on the lower surface of the slide. So long as microscopes of the same type are used, moving-stage coordinate references are sufficient to indicate the location of a grain. For use with different microscopes the microscope firms stock various devices (indicator slides) facilitating transformation from one coordinate system to another.

Even so, it is not always easy to locate a specific grain in a rich preparation. This is especially important in description of sporae dispersae, in which the holotype is one spore. One may therefore resort to preserving the actual specimen by picking the pollen grain out of the preparation and keeping it as a single-pollen preparation (cf. Fægri, 1939; Klaus, 1953; Gluzbar, 1968). With some practice this may be done relatively easily, either by 'fishing' the grain on to a piece (not a drop) of glycerol jelly on the tip of a needle or by sucking it into a microcapillary tube. Such microcapillaries are made by pulling out a commercially available capillary over a not too hot (spirit) flame to a bore *ca.* 100 μm; ideal cut-off angle 45°. Both processes must be carried out under low-power microscopic control. Glycerol jelly fishing works best in relatively thin slides in which parts of grains project out of the liquid film. If there is too much liquid in the preparation the pollen grain floats away; in such cases it may be collected in the capillary tube. Kidson and Williams (1971) have described a simple micro-manipulator for pollen grain fishing. This method can be used under an ordinary compound microscope. Alternatives are to use a stereo-microscope, an inverted, or a projection microscope, all of which correct the picture sideways.

For optical microscopy the pollen grain is subsequently transferred to a larger piece of glycerol jelly with the piece used for fishing, or blown out from the capillary on to the larger piece, which is afterwards melted and covered (preferably with a very small cover-slip) and sealed with a paraffin seal underneath the cover-slip. Care should be taken that the grains do not shrink or crumple when being transferred; the glycerol jelly should be rather soft, and the whole left to dry for a day or two before sealing.

If it is desirable to preserve a pollen grain found during ordinary analysis, the cover-glass must be carefully *pushed* off under low-power microscopic control. Lifting the cover-glass creates currents that displace pollen grains. The seal of lacquer-sealed preparations must be broken away round the edges first.

If silicone oil or some other non-volatile embedding medium has been used during analysis it may be unnecessary to make single-grain preparations. However, the difficulties of finding, or the danger of losing, a grain are always much greater in a whole mount than in a successful single-grain preparation.

Microphotography (light and SEM) also provides good documentation and has the great advantage that photographs can be reproduced. SEM gives an adequate picture of the general shape and surface ornamentation, but does not penetrate. High magnification light microscopy focused at successive levels through the grain brings out structural details, but it is a cumbersome method and the result is inferior to that of special SEM photographs of the interior, producing after opening the grain (generally not feasible if for documentation).

Modern microphotographic equipment with

electronic light-measuring and almost vibration-free shutters makes it easy to take excellent photos on the routine microscope, even better on a special photomicroscope.

Digitalized image analysis by computer is a novel technique, which has so far not been used in pollen studies, but which seems promising. It does not bring out anything that is not already there, but emphasizes and brings out features otherwise obscure, and produces a real data for further statistical analysis. A new microscopic technique (confocal microscopy) may prove to be the solution to photographic documentation. So far, we have not had the opportunity to test such methods.

Some pollen grains are impossible to identify and document using the light microscope, because the identification criteria are at or outside the resolution power of the objective. It is possible to document a limited amount of such pollen per preparation by producing single-grain preparations for the scanning electron microscope.

If a problem pollen is frequent in a sample, it is possible to make a total SEM preparation and locate specimens afterwards.

Drawings are mostly less useful for documentation. It takes very keen observation and artistic skill to make a good drawing of a pollen grain. A good drawing is equal to, or often surpasses, photographs with regard to information content (Wodehouse, 1935; Erdtman, 1962) but indifferent or poor drawings are of very little value for future identification or reference. Literature abounds in pollen illustrations not satisfying the demands of Article 38 of the Code of Botanical Nomenclature, that a valid figure should show 'essential characters'.

6. Presenting the results. Pollen diagrams and maps

Motto: What do we want to show?

6.1. The stratigraphic pollen diagram

The immediate result of a pollen analysis, the *pollen record*, is usually a very large matrix of raw data. This matrix is the basis for any subsequent discussion and numerical treatment, and must therefore be available, but it is much too unwieldy to be of direct use to a reader. Results are therefore presented in simpler, graphic forms: *pollen diagrams*, introduced already by von Post in 1916. The construction of the diagram is the last link in the chain of technical procedures that started in the field. The objective of the diagrams is visualization and simplification. The diagram should give a visual picture of the material of the data matrix, and present it in a simpler way so that the reader can grasp the salient features with less difficulty and less time-consuming work than by studying the data matrix directly. One should therefore be careful not to put too much information into one diagram, losing the salient points in a maze of less relevant data—relevance to be understood in relation to the objective of the investigation.

In the planning of a diagram layout there are two opposing considerations: documentation vs. legibility. A diagram of a metre's width with more than 150 columns is so unwieldy as to defy its own purpose. Either less important curves should be suppressed or added up to a common sum, or the diagram should be divided into partial diagrams showing various aspects, all bound together by one or two index curves repeated in each partial diagram. It seems unnecessary to point out that the vertical scale should be the same in all partial diagrams. Experience shows that it is not.

The pollen diagram (Fig. 6.1) summarizes the results of field and laboratory investigations and contains different sets of data, which should always be presented in a more or less comprehensive state: chronostratigraphic, lithostratigraphic and biostratigraphic (cf. Hedberg, 1972). In the diagram they should be presented side by side, by preference in that sequence. Berglund and Ralska-Jasiewiczowa (1986) present an example of a very comprehensive pollen diagram. Usually, problems do not demand all these data in the same presentation.

The chronostratigraphic part of the diagram should contain a depth scale, ^{14}C dates or other absolute dates and relative datings by conventional archaeological or geological period designations or pollen zone types. In very extensive diagrams some of these data should be repeated at the right-hand side of the diagram for easier reference.

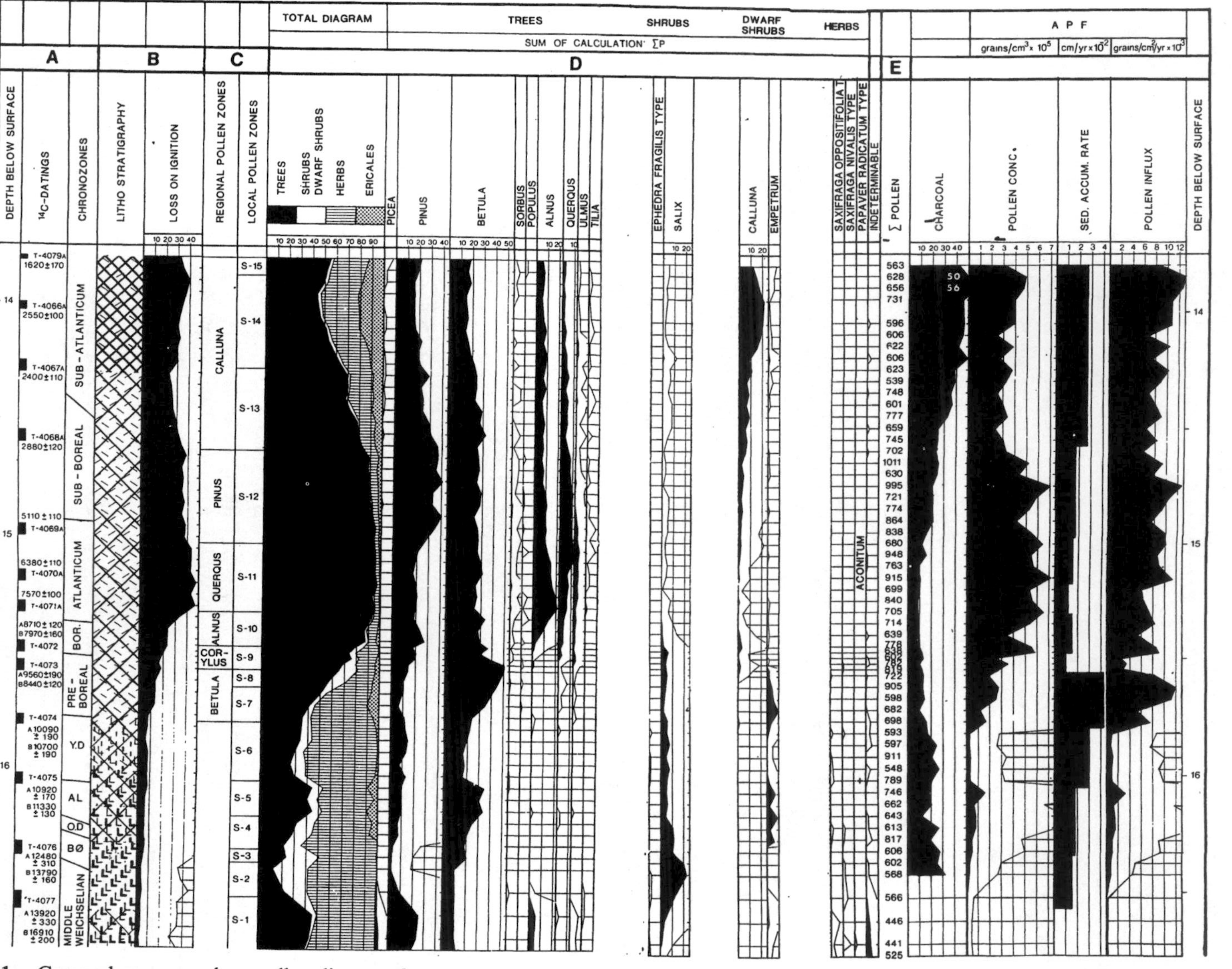

Fig. 6.1. Comprehensive modern pollen diagram from western Norway. A and B: chrono- and lithostratigraphic columns; C: regional and local zones; D: biostratigraphic diagram; for reasons of space, most taxa have been omitted (open columns, the diagram as published comprises *ca.* 120 taxa); E: pollen sum for statistical control. Note curves for charcoal and data for calculation of influx. Depth scale at both ends. From Eide and Paus, 1982; cf. Paus, 1988; redrawn.

The lithostratigraphic part should give a graphic summary of the data on composition of the deposit as identified in the field and laboratory (Troels-Smith data) and/or a genetic interpretation of the deposit (Fægri–Gams notation). In this part also belong results of chemical and physical tests not included in the above data: loss on ignition, elementary composition, etc.

The biostratigraphic part is the pollen diagram proper, summarizing pollen (and spore) counts as influx figures or percentages, according to the demands of the problem and the availability of data. If and which non-pollen microfossils should be entered in the general diagram depends on their character and on the objective of the investigation. They should be presented in the same way (influx or percentages) as other constituents of the diagram. Since that is usually not possible with macrofossils, these should not be included in a regular pollen diagram, but should form a diagram of their own. Local zoning is also presented in the biostratigraphic part. The results of further numerical transformations of the raw data of the count may also be entered in this part, but by this time the diagram may have become too long and may have to be divided (between two or more pages). For ease of comparison, some basic part or curves of the original and the zoning should be repeated for reference with each part diagram.

The *minimum contents* of an ordinary pollen diagram should be depth scale and some relative or absolute time scale, a sediment column and pollen curves. Modern, comprehensive diagrams necessarily are very extensive and present serious printing problems. Unfortunately they are often printed too small to be useful. For practical reasons we have abstained from presenting a complete diagram of this type in this book. It will also be noted that many of the small diagram pieces published here have been stripped down to demonstrate salient features more clearly. We strongly advocate that this is not done when presenting original material.

The vertical axis of the pollen diagram represents time, usually expressed as stratigraphic time, i.e. depth of deposit. If the deposit has been thoroughly dated, absolute time may be used. Such time-compensated diagrams are essential for comparison between sites and even more between regions.

For general use the drawing principles and symbols are quite well standardized. From a certain confusion of aberrant types in the early days of pollen analysis, two or three are left and generally used today. Two principles should be kept in mind in the construction of pollen diagrams: clarity and comparability with other diagrams. Changes in the more or less standardized forms may be necessary, but they should never be made lightly, though they may appear desirable.

The pollen diagram is a conventional aid, and its usefulness is directly proportional to the universality and acceptance of the conventions. The classical diagram was intended, above all, for studies of the development of vegetation under the assumed influence of climatic change. It is also useful in many other contexts, but the set of conventions behind it must not become a straitjacket: other problems may demand other diagrams. The diagram must always be problem-oriented.

6.1.1. *Diagram types: cumulative diagram*

The three types of diagrams generally used today are referred to as *cumulative, composite* and *resolved.* They can be used both to represent the result of percentage calculations of various kinds and for 'absolute' diagrams: influx, concentration etc.

The *cumulative* diagram (Fig. 6.2A) is used only for percentage data which for each level (sample line) add up to 100. Pollen curves are thus added to each other and circumscribe areas corresponding to the running percentages of the taxa. The areas of different taxa may be set apart by differential shading. In the *composite* diagram individual pollen curves are set out from the same (or two) base line(s) (Fig. 6.2B) and the curves are marked by pollen symbols (Fig. 6.3). In the *resolved* diagram (Fig. 6.2C) the curves are displayed separately, each from its own baseline. Special signatures are usually not necessary.

The three diagram types are not mutually exclusive. Very often diagrams are presented which

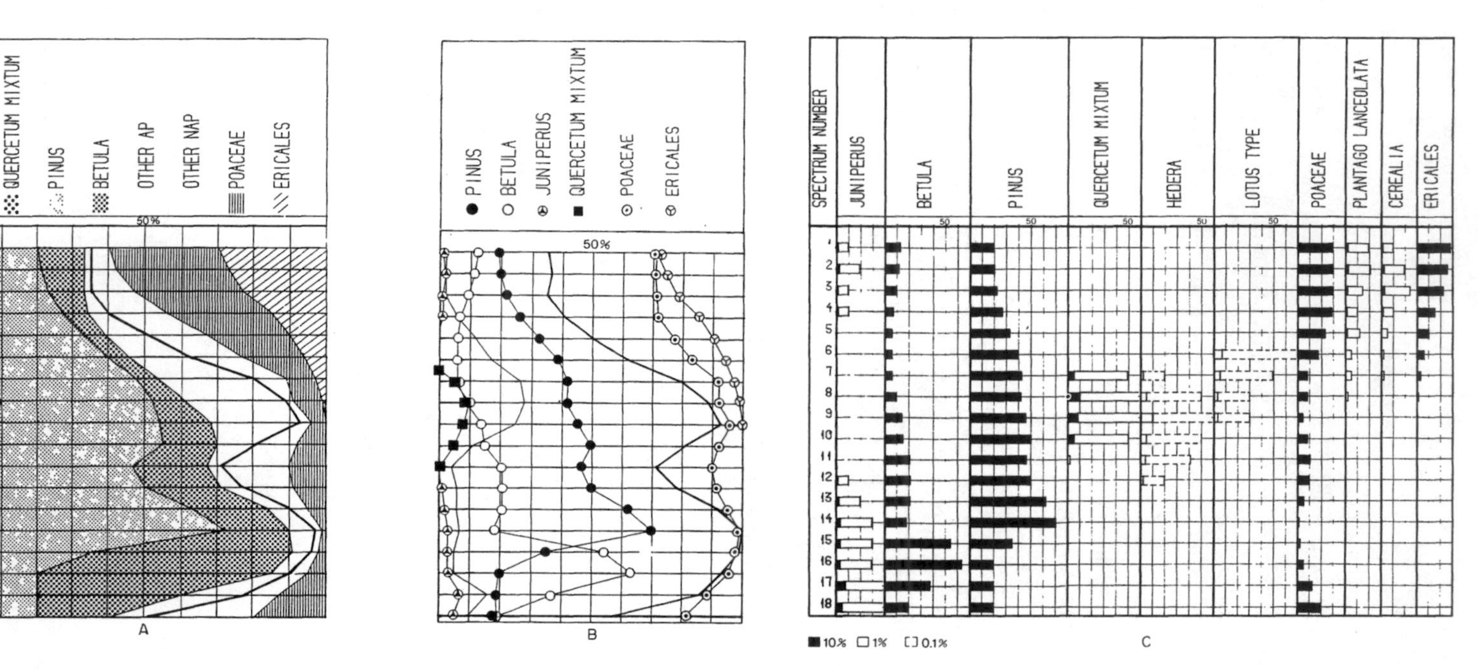

Fig. 6.2. A model biostratigraphic test diagram displaying the results of a (fictive) pollen count by different diagram types. A: Cumulative diagram —curves are set apart by differential shading (area signatures). B: Composite diagram—the heavy line indicates the total percentage of AP. Baseline for AP curves at the left, for NAP curves at the right edge of the diagram. C: Resolved diagram. Minor curves have been added which could not be displayed in diagrams A and B. 10 × and 100 × exaggerated curves have been used (added) for some taxa.

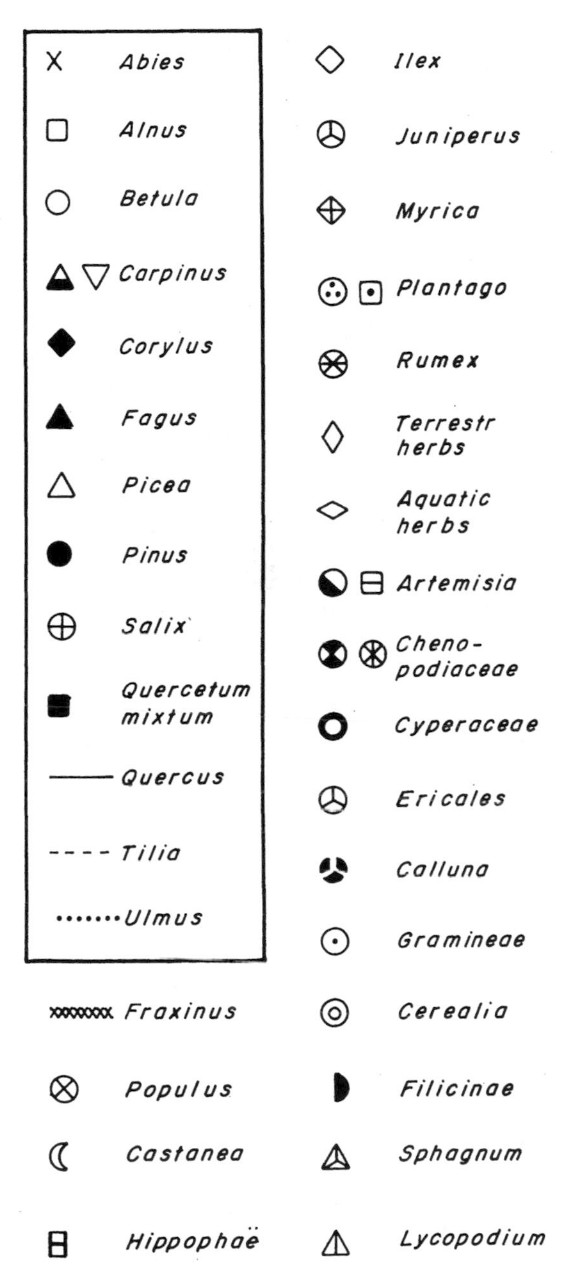

Fig. 6.3. Pollen symbols used in diagrams from northwest Europe. Those in general use are framed.

embody elements from two or all three types (cf. Fig. 6.4).

The cumulative diagram is the simplest, but it is useful for only a very small number of taxa, best for alternative variation. Even in such a simple diagram as Fig. 6.2A some of the curves become contorted and are difficult to follow.

The most frequent use is for the AP/NAP curve, as in Fig. 6.4, or in the total diagram of Fig. 6.1.

With regard to pollen spectrum lines, percentage grid and other technical details the same principles are valid as with regard to composite diagrams (q.v.), but in many cases the percentage line network may be simplified or omitted—depending on what use is to be made of the diagram.

6.1.2 *The composite diagram*

The composite diagram was introduced by von Post in his first presentation in 1916 (Fig. 6.5), and has not changed very much since then. Today it is rarely used alone, but must share the field with other presentation types. Many of the general principles of diagram construction are best discussed in connection with composite diagrams. The type is therefore treated at some greater length than its actual present use would otherwise justify.

The original composite diagram is exclusively biostratigraphic: each sample is represented by a horizontal *spectrum line* at the corresponding level. The percentages are recorded on the sample line and each species is indicated by means of a conventional symbol (Fig. 6.3). Those generally in use in Europe were originally chosen by von Post (cf. 1929: 556) to indicate the order of immigration of the respective species in middle and south Sweden. Outside that area the signature system is purely conventional, since the order of immigration changes from region to region, but the symbols are generally accepted. It is self-evident that the use of these symbols in an accepted sense in European literature does not preclude their use in a different sense in another floristic area. The same signatures can be should be used to represent the same taxa also in other graphic presentations (cf. Fig. 2.8).

Each horizontal line with its symbols is a *pollen spectrum*, and the diagram consists of a number of such spectra from different levels. To demonstrate more clearly the trend of vegetational development

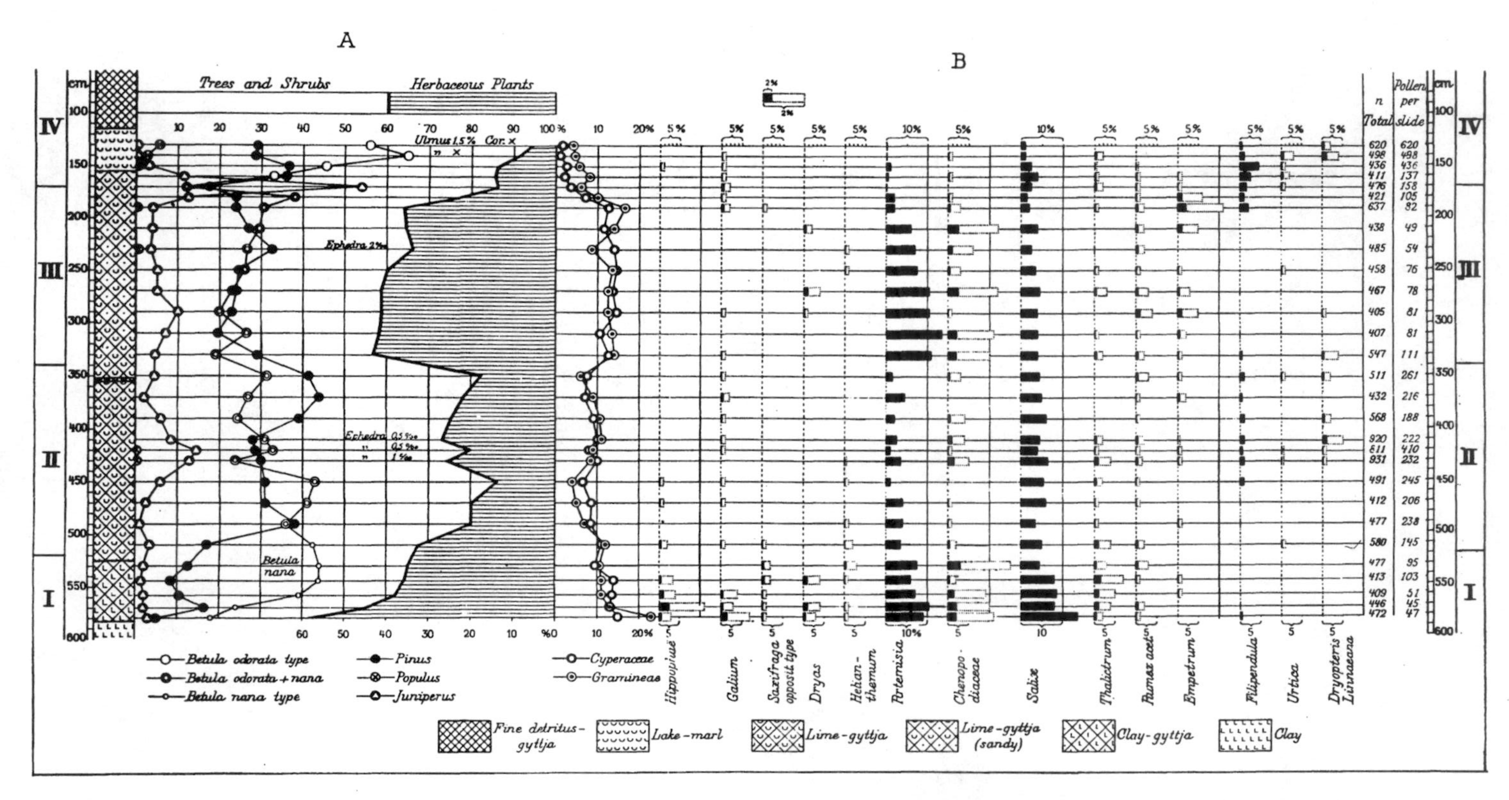

Fig. 6.4. Late-Glacial total diagram (aquatics not included) from Bornholm, Denmark. Diagram A contains a (cumulative) AB/NAP diagram with (in the AP part) the (quantitatively) most important taxa in a composite diagram. Important NAP taxa are shown in a composite diagram in part B. In this part are also shown (resolved) minor constituents, partly in exaggerated scale. Some taxa are indicated in lettering (*Ephedra*). Zones according to Jessen, 1935. From Iversen, 1953b.

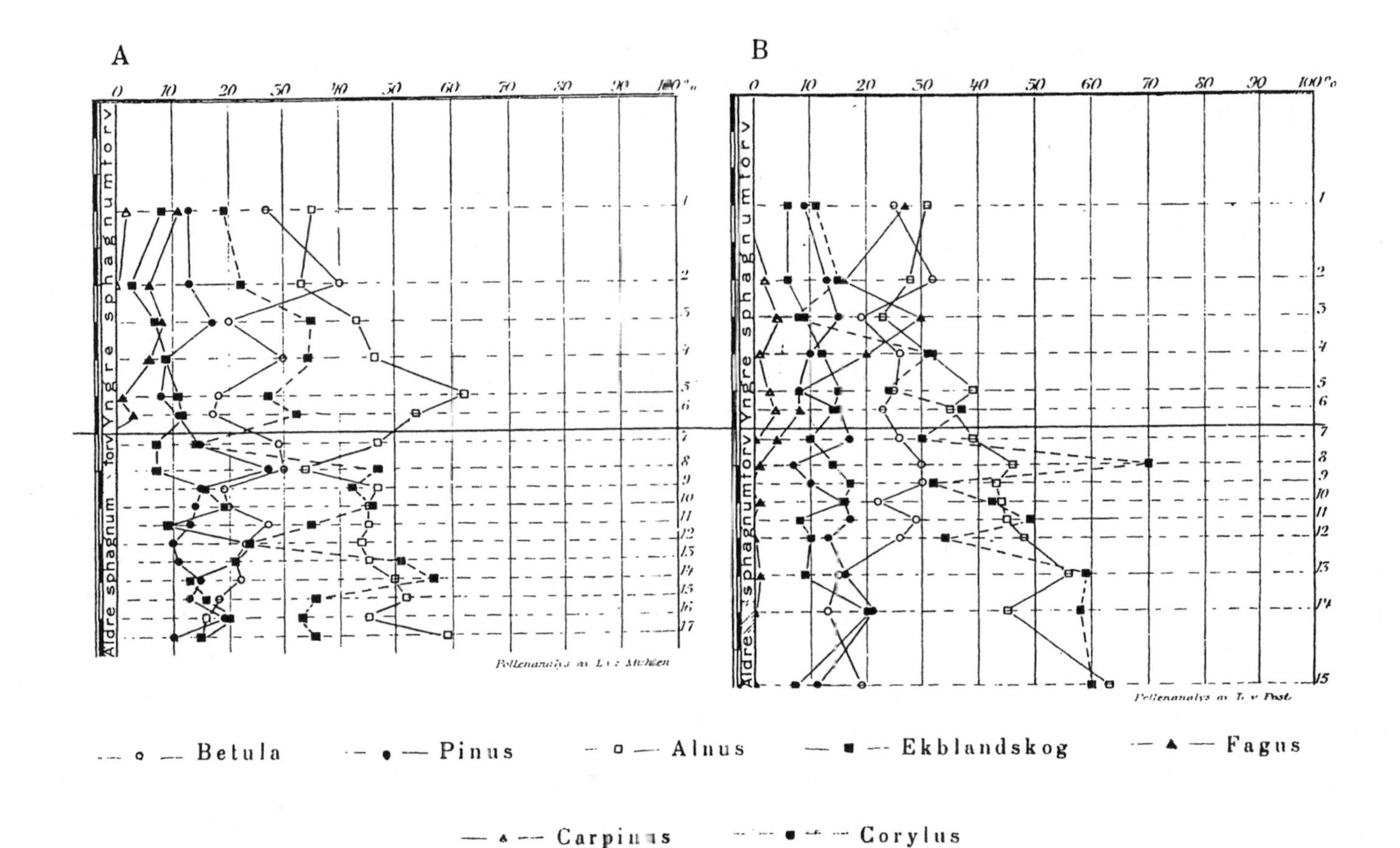

Fig. 6.5. Two composite pollen diagrams from a southern Swedish mire, published by von Post in his 1916 lecture. Diagram A represents a point 55 m from the edge, point B 300 m. The aim was to demonstrate the similarity of curves in sample series taken at some distance from each other and counted by different persons. The higher *Alnus* curve in the younger part of diagram A suppresses the other curves. The high *Corylus* curve in the older part of the diagram does not have the same effect, as *Corylus* was, at that time, calculated 'outside'.

the symbols for each taxon under consideration are connected from one spectrum to another by *pollen curves*.

The diagram is, above all, an instrument of documentation. To achieve this effectively it should enclose a *reference grid* consisting of spectrum lines (for all spectra) and (vertical) *percentage lines* indicating a suitable interval of percentages, in an ordinary diagram usually 10%. If spectrum lines are indicated a general depth grid is redundant, but there should be a depth scale somewhere, best at both ends of the diagram. Grid lines should be drawn thinner than baselines and pollen curves etc. (not always done in the illustrations of this book).

The absence of reference grids detracts from the usefulnes of a diagram. In some cases a diagram is not readable until a grid has been added. It is almost unbelievable to what extent and how often authors, even in proposed standard diagrams, neglect this simple rule, which, as shown by some of the following examples, is of paramount importance for the legibility of the diagram: if it is not adhered to, the reader experiences a great deal of trouble attempting to find out what the author had in mind, or in despair gives it up altogether (see Fig. 6.9).

Frequently some, or even all, curves of a composite diagram are crowded together and are difficult to follow individually. In some cases this inconvenience is not too serious; if one of the curves is of special interest, it may be set off from the others by hatching of the area between it and the zero line, or in some similar way (cf. Fig. 7.1). It may alternatively be given in a separate diagram. The latter method may also be resorted to for those constituents the quantities of which are so small that the curves would not be legible if drawn to the same scale as the others. If statistically significant the change of a curve from 1% to 2% may be of as great importance as a change of another curve from 30% to 60%, but it is obvious that they cannot be represented on the same linear scale.

One possible way out is to present one minor, but important, constituent on an exaggerated (10 ×) scale. Since this is usually done by a separate curve, it leads to the resolved diagram.

Another alternative to bring out minor constituents is to use a logarithmic scale. This is useless for values below unity, but can be remedied by adding 1 to all values before plotting. However, on the whole logarithmic scales are not recommended —especially not for percentage diagrams, where the logarithmic expresion suppresses the variations of high-value curves even more than the already existing similar effect of the percentage calculation itself (cf. Section 7.1.2.2). For 'absolute' data this is not a consideration, but even in such cases auxiliary scales (cf. Fig. 6.2C) are more recommendable.

The total area of a (main) composite percentage diagram should extend to the 100% line, even if no single curve reaches above 50%. Also in resolved diagrams one should not vary the range of the individual diagrams too much. The extra space this demands is a low price to pay for greater legibility and greater comparability with other diagrams. The main diagram should comprise all types included in the basic sum, but care should be taken not to include too much in one diagram, as legibility suffers if diagrams are too crowded. The number of auxiliary diagrams that can be attached to the main one is almost unlimited. There should be a certain correspondence between the breadth of a partial diagram and the scale adopted for the same.

Forested and non-forested areas have different dominating plant species; they produce different pollen taxa, and their dispersal is different. It is therefore customary in percentage diagrams to distinguish between tree pollen (arboreal pollen, AP) and non-tree pollen (NAP). Dwarf-shrubs go with the NAP, shrubs sometimes with AP, since the pollen types (e.g. in *Betula*) are not distinctive for arboreal vs. shrubby species. According to the problem at hand the basic sum is constructed so as to represent the vegetation of the one or the other landscape type, or both: a forest tree (AP) diagram or a non-forest diagram (NAP) or a *total* diagram (cf. Fig. 6.6) in which percentages are calculated on the basis of AP plus NAP.

Separate AP and NAP diagrams present, respectively, the changes of the vegetation inside the forested and the non-forested parts of the landscape. The AP diagram from an open site is less influenced by local circumstances, and a pure AP diagram, calculated on the $\sum$AP basis, therefore suitable for regional comparisons, presuming a

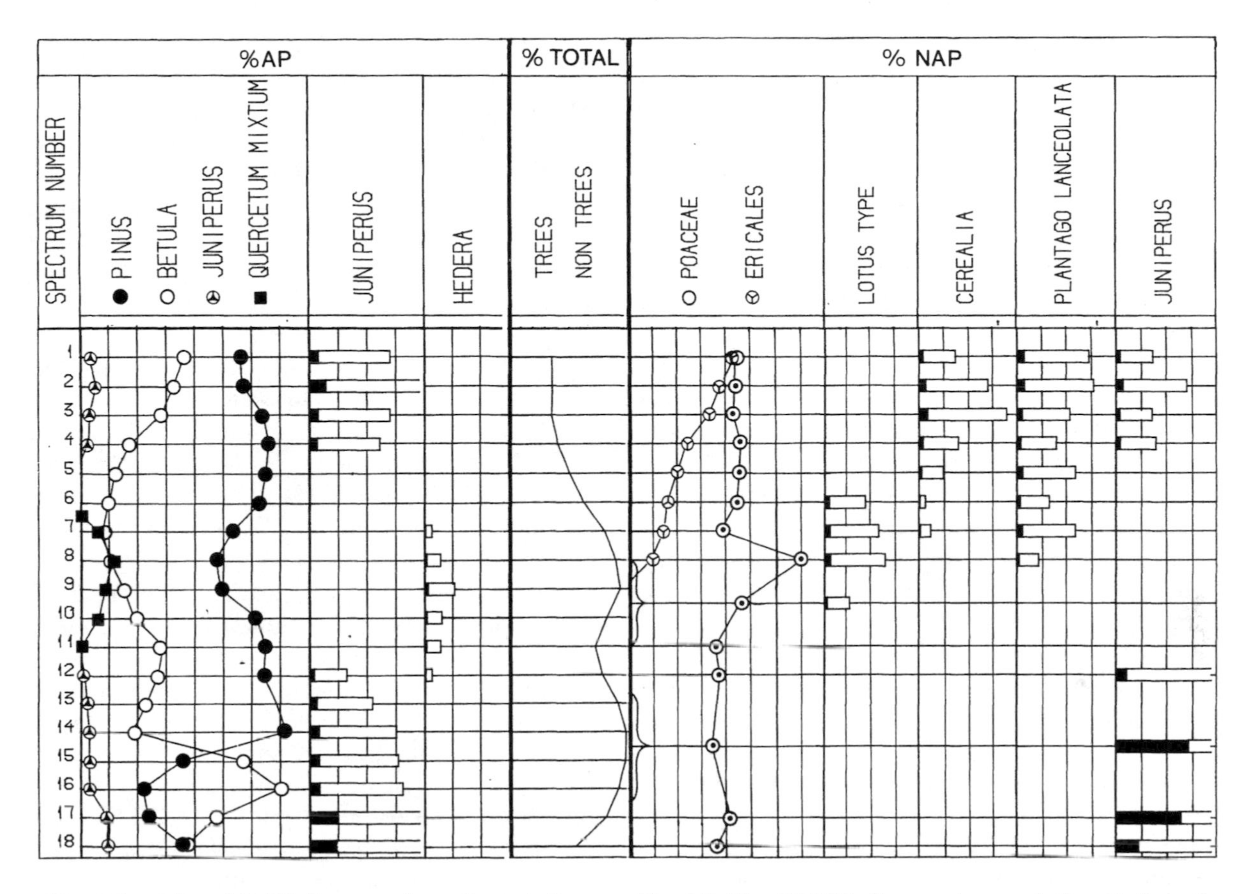

Fig. 6.6. AP and NAP diagrams from the test diagram, Fig. 6.2. The TOTAL diagram is cumulative. Both in the AP and the NAP parts of the diagram composite as well as resolved part diagrams have been used. Only some taxa are included. *Juniperus* is included in both the AP and the NAP diagram. The latter shows the much greater role played by *Juniperus* during the older part of the diagram. In the NAP part material from several counts has been pooled to obtain sufficient material for a meaningful percentage calculation.

reasonable forest cover in or near all localities.

The principle of the NAP diagram follows that of the AP diagram: the basic pollen sum is made up of the pollen of wind-pollinated herbs, grasses and shrubs constituting the vegetation. Insect-pollinated species, the pollen of which forms part of the pollen rain (e.g. *Calluna*), should, of course, be included, but not those that disperse their pollen unevenly and aberrantly, both in space and over time. They may be locally important or of importance for the special problem under consideration. Usually it makes no real numerical difference whether they are included in the basic sum or not, but it is characteristic that their curves sometimes have freak maxima that, percentage-wise, influence all other curves. Such taxa are therefore better calculated in percentages of the general NAP sum plus themselves. This pertains both to pure NAP diagrams and to the NAP part of total diagrams.

As the R values in influx diagrams are more closely related to actual representation in vegetation, as will be discussed later (Section 7.1.3.1) such diagrams are easier to understand and to visualize in terms of vegetation. However, the NAP diagram suffers from more complications than the AP diagram (or the AP part of the diagram), and demands greater botanical knowledge. Because of the ineffective dispersal of pollen from the lower strata of vegetation the NAP flora is more dependent on the mother formation than the AP. Also, mother formations produce more NAP than AP. In, for example, a heather peat, *Calluna* may contribute 80–90%, even more, of the total NAP deposition. NAP diagrams from deposits the mother formation of which produces NAP are generally useless for the purpose of regional analysis, but they may be of decisive value for the study of local conditions, according to the problems at hand. Such diagrams are often recognizable by the excessive dominance of one pollen type and by violent changes of dominance (see Fig. 7.19). Resulting statistical depression of other curves must be taken into account.

Diagrams from sediments are usually less exposed to such distortions and are safer when the problems calls for regional comparisons. In other contexts peats are more suitable, even including the excessive variations. It is part of the field technique to evaluate the contamination likely to come from local vegetation in the selection of sampling sites.

Sometimes it is possible to eliminate the locally produced NAP, but all such procedures imply more tempering with the data than safety would accept. With some reservation, it can be done in limnic (and marine) deposits in which the locally produced pollen, viz. that of the aquatic species, is more or less recognizable and can be excluded. Expecially in oligotrophic lakes in which the surrounding grass and sedge belts are insignificant, the NAP flora of the sediments will represent the vegetation of the surrounding open land without much distortion deriving from locally produced pollen.

The NAP diagram (part) is always more locally influenced than the AP diagram, and in a forested, or mainly forested area, it does not represent much more than the open land bordering on the basin (cf. Borowik, 1963, 1966). The NAP diagram may be important for the solution of special problems, but it is of only limited interest if general problems are to be dealt with.

The quantity of NAP in a forested area is generally small (*ca.* 10% AP or less in limnic deposits). Under such circumstances an ordinary count does not give sufficient material for a separate NAP study without much extra work. It may be necessary to combine the NAP counts from several samples to obtain a satisfactory basic sum (Fig. 6.6). Samples should be combined from biologically uniform sections of a diagram only. In areas with little or no forest cover the relative quantity of NAP is so high (50% of the total or more) that sufficient material is generally obtained in connection with the total or AP analysis.

When forest covers all the upland the total diagram will be more or less identical with the AP diagram (consequently one diagram is sufficient), since the ground flora of the forest scarcely contributes anything to the general pollen rain. The NAP present registers the vegetation of wet ground; fen, rush thickets, etc. A difference between the AP diagram and the total diagram is found in regions where, and at times when, forest-less dry-ground areas occur; in these cases the latter must be represented in the diagram, too. This is particularly imperative in Late-Glacial diagrams and in sub-alpine or sub-arctic areas. The total diagram is also of great importance if the disappearance of forest is due to cultivation, among other reasons because it serves as a warning that the AP curves must be regarded with a certain suspicion because there has been human interference with the forest. In north-western European Late-Glacial diagrams the difference between AP and NAP is more or less illusory (*Betula nana, Salix herbacea, Juniperus*) and the total diagram is the only means of adequate representation. An AP percentage diagram would be misleading and should be avoided altogether, even if the post-Glacial part of the same deposit is to be represented by such an AP diagram.

The advantage of the composite diagram lies in the wealth of information given in a compact form and in the clear demonstration of the interrelationship of the curves. The disadvantage is that the diagram easily becomes so crowded that it is difficult to follow the course of the individual curves.

6.1.3. The resolved diagram

The easiest way to surmount the latter difficulty is to separate the individual curves, which is done in the 'resolved' diagram, (cf. Fig. 6.2). What is combined in one composite diagram is in this case given in as many diagrams as the former possesses curves. A resolved diagram is obviously easier to read, most symbols are redundant, and there is no crowding of curves; on the other hand, it requires much more space and it is difficult to follow the interrelationship between the individual constituents.

Figure 6.7 demonstrates the use of resolved diagrams for a regional comparison; the appearance of new species (*Picea, Fagus*) is particularly clearly brought out. The curves of the resolved diagrams may be given as 'saw-blades' or as histograms (Fig. 6.8). The former are easier to draw, but the latter give a more objective presentation of the material. If small samples are taken at intervals the individual columns of histograms should be kept separate; if the whole profile has been analysed (continuous sampling) a contiguous histogram should be made (Fig. 6.8).

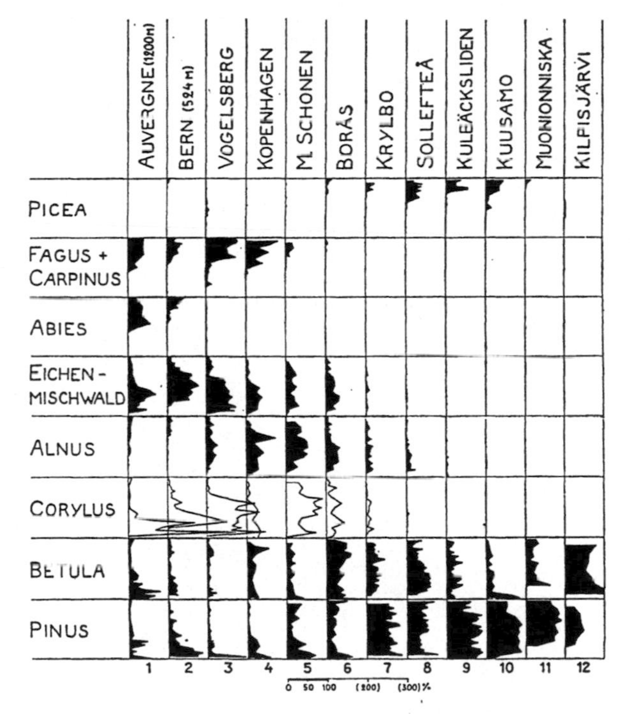

Fig. 6.7. A series of resolved diagrams ('sawblade' presentation) illustrating the main features of the vegetational development in a transect through Europe from south to north. The climatic optimum is registered in pure broad-leaf forest in the south, passing through mixed forests and birch to pine forests in the north. *Corylus* has been drawn as a line, not as a silhouette, because the taxon was at that time not included in the 100% sum. From von Post, 1929. The absence of a grid is acceptable in a presentation like this, which does not present primary material.

In many cases it is advantageous to combine a main composite diagram with resolved auxiliary curves for minor constituents. Spectrum levels and percentage lines should be included in resolved diagrams as well as in the composite ones. On the other hand, one should not reproduce too much of the grid, as the general picture is rather unsatisfactory if the grid is too compact. It is unpractical to let the area of all partial diagrams cover 100% but ranges should allow sufficient 'air' between curves. Especially, it is recommendable that a 100% range be used for curves exceeding 50% in more than one or two values. Only in that way can the reader immediately see the actual values of the percentages directly from the visual image. The range of each partial diagram should be clearly indicated. The main scale should be consistent throughout; auxiliary scales for exaggerated values should also be indicated (cf. Fig. 6.2C).

The more complex the diagram, the more important becomes the reference grid. Pollen-analytic literature contains many resolved diagrams without a grid, the curves of which are consequently too difficult to identify both as regards pollen category and depth level of individual features of the curves. It is unfortunately easy to quote many publications in which the absence of zero lines for the individual

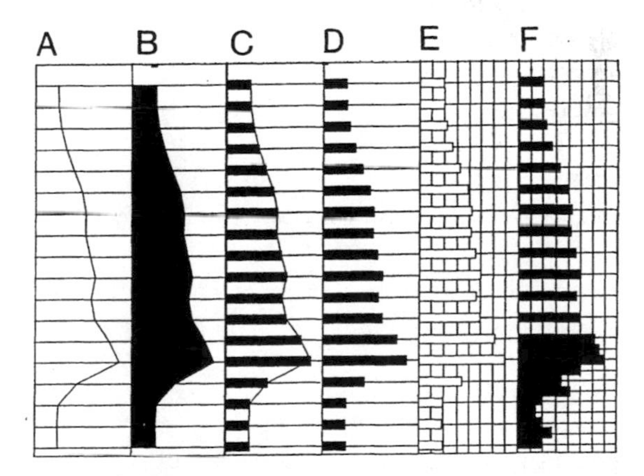

Fig. 6.8. Presentation of the same curve (*Pinus* from Fig. 6.2) by different techniques. A: open continuous curve; **B.** ditto, blacked; C: curve plus histogram (redundant); D: ordinary histogram; E: open histogram; F: histogram with continous sampling indicated in the lower part (contiguous curve). In A–D there are spectrum lines only. In E and F there is also a percentage grid.

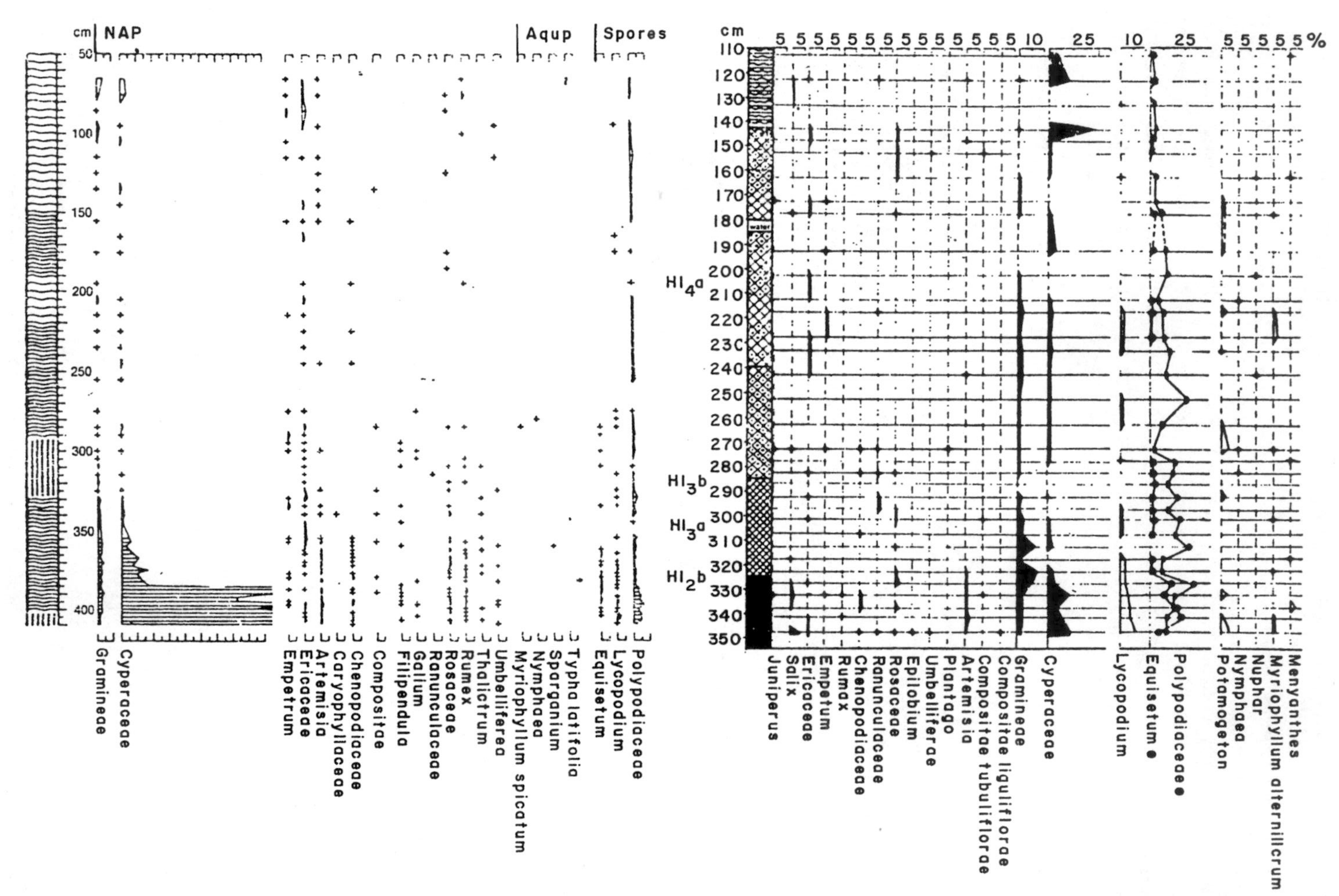

Fig. 6.9. Two (NAP) diagrams quoted from literature. The left-hand one is very difficult to read due to the total absence of guiding lines both vertically and horizontally, and a percentage scale. Compare the right-hand one. Both diagrams redrawn from the originals.

species makes it impossible to read a complicated diagram (cf. Fig. 6.9).

The grouping of curves in resolved diagrams is often more or less arbitrary or conventional, frequently alphabetic. In ecologically oriented diagrams curves of those species which belong to the same major or minor plant community are (or should be) grouped together. Not all curves can be treated in this way: many of them represent taxa which occur in a variety of plant communities, e.g. *Pinus sylvestris*. Others, especially local NAP, may be highly indicative, even of narrowly circumscribed plant communities. Again, this varies according to the geographical position. The same taxon may be widely distributed in the central parts of its distribution area, but possess great indicative value in marginal places, e.g. *Quercus robur* in southern England or western Norway.

The representation of the different species varies greatly with pollen productivity and dispersal of the species. Curves cannot, therefore, be immediately combined, since one of them may be situated around the 2% level, another at 25%, both being equally important. However, by mathematical transformation (autoscaling, cf. Fig. 6.10) variations in pollen productivity, dispersal and preservation can be compensated for. Autoscaling can be carried out in various ways, e.g. by setting the area below each curve = 100 and calculating new values on that basis, or by calculating the values in percentages of the highest value of each curve. Usually there is not much difference between the results, except that the first method is more dependent on continous curves. It should never be forgotten that all such transformations represent artificial numerical processes, by each stage removing data further from botanical reality.

There is a limit to what can be entered in a diagram. Very rare types, which occur as single grains in a few spectra, are often more profitably entered in a table or set out under name in the diagram, (cf. Figs 6.4 and 7.17). It is recommended that the original data be placed in a scientific institution (preferably the one in which the work was performed) or library, where it is available for consultation by other workers.

Different purposes demand different diagram types, and strict rules cannot be laid down. All representations must appear as compromises between practical considerations and theoretical demands, between the demand for clarity and visibility on one side and for documentation on the other. It may be desirable to present the same material in different ways even in the same paper, or to divide the diagram in parts, stressing various indicative features. However, original material should, if

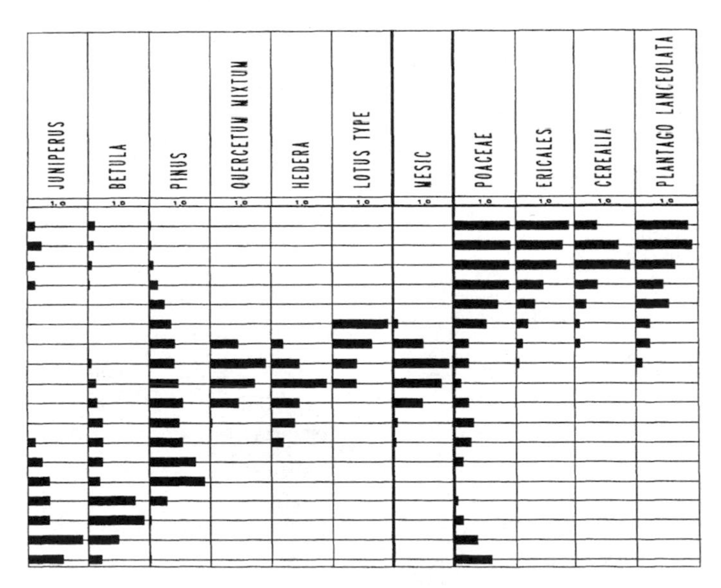

Fig. 6.10. Autoscaled version of the test diagram, Fig. 6.2.

possible, be published in one of the customary ways, whereas diagrams summarizing the results of one, or more, investigations may be treated more or less freely (cf. example in Fig. 6.11). Such diagrams serve to illustrate an opinion, but they are inadequate as representations of the material upon which that opinion is founded.

The manual calculation and production of a good diagram is a time-consuming and often boring job, needing technical skill and much patience. Luckily, most of this can now be left to the computer which can be programmed to calculate percentages and to draw the diagrams. It represents saving of time (the training of assistants is also time-consuming) which may be important, even if the saving is usually fairly small in comparison with the time necessary to produce the primary raw data. Once data have been entered into the computer it can test different calculation and presentation forms without any extra effort. Early (and some contemporaneous) computer-drawn diagrams were so definitely inferior that they should be accepted only as a very last resort. Modern computers produce diagrams of excellent quality. The test diagrams and Fig. 6.1 in this book were made by the program CORE SYSTEM, developed at our institute. Computer treatment should also minimize the danger of accidental errors in drawing. A danger of too much reliance on the computer is the loss of immediate feeling of the features of the diagram. Except for routine work the rough hand-drawn diagram sketch developed during the analysis work is an essential stage in attempts to discover where problems are and how to solve then, and where information is —sometimes—concealed.

6.1.4. The influx diagram

The general problems and use of influx data are discussed in Section 7.1.2.1. The presentation of such data is, on the whole, equivalent to the presentation of percentage data, except that the numerical variation is much greater, since values are not confined within a 100% universe. As a consequence, influx data must be presented in resolved diagrams, as sawblades (Figs 6.1, 7.3 and 7.4) or as histograms (Fig. 6.12).

Concentration and rate of growth of the deposit are necessary parameters for the calculation of influx. They are represented in Figs. 6.1 and 6.12. Usually they are omitted in ordinary presentation. This makes it difficult for the reader to evaluate the data presented, especially the degree of interpolation of deposit accretion values. The very irregular influx curves sometimes seen in the literature may be due to spasmodic growth of the deposit between dated levels, as in regenerative peat, unless dates are very closely spaced.

6.2. Zoning of diagrams

A pollen diagram covering thousands of years is generally rather complicated, and in order to deal with it more easily the investigator has to subdivide instruments both for the study of internal variation in the one diagram and also of correlation between diagrams.

The stratigraphic column of the diagram represents the *lithostratigraphy* (Hedberg, 1972) of the deposit. It is of eminent importance for local stratigraphic correlation within a basin of deposition and also for an ecological interpretation, but it is obviously under such strong local control that it is of very restricted use for comparison between regions or even between basins within the same region (cf. the occurrence of marine sands at varying age levels in basins at different altitude). Even within the same basin lithostratigraphic units may be metachronous. The concept of the lithostratigraphic type sequence and locality is untenable in pollen analysis and can only serve as a straitjacket.

Chronostratigraphic units are the primary material for worldwide correlation, but they are practically never immediately evident from a pollen-analytic study of the deposits, and can only be established by indirect methods such as radiocarbon dating, and then they are in themselves redundant, being replaceable by actual dates.

The *zone* as used in pollen analysis represents the *biostratigraphic* unit: 'that element of stratigraphy which is concerned with . . . units based on

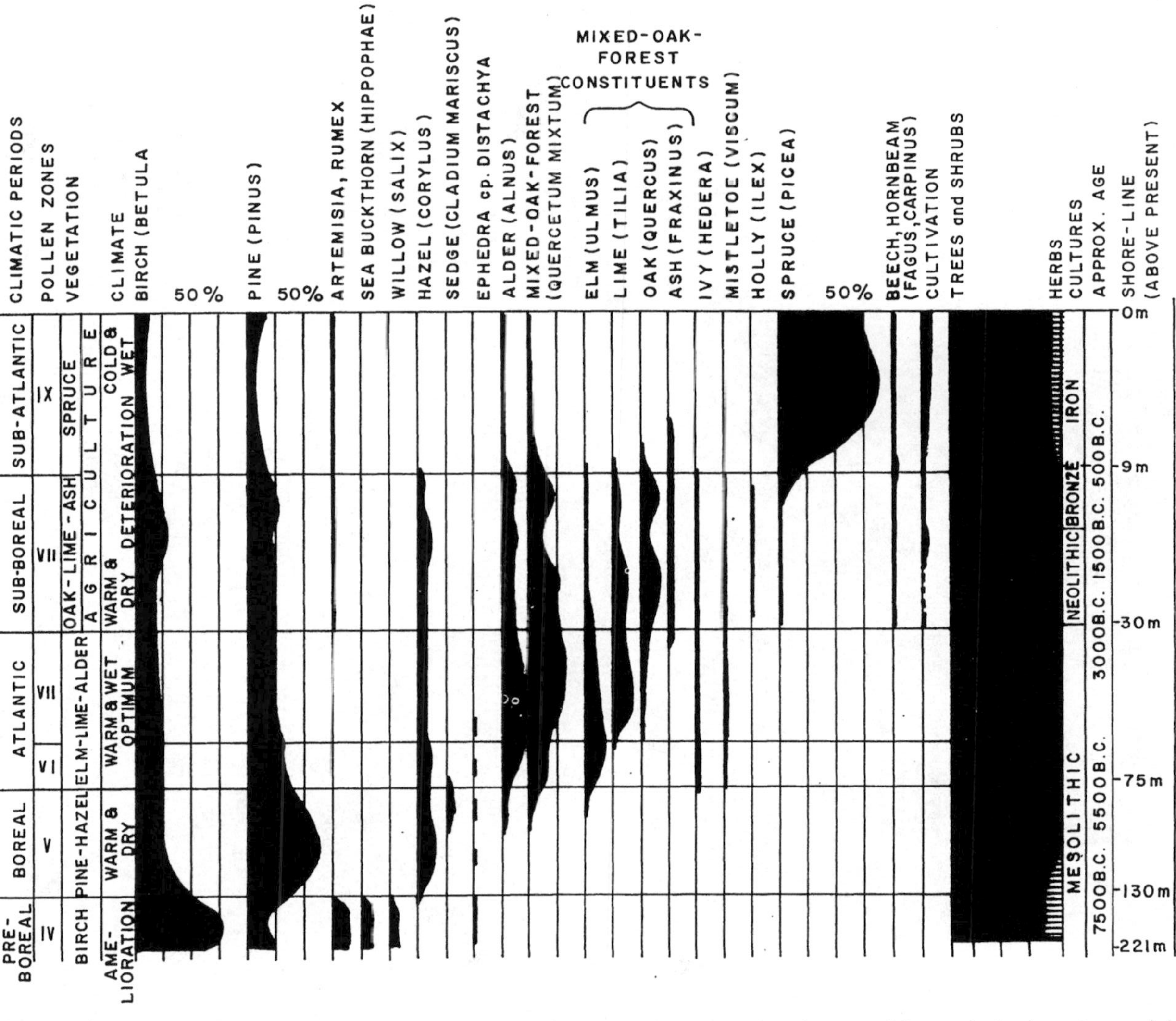

Fig. 6.11. Schematic diagram summarizing in results of investigations within a large area (southeastern Norway). At the extreme right an AP/NAP diagram. Such diagrams are meaningful only if based upon material from an ecologically uniform area. From Hafsten, 1960b, modified. Reproduced with the permission of the copyright-holder: author.

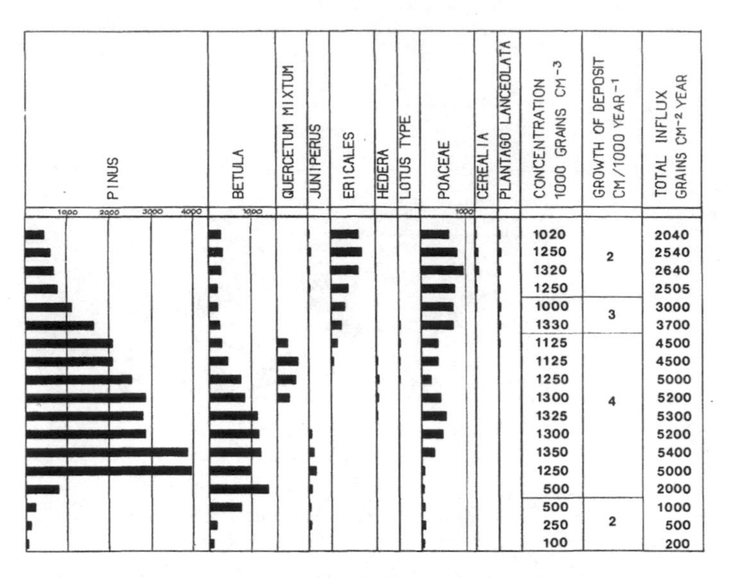

Fig. 6.12. The test diagram, Fig. 6.2, represented as influx diagram.

their fossil contents' (Hedberg, 1972: 16), i.e. pollen flora. By modern definitions 'zone is a common, usually informal, term for a minor stratigraphic interval in any category of stratigraphic classification'. Pollen-analytic zones are therefore biozones, a precision that is unnecessary in ordinary use. Whereas at any rate most pollen-analytic zones will be assemblage zones (cenogams), it hardly pays to go too deeply into the philosphy of stratigraphic classification, which is currently based on a different type of deposits where zones are roughly defined by presence/absence relations and not by the punctilios of pollen-analytic percentage calculations.

A pollen-analytic zone is a sequence within a diagram,* characterized by its flora. The flora may it into parts, usually referred to as zones. Zones are be more or less uniform throughout the zone, or it may undergo a definite change; a constituent of the flora increasing or diminishing, passing a maximum. It goes without saying that the composition of the flora, i.e. the definition of the zone, is dependent on (1) the geographical, i.e. general environmental situation; (2) the environmental history; (3) the successional stages of the vegetation. Strictly speaking biozones are therefore local, and even if two strata in two deposits in geographically remote areas may represent the same period of time and the same climatic development, they should not be considered biostratigraphically identical if their floras are different. On the other hand, the same or nearly the same flora may appear in different latitudes at different times, i.e. different stages of the climatic development (cf. Fig. 7.12). In spite of a possible floristic similarity, such packages of deposit should not be considered as representing the same zone.

Those changes in diagrams on which definitions of zone borders are based should preferably be unequivocal, providing clearly defined borders. Obviously, they are practically never instantaneous. Whether or not ecological reasons can be given for the changes observed has nothing to do with zoning, and neither validates nor invalidates the zonation.

Some authors define zone borders by the intersection of two pollen curves. However easy and distinct this is in the individual diagram, it is not to be recommended, since the level of crossing between two empiric curves, each with a built-in statistical error, is subject to fortuitous and

* The one thing a pollen-analytic zone is *not*, is a body of sediment, as confused by Cushing (quoted from Birks, 1986b: 746). A body of sediment may be defined by its pollen content, but the pollen zone is a feature of the diagram, not of the deposit.

sometimes strong fluctuations. Even within the individual basin such a definition may cause confusion as the floral composition of littoral deposits may vary from that of profundal ones. Or the curves from two closely adjacent basins may for some, e.g. edaphic, reason be higher or lower, and the crossing will take place at different levels, even if the trend of the curves, which is the important part of the diagram, is the same (cf. Fig. 6.13).

As the zone is an instrument for working with the diagram it should be defined by diagram characteristics alone, without any climatic or ecologic argument being brought in. At the next step the diagram may be used as a basis for climatic or edaphic inferences, and the danger of circular reasoning should always be kept in mind.

The concepts of zones and zoning are very old in pollen analysis, and although the basic concepts were sound, some of the applications are open to criticism. As biostratigraphical units, zones are depositional sequences of spectra showing a common characteristic composition, partly through the absence or presence of certain elements, partly through their mutual quantitative relation. Thus, the first appearance of *Corylus* or *Alnus* defines the beginning of another zone, i.e. another pollen assemblage. The change-over from equal representation of *Ulmus* and *Quercus* to the dominance of the latter indicates the beginning of another zone.

Pollen zones are nothing more than that. Locally they characterize a certain depth of deposit, but the deposit itself does not form a zone. The pollen zone is a feature of the diagram alone and the *primary assemblage zone*, being a feature of the individual diagram, is of limited general interest. It is a matter of opinion how far one should go in formal designation of such zones.

When more than one diagram is available from a deposit, or from a restricted area, equivalent pollen assemblage zones are usually found in the majority of them, and by integrating the primary assemblage zones one may establish a local *assemblage zone pattern* (to avoid the equivocal term 'type').

As long as distances between sampling points are small, and habitats uniform, this procedure should be fairly safe. However, if one tries to proceed to the establishment of a *regional* assemblage zone pattern one has to be very careful and realize the *a priori* restricted significance of the resulting patterns: they do only comprise certain diagram sections of similar composition, but they may be completely metachronous: shorter and longer due to climatic conditions (cf. Figs 7.11 and 7.12) or may even have varying modal values because of

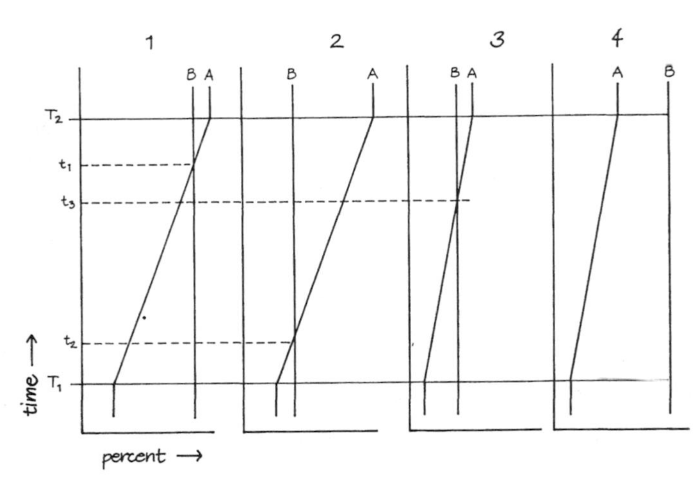

Fig. 6.13. Metachrony of the intersection between two curves showing the same trend, but different values. Schematic pollen diagram showing the development of taxa A and B between time T_1 and T_2. Taxon A increases by a factor of 4, B is constant. According to the quantity of A and B the curves intersect at different times ($t_1 - t_3$) or not at all (diagram 4).

immigration delays. Mathematical control can assure that conclusions do not go further than the numbers permit, but can in no way guarantee against such pitfalls as these.

A certain confusion between biozones and the corresponding time units has led to loose usage in many pollen-analytic publications. Zoning systems are often used in the wrong context. In particular, the Blytt–Sernander periods, which have played such a great role in chronologic discussions in northern Europe and elsewhere (Boreal, Atlantic, etc.), never were lithostratigraphic, and if they at any time possessed a biostratigraphic meaning (in southern Scandinavia), it was lost long ago. Consequently, they can only be used as chronostratigraphic, i.e. chronologic units (cf. Mangerud *et al.*, 1974). Originally they were defined as periods (of immigration of certain groups of plants), i.e. as chronozones. Their use as biozones is not justified. The corresponding biostratigraphical units in southern Scandinavia would be the 'oak zone', 'spruce zone', etc. used by some of the older schools. Similarly, Jessen's original division of Danish pollen diagrams was clearly biostratigraphical, and the use of 'Jessen's zones' outside Denmark, i.e. as chronostratigraphic units, is open to the same criticism.

One may ask why descriptive terms for chronozones should be used at all, considering that absolute dates are available for the late-Quaternary. The answer is that such terms are convenient and of great mnemonic value, like the Middle Ages, the Elizabethan Period, etc. Like such period names, the old zone designations also carry additional information with reference to climate, vegetation etc. but like them too, they are either quite local (Elizabethan) or they become metachronous (Middle Ages). It is therefore extremely dangerous to use them outside their area of origin, especially because the climatic, etc., connotations, may become misleading. On the other hand, (Quaternary) geology abounds in more or less local zone names which indicate the same time period, and the existence of which makes literature more complicated to read, unless actual dates are added for the benefit of non-local readers.

It is also not recommendable to try to establish such old zones as strict chronologic units, because errors in dating may give the name a completely new and confusing connotation.

With zones reduced to local phenomena, most of the importance previously attached to their establishment, and to the comparison with standard zonations in others parts of the world, has diminished appreciably. The zone remains an important means of describing the local vegetational development and relating this development to various physical factors. To describe solemnly type localities for such local phenomena is hardly necessary, especially in view of the metachrony of identical pollen assemblages.

Zones may be subdivided. If the curve of a forest type within a zone gradually rises to a maximum and then declines, subdivision should embrace three parts: (a) rising curve, (b) culminating, and (c) declining curve. Whether such a course of the curve is conditioned by climate or by the immigration of another forest type, or both, it is obvious that such changes of the diagram have a definite ecological background and therefore may be common to the vegetation of a larger region, though local conditions may modify the picture in each individual case. It may be assumed that such changes of the curves are synchronous within a fairly large area, even if absolute values differ between diagrams.

It is much more difficult to interpret the short-term changes of the pollen diagram which manifest themselves as a minimum or maximum of only one, or very few, samples. What do such short-term fluctuations mean in terms of vegetation? Is it possible that the whole vegetation of an area is subject to small *synchronous* variations? Certainly, a few years of meteorologically conditioned profuse flowering of one species may theoretically be sufficient to cause an appreciable synchronous maximum of the corresponding curve. But a maximum may be due to quite different causes, e.g. if the pine forest surrounding our basin has become over-ripe and blows down in a storm, to be succeeded by a generation of birch, producing a birch maximum before the reappearance of pine. Maxima of the latter type are isolated features of distinctly local distribution; unfortunately they cannot easily be

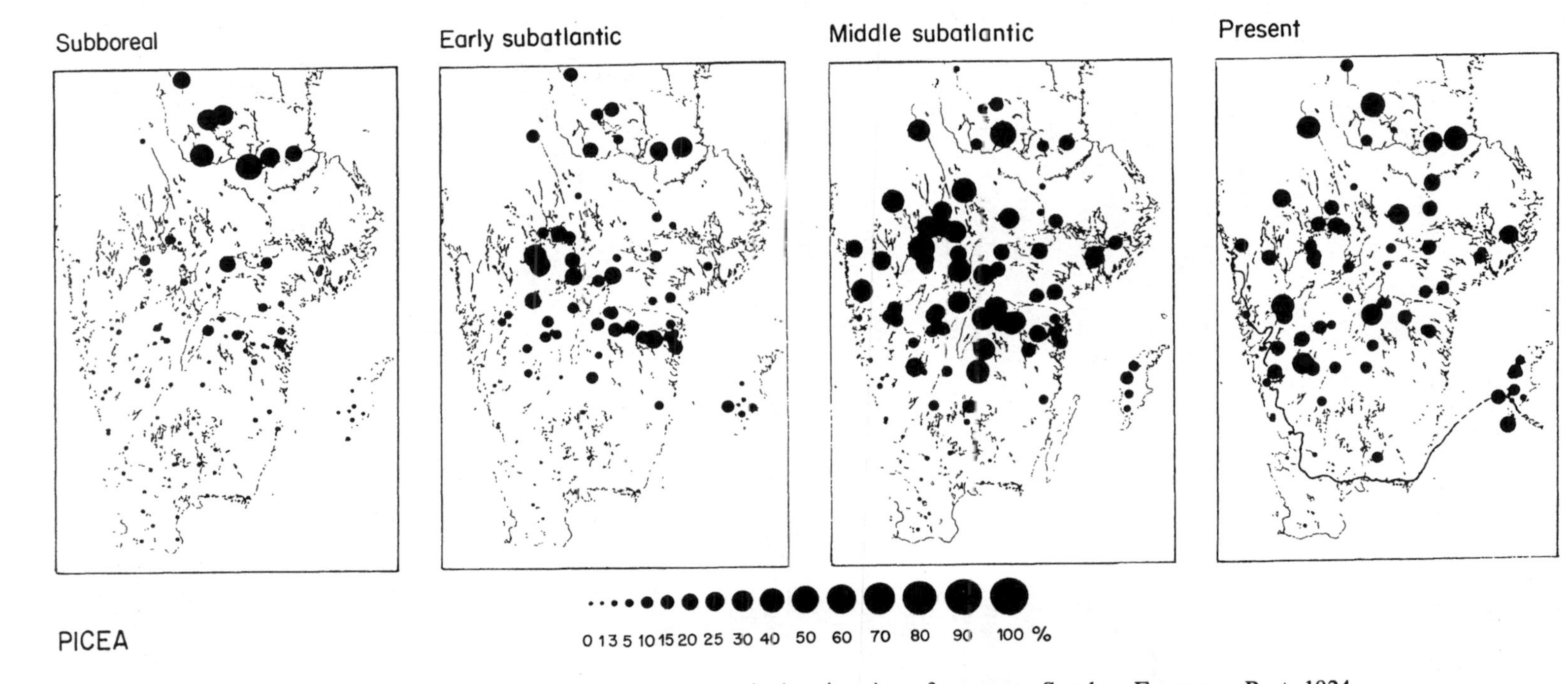

Fig. 6.14. Resolved pollen map showing the immigration of spruce to Sweden. From von Post, 1924.

distinguished from the former. Even the fact that a corresponding maximum is found in another diagram from the same area is no proof, since the succession between birch and pine may be a regular feature in the cyclical regeneration of the forest, and we have no reason to presume that this regeneration is simultaneous throughout the area; the contrary is indeed more likely. It is hazardous to correlate diagrams by means of short-term fluctuations without adequate dating control, and to establish synchronous levels the ecological significance of which is not clearly understood. A long-term change of dominance, a slow rising of a curve with a subsequent decline, may be interpreted ecologically with a certain probability, and so may a sudden but irreversible change. Such features of the diagram are suitable for local correlation. After the advent of agriculture, and consequent clearing and destruction of the forest, correlation presents even greater difficulties than before; compare the very characteristic but metachronous 'clearance phase' in Danish pollen diagrams (Iversen, 1941). Pollen curves should not be regarded as phenomena *per se*, as a kind of index fossils, and then used independently of their botanical background. Such a procedure is bound, sooner or later, to lead to incongruous results.

6.3. Maps

The ordinary pollen diagram contains two sets of variables: time (i.e. depth) and species composition, the latter again containing two variables: the number of species present and their quantity. A map would introduce another variable, viz. spatial arrangement, also with two possible subvariables: geographical location and altitude. It is not possible to bring all these variables into one diagram that can be visualized.

Already in his 1916 lecture von Post (1918) mentions a map, which was, however, never published, but in 1929 he demonstrated the spatial distribution by a set of resolved diagrams ranging from Auvergne to the Sub-Arctic (Fig. 6.7). In 1924 he had introduced truly cartographic representation in resolved maps (Fig. 6.14).

These leads have been followed in various ways, but it has proved very difficult to produce a synoptic map. Figure 6.15 shows how far one can go, provided the number of species is small and the spatial arrangement is laminar.

Welten (1952) has given a similar presentation for an altitude series within a limited area (Switzerland) (cf. Fig. 7.8). He has reduced the complexity of the sectorial notation in fig. 6.15 by using forest types, i.e. plant communities, as mapping units, in this case represented by the dominant tree(s). Simonsen (1980: 53) has published a similar diagram for Norway (cf. Fig. 6.16). Both suffer from the absence of (then unavailable) dating.

Usually problems are too complicated, and one must choose between presentation of a time sequence in one locality, i.e. a pollen diagram, or the composition of vegetation at one time level within an area, i.e. a fossil vegetation map, which should then correspond, as closely as possible, to recent vegetation maps. The validity of fossil vegetation and flora maps is limited to one

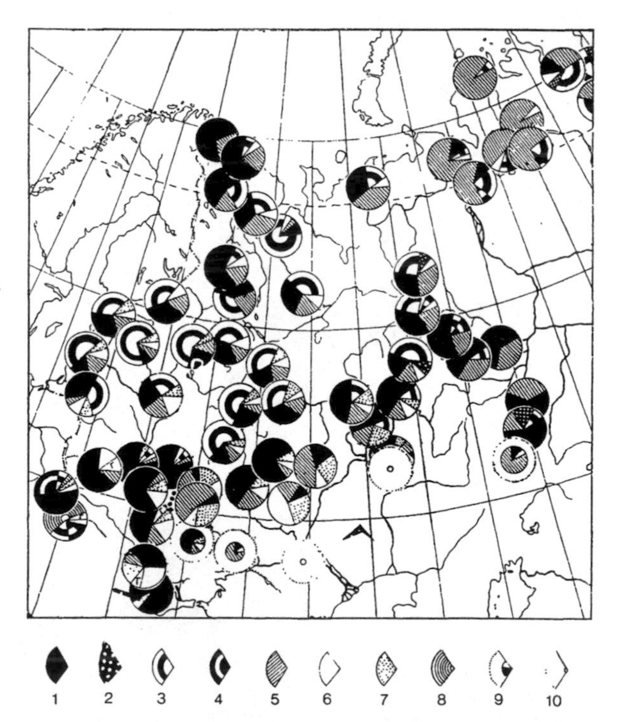

Fig. 6.15. Early Holocene synoptic pollen map from Eurasia. After Nejshtadt 1957, redrawn. (1) *Pinus*; (2) *P. cembra*; (3) *Picea*, (4) *Abies*, (5) *Betula*, (6) *Alnus*, (7) QM, (8) *Fagus*, (9) forest steppe, (10) steppe.

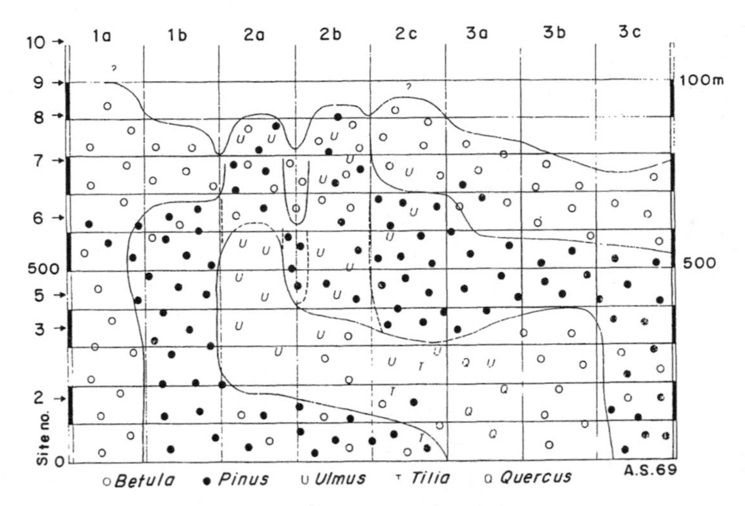

Fig. 6.16. A 'vertical pollen map' from western Norway. Forest types are indicated by the pollen signature of the leading taxon (cf. Fig. 6.3). Large figures at left edge indicate station number. Figures at the top indicate time, from Preboreal to recent. From Simonsen, 1980.

particular period only, and it is imperative for their usefulness that the diagrams on which they are based are correctly dated. A metachronous map is not only useless, but misleading as well.

Like diagrams, fossil vegetation maps may be presented as composite or resolved types. Composite maps (Fig. 6.15) indicate the total composition of the vegetation at the time in question, whereas the resolved type indicates the distribution and importance of each individual constituent in a separate map (Fig. 6.14). A whole series of maps is then necessary, one for each taxon, to give a complete picture of the vegetation of the area.

The usual type of composite map has circular symbols, in which the relative importance of the constituents is shown by sectorial division of the circle, (cf. Fig. 6.15). As this type of map has not been widely used (it is not always easy to read), the use of symbols has not been consistent. Gams (1937 —cf. 1938) has proposed a symbol system for Europe.

Unless the composition of the vegetation is very simple, better representation of facts is achieved by the use of a resolved map, in which the part played by one species at a time in contemporary vegetation is indicated by the size of the symbol, usually a black circle (cf. Fig. 6.14). This map type is especially useful for detailed maps of smaller areas (cf. Birks, 1985: 12). If the number of localities is very great, these symbols may become too crowded and overlap. Firbas (1939) has used small lines instead of circles, but even such maps are rather complex and difficult to read.

A major improvement was introduced by Szafer (1935 a, b). His isopollen lines (unfortunately dubbed isopolls) are synchronous lines delimit-

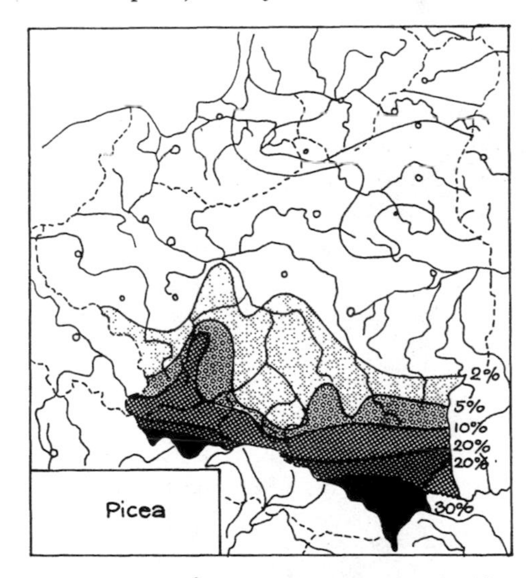

Fig. 6.17. Isopollen map showing the incidence of spruce pollen in Poland at the end of the post-Glacial warm period. From Szafer, 1935a.

ing areas with the same percentages of a certain species, in the same way as isotherms delimit areas having the same temperature. Isopollen lines give a cartographic presentation which is equivalent to that used in recent vegetation maps—though, of course, the figures must be read in a different way (cf. Fig. 6.17). The isopollen maps give excellent comprehensive summaries, but owing to the unavoidable generalizations in their construction they are unsuitable as primary documentation of pollen-analytic data.

Apart from small-scale, very generalized maps it is important that all stations taken into account for construction of the map (also zero values) are distinctly indicated at least in one map of a series.

Isopollen maps are more versatile than the previous types, and have been constructed in various variants, for various purposes and also in various scales with concomitant degrees of generalization. The simplest type is the classical isopollen map showing the distribution of one taxon at one time, graded after percentages and suffering from the general weaknesses of the percentage calculations. Corresponding maps based upon data presumed to give better estimates of real abundance (corrected percentages) are called *isophyte* maps (H. H. Birks, 1985). The ideal would be to base them on pollen influx counts, but such data are still too scarce to permit construction of maps. Also, the inherent weaknesses of influx data make such maps, for the time being, unpracticable.

By subtracting the data of a younger map from those of an older one *difference maps* can be established, demonstrating the advance or retreat or,

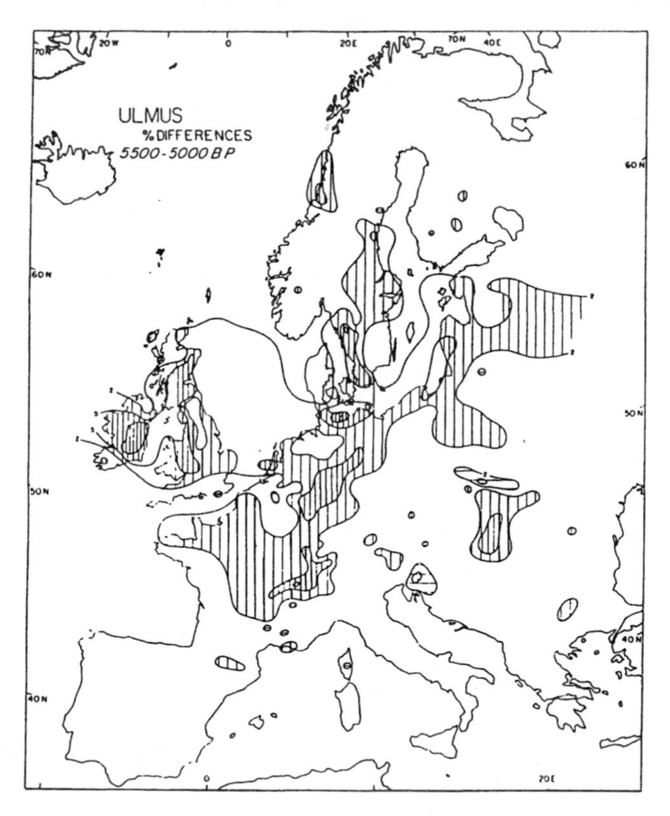

Fig. 6.18. Difference map for *Ulmus* values in Europe 5500 and 5000 *BP*. Vertical shading indicates decline in pollen values. From Huntley and Birks, 1983. Reproduced by permission of the copyright-holder: Cambridge University Press.

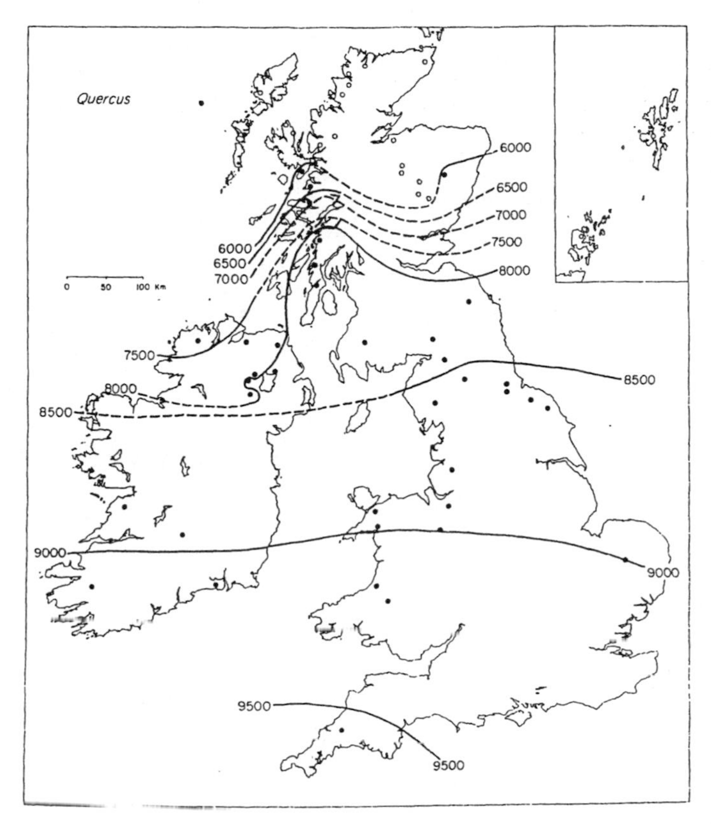

Fig. 6.19. Isochrone map for the first rise in pollen values for *Quercus* in the British Isles during the Holocene. From Birks, 1986a.

less drastically, change of abundance of the taxon during the period concerned (Fig. 6.18).

Isochrone maps indicate the time when a taxon reaches—or leaves—the stations mapped. Here the difficulty is to define a meaningful criterion for mapping, i.e. a fixed percentage which is taken as the zero value for local occurrence or establishment (cf. Fig. 6.19).

With a very detailed mapping, a good knowledge of recent vegetation types and a thorough ecological analysis of pollen diagrams it is possible to construct palaeovegetation maps. Isochronic *palaeovegetation* maps of the borderline between major plant communities have been constructed to show, for example, the equilibrium between prairie and steppe in the USA.

7. Where did the pollen come from? Interpretation the findings

Motto: Never take anything for granted.

The interpretation of a pollen diagram comprises two steps: (1) establishing the composition of the vegetation that delivered the pollen rain registered: reconstructing; (2) drawing inference from the vegetation data back to the agents behind them: climate, ecology, human interference, etc.: interpretation sensu stricto or application of the data. This chapter chiefly deals with the first step. The second is dealt with in the chapter on application. The special interpretation problems of archaeological sites are discussed in Chapter 9.

7.1. Pollen data and vegetation

The pollen spectrum is a record of contemporaneous vegetation within a certain area, but due to differences in pollen productivity, transport and preservation there is no one-to-one correlation between the representation of a species in the vegetation cover and in the pollen spectrum. The translation from a pollen spectrum—or a sequence of spectra: the diagram—to the vegetation is a process that demands understanding of the effect of all the processes detailed in Fig. 1.1, and the ability mentally to run these processes backward. This is the normal first step in the interpretation process of the normal diagram.

However, in many cases the reconstruction cannot follow the normal pattern: the deposit is abnormal, the pollen rain influenced by extraordinary forces, conditions of sedimentation and preservation deviant, etc. This introduces various 'errors' in the diagram. The diagrams are not erroneous in themselves, but if they are interpreted according to the ordinary rules the interpretation certainly will be. On the other hand, the existence of such aberrations is in itself a registration of conditions differing from the normal, and is therefore of primary value in understanding what has happened. From having been, in the days of classical analysis, considered errors and obstacles, such aberrations are the primary material for interpretation in many modern problem approaches.

The analysis itself, and the preparation of the diagram, is all technical: much of it may be left to technically trained personnel without adequate scientific qualifications—and to computers. Even in routine analyses, however, the investigator should personally count some samples to check upon irregularities which might pass unnoticed by an assistant. If there are irregularities of any kind, routine work is very dangerous, and should never be left to assistants without scientific background.

Schematic routine work is even more fatal when it comes to the interpretation of diagrams. One must not forget for one moment that the pollen diagram represents nothing but those parts of the pollen rain than have been recovered after fossilization. The primary objective of the diagram is to give a record of the whole vegetation of the locality in question. The pollen diagram has no direct bearing on any other factor, whether climate, soil, exposition or anything else. The first step is reconstruction of vegetation from the pollen data presented in the diagram. Only thereafter may an attempt be made to find any reasons for the vegetational changes observed. A corollary to this is that the pollen analyst should be a trained field botanist. Pollen analysis without a firm botanical foundation is certain sooner or later to produce incongruous results.

Consequently, the first conditions to be taken into consideration are the botanical ones, affecting the life of the plants, the pollen of which we have registered.

The simplest case is that of the classical diagram from a large basin, supposed to represent the regional pollen rain of a forested region only. The data, whether influx or percentages, are based upon assumption of the existence of a general pollen rain, containing pollen in percentages commensurate with the representation of the respective pollen emitters in the regional vegetation. We shall later discuss special diagrams where the concept of a general pollen rain is invalid.

Pollen diagrams represent only flowering. If a species is regularly prevented from flowering, e.g. because it is regularly harvested before reaching flowering age, it will not show in a pollen diagram, even if it may be frequent in the area. When there is evidence that humans have entered into the landscape, such contingencies must always be kept in mind.

As long as the main object of pollen analysis was to infer climatic changes it was of prime importance that the investigator should be able to recognize and discount such non-climatic features as were likely to occur in pollen diagrams, e.g. due to the ordinary succession in a fen area (cf. also the almost identical, but highly metachronous regression diagrams published by Florin (1945), in which a certain stage in the isolation of basins from the Baltic is characterized by a distinct maximum of *Alnus*). The interpretation must recognize that this maximum represents the alder carr growing round brackish-water basins, and therefore carries no climatic or chronological information whatsoever (cf. Fig. 7.1). This is a case in which *reconstruction* is straightforward: The *Alnus* carr certainly was there, but the *interpretation* is complicated by the fact that it carried no climatic information.

Some of the objections raised against the application of pollen analysis with regard to inferred effects of successional or pre-climax vegetation belts on diagrams demonstrate the necessity for botanical training in the interpretation of the diagrams, not imperfections of the method. The paper by Couteaux (1977) gives several examples of this: the apparent weaknesses pointed out are nothing but too schematic use of data by persons not understanding the ecological background.

Differences in pollen productivity cause some of the most persistent problems in pollen diagram interpretation. In a completely forested area all information is contained in the AP diagram, and the pollen-emitters are, on the whole, comparable.

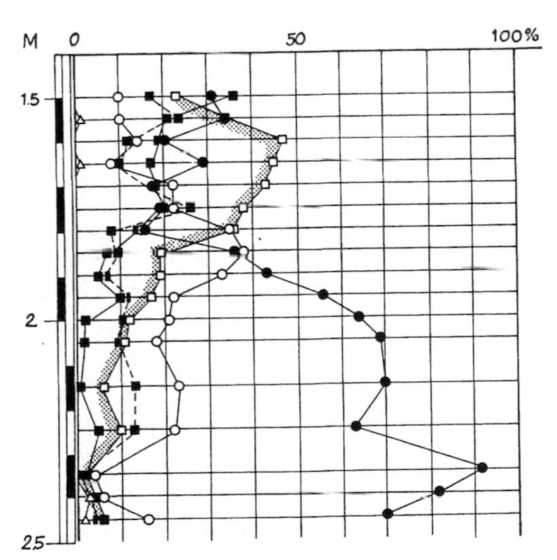

Fig. 7.1. Overrepresentation of *Pinus* in the marine phase (up to 1.80 m) followed by overrepresentation of *Alnus* from the shore vegetation surrounding the brackish-water basin. After Florin, 1945, redrawn. Reproduced by permission of the copyright-holder: M.B. Florin.

If the diagram is to be translated adequately into terms of vegetation, the varying pollen production of different species must, nevertheless, be considered. As pine produces great quantities of pollen 10 % pine pollen (or a corresponding influx value) is almost of no account: pine hardly grew at or near the site at the time of deposition of the sample. As beech produces little pollen, 10 % beech (or corresponding influx) is highly significant and generally indicates that beech has played a substantial part. If we use Pohl's (1967) production figures (their significance and applicability must not be exaggerated) a pollen spectrum registering 33 % each of pine, oak, and beech should represent the production of a vegetation of 5 % pine, 35 % oak and 60 % beech. This is a theoretical model only; in nature conditions are much more complicated, for one thing because of transport distance differences.

The quantity of pollen of a particular species deposited per unit area depends upon a number of factors:

1. The frequency of the species in the region, which is usually the problem under study.
2. Its absolute pollen production, which varies both specifically and individually according to the conditions under which the specimen grows: open position or closed stand. Pollen production is also dependent upon the frequency of flowering years. Some forest trees flower freely at intervals only. This may partly be due to morphological factors: flowers or influorescences being situated in such a position that no flower can be formed in the same place in the following year; and partly to physiological factors: a period of 'rest' being necessary before the next profuse flowering. The nearer one comes to the limit of the distribution area of the species, the longer the periods of 'rest' between flowerings. Dendrological literature gives information about the years of rich seed production, which is, however, not necessarily the same as years of rich flowering. Authorities do not always agree, but there seems to be a general agreement that *Betula* and *Alnus* flower and fruit freely every year, and that there are intervals

between flowering years of *Picea*, *Fraxinus*, *Fagus* and *Quercus*. The position with regard to *Pinus* is more obscure; at any rate in *P. sylvestris* flowering seems to be more regular than seed-setting, which is less certain because of the long ripening period.

3. The dispersal mechanism of the pollen also influences the deposition pattern. Dry pollen is better dispersed than pollen grains that form lumps, both because of the smaller size of the particles and because of the greater number of units.

The factors and processes dealt with under 2 and 3, above, express themselves in the R value mentioned in Chapter 1, i.e. the relation between actual occurrence of a taxon in the vegetation and its pollen registration. If the latter is an influx registration the R value is fairly simple, and there is (within limits) a linear regression between the two sets of data. This representativity factor, R_{abs}, is still dependent on ecologic parameters and takes on different values in different climatic regions, but it can be used, with reservations, within fairly large areas.

Provided we can compensate for the factors mentioned under 2 and 3, above, the quantity of pollen per unit area is a measure of the frequency of any flowering wind-pollinated species in the region. Most northern temperate forest trees are wind-pollinated, which facilitates the reconstruction of the forest composition. During the flowering season pollen is scattered over the whole area and may be recovered from the contemporaneous surfaces of bogs and sediment deposits. A few northern temperate forest trees are pollinated by insects, the most important being *Tilia*, *Acer* and *Salix* species, most of which, however, produce relatively great quantities of pollen. Their pollen dispersal is not as effective as that of anemogamous pollen, and *Salix* and *Tilia* are seriously underrepresented in the regional pollen rain from forested regions. With some caution one may treat them in the same way as wind-pollinated species, but *Acer* is in a different position, and its occurrence is very difficult to evaluate by pollen analysis, at any rate in Europe. This is aggravated by the exceptional susceptibility to degradation (Sangster and Dale, 1964).

The customary percentage calculations produce some additional effects, which are not always realized. Because the total sum of percentages always amount to 100 (a closed universe in the statistical sense), there is a strong negative autocorrelation between numbers, and consequently there is no linear regression between pollen percentages and actual quantities recovered. The incidence of a species is a function not only of the frequency of that species, but also of the composition, i.e. the total pollen productivity of the forest. The percentage calculation introduces a series of new problems, the influence of varying pollen productivity dealt with here being only one of them. Partly, this can be compensated for by numerical methods, but on the other hand each step in transformation (mathematical or otherwise) interposes a barrier between data and the biological reality.

7.1.1. Pollen deposition in relation to composition of vegetation

Fagerlind (1952) was one of the first to discuss the relation between (recovered) pollen rain and actual vegetation by means of a mathematical model. The ideal situation that the pollen spectra should be influenced by quantitative composition of vegetation alone he finds greatly modified by differences in pollen productivity and dispersal in various plants. The comparison between two values on a pollen curve suffers from great mathematical uncertainties but, even so, Fagerlind concludes that some of these may be self-correcting, others compensated for. If the whole trend of a curve, represented through a series of samples, is taken into account, many of the mathematical objections are more or less invalidated.

Because of the inherent non-linearity of the relation between actual presence in vegetation and representation in a percentage diagram, the corresponding R value, R_{rel}, depends on so many other factors that its general usefulness is limited. There have been many determinations of R_{rel} values from various parts of the world; Table 7.1 gives some of

Table 7.1. R values in surface samples from various plant communities

	I	II	III	IV	V	VI
Pinus	6.6	7.7	22.4	—	15.8	200
Betula	6.0	7.9	1.5	3.0	13.6	5.6
Alnus	—	—	5.0	4.3	17.7	21
Corylus	—	—	—	—	13.7	—
Quercus	2.8	1.8	0.6	—	1.6	240
Picea	0.9	1.1	0.6	—	13.4	0.6
Abies	0.8	0.1	—	—	—	0.1
Carpinus	—	—	—	—	7.7	—
Fagus	0.5	0.2	0.9	—	1.0	1.5
Pterocarya	—	—	1.1	—	—	—
Tsuga	—	—	0.8	—	—	1.2
Thuja	—	—	0.6	—	—	0.3
Tilia	—	—	0.13	—	—	0.4
Acer	—	—	0.03	—	—	0.2
Larix	—	—	—	—	—	0.01
Salix	—	—	—	0.4	—	—
Empetrum	—	—	—	1.0	—	—
Cyperaceae	—	—	—	0.5	—	—
Gramineae	—	—	—	0.3	—	—
Ericaceae	—	—	—	0.2	—	—

From P. Müller, 1937 (I), Steinberg, 1944 (II), Tsukada, 1958 (III), Iversen, 1947b (IV), Pohl, 1937 (V) and Davis, 1963 (VI).

them. Various aspects of calculation and use of R_{rel} values are discussed by Birks and Gordon (1985) and by Prentice (1986), who indicate methods of standardizing R_{rel} values for comparative use. However, even such calibration is so plant community-dependent, and dependent on the climatic response of the individual taxa, that the use is also in this case restricted. The main importance of the R, especially the R_{rel} concept, is as an abstract concept embodying the modifications data undergo between the pollen diagram and the vegetation the diagram represents. Figure 1.1 shows how many modifying factors are involved and, unfortunately, we do not possess the data necessary to make a quantitative conversion.

The columns of Table 7.1 represent different species within the various genera; on the whole there is a rough, but fairly good, correspondence between figures. There is a contrast between a group of high producers and a group of low ones, with some unaccountable exceptions such as *Picea* in column V and *Betula* in column VI. *Quercus* is another type, the R_{rel} values of which come out in different positions in various investigations. This can hardly be an effect of taxonomic differences alone.

Whereas the ranking order in Table 7.1 is fairly consistent, the absolute values are rather different, and will certainly tend towards greater differentiation in richer forests, the components of which differ greatly in pollen output. In Finland the subrecent pollen spectra reproduce the actual composition of the forest comparatively well because they are totally dominated by *Betula* and *Pinus*, which are equally great pollen producers. However, *Picea* is underrepresented as usual (Aario, 1940: 56 *et seq*.). The situation is more complicated if main forest trees are seriously underrepresented in the pollen rain. Only by very careful work and great caution in interpretation may mistakes be avoided under these conditions (cf. Potzger *et al*., 1957).

Potter and Rowley (1960) have compared pollen rain and vegetation in an arid grassland–woodland region; they found overrepresentation of *Juniperus* over *Pinus* and *Quercus* in atmospheric pollen, while *Juniperus* was seriously underrepresented in moss and lichen cushions as well as in the alkaline clay soil. The effect of selective pollen destruction in cushions appears clearly. Pollen of grasses was underrepresented both in the atmospheric pollen rain and in surface samples, while pollen of Chenopodiaceae–Amaranthaceae was found abundantly.

With more data becoming available, numerical studies have been made of correlations. The ideal case conditions would be a 1 : 1 relation, i.e. a 45° slope and 0 *y* intercept (cf. Fig. 7.2). Unit regression (*Quercus*, *Pinus*) indicates proportional representation, i.e. a 'normal' (for the region!) pollen productivity, whereas the productivity of *Betula* is evidently supernormal and that of *Larix* and *Acer* subnormal. The high *y* axis intercept of the *Pinus* curve indicates that the asymptotic value of this pollen type is above normal, i.e. that *Pinus* is overrepresented in the general pollen rain.

In the cases of *Sorbus* and *Ilex* it is obviously not meaningful to calculate a regression. *Sorbus* represents a type, the pollen of which is to a certain extent dispersed in the air, but which produces so little pollen as to be almost absent in the registra-

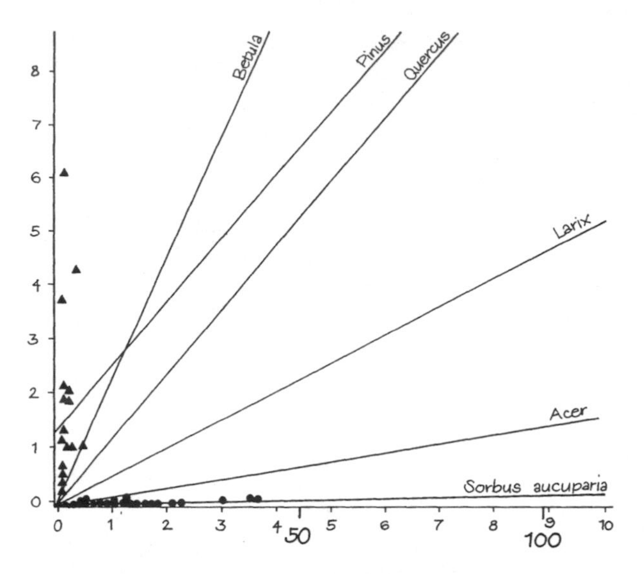

Fig. 7.2. Regression of pollen percentages in surface samples on actual representation in the vegetation. Data for *Sorbus* after Birks, 1980; *Ilex* (triangles) and the other regression curves after Delcourt *et al.*, 1984. Small numbers indicate the scale of the *Ilex* diagram, large numbers (*X* axis only) that of *Sorbus*. The scale of the other diagrams is 1/10 of the *Ilex* values. Redrawn. Reproduced by permission of the copyright-holder: American Association of Stratigraphic Palynologists.

tion even if it represents a significant part of the vegetation cover. *Ilex* represents an opposite case: pollen dispersal is so bad that, notwithstanding its coverage, no pollen is dispersed in the air, all falls down at the foot of the mother tree. The quantity registered is an indication of the distance to the nearest pollen producer and tells nothing about the part played in the composition of the vegetation. Obviously, neither has any real place within the concept of the pollen rain, *Ilex* because it never enters it, and *Sorbus* because the registration is too erratic for regression calculations.

7.1.2. *Presentation of data*

The technical problems of data presentation have been dealt with in the preceding chapter. Here we shall discuss presentation on relation to information content about the vegetation concerned.

7.1.2.1. *Influx diagrams*

The use of absolute pollen numbers would compensate for some of the problems inherent in relative percentage calculations, as the data from each pollen source can be evaluated separately, independently of the other pollen producers occurring at the same time in the same general area. Theoretically, the translation back to vegetation data should be easier and safer. Unfortunately, in actual practice this is not always so.

One reason for this is that we do not possess adequate data for the pollen production of individual species per area unit. The same pertains to the attenuation of pollen quantity with increasing distance from the source. Both quantities vary greatly with climatic conditions. The same number of pollen grains deposited on a surface unit may therefore mean very different things with regard to vegetation.

It is in reality very difficult to establish meaningful attenuation curves, as one will in most cases very soon reach the area in which the production of the individual stand is indistinguishable from the general component of the species in question within the pollen rain as a whole. Only very rarely is a stand so isolated (and at the same time flowering naturally) that this can be avoided, and the attenuation curve pursued beyond the distance of asymptotic values, i.e. a couple of hundred metres. Data from such stands are very rare, if they exist at all. Also, they must be collected over a sequence of seasons.

Nevertheless, absolute pollen frequency (APF) data are of great value in interpreting doubtful cases. We refer to M. B. Davis (1967b) who, by means of APF, for the first time could demonstrate beyond doubt that the so-called pine period in eastern North America really reflects a pine forest, and not statistical overrepresentation of pine pollen in a vegetation without appreciable representation of that genus (cf. Fig. 7.3). When issues are as clear-cut as this one, comparative influx data from recent vegetation are redundant.

The diagrams in Fig. 7.4 represent a case in which forest conditions could hardly have been elucidated without influx data. Neither the rise of

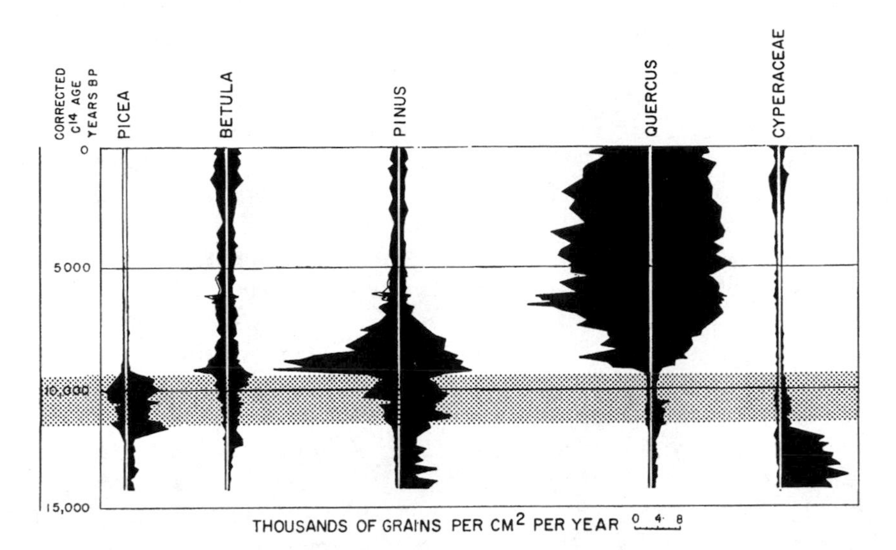

Fig. 7.3. M. B. Davis's (1967b) registration of pine forest at the transition between Late and Post-Glacial. Three stages are evident: tundra, forest–tundra (shaded) and forest. The NAP percentage values (cf. Cyperaceae) decrease already in the forest–tundra, being statistically depressed by the pollen production of the sparse tree cover. This effect is probably enhanced by the size of the basin (Rogers Lake, 140 ha, cf. Davis and Deevey, 1964) in which the regional pollen is more effectively registered than the local. The close correspondence between percentages and influx in the forest phase corroborates Middeldorp's assertion (cf. p. 122 *seq.*) about the constancy of the total pollen influx. The curves demonstrate different reactions: *Picea* belongs to the forest–tundra and disappears when the forest closes. Influx and percentage curves are very similar. *Betula* is the same, but (the same taxon?) continues also in the forest. *Pinus* is overrepresented in the tundra phase, forms the first forests and then recedes. *Quercus* shows a slight increase in the early forest tundra phase (climatic oscillation?), but only expands in the forest. Cyperaceae produces the same amount of pollen the whole time: the variations of the curve are purely statistical effects of the percentage calculations. Data from Davis, 1967b. Influx left, percentages right.

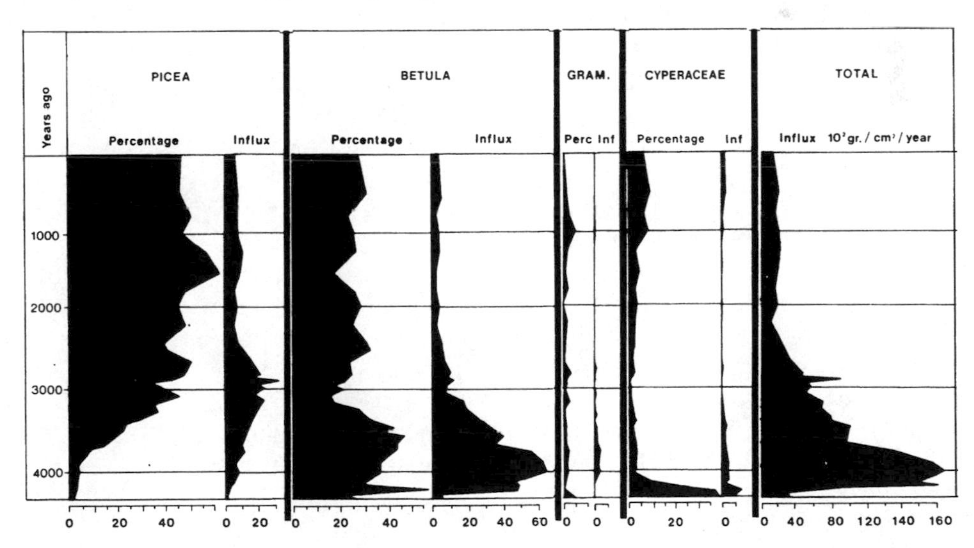

Fig. 7.4. Percentage vs. influx data. From McAndrews and Sansom, 1977, redrawn. Only some taxa have been included. Cf. text. Reproduced by permission of the copyright-holders: Les Presses de l'Université de Montréal.

forest in the bottom samples, nor the deforestation at 3000 BP, nor the small *Picea* maximum at that time are apparent from the percentage diagrams, even if the first-mentioned could be deduced from the Gramineae curve.

In the older literature APF usually meant absolute pollen number per sample or per slide only; later per volume of material. As these numbers are related to the pollen deposition via the then generally unknown accumulation rate, the figure was of limited value and was on the whole neglected or established in a very rough manner which could only indicate the most obvious changes of magnitude. Influx diagrams have been made possible and eminently meaningful by modern, independent, dating methods.

The technique of APF analysis has three phases: (1) the establishment of the absolute pollen number on the slide; (2) the establishment of the absolute pollen number in the deposit (concentration); (3) the establishment of the absolute pollen number within a time interval (influx). The last figure is the meaningful one in an ecological context.

The problems of the first phase have caused some difficulties, which have now largely been overcome; they are discussed in Section 5.3.3. The problem in the second phase is mainly to define the volume of the sample with sufficient accuracy. This is a sampling problem and has been dealt with in Section 5.3.1.

Phase 3 is the most problematical. Geologically, a time sequence is equivalent to a stratigraphic sequence: vertical distance equals time. The transfer data, i.e. the rate of deposition, can vary enormously, even within the same basin and short periods. It is therefore necessary to obtain dense datings to be able to draw inferences about rate of deposition. Such datings will in most cases mean radiocarbon dates, either directly or indirectly by inference from other dated deposits. The inherent weakness of radiocarbon dating makes a certain caution necessary, especially if there is a close correlation between assumed rate of deposition and calculated influx values (Liu and Colinvaux, 1988). Only rarely is one able to utilize a varved, self-dating sediment (Welten, 1944) or other types of absolute dating.

Modern influx data for the calibration of the influx pollen diagrams can, in theory, be obtained by trapping deposited pollen grains. As discussed in Chapter 2, trapping is subject to many difficulties, the most important of which is that pollen records from moss cushions, etc., are usually not dateable in number of years of deposition, nor are the effects of degradation of pollen known. Alternatively, yearly trapping in artificial traps fluctuates so much that one must integrate over a sequence of years. To control the effects of different dispersal one must also establish gradients from the pollen source outwards. Such data are not freely available anywhere. They may become reality with the increasing need in the future, because of the development of influx diagrams. A third method is to calculate, like Pohl, the actual production of individual trees, and calculate deposition data from the absolute productivity figures.

However, Davis (1976b: 227) has already expressed some doubt about the consistency of APF disposition and influx dates in lake sediments. These doubts have been amply justified: the various processes taking place during sedimentation in lakes and redeposition of sediment before final settlement: water-borne influx, wind-generated resedimentation, etc., cause great and for the time far-from-controlled irregularities (cf. Seiwald, 1980), some of which may be due to deficient control of short-period fluctuations in sedimentation rate.

Pollen diagrams from peat (whether influx or percentage) are almost impossible to interpret in all details if they cannot be checked by comparison with sediment diagrams. The local pollen production of peat bogs does not influence NAP only, but may also influence the AP data in many unforeseeable ways. In ordinary work (percentage diagrams) peat should not be analysed if sediment is available, notwithstanding the stratigraphic problems encountered in sediments. In some spectra influx diagrams from peat bogs are more regular than those from sediments, as no or very little resedimentation takes place. Middeldorp (1982) has reversed the use of APF by assuming that influx is relatively constant as long as no major vegetational changes take place. In this way the main changes in pollen deposition register differences in sediment accretion. Relative figures can then be translated into absolute ones by more far-spaced datings than without this help.

Obviously the exotic marker method (Section 5.3.3) can be used for quantitative pollen counts in all of the many substances yielding pollen data. In ordinary pollen analysis this automatically gives absolute pollen numbers per slide. As it is also becoming increasingly customary to use a standard volume of material for analysis, concentration values are also established in routine analysis with very little extra labour. The point outside direct control of the pollen analyst is the establishment of sedimentation rate, i.e. dating. In most cases one must contend with ^{14}C dates at some distance from each other, and establish average rates from such data. This assumes uniform sedimentation rate, which is unproven. Discrepancies at this point may explain the unexpected variations observed in some influx curves. A control such as Middeldorp's (above) might help in such cases.

Pollen influx diagrams are especially useful under marginal conditions: arctic–alpine timberlines, desert–steppe–forest gradients, etc. Many of the difficulties earlier authors have encountered in interpreting such diagrams are results of insufficient biological training and treatment of pollen data as mathematics, not as biology. However, in other cases influx diagrams are necessary to decide questions that have been difficult to answer from percentage diagrams, like the ones quoted above.

The APF gradient across the timberline, which is a parameter of great importance, is usually distinct. In northern Canada, Ritchie and Lichti-Federovich (1967) found between five and 700 grams per cm^2 a year in the tundra as against $2\frac{1}{2}$ thousand to $11\frac{1}{2}$ in conifer forest. Within a vegetation type, individual registrations vary by orders of magnitude. Values corresponding to the above have also been registered elsewhere. Differences of pollen production in uncultivated and cleared land represent another area where APF data may be important in the interpretation.

Other data indicate that the pollen deposition in lakes in the tundra is of the magnitude of 10^3 grains per cm^2 and season, whereas it is about ten times higher in the temperate forest (M. B. Davis, 1969; Craig, 1972). With due consideration for the paucity of data, these figures are a little higher than one would expect from the registrations recorded by Ritchie and Lichti-Federovich (1967). However, data are still too sparse for a global synthesis, and most of those available have not been collected especially for this purpose. The presentation of influx data in diagrams is discussed in Chapter 6.

7.1.2.2. Interpretation of percentage diagrams

An all-important element in the interpretation of percentage pollen data is the *basic pollen sum*, mentioned already in Chapter 1. As stated there, there is nothing like a single, 'correct', basic sum. The pollen sum, the basis for the calculation carried out on the sample, is an instrument for the study of a question at hand and must be constructed accordingly. The basic sum for the study of regional forest development is different from the one needed to study the agricultural occupation of the inhabitants of an archaeological site, and that is again different from the one forming the basis for a study of the development of a small forest area under various influences. Even the last-mentioned case can be studied under various angles, each of them represented by a separate basic sum (cf. Fig. 7.5). Especially in an area the vegetational development of which has not been previously established, and where methods of study are untested, it is a sound idea to start with an all-inclusive sum and reduce according to experiences.

Only one thing is certain: by mechanical adherence to one basic sum concept one is bound sooner or later to lose information which would have come up if another presentation had been chosen; compare the role of *Juniperus* in the bottom part of Fig. 6.6, compared with other diagrams representing the same analyses, but calculated on different bases.

Some taxa are unsuitable for inclusion in any basic sum, especially those that show extreme and

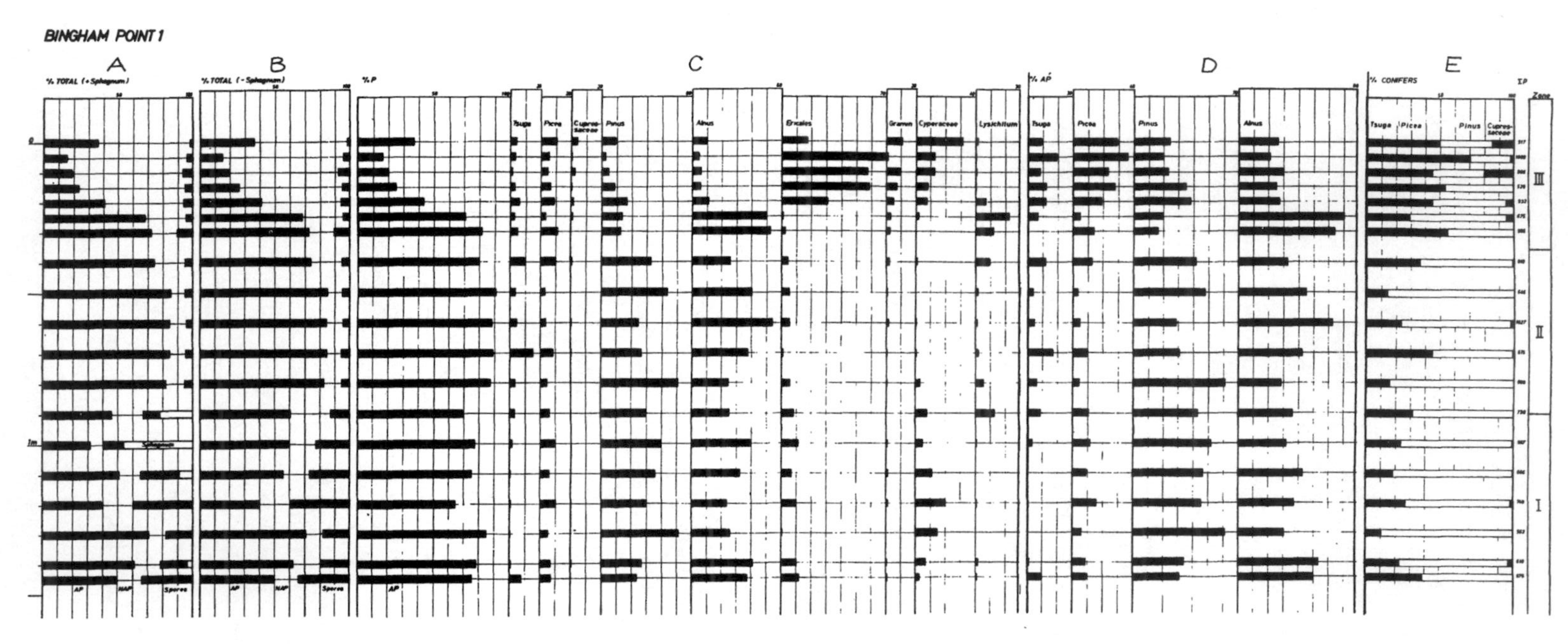

Fig. 7.5. Use of various basic sums in the analysis of development in a small, forest-less area at the Alaskan coast. A: All spore and pollen taxa included. *Sphagnum* is decidedly local and taken out in B. The other (fern) spores are probably forest-bottom produced and of little interest for the problem at hand. Sphagnum has been taken out in C, which is a regular biostratigraphic diagram. The top five samples show the disappearance of local forest and replacement with heather and rushes. The forest-less area is established. D is an AP diagram to study the change in composition of the forest surrounding the area. The increase of *Picea* (*sitchensis*) in the topmost five samples is obvious. *Alnus* is a forest-edge tree bordering the area. Taken out, E shows the composition of the conifer (i.e. regional) forest, and shows recession of *Pinus* (*contorta*) in the top samples. Diagrams A, B and E are cumulative, C and D resolved. The top frame is stepped to separate the individual curves. From Fægri, 1968.

abrupt variations, occurring in (single-sample) maxima which depress and distort all other curves. If this is a unique case, and especially if its origin can be inferred, it can be compensated for by a simple calculation like the one discussed p. 142, again keeping in mind the danger inherent in all tampering with data. More persistent cases are better treated by calculation on the basis of a special sum. The old way (already in von Post, 1918) to calculate percentages on the basis of a sum in which the relevant taxon did not form a part, is not to be recommended.

Figure 7.6 illustrates a very simple, but frequently overlooked, arithmetical consequence of the percentage calculation: if two constituents, A and non-A, produce equal quantities of pollen per area unit ($x = 1$), and they constitute the pollen rain in the proportion of 90 % A and 10 % non-A, the area representation is obviously the same: 90/10. If A doubles its output while that of non-A remains constant ($x = 2$), the pollen representation of A rises to 95 %, while B is reduced to 5 %. Thus, the rise from 90 to 95 % represents a doubling of the output of A. Figure 7.6 shows that this

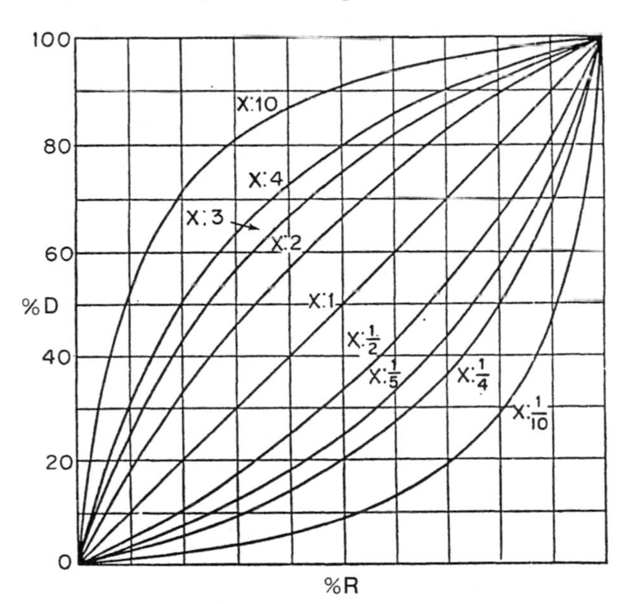

Fig. 7.6. Percentage representation, D, of a constituent, A, in pollen spectra corresponding to the representation, R, of the same in the vegetation provided A produces x times as much pollen per area as non-A.

'law of diminishing return' is most pronounced near 100 %. The ecological counterpart of this problem would be if species A began to flower more profusely without changing its area, or if it colonized an area which previously did not produce any pollen, or the pollen of which did not enter into the calculation. This would be a rather exceptional ecological situation. More frequently the expansion of A takes place at the cost of non-A. In this case relations are more complicated. As long as the ratio between pollen production of A and non-A remains low (below 1 : 4), an increase in area (with the same productivity) of A will give a substantial increase in the diagram percentages too. If the ratio is higher, increases above 20–25 % are poorly registered (cf. Fig. 7.7). As discussed later, numerical reduction to a certain degree compensates for this.

Actually, conditions are more complicated since at least non-A, and possibly A as well, are usually heterogeneous mixtures, the composition of which changes simultaneously with changes in the A/non-A ratio. The general effect of this will usually be to lessen discrepancies deriving from the percentage calculation.

If we visualize a forest consisting of equal parts of oak and pine, and we use the pollen production figures quoted, we find that the corresponding spectrum will contain 15 % oak, 85 % pine. If beech is substituted for pine (apart from the botanical improbability of that succession) the same quantity of

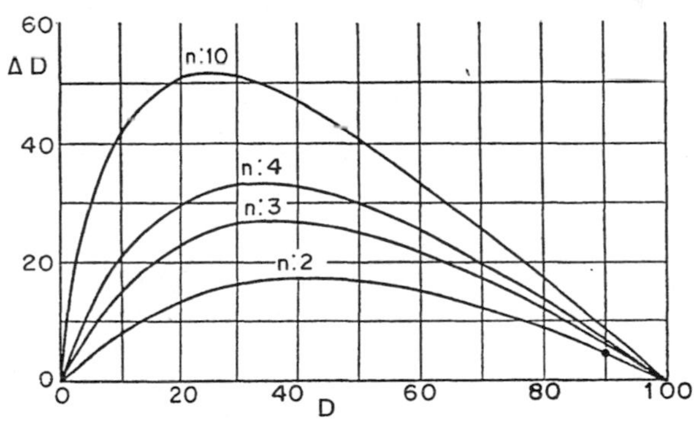

Fig. 7.7. Increase in pollen percentage, ΔD, when the pollen output of a constituent previously contributing D % increases by a factor n, other conditions constant. The example quoted in the text (doubling of the influx of a 90 % constituent) is indicated by the circle on the $n = 2$ curve.

oak will give 60 % of the pollen as against 40 % beech. If the beech is then replaced by a tree, e.g. *Acer* spp. or *Populus balsamifera*, which is scarcely, or not at all, registered in the spectra, we shall find almost 100 % oak pollen, although the quantity of oak has not changed at all. It is necessary to take into account not only the curve under discussion, but the others as well.

7.1.3. Reconstructing the vegetation

7.1.3.1. Adjusting percentage diagrams for better correspondence with vegetation

As indicated in the preceding section the reconstruction of vegetation from absolute data is no easy task in spite of the 1 : 1 correspondence between pollen deposition and occurrence of the species in the vegetation. In percentage diagrams this problem is aggravated by the numerical effects of percentage calculation. If percentage pollen data are to represent adequately the percentage composition of the corresponding vegetation, the variation of R values should be taken into account. Because these values are not consistent, but vary with the external conditions, there is no point in going into too much detail. Also, some R values are so excessive one way or the other that their introduction would completely upset the diagram. What can be done is to use a rough-and-ready scale with one correction figure for the large producers, another for the very small ones.

Iversen proposed (in lectures, cf. 1947a: 242) a numerical reduction based upon a division of pollen types into three groups. A, B and C. The A group comprises species that contribute great quantities of pollen, i.e. (in northern Europe) *Pinus*, *Betula*, *Alnus*, *Corylus*. The B group comprises those that contribute moderate quantities: *Picea*, *Quercus*, *Fraxinus*, *Fagus*, etc. Some of the species of the B group (B_2), while still undoubtedly belonging to the diagram proper, e.g. *Tilia* or *Hedera*, are exceptionally small producers or their pollen is ineffectively spread because a majority drops to the ground with the flowers and is never released into the pollen rain. The C group comprises pollen types that are so scarce that it is immaterial whether they are included in the pollen sum or not. The species in question produce and/or release very little pollen, many of them being strictly entomophilous or even autogamous. The fact that many of them are also very scarce contributes to their numerical insignificance in pollen analysis. Nevertheless, many of them may be of great indicator value, e.g. (in northern Europe) *Ilex*, *Viscum*, *Vitis*, *Lonicera periclymenum*, and if used as indicators they should always be presented in special diagrams. In other regions various other pollen taxa play a similar role. As previously mentioned, many of them should, for theoretical reasons, not be included in the basic pollen sum, indeed hardly calculated as percentages at all.

If we want our pollen diagram to give a more adequate representation of the real composition of the vegetation the fraction of the A group can be reduced, e.g. by a division of the pollen numbers by a correction factor before they are added to those of the B group to make up the pollen sum. The B_2 should be multiplied in the same way. The principle is illustrated by Table 7.2.

When discussing the position of a pollen type in relation to the groups mentioned above (A, B, B_2, C) one must, besides pollen productivity, also take into account the effectivity of dispersal of pollen (*Tilia*, *Abies*), the periodicity of flowering, the ratio between male and female flowers, etc., especially if NAP is also included. The aforesaid factors may counteract one another, e.g. *Calluna*: great production, ineffective dispersal. Whereas the reduction figure 4 has proved adequate for dealing

Table 7.2 Recalculation to compensate for pollen productivity.

	Number of grains	Original percentage	Number (and percentage) after conversion
Pinus	120	54.5	30
Betula	24	10.9	6
Quercus	16	7.3	16
Tilia	4	1.8	16
Fagus	24	10.9	24
Corylus	32	14.6	8
	220	100.0	100

with Danish forest trees, there are of course species or genera that would require different treatment. Such a converted spectrum gives a more adequate representation of the actual composition of the vegetation (cf. Iversen, 1949). Fagerlind (1952) has pointed out that such conversions will in practically all cases reduce the mathematical errors inherent in the percentage calculation. To what extent such converted spectra should replace ordinary ones is a matter for discussion. In any case the original counts must be made available to the reader. The main use of diagrams with converted values would be for demonstration purposes, or in papers intended for non-specialists, including pollen analysts not conversant with the vegetation under scrutiny.

Pollen production also varies with the density of the stand; one tree in an open position may produce more pollen than dozens of trees of the same species in a dense stand (cf. Borse, 1939: 129). It is probable that area production figures are more constant than those of production per tree, but this subject is not sufficiently known yet. Roadside trees in cultivated areas, especially elms, produce enormous quantities of pollen and distort the composition of the recent pollen rain. Surface samples from cultivated areas frequently show too high percentages of anemophilous roadside trees in the region. Heusser (1978) has registered excessive *Alnus* (*rubra*) values in Oregon, and Fægri (unpublished) the same in Washington. In both cases *Alnus* was a very subordinate constituent of the forest, but flowered extremely well in open positions, forest edges and isolated trees in the fields.

Human methods of husbandry may both prevent and increase flowering. However, also under more natural conditions perennial plants may persist for very long periods in a vegetative state outside the area where they flower. Fröman (1944) has maintained that *Hedera* in middle Sweden has persisted in a vegetative state since the post-Glacial climatic optimum.

Pollen of more or less cleistogamic plants is also badly underrepresented, if represented at all. Unfortunately, some of the most important cereals belong to this group and disperse pollen for a very short time only (Pande *et al.*, 1972). A great part of their pollen output is spread at harvest time (Vuorela, 1973), and may be recoverable then, or is recoverable in cereal products (Greig, 1982: 58). It is not known when the ancestral forms of modern cereals became cleistogamous, probably fairly early, as their registered pollen outputs have never been comparable with that of the obligately outbreeding *Secale*.

The problem of defining the composition of the reference vegetation also interferes with calculation of R values. Hansen (1949) has shown that basal area, number of trees 1 inch and over DBH, and number of trees 10 inches and over, give completely different R values when compared with recent pollen from moss cushions in a *Pseudotsuga–Quercus garryana* forest. Oak is in one case underrepresented by 43 %, in another overrepresented by 19 %. *Acer* is permanently underrepresented, however.

If the vegetation is to a great extent composed of species the pollen of which is not found in fossil deposits (e.g. because the species in question are exclusively insect-pollinated), the pollen flora will give not only an incomplete, but also an essentially deceptive, picture of the composition of the forests. In Europe this source of error is negligible, but in North America the difficulty of recovering the pollen of some *Populus* species from the deposits represents a serious imperfection (Erdtman, 1931). In tropical regions the greater number of zoogamous species presents great difficulties. The theory of tropical pollen analysis is still far from understood. If the vegetation of an appreciable, but incalculable, part of the area is not represented in the pollen rain, the pollen-analytical discussion of the remainder of the vegetation on the basis of percentage records is of limited interest only. Influx dates, if available, would be meaningful for those species that do disperse their pollen.

Cultivation of an area may distort diagrams and substitute, e.g., a long-distance transport AP diagram instead of a local one if cultivation is sufficiently intensive. Change in sediment or peat type may introduce changes in the curves—particularly those of NAP, but even in the AP curves, if trees are constituents of the mother formation. Forest fires may cause changes under natural conditions. These are probably reversible and of short duration, except in regions of vegetational instability in which a forest fire may lead to the realization of inherent tendencies in the development. All these and many other factors of local significance should be taken into account in the interpretation of a pollen diagram.

As a consequence of all these complications it is impossible to establish R_{rel} values of general validity. They are only valid for the actual situation from which they were calculated (cf. Comanor, 1967).

The best check on the representativity of a pollen diagram is obtained by comparing the topmost sample with recent conditions. Care should therefore be taken to obtain samples from the topmost, recent and sub-recent, layer of the deposit, even if recent spectra suffer from the drawback that contingent differential destruction of pollen grains with incorporation in the sediment has not yet taken place. Ritchie and Lichti-Federovich (1963) have demonstrated that such spectra may contain pollen grains which would not have been recovered if the sample had been left to fossilize. The recent pollen rain contains *ca.* 30 % *Populus*, in good accordance with the composition of the forest, but less than 2 % is recovered from recent sediments. However, if such cases of non-preservation (or nonrecognition) are taken into account, recent pollen rains, as preserved in sediments under formation, are fairly good indications of the conditions as found by analysis of fossil deposits. On the other hand, recent samples collected from moss cushions, etc., represent a much drier substratum than peat or sediment, and differential destruction may have advanced further than in deposits used for regular analysis. Consequently, accumulation of relatively resistant grains may take place. Besides, recent moss cushion spectra will in many cases be subject to local NAP overproduction, different from that of the deposits, and may be unduly influenced by the nearest trees (Potzger *et al.*, 1957).

The occurrence of any taxon is subject to constraints: climatic, edaphic, historical. The presence of a taxon therefore indicates that such constraints are not operative at that place and that time. On the other hand, non-occurrence does not immediately indicate anything about which constraint is effective. For example, absence may be due to historical factors, while climate would have permitted the taxon to grow there. With this reservation every taxon is an indicator, the usefulness of which varies immensely. It may be highly significant in one context, of little interest in another. *Hedera* or *Quercus* mean something quite different in southern England and western Norway. A birch forest is not particularly interesting in Scandinavia, but significant in southern Europe. The latter example also demonstrates that whereas a taxon (birch) may be of little indicative value if represented by individual specimens, it may be significant if it dominates a plant community. The general rule is that a taxon or a plant community is more indicative the nearer one is to the limit of its area of occurrence.

In rare cases pollen (and macrofossils) indicate the presence of character or differential species (in the phytosociological sense) which signal more or less well-defined plant communities (cf. e.g. Rybničková and Rybniček, 1972), the climatic and edaphic demands of which are well studied and defined. Usually, one must contend with less precise indications, but any plant assemblage defines its ecological niche by a combination of the constraints of its constituents (cf. Fig. 9.4). If the assemblage registered by its pollen deposition can be shown to be identical with a living plant community, it may be taken to represent the same habitat type—climatic and edaphic. However, many assemblages have no known modern analogue, which has disturbed many investigators. We cannot accept this as anything remarkable: the development of soil profiles and plant communities (by immigration of new species) present so many variables that the probability of finding analogues in time or in space must remain small. One is tempted, with Janssen (1980), to speak about 'the type of objectivity that consciously does not want to take into consideration the (admittedly poor) insight we have gained in more than 60 years of palynology', and as a result take refuge in 'numerical analyses which show the obvious or arrive at a zonation that for our purposes is unsuitable'. Interpretation of a pollen diagram is botany: it cannot be replaced by mathematics, and it needs the courage to make extrapolations.

Ecological groups in pollen diagrams are often

more clearly apparent after autoscaling (cf. Fig. 6.10). After autoscaling, curves are directly comparable and several curves can be added up to a composite (average) curve for a plant community or another ecological grouping, notwithstanding primary differences in pollen productivity of the individual taxa.

In the autoscaled diagram all pollen types are presumed to be equally important, which they certainly are not, and the diagram is useful only for discussions of trends, not of quantities. Autoscaling gives best results in fairly complete curves, based on values with low statistical noise, but can be used also with fragmentary curves of large relative errors, even if results are then rather crude (cf. Fig. 6.10). Unless one is very careful, autoscaling is a fertile field for circular reasoning.

7.1.3.2. *From diagram to vegetation*

As seen from Fig. 1.1, the vegetation is the central object in a pollen-analytic investigation. The pollen is produced by the elements of the vegetation and the presence of pollen permits inferences back to the vegetation that produced it. For further conclusions about ecologic (s.l.) relations, understanding of the vegetation is essential.

The information derived from a pollen analysis is a definition of the relevant vegetation by means of pollen spectra and diagrams. The various vegetation elements may be defined at different levels.

The immediate result is a record of the flora, of the species that grew at or near the site when the deposit was formed. If the ecologic demands of the individual species are known it is possible to draw inferences from this simple list. One limiting factor is the taxonomic level down to which the pollen can be identified (cf. the keys). By morphological criteria alone it is often not possible to come down to an ecologically meaningful level. If other criteria are brought in to sharpen the taxonomic identification, one may obtain more precise results, but then the danger of circular reasoning is imminent.

Also, in a simple flora list the indicative value of all species is equal. Weighting must be based on other experiences and criteria, again with the danger of circular reasoning. By introducing a quantitative element one may get a better picture of the composition of the vegetation. The area covered depends on the deposition conditions of the pollen, as discussed in previous chapters. At any rate the similarity or dissimilarity with recent vegetation adds to the indicative value of the flora as a whole.

Again, a more penetrating (syn-)taxonomic identification of vegetation units makes the record more indicative. This is especially the case if individual plant communities (phytosociological units) can be identified, which usually depends on the finding of character or differential species. If a plant community has been identified by means of such species one may—with reservations—postulate also the presence of such species that belong in the community but have, for one reason or the other, not been recorded in the pollen data. The ecological indicator value of the vegetation record is correspondingly increased (cf. Rybničková and Rybniček, 1971, 1972, 1985, and Smit and Janssen, 1983). The sharper the vegetation unit can be defined, the more indicative it is, and also more indicative than the combined record of all species (without weighting).

However, for a numerical analysis it is safer to base inferences on the wider ecological demands of individual species than on the more specific ones of plant communities. The former only presumes that the species in question is found within its whole potential area, and that it has not changed its ecologic characteristic in the meantime. The latter in addition involves so many qualifications that the possibility for a biased selection is a real danger (cf. Birks, 1981) and also the lack of a modern analogue which can act as a calibration is a drawback.

7.2. Migration of plants and communities

So far, the reconstruction and interpretation of vegetation has considered static, regular conditions. In nature this is not so. Dynamism is always present. Aberrant conditions are frequent. In the following sections we shall first deal with some of the effects of the inherent dynamism of the vegetation cover and its expressions, then with more aberrant cases.

Migration of plants and plant communities

may have many causes, most of them ultimately going back to effects of climate or to human interference. As these two forces are the ones usually standing in the focus of interest in pollen analysis, it is important to find out how much may be due to the one and how much to the other.

Immigration of flora and vegetation elements to a region is one of the most obvious effects of migration processes. Such immigration may be recorded directly by finds or surmised via phytosociologic reasoning (cf. the preceding section). Registrations of immigration can only be interpreted in the positive mode: the find of a pollen type means that the taxon in question occurred inside the area from which such pollen could come. The opposite conclusion is not permissible. However, there are great differences with regard to individual taxa, apart from those of taxonomic attribution. Pollen of common and widespread taxa indicate an *earliest* date of immigration, but it is difficult to decide if the first faint traces are due to long-distance transport or to a sparse local occurrence (or contamination, cf. Zoller, 1960a: 63), alternatively when the representation in the diagram is sufficiently strong to give positive indications of local occurrence. More or less arbitrary figures have been quoted and, while they are undoubtedly valuable as indicators, their value must not be overemphasized. They depend on the whole context, statistical competition with other pollen producers in the same universe, etc. In a forested area the threshold value for *Pinus* is tentatively supposed to be 5–10 %, for *Quercus* perhaps 1 %. In a non-forested area the same percentages would not have any indicative value regarding local occurrence.

Taxa which are rarely and/or sparingly represented in the ordinary pollen spectrum represent the opposite case: their pollen is so sparingly produced and/or badly dispersed that one may be fairly certain that, if their pollen is found, they did occur locally. On the other hand, its presence is so fortuitous that nothing can be concluded from absence. Thus the find of such pollen grains establishes the *latest* date of immigration.

In comparison with that furnished by common pollen types the 'incidental' information of the presence of species that are not regular contributors to pollen rain suffers from many of the disadvantages that characterize the evidence provided by macro-fossils (cf. Section 1.1.1), but it may nevertheless be of great value. This refers particularly to species that are otherwise rarely met with as fossils—but the pollen of which can be expected to appear if the species has been present in reasonable quantities, and if a sufficiently large number of pollen grains has been counted. *Viscum* is a good example. Once the presence of a species outside its present area today has been ascertained (through macro- or microfossils), conclusions may be drawn from this fact, though not necessarily climatic conclusions.

Especially in the Late-Glacial and early post-Glacial the slow rate of immigration of many tree species has caused a great deal of doubt if the well-known dominance sequence should in principle be interpreted in terms of immigration succession or in terms of climate. Particularly in northwestern Europe there are indications that migration effects play a greater part than usually surmised, and may not only conceal, but also imitate, the effects of climatic changes.

As long as taxa have not reached their climatically conditioned limits, their presence and especially absence cannot be used in a climatic argumentation. The poverty of the northern European early post-Glacial forests may be more due to succession phenomena than to climate (Fægri, 1940: 46; Iversen, 1954b). The varying distance of the areas of investigation from the glacial refuges of a slowly or intermittently spreading species results in a highly metachronous arrival in these areas. The beginning of, for example, the *Picea* curve in European diagrams, locally a highly important indicator level, is thus of greatly differing age in different areas. Radiocarbon dating has definitely confirmed earlier suspicions, and shown that earlier concepts were fallacious (Hafsten, 1985).

Sometimes reaction lags are unexpectedly great. Between two lowland stations in the isle of Skye, some 10 km apart, Birks (1973: 366) postulates a delay of more than 2000 years in the establishment of birch forest. If the establishment of birch forest had been unreservedly accepted as a chronologic marker, rather serious mistakes would have resulted. In this case, conditions were apparently just balancing at the edge of the critical level between birch forest and no birch forest.

A more large-scale demonstration of the same kind is given by Smith and Pilcher (1973), who show that most established pollen-analytic marker levels are metachronous with up to a couple of thousand years of difference the length of the British Isles. Characteristically, the only ones that are synchronous are negative: decline of *Ulmus* and *Pinus*. Obviously the successional lag does not have the same effect on retreats as on advances. Both these examples show the danger in relying on vegetational data alone without any control from (semi-)independent evidence: absolute chronology, glaciological data, land-sea relations, historical dates, etc.

A very convincing example of the effect of delayed immigration of a tolerant species has been given by Welten (1952), from Switzerland. The catastrophic change-over from QM to dominance of *Abies* in the 600–1600 m altitude interval during early Neolithic time (cf. Fig. 7.8) cannot easily be explained as the effect of a climatic change of corresponding magnitude. As long as there was no competition with *Abies* the QM components dominated all the way up to 1600 m, but after immigration of *A. alba* they could no longer compete, and were crowded out from an area which they might, for climatic and edaphic reasons, still have inhabited (cf. Zoller, 1964). This does not preclude the fact that migrations and successions observed, at any rate in some cases, may have been released by climatic or edaphic changes that have disturbed the ecological balance and accelerated migrations; but their vegetational effects may be out of proportion to the magnitude of the releasing factor.

A case similar to the one described by Welten (1952) may be that of *Picea* in Scandinavia. During the hypsithermal period this tree was 'held back' at the northeastern border of Fennoscandia (Nejshtadt, 1957); later, it expanded vigorously. However, to what extent the expansion was due to climatic change or merely reflected a change in competition has not, so far, been ascertained (cf. Moe, 1970).

The decline of a pollen type, even if synchronous within an area, is not necessarily a direct climatic effect. Statistical effects may mask vegetation development, agricultural measures and suc-

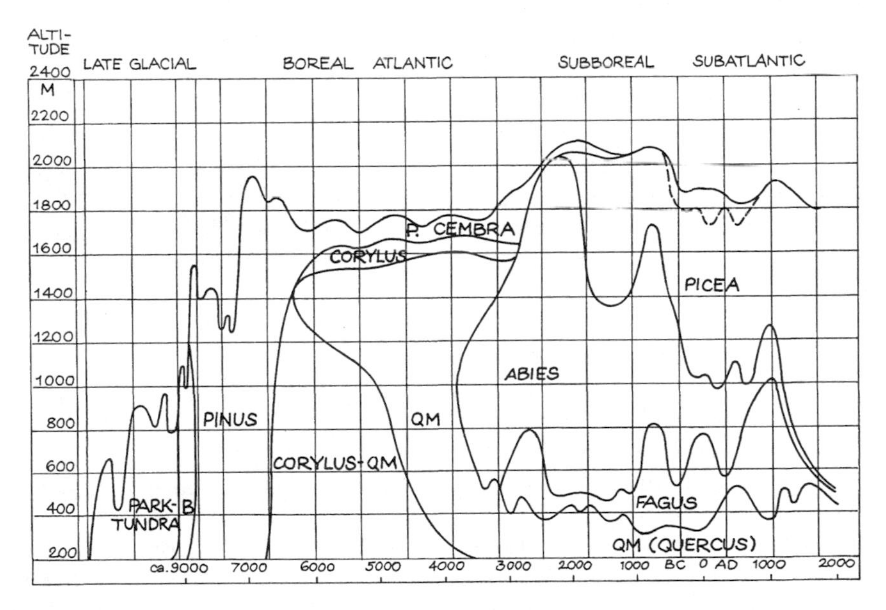

Fig. 7.8. Change of dominance of forest trees with altitude during Late and post-Glacial in Switzerland. After Welten, 1952. Redrawn. Reproduced by permission of the copyright-holder: Geobotanisches Institut ETH Stiftung Rübel in Zürich.

cessional phenomena may influence the vegetation picture, and the introduction of a new pathogen may establish a new equilibrium in which a former dominant may be repressed. Especially if the pathogen is transported outside the area where potential hosts have acquired a certain immunity the effects can be widespread and almost instantaneous, as shown by recent evidence in plant pathology (elm disease, chestnut blight). Ever since humans started to move 'faster than nature' there has been a possibility that pathogens could be transported that way with or by humans.

7.3. Altitude effects—tundra and desert

Still within the conceptual framework of the general pollen rain is the influence of elevation on pollen recruitment in mountains, since the dispersal of pollen across a short distance may bring it over and into a different vegetation area, with the possibility of deposition in an assemblage of completely different composition. Vertical pollen transport has been discussed since the days of Rudolph and Firbas (1927). Since plant communities of higher elevations usually produce less pollen than those of the lowland, the lowland pollen, even if attenuated, may make itself strongly felt. Markgraf (1980) found that during daytime there was a positive transport of lowland pollen to higher altitudes, but found no appreciable downward transport. Summit areas are under the influence of gradient winds, not of local updraughts. Other investigators (Hamilton and Perrott, 1980) postulate a downdraught as well. Probably, conditions vary from one mountain massif to the next, both with orography and exposure to gradient wind direction. According to Flenley (1973) conditions in tropical mountains are similar.

For reasons given above, one cannot expect any correlation between the registration of (arboreal) taxa in the pollen rain and the actual (alpine) vegetation. Frenzel (1969) summarized observations on the registration of arboreal pollen in, respectively, alpine, tundra and arid (steppe and desert) vegetation. In the tundra, like in the alpine region, there is no correlation at all between AP percentages and distances to the nearest forest, with respect to altitude (Fig. 7.9) whereas in arid communities there is a distinct negative correlation (Fig. 7.10). This would indicate that the latter communities produce more local (NAP) pollen than the tundra, and therefore depress the AP values. This tallies with the observation of not inconsiderable local pollen registrations even in Sahara across a 1300 km traverse (van Campo, 1975).

The influence of locally produced NAP in alpine deposits varies both with the character of the alpine vegetation, and with that of the lowland underneath. According to the variable local pollen production of the alpine zones the NAP percentage in alpine deposits is differentiated. In the low-alpine belt it is usually so high that a distinct dip in the AP/NAP ratio characterizes the timberline. The presence of special alpine indicators may give additional information distinguishing between alpine and inferalpine vegetation; so do APF data if available. Nevertheless, a certain circumspection is necessary in the interpretation of diagrams from mountainous areas. In the very low-productive high-alpine zones pollen data are distinctly misleading (Fig. 7.9). As long as one moves within and

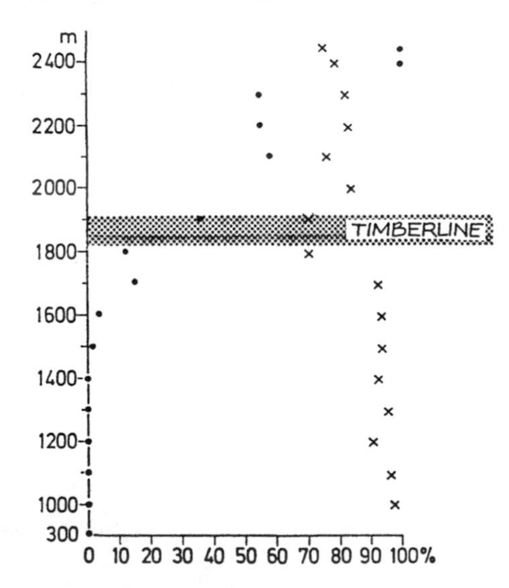

Fig. 7.9. AP percentages (crosses) and percentage of long-distance transported pollen (dots) in relation to altitude and timberline in a temperate mountain (Roumania). After Frenzel, 1969; redrawn. Reproduced by permission of the copyright-holder: Springer (Heidelberg).

between vegetation types with comparable pollen production, conditions are usually not serious and vegetation belts fairly well differentiated. If one community produces much less pollen than another, the former may be difficult to distinguish in the pollen percentages.

7.4. Regional parallelism

Even within the same climatic area there is a regional parallelism between diagrams influenced by different edaphic conditions. The former vegetation types of a sand and of a clay area differed, just as the recent vegetation differs, and their pollen diagrams differ correspondingly. If, however, a (climatic) change has set in during the period covered by the diagrams, the corresponding effect on vegetation will show in both areas, though not necessarily in an identical manner.

The trend of curves will be identical in regional parallelism, but actual values differ. If edaphic differences are very strong, trends may even reverse and become mirrors of each other, e.g. different reaction on humidity changes between a dry and a wet habitat, or between a super-humid and an arid climate (sun-exposed and shaded valley side). The more local are such diagrams the more distinct are the differences.

Though obvious to anybody who realizes that the interpretation of pollen-analytic data must be in terms of vegetation, investigators sometimes overlook the fact that the same climatic event may be reflected in quite different vegetation developments in pollen diagrams from different areas.

Parallelism is illustrated by Figs. 7.11 and 7.12; Fig. 7.11 presents a model of the equilibrium between two plant communities: the more exigent x (shaded) and the less exigent y (open). It is presupposed that they can be recognized pollen-analytically, e.g. like nemoreal broad-leaf and boreal conifer forest. The model also presumes that one (simple or compound) climatic parameter, e.g. temperature, regulates the equilibrium between the communities. Figure 7.11 illustrates what happens during a positive oscillation in respect of the decisive parameter: with amelioration the critical

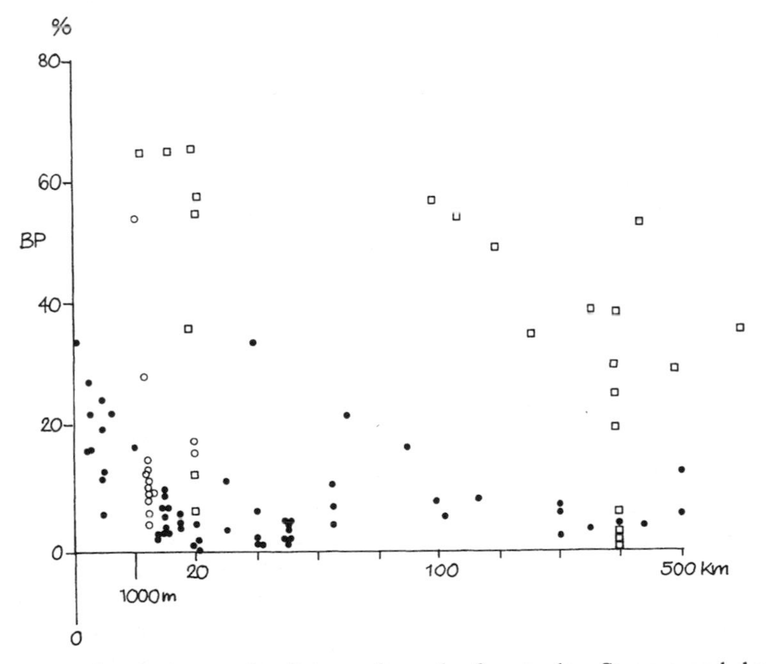

Fig. 7.10. AP percentages in relation to the distance from the forest edge. Steppes and deserts: dots and rings, respectively (water-tank registrations). Tundra: Quadrats. After Frenzel, 1969; redrawn. Reproduced by permission of the copyright-holder: Springer (Heidelberg).

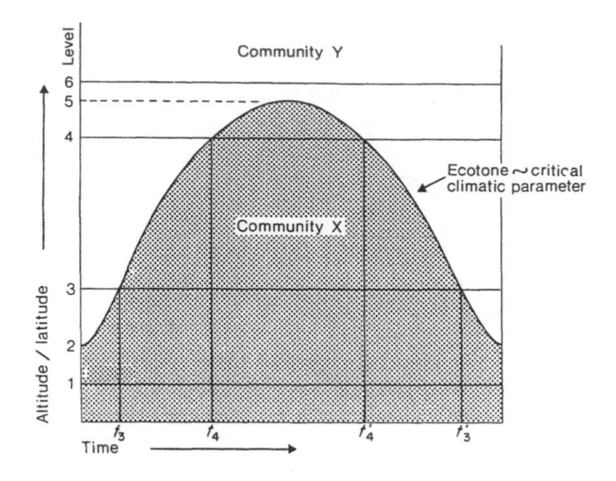

Fig. 7.11. Vegetational development in a latitudinal or altitudinal cline during a climatic oscillation: temporary expansion of community x into the area of community y. Cf. text. Reproduced by permission of the copyright-holder: CNRS, Paris.

value, which separates between the two communities (and which is in the field registered as the border-line, the ecotone, between them) moves towards increasingly higher latitudes (or altitudes). Community x expands. The ecotone which was originally found at latitude (level) 2, is found at increasingly higher latitudes (levels), passes 3 and 4, until community x has expanded to latitude 5. Then recession sets in, and the ecotone moves back to latitude 2 with community y expanding. No account has here been taken of the delay caused by the long reaction time before vegetation is adjusted to climate. Exactly the same model can be used if the decisive parameter is one of, e.g., humidity.

Pollen-analytically, the picture is more complicated. At level 1 nothing is registered: the station is all the time inside community x. At level 3 a long favourable period (t_3–t_3') is registered, when the original community, y, was replaced by the more exigent community x. The same is the case at level 4, but the favourable period is registered for a much shorter period (t_4–t_4'). At level 6 nothing is registered again: the station was all the time inside community y.

The model illustrates how the pollen-analytic registration of *one and the same* climatic event differs in different climatic positions. An investigator working at station 1 or 6 will maintain that there has been no climatic event, whereas investigators at 3 and 4 will agree that there has been one, but will disagree as to its duration and importance.

An important corollary is that the pollen-analytic registration of a climatic event in different climatic zones *is in principle metachronous*, even if we presume instantaneous reaction.

In Fig. 7.12 the model is extended to comprise a double climatic oscillation, and several plant communities (climax zones), defined by their pollen-analytic record.

The vegetational development in the stations during the time t_0 to t_4 can be summarized as shown in Table 7.3.

Whereas both oscillations are registered at stations E and B, none are at A, G, H, I. At stations F and D only the younger oscillation is registered, at station C only the older. The same vegetational change, viz. birch forest–taiga–birch forest, represents the older oscillation in C, the younger in D. As the other oscillation is not represented in either of these diagrams, the danger of erroneous identification is obvious.

Nevertheless it is the same climatic double fluctuation that is reflected so differently. An instructive demonstration of this principle is given by Grichuk and Grichuk (1960, Fig. p. 79) from the USSR.

This model is based upon the strong reactions due to changes in the local pollen rain component. However, when the whole system of vegetation is displaced this has a certain, though minor, effect on the regional pollen component, and may therefore be observed by slight changes in the registration also at stations where no appreciable change took place in the local pollen rain (cf. Fig. 7.13). An actual case is shown in Fig. 7.14; the landnam changes in the *Ulmus* curves further inland are registered also in the local pollen rain at an isolated exposed island where there is no reason to presume that any landnam took place.

Figure 7.12 may also serve as an illustration of the Allerød and Bølling oscillations in European pollen diagrams. Curves F and E show the classical case, Denmark, whereas conditions farther south are represented by curves D–A. Two conclusions

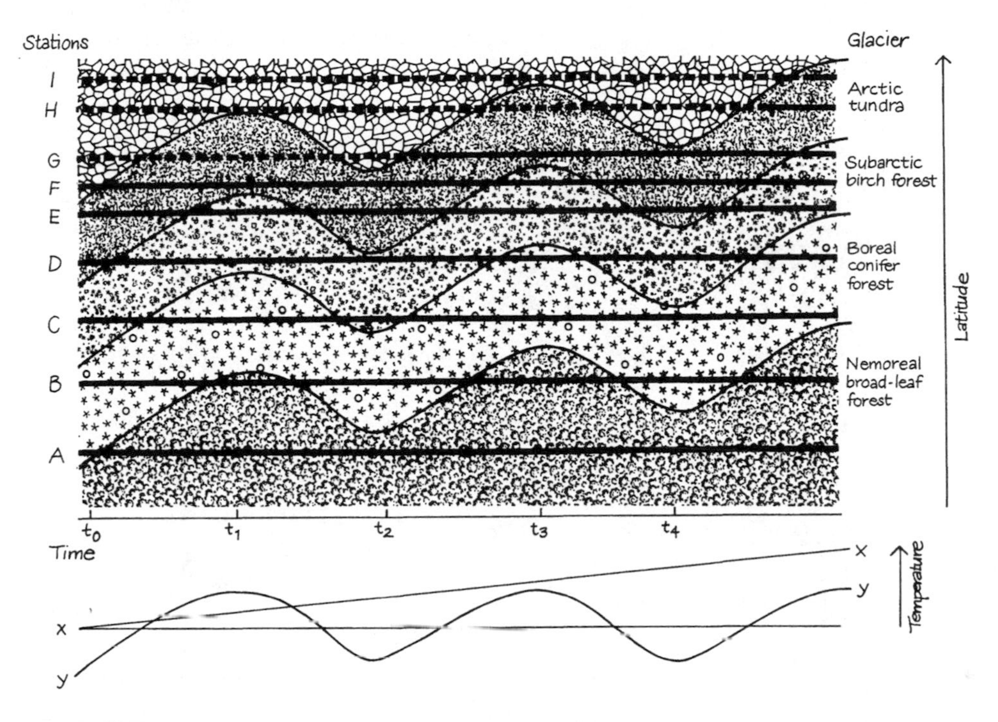

Fig. 7.12. Different registration of the same climatic events in a schematic transect from the centre of a glaciation (top) outwards. Vegetation zones correspond to conditions in northern middle Europe. Heavy horizontal lines: stations at various distances from the glaciation centre. Shadings represent plant communities migrating towards or away from the glaciation centre with changes in climate. The intersection of line (station) and shading (plant community) indicates the plant community at the station at the relevant time, e.g. at station D, there is boreal forest at time t_3. Dashed lines represent periods from which there is no deposit because of simultaneous or subsequent ice-cover. Bottom figure: analysis of the climatic development registered by the migration of communities of the top figure. It presumes a long-time trend (curve *x–x*), upon which are superimposed two oscillations (curve *y–y*).

Table 7.3. Vegetational development in stations A–I, above.

Station	Development	Oscillation(s) registered
I.	Glaciated (no organic deposit) until time t_{4+}	None
H.	Glaciated until t_{2+}, glaciated again at t_{3+}, no deposit until deglaciation at t_4	None
G.	Ice-free at t_{0+}, glaciated again at t_{1+}. Deposits since t_{2+}	None
F.	Tundra–birch forest at t_{2+}—tundra at t_{3+} —birch forest at t_{4+}	Younger
E.	Tundra–birch forest–tundra -birch forest–tundra–birch forest	Both
D.	Birch forest–taiga–birch forest–taiga	Younger
C.	Taiga–nemoreal forest–taiga–nemoreal	Older
B.	Taiga–nemoreal forest–taiga–nemoreal–taiga–nemoreal	Both
A.	Nemoreal all time	None

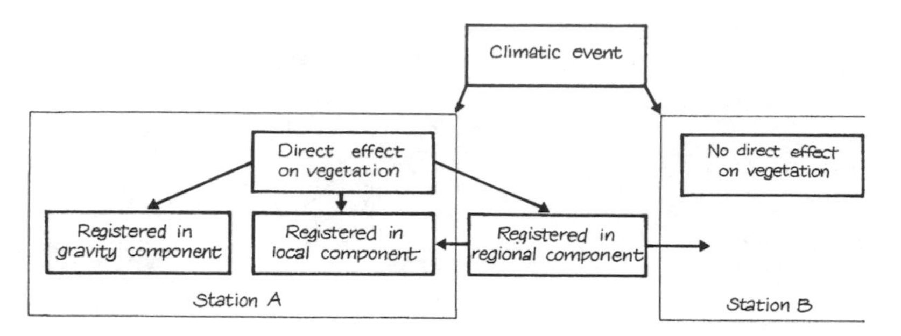

Fig. 7.13. Indirect registration of a climatic event, which produces no direct effect at station B. However, the composition of long-distance pollen from station A changes, which may—or may not—influence also the registrations from station B.

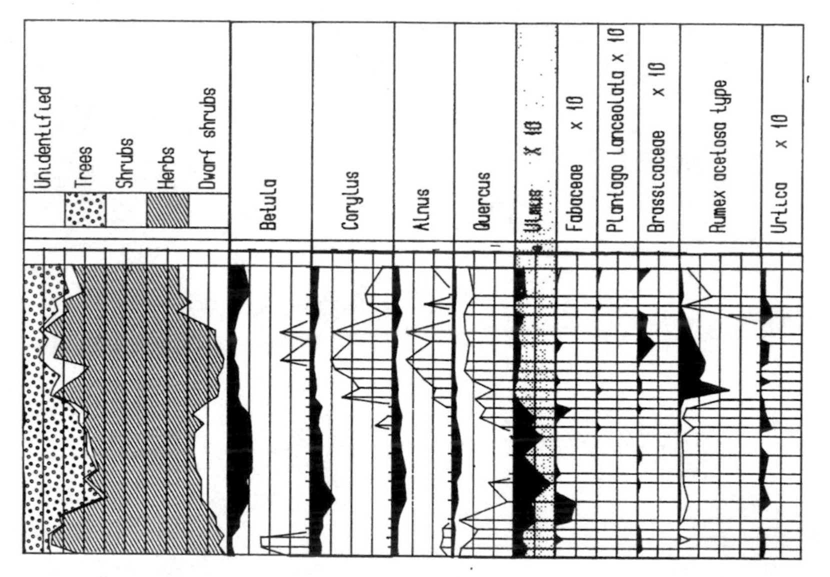

Fig. 7.14. The *Ulmus* curve in relation to some cultivation indicators and/or hemerophilous species at Fedje, an exposed island off the Norwegian west coast at *ca.* 60° N. lat. The elm decline is registered although there is reason to believe that there was no flowering elm at or near that island any time. From R. Danielsen, 1986; redrawn.

may be drawn from this: (1) that the pollen-diagrammatic representation of the Allerød, or any other climatic oscillation, must differ in different climatic areas; and (2) that if an identical development is found in two sufficiently different climax areas it *cannot* be the same fluctuation. Thus, if a 'classical' Allerød (tundra–birch forest–tundra) is found, e.g. in the lowlands of southern central Europe, it must represent a climatic fluctuation older than Allerød (cf. curves C and D).

Figure 7.12 also demonstrates the importance of critical levels. Unless the change in vegetation conditions transgredes a critical level there will be no registration. On the other hand the absence of a registration in a diagram does not mean that no changes have taken place. It is simply that the changes have not been sufficient to cause reaction; alternatively that the vegetation has not been sufficiently sensitive to react in such a manner that the reaction can be recorded.

It is easily seen that a discussion of whether there has been no, one, two or even three Allerød phases is meaningless, even if the dating is correct: climate fluctuates continuously and the only problem is if these fluctuations transgrede across a critical level. The so-called Allerød oscillations were so large that they registered across a large area, but any much smaller oscillation might have given the same registration at any station, provided that it transgreded across such a level. There is a fine parallel in the middle eastern Swedish bog (Snöromsmossen) that registered five Grenz-horizons instead of the usual one or two (Granlund, 1932: 82): it just happened to be in a very labile, reactive state with regard to the triggering effects (humidity variations).

7.5. Some interpretation problems: aberrant registration

In the preceding discussion the underlying assumption has been that the pollen record gave a 'true', undistorted picture of the contemporaneous vegetation within the area under study, i.e. that the numerical relations between pollen registration and the part played by the taxon in the vegetation cover tallied, with reservations for differences in R values.

There is only one fact in pollen analysis that always holds true: a pollen grain of a plant species came from a specimen of that species. What has happened between the time it left the anther until it was recovered on a microscope slide may be anything from just dropping down to a complicated story of transport and deposition and possibly re-deposition. In this sequence of events the 'normal' process may have been distorted in various ways, producing pollen diagrams which, if the abnormalities are not discovered and compensated for, will lead to erroneous interpretations.

In the following, we shall discuss some such potential sources of errors, always keeping in mind that these are sources of *interpretation errors*–Nature does not commit errors. We shall neglect trivialities like errors caused by wrong identifications, wrong arithmetic, contaminated field or laboratory equipment and other unforgivable effects of sloppy and careless work. If properly understood, the other 'errors' in reality carry independent information, not contained in the 'normal' diagram, or at any rate less obvious there.

7.5.1. Stratigraphic problems

Problems are partly stratigraphic, partly connected with the deposition of pollen in the sediment or peat. Stratigraphic problems discussed here are those common to all stratigraphic work, viz. hidden erosion contacts and redeposition.

One deposition problem deals with *fractionation*, which exists everywhere, but is most important in minerogeneous sediments. Practically all minerogeneous sediments (with the exception of tills) are fractionated to some degree, and pollen grains accumulate in the fraction having the same rate of sedimentation. This means that it is futile to search for pollen in sand, because grains of sand sink much faster than pollen grains, even if they are of the same size. If there is any pollen in sand it will usually represent a small, fortuitous fraction of the total pollen rain, probably deposited with turbulence in the bottom layers. In predominantly minerogeneous sediments pollen should only be expected in (clay and) very fine silt fractions. Especially in marine sedimentation this can lead to large-scale differentiation (Traverse and Ginsburg, 1966).

As early as in 1924 Lundqvist demonstrated that an apparently continuous lake deposit may conceal a *hiatus* of great duration: the sediment package in question had been removed by some change of the current system of the basin. Such cases are rather frequent in large, shallow basins, in which the mechanism of filling-up is very complicated, and may also lead to other irregularities, e.g. by temporary drying-up of the basin so that no sediment is formed. Usually a hiatus is recognizable by very sharp contact between two types of deposit (and conversely sharp contacts may always be suspected) but occasionally it may not be discernible, at any rate not in a sampler. In other cases erosion contacts are indicated by thicker or thinner—in some cases exceedingly thin—layers of sand or silt.

Even in apparently homogeneous sediment series, erosion may have caused great lacunae (cf. Lundqvist, 1924) especially in the littoral parts. Samples should be collected centrally, where the sequence is presumably most complete. The investigation of more than one series from the area under consideration will in most cases disclose such undiscovered breaks of the sequence.

As the deposition of sediment is dependent on wind-generated currents, changes in wind conditions may also cause irregularities in sediment sequences. The configuration of the basin is important: redeposition from natural causes is less imminent in deep than in shallow lakes, and less in a basin with a complicated contour.

The sedimentation depth varies with orographic and climatic conditions. Pennington *et al.* (1972) have described lakes with rocky shores in northwestern Scotland in which no sediment was deposited above 50 m depth. Most redeposition material comes from shallow-water deposits that are eroded and washed out. A major part of the material, especially pollen, drifts out into the central part of the basin and is redeposited in deep water. In this way pollen grains are focused into the sediments of the central part of the basin, where sequences are usually the most complete. Changes in the pollen record due to redeposition going on continuously are usually negligible, amounting to a certain smoothing of the curves only. Dangerous redeposition occurs in the wake of unique, often catastrophic events causing changes of sedimentation conditions.

In a small lake of regular shape M. B. Davis (1968) found that 80 % of the pollen recovered from traps during the year was redeposited, chiefly being dislocated from the surface of the bottom deposit during periods of overturn in autumn and spring. Only 20 % represented new material entering the water from the air. This would mean that there is an average period of some 4–5 years before a pollen grain or similar particle is definitely incorporated in the sediment. This effects an integration over time, largely obliterating the yearly variations in pollen productivity between species and of the vegetation as a whole. Depending on the dimensions and configuration of the basin, prevalent currents, etc., there may also be a more or less effective integration over area, so that the initial variations due to differences of buoyancy may also be obliterated. The experiences quoted above show that this is not always the case. Some lakes register annual deposition varves. In these there is obviously not sufficient overturn to influence bottom deposits once they have settled.

Various additional irregularities in the deposition of lake sediments are enumerated by Nichols (1967a, b): mixing of sediment (bioturbation) by burrowing fauna or in shallow lakes by wading animals (probably a very important factor under subarid conditions), sublacustrine landslides due to gullying or earthquakes, changes of water level with concomitant drying out and destruction of sediment, and displacement of sediment blocks picked up by rising bottom ice in spring, the same possible also in terrestric peat inundated in winter (Låg, 1949). All these dangers are pertinent, but on the whole very little common sense is necessary to evaluate their possible role in the actual context, once they have been pointed out.

Rafts of allochthonous material can be embedded in a sequence, and will bring with them their own pollen flora. If the raft is approximately contemporaneous with the rest of the deposit this will not mean very much, but a raft that was fossil already causes difficulties. Birks (1970) has described a raft of moss peat, probably from a spring, floated into a basin and deposited in an otherwise minerogeneous deposit, masquerading as an Allerød sequence.

Macroscopic resedimentation like this is in most cases relatively easy to discover, at any rate if open sections are available. If the original sediment has been ground up, and the pollen grains contained therein are redeposited individually in the younger deposit, the situation is more difficult. Such redeposition takes place where former peat bogs or littoral accumulations are eroded, either due to changes in water level or due to changes in the prevailing wind direction. Redeposited sediments contain a mixed pollen flora, partly deriving from contemporary, partly from an older vegetation. Typical redeposition sediments are often recognizable by an abnormally high content of coarse particles—the finer grains having been carried further away. In the extreme-oceanic blanket bogs small peat slides frequently lead to small-scale redeposition; however, since blanket peats are poorly suited for pollen analysis this inconvenience is of less importance. The deposits under floating quagmires also belong to the types in which redeposition may be suspected, as demonstrated by Troels-Smith (1956); cf. Fig. 7.15.

Redeposition and other displacement of pollen have been summarized by Jones (1958). The most frequent cause will be the release of pollen from

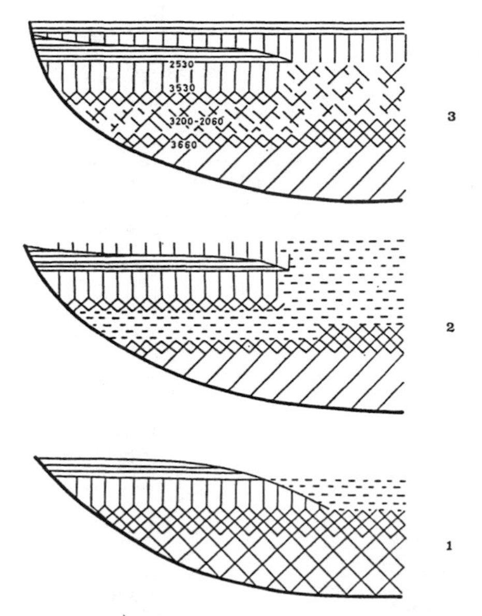

Fig. 7.15. The formation of an inverted sequence through water-level changes. (1) An ordinary hydrosere has led to the formation of marginal telmatic and terrestric peat over allochthonous sediment. (2) A rise in water level has caused the peat to float up and tear the sediment pack apart. It has also flooded the former terrestric peat. (3) The hydrosere is completed, partly for the second time, and the marginal terrestric peat zone is re-established. In the separation zone a metachronous sediment has formed through mixing of parts of the original with new material. Modified after Troels-Smith, 1956. The dates are some of the ^{14}C dates from the actual example.

older deposits by erosion. If Carboniferous spores are re-embedded in Quaternary deposits the damage is not serious, but if metachrony is less pronounced, results may be very confusing.

After cultivation of the soil has set in, man-made redeposition is an important problem; open fields are subject to sheet erosion and if the surface layers incorporate older pollen material it may be redeposited in adjacent basins.

Changes of water level frequently cause erosion of older littoral deposits, which become exposed to a hydrologic regime not previously encountered and against which the deposit is not sufficiently stable. Marine transgression, when salt water intrudes into a former freshwater basin, is often rather erosive even if the change in water level is slight, since the freshwater vegetation is killed, leaving the deposit exposed. Redeposited pollen from peat or acid dry land deposits may completely distort the pollen flora in the first salt water deposits above the limnic/marine contact, as demonstrated in Fig. 7.16.

Material removed by erosion will generally be distributed over such a vast surface that the contamination of contemporaneous pollen diagrams is negligible, but in serious cases a layer may be dominated by redeposited peat or sediment and its pollen flora distorted (cf. examples in Ernst, 1934: 304). The remedy is the same as in the preceding case: more diagrams from the same area (plus a little common sense).

7.5.1.1. Secondary pollen

Deposits containing redeposited material contain both the primary, contemporaneous pollen flora and a larger or smaller component of older 'secondary' pollen derived from the redeposited material. This component may in many cases be distinguished in analyses.

Exines of fossil pollen change as a function of time, and old, redeposited pollen grains can be distinguished from more recent ones. Both the type and speed of such diagenetic changes vary with external conditions, and it is not easy to indicate characteristics of general applicability. Sometimes colour differences of unstained exines may be indicative; in others (Stanley, 1966) the capacity of absorption of stains may serve to distinguish more or less quantitatively. According to van Gijzel (1967) the autofluorescence of exines changes with time, and can be used to distinguish between native and redeposited material if the age difference is sufficiently great.

Iversen (1936) demonstrated that moraine (boulder clay) may contain great quantities of pollen from pre- or inter-glacial deposits. These pollen grains are redeposited in late-glacial clays., thus giving rise to a mixed pollen flora in which the 'secondary' pollen frequently predominates. In ice-marginal lake deposits in Poland, Jaworka (1971) found practically 100 % secondary (pre-Quaternary) pollen and spores. Any pollen spectrum from a glacial clay or a clayey sediment must

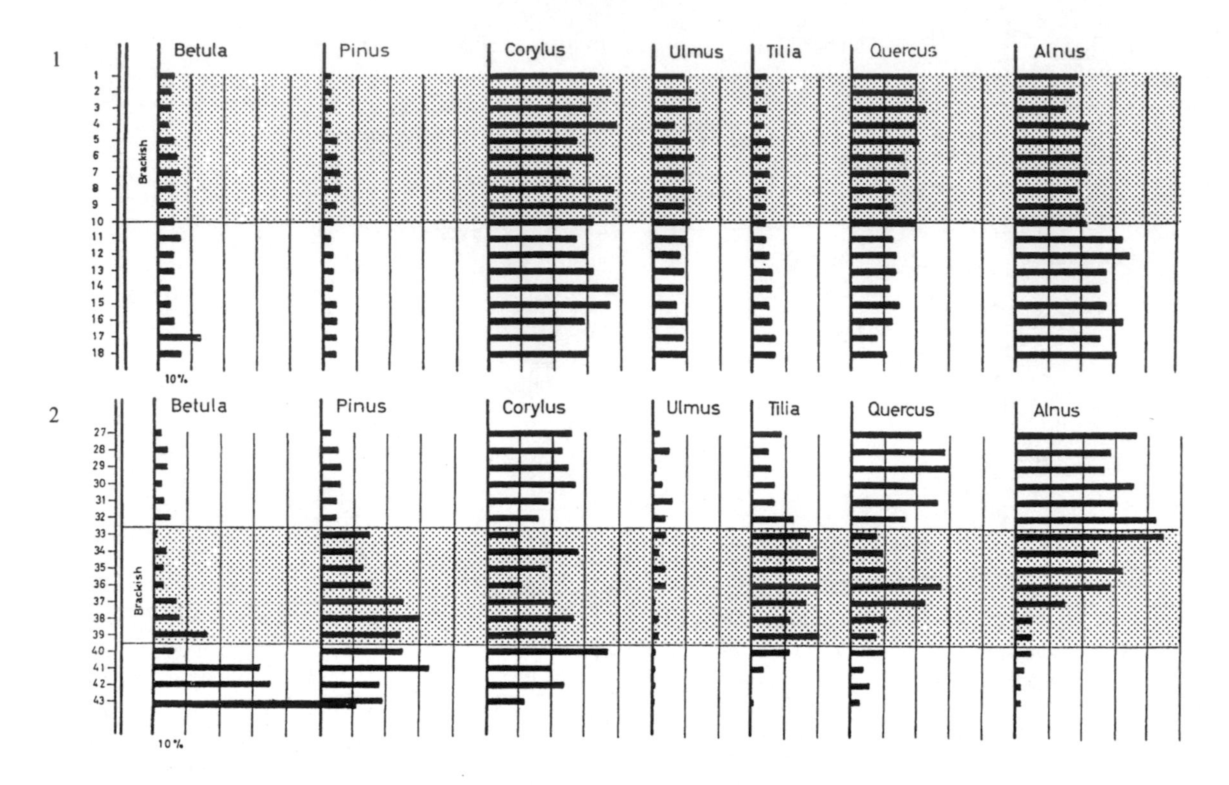

Fig. 7.16. Erosion and deposition caused by marine transgression: Two pollen diagrams from the same basin (Bundsø, Island Als, Denmark). Diagram 1 (top, A. Andersen, 1954) represents the deepest, central part. Diagram 2 (Jessen, 1938, both recalculated and redrawn) is marginal. The transgression took place when Baltic water flooded the lake. The marine deposit is indicated by shading. In diagram 1 nothing very serious happens: curves are smooth. In diagram 2 the deposit immediately below the marine phase is obviously much older (high *Betula* and *Pinus*, very low *Alnus*), and indicates a hiatus. The material eroded away has been mixed into the younger deposit giving, there, too high values for *Pinus* and *Betula*, too low for *Alnus*. All three samples immediately above the transgression level are distinctly contaminated. The high *Pinus* values higher up may be due to marine overrepresentation.

be regarded with suspicion until the presence or absence of secondary pollen has been thoroughly investigated (Fig. 7.17). The contaminated spectra may nevertheless give an adequate picture of the composition of contemporaneous vegetation if the secondary pollen can be subtracted from the total sum. The principle is in this case comparatively simple: the composition of the secondary pollen flora may be obtained by analysis of an unweathered sample of the boulder clay from which the secondary pollen is supposed to have come. It is necessary that it is unweathered, since the pollen content of moraines is easily destroyed by subsequent weathering. The secondary pollen flora consists partly (1) of pollen types which cannot be separated from those that might be expected in the clay sample; partly (2), however, of pollen types that are obviously secondary, either because the species in question did not grow in the area after the Ice Age, or because their climatic demands could not have been satisfied under the prevailing conditions, e.g. *Tilia* pollen in glacial clay. This second group constitutes N% of the pollen in the moraine. In a clay sample the same group constitutes n%. Since no pollen of this group is authochthonous in the clay, n/N of the total pollen in the clay must be derived from the moraine. This reduction can therefore be applied also to the pollen of the first

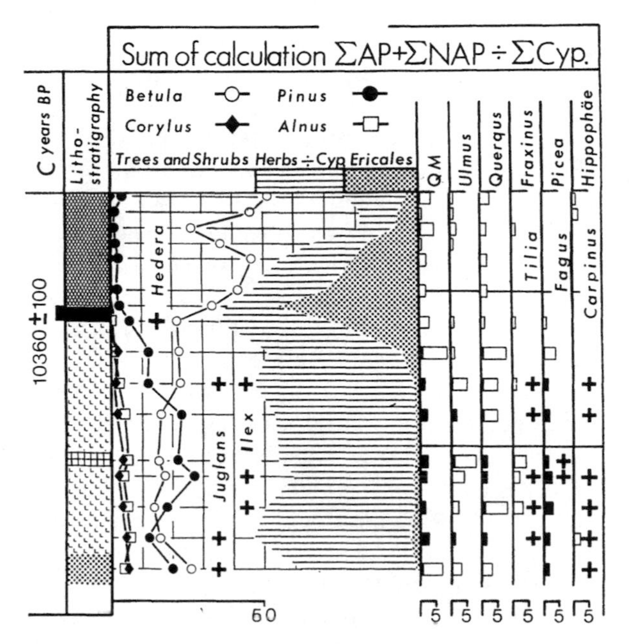

Fig. 7.17. Secondary pollen in a Late-Glacial marine clay in a deposit from the outer Norwegian coast at *ca.* 63° N. lat. The clay contains great quantities of resedimented material of Riss age or older. The lacustrine series shows the regular succession from forest-less (*Empetrum* heath) to birch forest. Since there was no pine growing at the shore, the pine curve probably represents secondary pine only, no marine overrepresentation. The decline of the pine curve (together with the other secondary constituents) indicates the stronger local pollen production (*Betula*, NAP), depressing the values. From Paus, 1982; redrawn.

group: the presumed autochthonous 'primary' pollen spectrum is left. The method is based upon the assumption that the relative frequencies of pollen types have not undergone change during lateglacial redeposition. Experience in Denmark indicates that this assumption is correct. Another point is the non-linearity of pollen percentages in relation to absolute values. However, as long as values do not exceed 50 % these mathematical errors are probably insignificant in relation to the statistical noise in the data. Differential stain acceptance can also in this case serve to distinguish between autochthonous and redeposited pollen. 'Secondary' pollen is not found in all glacial clays: in central glaciation areas the moraine may be sterile, not yielding any pollen by erosion. The effect is restricted to marginal areas (e.g. Denmark).

7.5.2. *'Sources of error'*

Whereas the problems dealt with in the preceding section are essentially the same in all stratigraphic work, the incorporation of pollen grains in a deposit also depends on special factors which, if not properly understood, make themselves felt as the 'errors' of classical pollen analysis.

Some of them were realized by von Post and discussed in his 1918 paper; others have been recognized later.

7.5.2.1. *Local overrepresentation*

Pollen registrations contain three not distinctly demarcated components: a local, a regional and an extra-regional. Roughly, they correspond to the descending branch of the deposition curve (Fig. 2.12), the asymptotic value and the upper air component. If the problem is the regional development of vegetation under the influence of climate or humans, the regional component is the one that matters. Of the two others, the extra-regional is usually not significant because it is so small, whereas the local component, the so-called local overrepresentation, is a disturbing element. Local overrepresentation is one of the classical problems in the discussion of the pollen-analytic method. The local component represents a large quantity of pollen and may influence and distort the regional diagram seriously.

Local overrepresentation is particularly likely to cause disturbances in pollen samples from deposits, in which the pollen producer is a constituent of the mother formation, e.g. pine forest peat formed on a dried-up bog covered with pine, or alder carr peat. In such cirmcumstances the deposit will in itself serve as a warning and, if possible, such deposits should be avoided. Alternatively one may have to establish a 'truncated' pollen sum from which the disturbing constituent has been omitted. Even so, there should be one total diagram, including this constituent, to make possible the evaluation of the effect of the elimination (e.g. Hafsten, 1951: plate 1). Other cases are more serious, e.g. when a tree on the bank has dropped pollen or whole catkins into the water, in which they have been deposited within a small area. The number of pollen grains left in falling catkins may be greater than that which has been dispersed in the air at flowering time (Rempe, 1937: 114). In such cases no macroscopic remains of the tree will be present in the deposit (profiles may be helpful), but usually the pollen grains will form lumps, thus indicating what has happened. Even if such pollen lumps are counted as units, the percentage of the species in question will in most cases be abnormally high. If this is the case in one, or very few, samples only, the overrepresentation may be compensated for simply by assuming that the percentage of the species in question has changed continuously between the adjacent 'normal' spectra. The other percentages may then be calculated from the following formula:

$$\% A_{corr} = A \frac{100 - \% B_{corr}}{S - B}$$

in which A and B are the actual counts of two taxa, S the total count of all taxa included; $\% A_{corr}$ is the calculated adjusted percentage of A after that of B has been normalized to $\% B_{corr}$ (Fægri, 1944b: 546).*

Owing to the small transport distance of most of the NAP the danger of local overrepresentation in the NAP diagram is greater than in the case of AP, but, on the other hand, it is generally easier to check the occurrence of pollen producers in the mother formation.

Local overrepresentation may seriously reduce the value of a diagram as an expression of the regional vegetation history, but may form the basis of a palaeophysiognomic analysis, which may disclose influences not revealed by ordinary methods. Sound botanical judgement is needed to determine, in every pollen diagram, the relative roles played by the two antagonistic components: local and regional pollen.

If the palaeophysiognomic changes can be referred back to a known vegetational and climatic sequence it may be possible to utilize such local diagrams for more general purposes, too (cf. Fægri, 1954a, dating the immigration of *Fagus* by means of local diagrams).

A very simple way of evaluating and, as the case may be, compensating for local overrepresentation is to identify macrofossils present in the deposit. The occurrence of pollen of the same species must then be considered with suspicion. A further refinement is achieved by the sociologic identification of the plant communities involved. From floristic analyses of present stands of the same communities one can with certain reservations also draw conclusions about the presence of species not represented as macrofossils (cf. Rybničková and Rybniček, 1971).

7.5.2.2. *Pine overrepresentation*

In north-hemispheric diagrams *Pinus* is often registered in suspiciously great quantities, which can in some cases be shown to be incommensurate with the actual occurrence of pines (*P. sylvestris* in Europe) in the area. There are two different types of pine overrepresentation.

The first type is due to the great pollen production of pine and the great ease with which pine pollen is dispersed. Where the local pollen production is weak the ever-present pine pollen comes to dominate the percentage picture. Influx data are not influenced. This pine overrepresentation corresponds to the high intercept of the pine curve in Fig. 7.2. It occurs outside or above timberlines, after deglaciation before forest has established itself, after deforestation due to cultivation, etc. The interpretation of these high pine values therefore varies from one diagram to the next, but usually it does not cause great difficulties as long as one keeps in mind that they do not indicate pine forest.

The other type of overrepresentation is due to differential buoyancy (cf. Hopkins, 1950). Pine pollen, in particular, floats on the surface for long

* E.g. *Pinus* 480, *Betula* 566 and the total 1070 grains. *Betula* is overrepresented; the correct percentage is supposed to be 10; the *Pinus* percentage will be :

$$P_{corr} = 480 \frac{100 - 10}{1070 - 566} = 85\tfrac{1}{2}\,\%$$

periods while non-vesiculate grains sink within a day or two, or even faster. When pine is frequent on adjacent shores, that part of its pollen output that falls on the water drifts with the wind, off or on the shore. In the latter case the pollen is trapped and the littoral sediments enriched in pine pollen. This effect is very pronounced on the sea because the collecting surface is so vast, and the pollen often drifts into shallow coves or lagoons, where it can be enormously enriched. There is then a dramatic drop in the pine curve at the transition from marine to lacustrine sediment (cf. Fig. 7.18).

The same process does take place in lakes, but the effects are much smaller, due to the smaller collecting surface. However, Iversen found that in offshore sediments in Danish lakes the pine percentages were lower than in littoral peat. In other cases pine has been described as being more frequent in the central part of the basin (Pennington, 1947). Such differences are probably due to physiographic differences between basins, and differences in the redeposition (focusing) mechanisms.

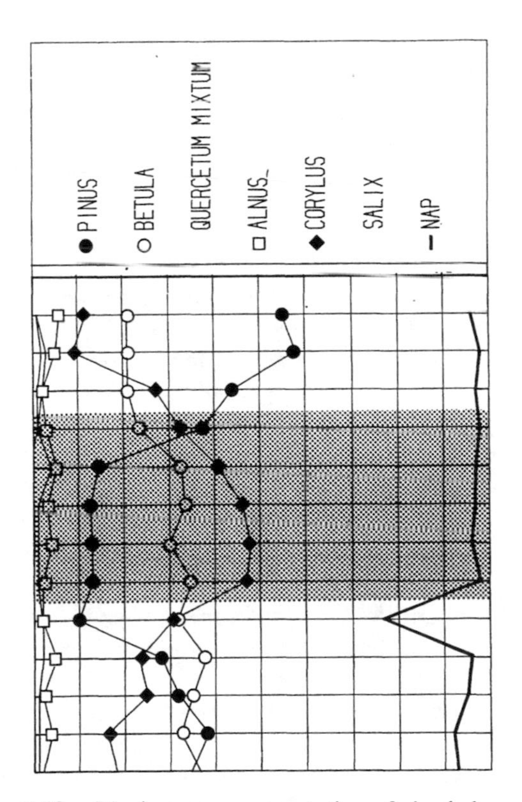

Fig. 7.18. Marine overrepresentation of pine below and above a brackish-water deposit. The absence of an *Alnus* registration (as in Fig. 7.1) is due to the absence of the tree, which had not yet immigrated during the Boreal. The great *Corylus* is a regular feature of the Boreal diagram of the region ('Aurebettjønnet', Fægri, 1944a; recalculated).

The above discussion has certain limitations. Pine is not invariably overrepresented in marine deposits, and other pollen may also float for a while (cf. Overbeck and Schmitz, 1931: 86). Irregularities in the pollen flora of marine deposits should always be expected. In eastern Sweden the *Pinus* overrepresentation in marine deposits is followed by that of *Alnus* in the subsequent lagoon sediments (Florin, 1945). Whereas the pine overrepresentation obviously does not express any vegetational fact, that of *Alnus* reflects an actual, but metachronous, carr vegetation bordering on the lagoon (cf. Fig. 7.1).

Hence, if pollen analysis is used for investigation of land/sea relations these sources of error must be kept in mind. Distorted diagrams cannot directly be used for synchronization. Usually, however, there will be other features of the diagram which can be used. In diagrams in which one pollen type is over-represented, correlation with other diagrams can be obtained through a study of the relationship between the other constituents of the diagram. In extreme cases one may be obliged to make auxiliary, 'truncated' diagrams from which the overrepresented species is omitted (cf. von Post, 1947: 234). It must be emphasized that such diagrams are even less representative of the actual vegetation of the area than the original ones: they should be used only for correlation purposes.

7.5.2.3. Long-distance transport effects

The second classical problem is *long-distance transport*, which in the first years of pollen analysis was taken very seriously (Hesselman, 1919b). Subsequent experience has shown that in most cases it is a subordinate factor. Considering the data in Chapter 2, especially Hesselman's results from the light-ships, this might seem somewhat surprising. However, the quantity of pollen produced by local vegetation is generally so immense that the quantities deriving from long-distance transport are relatively insignificant despite their absolute magnitude. It is self-evident that if long-distance pollen is caught in a place without AP-producing vegetation, the whole forest pollen sum will derive from long-distance transport and the AP diagram will be totally misleading as an expression of the local or regional vegetation (cf. Fig. 7.9).

If there is a local NAP production the AP/NAP ratio indicates if the AP is local or long-distance transported. However, there are completely unproductive plant communities (e.g. lichen heaths), and then there is no local pollen production for comparison. In western Norway the 'normal' NAP percentage in a forested area is *ca.* 5 %, whereas in a forest-less area it is, in the total diagram, *ca.* 50 %. This means that the local NAP production is of the same order of magnitude as the extra-regional, long-distance AP.

If there is the same ratio between NAP and long-distance transported AP in a forested area, the latter would also be some 5 % of the AP found. Such a calculation is, of course, highly conjectural, but it gives an idea of the magnitude of the long-distance contamination of the ordinary pollen diagram. This is not generally noticed, because most of the long-distance AP consists of the same pollen types as those produced locally, but in the case of species that did not grow locally the influence may be read directly from the pollen diagram. Results obtained by such calculations support the conclusions arrived at above.

In western Norway *Picea* and *Fagus*, practically speaking, do not occur. Thus, whatever pollen is found of these two genera must be due to long-distance transport. Careful analysis may give continuous pollen curves of these genera; these curves show a relation to the corresponding curves from the areas where the trees grew. Such curves may be used for two purposes, viz. to date otherwise uncharacteristic diagrams (Fægri, 1954a), and to give an indication of the strength of long-distance transport by comparison of the relevant percentages in both places. Again, the *Picea* percentages correspond to a total long-distance AP percentage of about 5. As the distances in question are very great, the differential rate of deposition is hardly of major importance in this case; the turbulence phenomena are of such a nature that they largely eliminate this factor. Similarly, in southern Greenland, where conifers (except *Juniperus*) do not occur, the long-distance conifer values (*Pinus*, *Picea*) in lake sediments amounted to *ca.* 1 % of the pollen total (Iversen, 1954a).

In the diagram as a whole, the long-distance pollen is of little significance, but it is a great hindrance when the question of immigration of a species arises. Does a small percentage of a pollen type derive from small local occurrences or is it due to long-distance pollen transport? The question cannot be answered by a single formula. Previously it was asumed that when the curve rose above a few per cent (the so-called rational limit) there was fair reason to think that the species grew within the area. In forested areas this may be correct in many, perhaps most, cases; but in this respect dogmatism is particularly dangerous. The association plays a great role: a curve in a diagram chiefly composed of prolific pollen producers (Iversen's group A) means something quite different from a curve in a group B diagram. The curve of an A species—in any association—may be continuous and even reach substantial values without any local or regional presence of the species. For *Pinus* in a spectrum chiefly composed of group A species the limit seems to be at about 10 %: generally values below 10 % cannot be taken to indicate the local presence of *Pinus* in the immediate neighbourhood of the deposit (cf. Fig. 7.2). On the other hand, a single grain of a group C species may be sufficient to indicate the presence of that species with fair certainty, e.g. *Ilex*, the pollen grains of which are produced in small quantities only, and are not wind-transported. This is based on the assumption that contamination may be excluded.

The smaller the total local pollen production is, the greater will be the demands on the frequency of a species in the pollen diagram to make it acceptable as indication of actual presence. On the other hand, in a densely forested area even low percentages may indicate presence of the species, but, unfortunately, it is impossible to fix general limits, and many cases will probably remain undecided. It is hardly possible, by means of pollen analysis alone, to date the immigration of a forest tree exactly. Absolute certainty that a species occurs (and may have occurred for some time!) in a station may be ascertained by the find of macrofossils only, and in some cases even they may have been brought a long distance by the wind, e.g. pine needles blown on and with snow above the timberline. Also, influx data calibrated against recent influx greatly improves the information content of such records—but not of the sparse producers.

The above pertains mainly to forested areas. With increasing distance from the forest the importance of long-distance pollen increases, and at a sufficient distance, some 10 km from the nearest forest, most AP must be counted as belonging to that category. This is where difficulties arise: the same pollen type may, even in the same diagram, sometimes be produced locally and indicate a temperate period and—in other diagram zones—be carried a long distance and indicate absence of forest, i.e. unfavourable conditions. A typical example is afforded by the *Pinus* pollen in northern Finland (Aario, 1940: 60): the pine representation falls from 80 % in the south to 50 % at the northern end of the pine area, and further to 20 % in the birch belt. Then it rises again to 80 % in the AP diagram of the tree-less tundra. Similar figures are found also for *Picea* and *Alnus*.

Until we have more experience each problem must be solved individually, using botanical discretion (and common sense!). Also, there may be a differential long-distance transport, due not only to differences in natural buoyancy of the various pollen types, but also to the conditions of turbulence in their areas of origin.

The occurrence of NAP follows quite different rules, if any rules can be formulated at all. These pollen types are less subject to long-distance transport than AP, simply because they are released at a much lower level above the ground, and their occurrence is therefore more local in character. Just at the border of the forest zone the NAP production per area is highest (apart from clearings within the forest zone), and then it falls off with increasing distance from the forest. Experience seems to indicate that this drop is often sharper than the corresponding drop in AP; accordingly AP dominance may be found again at a greater distance from the forest. The respective dominance of AP or NAP is conditioned by a complicated set of mutually interdependent and interfering factors, and no hard-and-fast rules can be laid down. An intimate knowledge of the vegetation types is the only safe foundation for the solution of these problems.

NAP curves in sediment and in peat are different. In the latter, NAP producers form part of the mother formation of the deposit, and so do the respective pollen grains, which are therefore strongly represented. Consequently, high NAP values (in peat) are of little importance as indicators of the absence of forests if they derive from one NAP type only or, in some cases, from the simultaneous occurrence of few types which together are characteristic of one plant community, e.g. *Empetrum* and *Rubus chamaemorus* in northern Finland (Aario, 1940: 69). The disharmonious high NAP values are characteristic of peat (cf. Fig. 7.19). The AP/NAP ratio gives a much clearer picture in sediments, and many of the difficulties met with in peats, above all the extreme variability of figures, will disappear if investigation is carried out in sediments. In typical sediments the NAP flora is harmoniously developed, containing representatives of all major

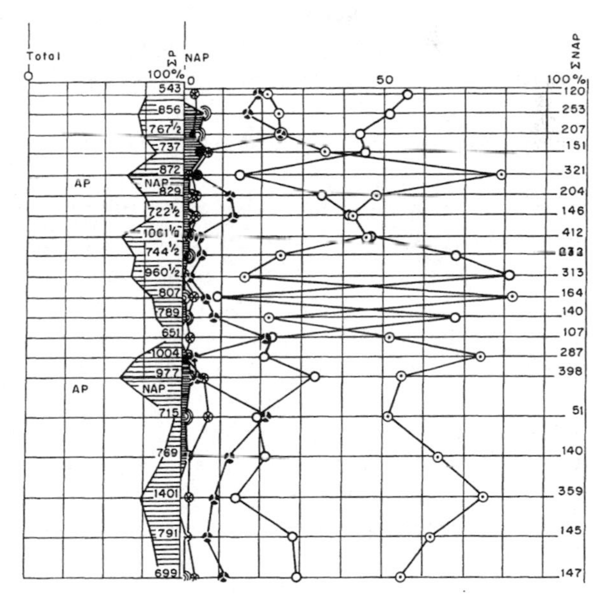

Fig. 7.19. NAP curves in terrestric peat showing very strong local dominance shifts. From Fægri, 1950b.

types. Even if their ratio varies, these variations are subordinate to those of the NAP/AP ratio as a whole. Local, single-type NAP maxima that dominate the NAP/AP relations are usually of no general significance.

7.5.2.4. *Preservation and degradation of pollen: differential destruction of exines*

The third classical 'source of errors', which has also been known and feared throughout the history of pollen analysis, is that of *differential destruction of pollen grains*, already mentioned. As the composition of the exine and its resistance to corrosion differ from one taxon to the next, it goes without saying that if pollen grains are subject to any notable degree of corrosion during fossilization or after, the composition of the pollen flora will change; the more resistant types will appear to accumulate owing to the disappearance of their counterparts, the less resistant grains. Fortunately it is, in most cases, very easy to check this error, since corrosion of pollen grains is easily observed whenever it occurs.

If corrosion has proceeded so far that some pollen grains have disappeared altogether, there will be many crumpled, broken grains, or grains the exine of which has assumed a peculiar, pitted surface which is easily recognizable as the first stage of corrosion. If more than half of the pollen grains of deciduous trees (conifer pollen is generally more resistant) show traces of corrosion, the sample should be discarded, since a distortion of the spectrum due to the total disappearance of part of the pollen flora must be suspected. In diagrams intended to be used as a standard for the area, corrosion should not be tolerated at all. In some cases, e.g. for dating purposes, it may be necessary to analyse deposits the pollen of which is more or less corroded; in such cases the diagram must be taken for what it is worth, and its features considered with due suspicion. The extent and degree of corrosion should be stated. In sediments corrosion will generally be of little importance, but in aerated terrestric deposits it is a dangerous source of misinterpretation.

Pollen exines (and spore walls) are generally preserved in peat and sediments even when almost all other organic constituents are reduced to a structureless and indefinable substance. There are, however, peat types in which pollen exines are more or less corroded. Corrosion seems to take place especially in aerated peats formed at or above the ordinary water level, near springs, etc. The acidity of the substrate is also of great importance for preservation of exines. In raw humus ('mor') pollen grains may be excellently preserved, even under relatively dry conditions, whereas they may be rather badly corroded in alkaline deposits, even if other preservation conditions have been favourable. This may be partly due to a bacteriostatic (and fungistatic?) effect of the acid medium. In minerogeneous sediments pollen grains may possibly be mechanically abraded. Exines are apparently not all (Fægri, 1961) or very slowly (viz. in the rumen of goats: Harris, 1971) broken down in the alimentary canal of pollen-eating animals. On the other hand there is an extraordinary report of ingested pollen being transferred to the bloodstream in a very short time (Jorde and Linskens, 1974) and being broken down and eliminated after a few hours. The pH of the bloodstream certainly has some effect but, even so, one almost suspects the occurrence of a sporopollenase in the human blood. Bee larvae, living on pollen, do not digest exines, which acumulate in their body, leaving a record of their diet (Fægri, 1961).

Cushing (1967) distinguishes between various, mutually not exclusive, types of deterioration of exines: thinning, corrosion, degradation, crumpling, and breaking, approximately in this sequence of seriousness for the interpretation of the diagram. However, a too finely cut classification of exine degradation hardly serves any purpose, especially as the causal factors are not too well understood.

Some exine types may be relatively resistant in one case, less so in another, depending on the medium. In addition to purely chemical corrosion, perhaps mostly by direct oxidation, some pollen-attacking micro-organisms may also be able to attack exines (Scott and Stojanovich, 1963), but the point is still controversial, and various patterns of erosion are not yet accounted for. Sangster and Dale (1961) conclude that some bacteria must possess enzyme systems equal to the task of degrading sporopollenin. Brooks and Shaw (1973a) registered growth of *Streptomyces* on pollen exines, but they found no bacteria able to degrade sporopollenin. The existence of sporopollenase in bacteria is unconfirmed. Skvarla and Anderegg (1972) conclude that if pollen-infesting fungi do possess sporopollenin-degrading enzymes, they use them only for the penetration of the exine. The same conclusion was drawn by Goldstein (1960; cf. also Elsik, 1966). Septate hyphae, representing higher fungi, seem to be restricted to entering pollen grains via the apertures.

In an experimental investigation Havinga (1971) observed strong degradation in sub-neutral medium (river clay, leaf mould) and very little in acid (*Sphagnum* peat, *Carex* peat, podzol). One of the reasons why tables of resistance to decay are contradictory may be that the concept of deterioration in itself comprises different processes (cf. the Cushing scale, above). A general thinning effect is prevalent in acid, peaty soil, whereas corrosion by perforation is found in milder, better-aerated soil. Various pollen types are more or less resistant to the one or the other of these types of decay. Harris (1971) quotes thinning of exines in the extremely acid milieu of a goat's rumen. This might suggest another purely chemical process.

7.5.2.5. Other problems

Errors due to non-random dispersal of pollen grains on the slide have been dealt with earlier. One of the sources of error seriously debated in the early days of pollen analysis was *vertical displacement of pollen* after primary deposition. However, Malmström's experiments (1923: 149) showed that mechanical downwash played no significant part in peat. Similar results have been obtained by Rowley and Rowley (1956). The very sharp and clear-cut variations in pollen curves from lake sediments of whatever type also indicate that the burrowing biota of lake bottoms do not appreciably influence the stratigraphy of the deposit, though cases of 'faded-out' curves are known (cf. R. Davis, 1967). The problems of redeposition are dealt with above.

Apart from the updraughts in mountainous areas (cf. above), there are also in such areas and on the coast *local winds* which may cause disturbances in the diagram (cf. Brinkmann, 1934: 338). This source of errors is in most cases insignificant, but may in particular cases prove significant if overlooked.

7.5.3. Forest bottom (local) analyses

In essence, forest bottom analysis is analysis of the composition of the local overrepresentation element or, what is the same, the gravity dispersal element. The smaller the basin—lake, pond, bog—in which pollen is deposited, the stronger is the local element. The pollen deposition from the regional rain is overwhelmed by the massive deposition from the nearest vegetation. The extreme is represented by analyses made directly in the forest bottom. In such cases another factor makes disproportionality between local and regional pollen even greater: Andersen (1970) has shown that the main pollen dispersal inside a forest is restricted to a few dekametres from the stem.

In nature, the transition between peat and forest bottom raw humus is gradual: pollen analysts rather early observed well-preserved pollen grains in dry-land humus deposits. Erdtman (1954b) published analyses of subarctic humus without being able to draw conclusions from the material. Pollen-analytical theory was not sufficiently advanced for such a task. Firbas and Broihan (1936) participated in a polemic about dating raw humus formation by means of pollen analysis. Beijerinck (1934) dated a podzol profile on the basis of pollen contents and reached conclusions so different from the usual concept of podzolization that the paper was largely overlooked, but the observations themselves were good enough (cf. below).

These three papers give examples of the use and non-use of terrestric, dry-land deposits in the days of 'classical' pollen analysis. Only later has the use of pollen data for direct elucidation of local ecological problems added a new dimension to such analyses. Fægri (1954a) analysed the humus inside a (beech) forest to obtain a picture of the vegetational changes that had occurred in the plot itself. On that occasion a continuous dry-land deposit was met with that dated all the way back to the Boreal.

In his studies in Draved forest in Jutland, Iversen (1964, 1969) went one step further inasmuch as the changes in the local forest composition were the principal objects of his investigations. Diagrams from closely adjacent stations differed considerably according to the treatment of the vegetation by humans (clearing, grazing, etc.—1969: 47). Pollen analysts have always talked about the possibility of using the changes of the pollen flora as the basis for more rational forest management. Apparently, this is the way.

The quantities of pollen found in dry-land deposits may be enormous; many such deposits would qualify as H 10 in the von Post-Granlund humification scale, and consist chiefly of unsaturated humus colloids (removed by KOH treatment) and pollen grains plus some fungus hyphae. Also the state of preservation may be excellent. As the pollen even from the trees of the canopy hardly travels more than 50 m inside a forest, and that of plants of lower strata even less (especially as insect-pollinated species preponderate) such spectra are extremely local. On the other hand they often contain pollen which is not, or only rarely, registered outside the forest, e.g. *Ilex* or *Impatiens noli-tangere*.

Not only forest produces raw humus: heaths also do, and pollen analysis of heath raw humus has proved valuable in the unravelling of the history of individual heath patches (Mamakova, 1968): afforestation, deforestation with concomitant changes in the composition of the field layer.

In dry-land deposits two factors are of predominant importance, more so than in ordinary peat: oxidation and vertical transport, which are to a certain degree coupled. If much air penetrates into the soil, degradation processes proceed fast. In such soil, pH conditions are usually favourable, and the soil fauna includes comparatively big animals with a great vertical range—above all, the big earthworms. Oxidation under mild pH conditions is destructive for pollen exines and the vertical displacement of the soil caused by the activity of the earthworms distorts the diagrams. With a lively soil fauna of this type pollen is displaced vertically, both upwards and downwards, and so far no proof has been given for the chromatogram-like differentiation assumed by some workers in the field, (cf. also the experiments by Walch *et al.*, 1970).

In the typical raw humus there are very few earthworms, and those that might be there do not to any extent migrate vertically, but stay in the topmost layer. There are no other fauna components that can displace pollen vertically, and the deposit therefore builds up regularly as in peat bog, only more slowly owing to initial oxidation, leading to much higher humification. There is no reason to suppose that pollen grains will sink down beyond the loose förna layer at the top, and their state of preservation is usually acceptable.

The scarcity of larger soil-inhabiting insects and worms in arctic-alpine soils creates conditions similar to those of lowland raw humus. Such deposits may also show a time sequence.

In one respect, dry-land deposits differ from peat: the diagenesis of ordinary peat is usually an extremely slow process, taking geological ages to change the chemical composition of the material. There is every reason to suppose that in the raw humus (slow) oxidation takes place all the time, which would mean that the older the humus, i.e. the deeper in the profile, the more oxidized it is. As pollen grains (together with sand) are among the least oxidizable components of the humus, they will accumulate in the bottom of the profile (Fægri, 1970, 1971). Iversen (1964, 1969) records several raw humus-like deposits ('mor'), the chief constituent of which is pollen, so-called pollen-mor. If all humus is oxidized away, the sand that might have come into the profile will remain as A_2 horizon, imitating a podzol that has been leached *in situ*. This would explain observations of pollen grains in podzol layers, including those of Beijerinck (1934).

No general consensus of opinion has been arrived at; we refer to discussions by Dimbleby (1957), Godwin (1958), and to Dricot (1961). In experiments Walch *et al.* (1970) have demonstrated vertical transport in both directions by rainworms, and have also shown that in their absence pollen grains do not pass a layer of leaves in their downward movement through loose soil.

If a vegetation on mild humus has been replaced by one that forms raw humus, the lower part of the diagram may have been subject to vertical displacement of pollen by the big earthworms living in the mild humus at that time. This part will then give a more or less generalized diagram in which the individual stratigraphic features have been obliterated in contrast to the upper part, where sequences are undisturbed. Gradual oxidation of the bottom layers with resulting concentration of pollen and sand will take place anyway.

The preceding discussion presumes that the soil profile develops on an existing minerogeneous base, which has already formed itself. If there is a continued, more or less regular accretion of mineral substance, e.g. in the form of blown sand, relations become more complicated and simultaneous deposition of mineral grains and pollen is a possibility.

So far, forest bottom analyses have rarely been systematically utilized for phytosociological purposes. Where the climax vegetation is so heterogeneous that it gives an undefined pollen picture, e.g. in the Tropics, the botanical analysis of what has happened in individual stations may be of greater importance than the general pollen spectrum integrating, as it does, a greater area of vegetation.

Whereas intermittent wetting and drying of a deposit under natural conditions generally causes rapid destruction of the pollen flora, exines seem to keep well under constantly dry conditions, e.g. in the desert. For pollen analysis this is, admittedly, of less significance as shifting winds and sands also mix recent and pre-recent pollen floras, allowing only very approximate conclusions—if any—about former vegetation. Occasionally it is possible to obtain more definite data: material of brickwork and pottery vessels (from Sudan) had been mixed with farmyard dung before firing, and contains appreciable quantities of recoverable pollen. Most of this probably came from the fodder ingested by the cattle shortly before. As cattle are, under those circumstances, fairly omnivorous, one may surmise that the dung contained a fair sample of the vegetation flowering at that moment plus, of course, what native pollen may have been in the clay itself (Bot. inst. Bergen, unpublished). Similar observations have been made in weakly fired brickwork (adobe) in other parts of the world.

7.6. When common sense fails: statistical and numerical analysis as aids for interpretation

Motto: Statistics can never prove that you are right, but may indicate that the opposite is improbable.

7.6.1. Statistical evaluation

In the classical days the designation for what is today known as pollen analysis was *pollen statistics* (used by Erdtman as late as in 1944). In essence, pollen analysis is a statistical technique, and as such it is subject to the rules that govern all statistical relations. However, statistical problems of pollen analysis have in the past often been neglected, both in sampling design and in the evaluation of data. Frequently, published results neither indicate confidence limits, nor do they publish data (i.e. pollen sums) that permit the calculation of such limits, which are essential for the evaluation of the carrying power of pollen-analytic statements. For definition and calculation of these statistical parameters we refer to any textbook of elementary statistics.

Data obtained by pollen counts are statistical estimates of the 'true' values, i.e. relations between pollen taxa in the preparation. The numerical validity of such estimates can be evaluated mathematically. The figures from the pollen counts are used to make inferences about the pollen rain, and finally, about the composition of the vegetation (Fig. 1.1). These inferences are not subject to mathematical tests of the same rigour. There is therefore a certain limit of rigour beyond which it does not pay to go into the statistical procedures.

The following discussion mainly deals with the problems of percentage analysis. Those of APF and influx are simpler, since the constraint of the 100 % universe does not apply. On the other hand, sampling errors in the R_{abs} values are more serious, since the transformation of influx data into quantitative vegetation data is a much more pertinent problem in that case.

Statistical and numerical analysis can elucidate various features of interest in the interpretation of the pollen diagram: (1) the confidence limit of individual values;* (2) the relations between observed values, e.g. the uniformity of curves, rise and fall of curves, correlation among species, etc. Whereas the first type of check is a purely numerical one, for which nothing but an elementary theory of statistics is necessary, the second check also implies theoretical assumptions, both mathematical and biological, the validity of which in individual cases may vary.

For this latter purpose, methods of advanced statistics (exploratory data analysis, pattern recognition) may be necessary. They will be dealt with in the second part of this chapter.

It should also be kept in mind that statistical testing in essence is hypothesis testing and cannot confirm a statement; it can only say that the hypothesis is wrong (if such is the case), or more correctly that it is improbable, and the improbability can be expressed numerically.

Statistical checks on the validity of statements presume the availability of numbers. In pollen analysis the relevant numbers are those of the samplings (and consequent sampling errors) between the vegetation and the pollen count; only the last one is usually available for statistical control. It is a weakness of pollen-analytic theory that there are practically no data permitting a check on earlier stages with regard to sampling errors. To provide such data is, at best, very laborious and difficult and, except for special studies on method, quite impracticable or even impossible. Empirically we have experienced that the concept of a uniform pollen rain is not contradicted in a general sense, but also that there are many aberrations, for which we have no control, statistical or otherwise.

Pollen counts comprise a sample only, and are therefore subject to a statistical uncertainty, which has been discussed by many previous investigators. No statistical training is necessary to realize that the more pollen grains counted, the more the sample is representative of the 'real' composition, i.e. for the universe of numbers as a whole. Especially in the early days of pollen analysis investigators made frequent tests to gain some experience about the size of the counts necessary to obtain reliable results. However primitive, these tests are important, because they have influenced all later pollen-analytic practice.

Based on practical experience Booberg (1930: 226) indicated that if *ca.* 150 AP grains were counted, the percentages of major constituents (> 25 %) would be comparatively safe to use as a basis for inferences. At that time minor constituents were not as important as today. One fact has, however, emerged from these and other experiences: if less than 150 pollen grains are counted per spectrum, the variations of the pollen curves must be very pronounced to indicate beyond doubt that the observations are 'real' and do not represent statistical uncertainties of the counts ('noise'). Nevertheless, some investigators still draw quite unwarranted conclusions in relation to the number of pollen grains counted. No statistical treatment can compensate for an initially too low pollen count, but it can—and should—warn against the abuse of such figures.

Based on twenty counts from the same preparation, each of 100 grains, Bowman (1931: 698) concluded that after 800–1000 grains had been counted, percentages, also of minor constituents, are fairly constant. Hafsten (1956) has published two curves of minor pollen diagram constituents (cf. Fig. 7.20). They show that with the count of 500 AP grains per sample fluctuations in curves of minor constituents are so erratic as not to permit any conclusions. When the number of grains counted corresponds to 5000 AP the variations* become interpretable.

Another simple test should be more or less automatic: if a curve maximum or minimum is formed by a single sample only, two neighbouring samples should also be counted. These two samples ought then to be intermediate or higher, giving a three-spectrum maximum. Single-spectrum maxima or minima should, also for statistical reasons, always be regarded with suspicion, and conclusions should not be drawn from them without further

* A confidence limit of e.g., 95 % means, practically, that in 95 % of all cases, the 'true' value is found within the interval (but in 5 % it is still outside!).

* For such counts it is obviously not necessary to count specifically all taxa. The non-relevant ones can be counted as one group. For statistical reasons one should not single out only one major pollen taxon and use that one as a marker; cf. the discussion of 'inside' and 'outside' percentages.

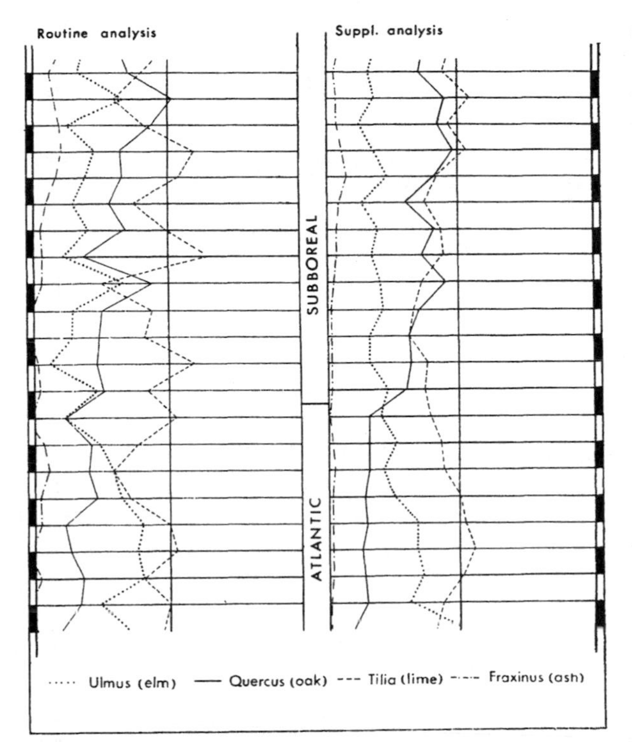

Fig. 7.20. Diagrams illustrating the effect of increasing the basic sum for percentage calculations. The diagrams show the same material, viz. the QM constituents of an analysis, but in the right-hand one the basic number has been increased about ten-fold. The distinct trends of the curves apparent from the right-hand diagram are lost in statistical noise in the left-hand one. Each diagram comprises 10 %. After Hafsten, 1956; redrawn. Reproduced with the author's permission.

control. A good plan is at the outset to count every 8th or 16th sample and intercalate samples until curves become sufficiently regular. This gives an automatic check upon all maxima except those becoming apparent at the last intercalation.

Practical experience gives certain indications as to how many pollen grains must be counted in order that contingent regularities shall not be hidden by sampling errors, but mathematical treatment of the material is the only way in which one can obtain more definite answers.

Ording (1934), Westenberg (1947) and Mosiman (1962, 1963) have subjected the statistical counting uncertainty to a formal treatment, presuming a binomial distribution of values. That this assumption could be regarded as valid was shown by Fægri and Ottestad (1948). Consequently the limits of statistical counting uncertainties are defined by the standard variation of the binomial function, which is dependent on the value of the observed percentage and on the number of grains counted (cf. Table 7.4, which shows the 95 % confidence interval). This means that if the observed percentages are as indicated in the left-hand column, it can with 95 % confidence be stated that the 'real' value will be found within the interval. The nearer the percentage approaches 50, the wider the limits of error, decreasing symmetrically towards both extremes.

In pollen analysis the numerical value of the confidence interval is less important than the *relative confidence interval*, i.e. the relation between the confidence interval and the observed value. If the count tallies 50 % and the confidence interval is 14 %, the relative confidence interval is

$$\left(\frac{14\,\%}{50\,\%} \times 100 = \right) 28,$$

and the 'true' value in 95 % of the cases presumably found between 43 and 57 %, which is tolerable. If the count gives 1 % and the confidence interval is 3.4 % the relative interval is 340 % and the 'true' value presumably between (− 0.7) 0 and 2.7 %, not a very useful observation: it gives too wide a berth.

Table 7.5 shows the counts necessary to establish relative confidence intervals of, respectively, 40 and 100 %, i.e. the confidence interval constituting 2/5 or 1/1 of the percentage under consideration. The table is not symmetrical on either side of 50 %, but for reasons detailed above (Fig. 7.6) the higher

Table 7.4. Width of the 95% confidence interval pollen percentages.

Percentage density	Number of pollen grains counted				
	50	100	200	500	1000
1(99)	9,2	5,4	3,4	2,0	1,3
3(97)	11,7	7,6	5,1	3,2	2,2
5(95)	13,9	9,3	6,4	4,0	2,8
10(90)	17,4	12,1	8,6	5,4	3,8
20(80)	22,2	15,9	11,2	7,2	5,1
30(70)	25,1	18,1	12,9	8,2	5,9
50(50)	27,2	19,2	14,0	8,9	6,3

Table 7.5. Relative confidence interval in relation to percentage and pollen counts

Pollen percentage observed	2/5 Relative confidence interval (95%)		1/1 Relative confidence interval (95%)	
	interval, by definition	Necessary pollen count	Interval, by definition	Necessary pollen count
50	40–60	100	25–75	16
20	16–24	400	10–30	64
10	8–12	900	5–15	144
5	4–6	1900	2.5–7.5	300
2	1.6–2.4	4900	1–3	780
1	0.8–1.2	9900	0.5–1.5	1600

values are of little practical interest. The table shows that even such moderate demands for statistic control are fulfilled only rarely, and as soon as percentages approach low values these demands are practically impossible to satisfy.

The data are based on the assumption of a simple binomial distribution. A more penetrating statistical treatment has been given by Mosimann (1962, 1963), who considered a multinomial distribution. His results have been used by Maher (1972) for calculating nomograms giving the 95 % confidence limits based on Mosimann's equation. Actually, values do not differ materially from those of Table 7.4 and, again, irregular variations are certainly greater than the difference between the two sets of statistics. Martin and Mosimann (1965) developed the statistical treatment further on the basis of actual analysis data.

Westenberg (1964, 1967) has used a different distribution function, viz. the hypergeometric distribution, and has published a set of nomograms for testing the significance of differences in a pair of pollen percentages. The statistics are of the same magnitude as in the preceding calculations. Such more sophisticated mathematical methods are interesting for showing the possible mathematical structure of the counting universe. They tell nothing more about the other sampling problems, and the small numerical refinements in confidence limits obtained by such methods in comparison with the simpler techniques (as above) are of little practical importance in analysis and interpretation.

An important practical use of confidence statistics is in the comparison between two percentages. If they are situated within the same chosen confidence interval (e.g. 95 %) there is no statistical reason to believe that they are different. If the difference between them is greater than half of the 95 % confidence interval, there is a 95 % chance that they are estimates of two different 'true' values. The chance can be expressed quantitatively, viz. as the confidence percentage, but it must never be forgotten that there is always a certain chance that this is the odd value that falls outside the interval. Counting of samples never gives absolute values.

One important, but frequently overlooked, point is that if variables, e.g. percentages, are chosen for comparison because the differences to be tested are greater or smaller than expected, sampling is not unbiased, and basic statistical assumptions are no longer valid.

The statistical counting errors dealt with above refer to the slide counts only. Two other questions of sampling errors are equally pertinent, viz. the question of whether values also distribute themselves binomially (1) if new preparations are made from the same major sample for each count, and (2) if samples are taken individually over the surface of the bog. Both questions have been investigated by Woodhead and Hodgson (1935)—who did not, however, test statistically the significance of their results—and by Fægri and Ottestad (1948) and later authors (see also Birks and Gordon, 1985: chap. 2). The answer to the first question is that no departure from the binomial distribution can be demonstrated.

The second question is more complicated; it pertains to the distribution of pollen grains over synchronous surfaces, i.e. the composition and distribution of the pollen rain. Woodhead and Hodgson (1935) analysed ten surface samples taken from places up to 60 ft (18 m) apart (p. 268). Fægri analysed fifty samples belonging to an apparently synchronous layer in an exposed cutting, the maximum distance between samples being 1 m. In both cases the observed data were consistent with hypothetical values for random distribution, and one may therefore conclude that over small surfaces the distribution of pollen grains in synchronous layers does not depart appreciably from the binomial: consequently the sampling error is in this case also defined by the standard variation of the binomial function. On the other hand it is self-evident that—especially in areas of variable vegetation—different parts of the bog may receive different types of pollen rain (some striking examples have been published by Lüdi, 1947), but this is a matter of representation, not of sampling error.

If the binomial counting errors were the only ones, the practical requirement would simply be that counts should be made as great as possible—or rather, sufficiently great to reduce the confidence interval to below a given level. The sequence of percentage figures would then approach the 'true' sequence with increasing size of the counts. However, it must be presumed that pollen curves—like most empirical curves—contain 'noise': irrelevant fluctuations which distort the picture; in other words, that variations which are in themselves 'true', are irrelevant for the investigation in hand.

In such cases the limit which the percentages approach with increasing size of the counts will contain maxima and minima that are irrelevant and misleading because they are consequences of irrelevant factors, the effects of which are restricted to short spans of time and space. It is evident that if there is any doubt about its validity it is more dangerous to use an irrelevant short-term maximum (or minimum) as a real one, e.g. for synchronization purposes, that it is to refrain from using a real maximum (or minimum). A special case is represented by freakish, but 'real', maxima probably due to aberrant long-distance transport outside the area of the taxon in question (e.g. A. Danielsen, 1970: 110).

Concepts like validity or dimensions in time and space, signal and noise, are not *a priori*, but depend on the problem under consideration. What is irrelevant for an investigation aimed towards the general climatic development may have high relevance in the context of an archaeological excavation. The concept of short-term fluctuations is a point in case. A diagram contains a certain number of details only, has a limited 'resolution power', and only a decrease of distance between samples can give more details. The distance in time between two pollen spectra is not a constant magnitude, either from diagram to diagram or even within the same diagram. It depends primarily on the method of sampling and on the rate of deposition of the peat or sediment to be analysed. In routine work an ordinary distance between samples is 5 cm; these 5 cm represent 50 years or less in a rapidly growing peat, 500 years or more in a slowly accumulating sediment. In any case, if a more or less pronouncedly cyclical development has taken place, and we want a satisfactory representation of the development around a maximum or minimum, the full recurrence should generally be represented by about eight samples (four samples per maximum or minimum). In the first of the two cases mentioned above, where distance in time between samples was on an average 50 years, the diagram cannot profitably be utilized for the study of a recurrent development of shorter total duration than 400 years: the 'resolution power' of such a diagram will be 400 years. If shorter fluctuations are involved, they will—even if they recur regularly—occur as irrelevant fluctuations which cannot be analyzed properly by means of the ordinary diagram. Similarly, the resolution power of the diagram in the second case mentioned above will be of the order of 2000 years per maximum or minimum (4000 years per full cycle).

Pollen diagrams often contain so much noise that it is not immediately apparent if there is also a trend or division of some kind (cf. Fig. 7.21). In many such cases simple statistical tools can help. Noise in trends can be practically removed by

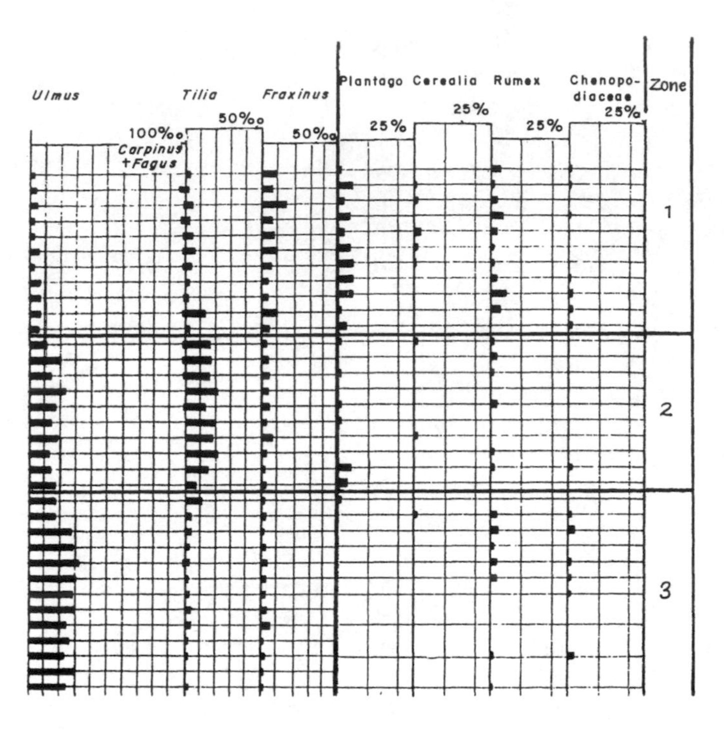

Fig. 7.21. The diagram tested for validity of differences in the *Fraxinus* curve, cf. text. From Fægri 1944b; redrawn.

smoothing the curve (running averages over three or five samples). However, it is a dangerous procedure, which can lead to loss of important information.

Parts of curves may be compared by contrasting variance within parts against that between parts. The following case exemplifies this (cf. Fig. 7.21). On the basis of the *Plantago* and *Ulmus* curves the diagram is divided into three (local) zones, probably indicating different levels of human impact. The *Tilia* curve obviously follows the same tripartite pattern. The question is whether *Fraxinus* does so as well. The differences are small, but a statistical test (Fisher's *t* test) gives the probability value of 1.2 % against the two lower zones being numerically different, and 0.22 % against the two upper zones being so. We may therefore confidently state that *Fraxinus* also exhibits different behaviour in the three zones, but this in itself *in no way permits conclusions* about human impact or other causal explanations. It does indicate that there is an effect, the causes of which may be interesting or not. Since the division of the diagram into three zones was made independently of the *Fraxinus* curve, the statistical test is valid; it would not have been valid for the *Ulmus* curve.

If the curves, especially those on the basis of which the subdivision of the diagram has been made ('indicator curves'), reach high values, they will be interdependent (negatively correlated) because of the closed universe (the 100 % sum), which goes against the theoretical basis for calculations like the above (cf. Adam, 1967). The problem has been taken up for theoretical discussion in modern numerical analysis (Aitchison, 1981).

If there is no independent indicator curve, but the diagram is divided into parts because they look different, the parts are not chosen at random and the within/between variance criterion does not, strictly, apply.

It has been customary in pollen analysis to calculate certain taxa 'outside' the pollen sum. The classical example is *Corylus*, which was from the beginning excluded from the pollen sum on the (fallacious) ground that 'die normale Rolle des Hasels nur die eines regelmässig vorhandenen Unterholzes innerhalb. . . . Baumverbände gewesen' (von Post, 1929: 550). Later on, NAP types, etc., have been calculated in the same way. The principles for this have been discussed in previous chapters. Logically the calculation of 'extras' is rather dubious, and mathematically it is even worse. From a statistical point of view the calculation of pollen types as 'extras' is not to be recommended in any case. In Mosimann's words (1963: 53): 'coefficients among inverse counts (i.e. calculating 'outside'). . . are neither estimates of corresponding coefficients among numbers of pollen falling nor among proportions of pollen falling. Rather, they are estimates of correlations among indices which are not useful quantities'. This statistical procedure is, nevertheless, used in the calculation of absolute pollen frequencies (cf. Section 5.3.3).

Nomograms for confidence limits of such data are given by Maher (1972). Indeed, the actual phenomena with which pollen analysis is concerned make the calculation 'outside' both mathematically and biologically unsound. As the species delivering the 'extra' pollen practically always compete for space with the 'ordinaries', the two groups are mutually exclusive. If there is a great deal of NAP this means that forest, i.e. AP-producing species, is excluded from a greater part of the area under investigation. As a negative correlation thus exists between the densities of 'ordinaries' and 'extras', the variations of one group in relation to the other will be greatly exaggerated. The calculation of pollen categories as 'extras' should therefore be avoided; it is merely traditional and has no background in vegetational fact. The traditional AP diagram is a means of studying the composition of forests; a NAP diagram is a means of studying forest-less areas, but the only adequate means of studying the whole vegetation is the total diagram. There is, of course, nothing to prevent an investigator excluding from the calculation any species for whatever reason—the diagram then tells what happened in those parts of the area in which the species excluded did not occur, but the calculation of such species as 'extras' is of very limited value. All pollen taxa should be expressed in percentages of a universe (a pollen sum) of which they form a part. If such a sum does not exist, a special one should be created by adding the values for the taxon (taxa) in question to the regular universe, e.g. spores added to the pollen sum to express the incidence of fern, etc., spores. The only exception is formed by those taxa which occur in such small numbers (below *ca.* 5 %) that inclusion or no inclusion does not materially change the numbers. If the number of grains included in the sum is fairly constant throughout a diagram, such taxa may also be represented by absolute figures from the count. The Bingham diagram (Fig. 7.1) is an example of a universe that is gradually stripped down for the discussion of increasingly narrower segments of the local vegetation.

7.6.2. Advanced numerical treatment of pollen data

Simple observation of correlation among pollen curves within a diagram and between diagrams is the basis of interpretation of pollen data. Some early attempts were made at a formal statistical approach, but neither statistical theory nor mathematical facilities were at that time adequate for coping with the complicated multiple correlations of pollen data. Also, the effect of the closed numerical universe (the 100 % sum) presented—and presents—serious constraints; so do also the fact that variables are locked into a strict time sequence which makes it meaningful only to compare a variable with its two closest neighbours. This imposes a serious constraint on many clustering techniques.

The introduction of the electronic computer has made possible the use of various numerical tools in pollen analysis, which were earlier unknown, or at any rate not practicable. It is a very large subject, and will not be dealt with exhaustively in this book, especially as a comprehensive survey of problems and methods has recently been published by Birks and Gordon (1985), which should be consulted. The purpose of the following paragraphs is chiefly to point out some of the possible fields of investigation with special reference to pollen analysis.

In palynology the computer has, until now,

chiefly been used within three fields: (1) calculation and construction of pollen diagrams (dealt with in Chapter 6). This chapter deals with numerical and statistical methods as aids in (2) morphological analysis and for recognition of microfossils ('palynomorphs'), and (3) for aiding the interpretation of pollen data.

7.6.2.1. Morphologic analysis

The problem of keying out an unknown pollen grain always remains with the analyst, and any help is important. However, no key is better than the morphologic analysis upon which it is based, and at present the lack of adequate descriptors for the identification procedure is the main bottleneck. The problem of computerizing the morphology of the pollen inventory is more pressing when sporae dispersae are to be dealt with, and its solution is thus of greater interest in applied and pre-Quaternary analysis. In order to be compatible with the demand of the computer, morphographic descriptions must follow a rigid system suitable for coding. Germerad and Muller (1971) have published a very comprehensive thesaurus which seems to be too complicated for ordinary use in Quaternary analysis. Simpler systems—manual or electronic—may be constructed *ad hoc* (cf. also Straka, 1986).

However, work along these lines is continuing, and whereas the general problem has not yet been solved, some very elegant special studies have been published, e.g. Birks' and Peglar's (1980) study of differentiation and taxonomic identification of pollen of *Picea* species in eastern North America (cf. also Jacobs, 1986). A more primitive numerical treatment of the same problem was published by Ting in 1966, a completely non-numerical one by Jentys-Szaferova (1959). From a general point of view the subject has been exhaustively treated by Reyment *et al.* (1984).

7.6.2.2. Diagram interpretation

This is the most promising and interesting field of use of numerical methods in pollen analysis, but also the most critical and complicated, both theoretically and (computer-)technically. We must refer to Birks and Gordon (1985) for all detailed discussion, and shall restrict ourselves to pointing out some basic principles.

Interpretation problems can be seen from two sides; the pollen spectra, i.e. chronologic samples, and the pollen taxa, which in this context are ecologic samples. Referring to the pollen diagram, the one may be considered a comparison between horizontal, momentary samples and the second between vertical, sequential data (the pollen curves of the individual taxa). The two sets of data are combined in modern analysis (cf. Fig. 7.24).

The study of samples deals with the homogeneity of the diagram, its division into more or less homogeneous parts, i.e. zoning. For this purpose there are various similarity and dissimilarity functions, which are fairly easy both to understand and to apply (cf. Birks and Gordon, 1985: Chapters 3 and 4). They are not far removed from biological reality, and the numerical assumptions and procedures behind them are simple and easily controlled. Figure 7.22 shows the results of various numerical zonings of the test diagram.

Birks (1974) carried out numerical analysis on two diagrams, comparing the results and also those of the zoning previously carried out by visual inspection. As might be expected there is very little difference between the results of the various methods and between these and the result of visual inspection. Mostly it means moving a zone limit up or down a spectrum, or establishing an extra zone which is apparent also on visual inspection, but then neglected as being ecologically insignificant, even if its mathematical validity is established.

However, these simple experiments show some of the problems inherent. The computer does not recognize the indicative value of the individual components of the diagram. A change in one curve between 1 % and 4 % may indicate more, ecologically, than the change in another between 10 % and 40 %. This can, of course, be compensated for by loading the values of certain components of the pollen rain, but then we are already moving off the objectivity which should be the hallmark of numerical methods. This is even worse when curves of the same taxon, to be ecologically meaningful, should receive different loadings in different zones/

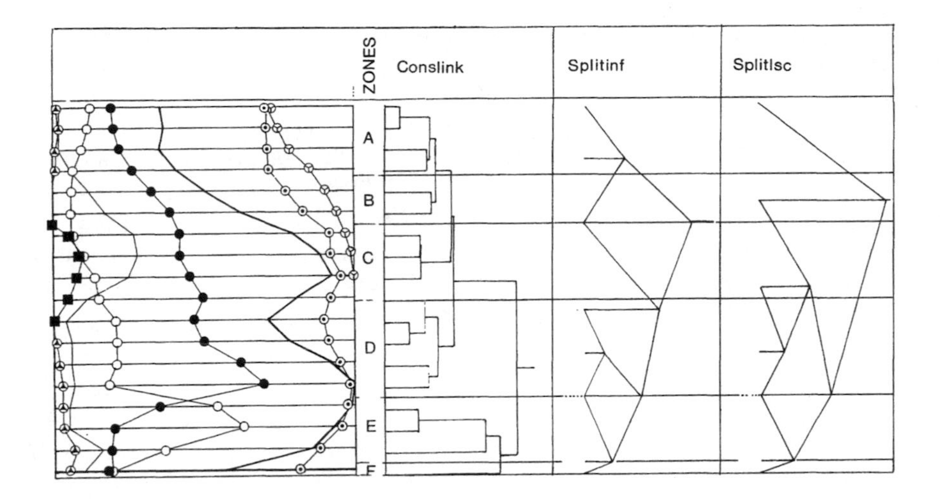

Fig. 7.22. Numerical zoning of the test diagram, Fig. 6.2. The material for this and Fig. 7.23 has been prepared by John Birks and placed at our disposal.

vegetation types of the diagram (cf. the *Artemisia* example in Birks and Gordon, 1985: 88). A more objective way would be to resort to autoscaling (Section 6.1.3), but then again, this presumes all taxa to be of equal importance, which they certainly are not. In numerical pattern recognition procedures all curves are also equal, notwithstanding their original absolute value.

Birks and Gordon (1985: 89 *et seq.*) set out some of the reasons why numerical and visual zonations do not always agree. Both are 'right'—on their own premises. Those of the computer are strictly numerical, and therefore presumably repeatable ('spurious objectivity'—Birks and Gordon, 1985). The zonations based upon visual inspection should, ideally, include a measure of *informed* subjectivity.

The objective of multivariate correlation analysis is to reduce the multidimensional universe of the original data to a low-dimensional set, which is more easily presented and perceived. There are several methods for which we refer to Birks and Gordon (1985), quoting their p. 167: 'ideally, we would like to have theoretical reasons for specifying a single most appropriate method of analysis'. Unfortunately, this is not (yet?) possible: various methods are differently suited to bring out the properties of the material at hand and also to answer the questions immediately pertinent. All numerical procedures are based on certain premises, and no result is more valid than are the

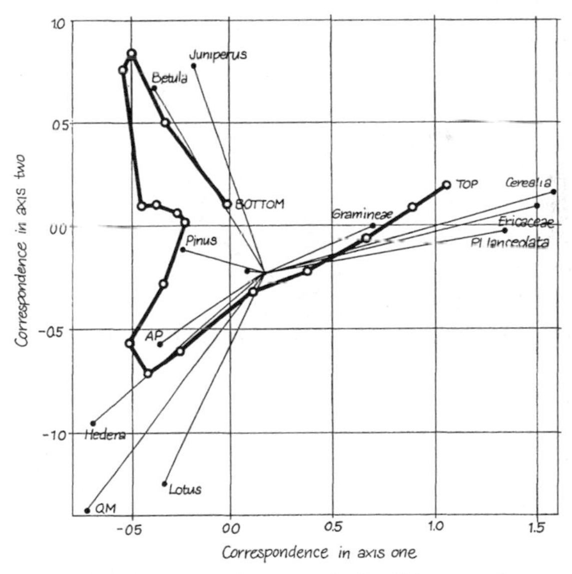

Fig. 7.23. Correspondence analysis of the test diagram, Fig. 6.2. Length of vectors indicates variability, angle between vectors indicates correlation e.g. a very strong correlation between *Lotus* and QM and no correlation between *Lotus* and *Plantago*. Samples divide themselves between a protocratic, a mediocratic and a terminocratic phase (Iversen's terminology).

premises on which it is based: whether or not the mathematical premises correspond to a biological reality. The result of a numerical analysis test should always carry the qualification that results are compatible—or not compatible—with the basic premises.

Pattern recognition analysis permits the grouping of pollen taxa into groups exhibiting the same behaviour in the diagram. Whether the reason for this behaviour is historic, climatic, ecologic or due to human impact, cannot be decided by mathematical methods. That needs biological (ecological) insight. It is therefore important that the results of such analyses be presented in a biologically meaningful form. Unfortunately, the most frequently used method, principal component analysis, does not give results that can be immediately read as ecology. Figure 7.23 gives the result of a correspon-

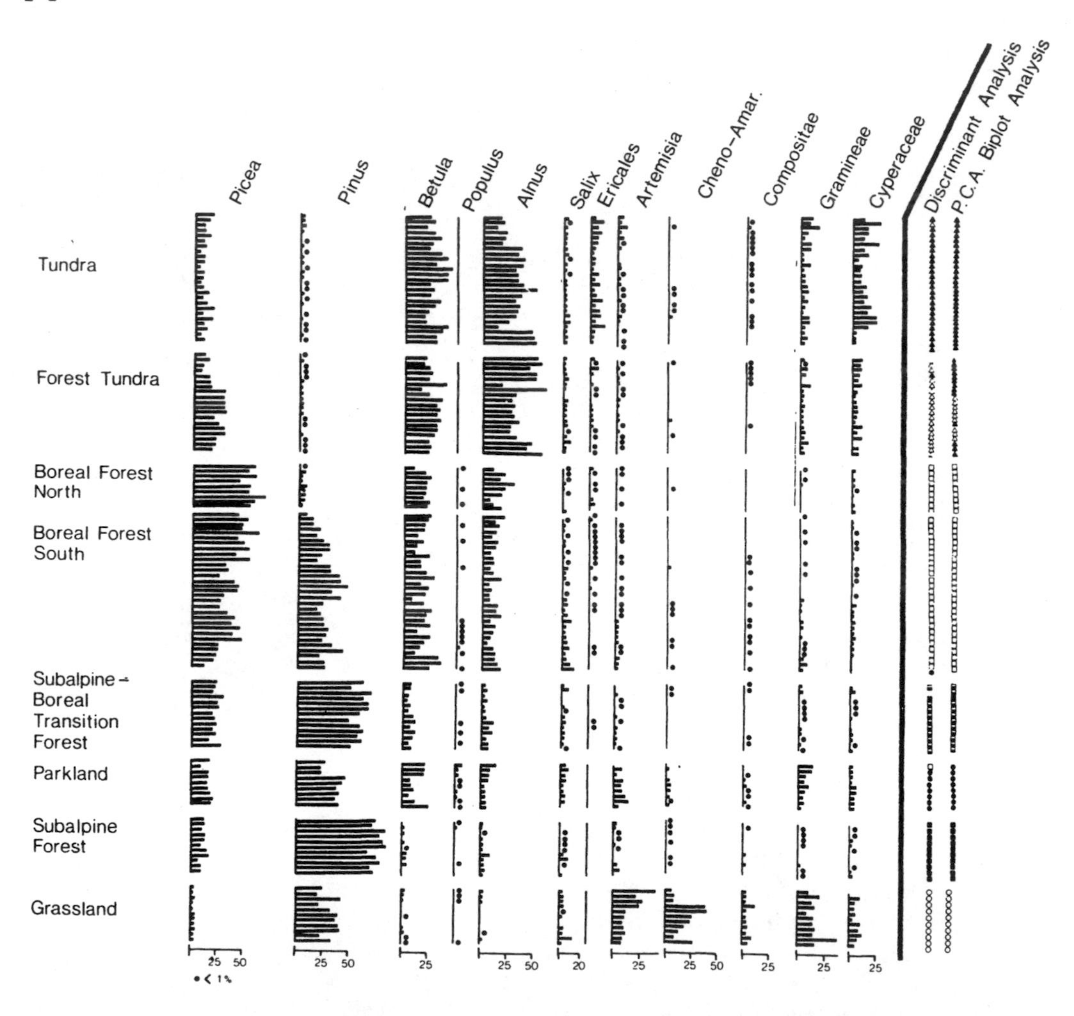

Fig. 7.24. Modern surface pollen spectra arranged according to the observed vegetation at the sites (A). The classification of each site according to PCB ordination and DA results are indicated at the right-hand margin of (A). At right (B, opposite) principal component biplots ordination of the pollen taxon loadings and modern pollen surface sample scores on the first (horizontal) and second axes. The observed vegetation setting of each sample is indicated by the appropriate symbol. From MacDonald and Ritchie, 1986; redrawn, subordinate taxa omitted from (A).

dence analysis of the test diagram, in which the biological significance is displayed.

Since the test diagram is small and contains a low number both of spectra and taxa recognized—so are many 'real' diagrams, too—it is easily appraised by visual inspection, and a complicated numerical procedure would be redundant. However, in cases of doubt a mathematical analysis may help to corroborate or refute the immediate impression.

On the other hand, it is very difficult to appraise directly large modern diagrams where the number of taxa recognized is of the magnitude of a hundred and the number of spectra may approximate the same magnitude.

In the absence of a computer this was done by cutting up the (resolved) diagram and shuffling the curves until they formed more or less natural groups. In such cases statistical ordination procedures are great time-savers and give an essential check on a meaningful arrangement of data, e.g. by bringing together curves of taxa showing the same development, and therefore probably being ecologically related. Correspondence analysis is a particularly useful technique for this purpose.

Mathematical pattern recognition and ordin-

B

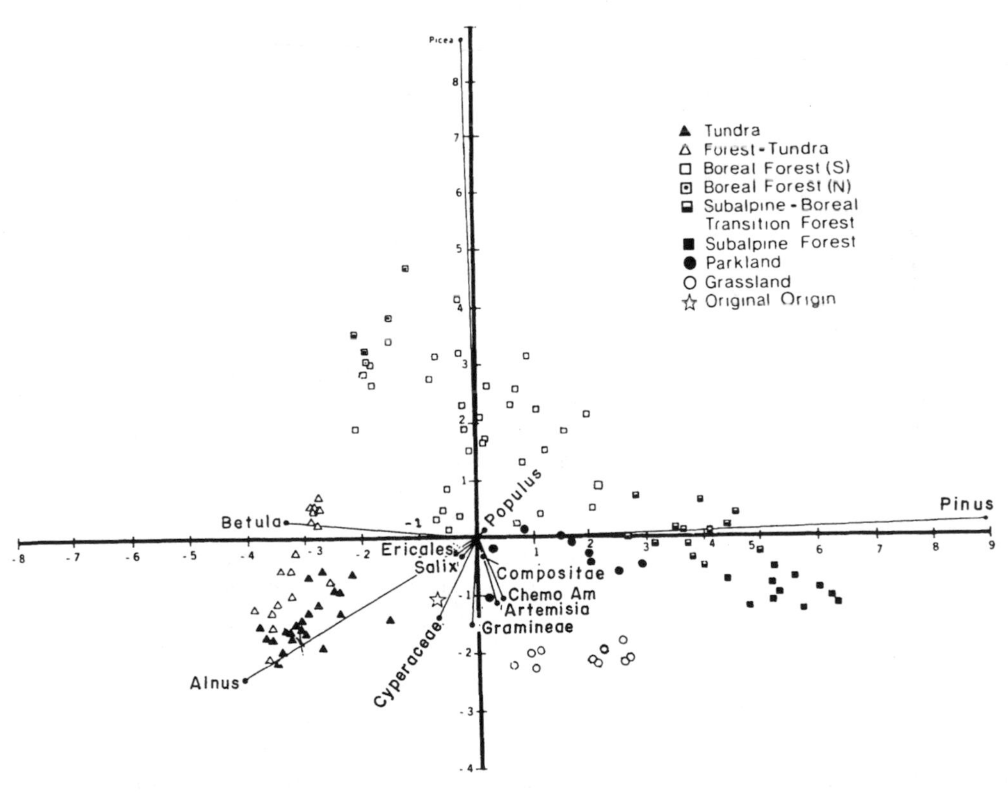

Fig. 7.24. *(continued)*

ation are also indicated in analysis of the relation between the pollen record and the actual vegetation. Figure 7.24(A) shows modern (percentage!) surface pollen spectra from a tundra–forest–grassland transition area in Canada. They have been subjected to discriminant and PCA biplot analysis. The results of the numerical analyses are, on the whole, identical with the vegetation observed in the field except for some transitional tundra–forest–tundra samples where PCA analysis disagrees with discriminant analysis and observation. Apparently, this is due to (super?) sensitivity of the PCA analysis towards the *Alnus* curve. By placing a fossil spectrum in plot B one may assess its modern equivalent provided other conditions are the same as in the test area. The same can also be achieved by direct ocular comparison between the fossil spectrum and those of (A). In cases of non-correspondence between the result of numerical analyses and direct vegetation observation, the latter should be given priority (cf. also the caveats given by MacDonald and Ritchie, 1986). The most important difficulty is the unavoidable lack of real analogues between, especially, Late-Glacial vegetation and the present one, due to effects of soil maturation and plant migrations. For the interpretation of the visual picture one can use imperfect analogues; the computer has difficulties in abstracting from them and, again, the necessary corrections may make the original objectivity of the mathematical methods rather spurious, to quote Birks and Gordon (1985).

Uncritical application of statistics or other numerical methods does not serve any purpose. Unless an investigator possesses a certain knowledge of the principles behind the procedures it can be very dangerous to use statistical or other numerical formulae. The pitfalls are numerous and treacherous, and an unskilled worker can make tests of no value whatever, or even worse: that are directly misleading. This also applies to exercises of mathematicians without the necessary biological background. If not properly understood a mathematical treatment may give a false sense of security: it can only give information about the figures on the tally sheet, but gives no control over all the other irregularities present in the material and its treatment. The value and very important mission of statistical considerations is to demonstrate the uncertainty built into the numbers obtained by analysis.

The easy availability of computer programs effortlessly performing various statistical tests on pollen data embraces some danger. 1. The danger of thoughtlessly pushing the same button each time without really reflecting if the method involved is meaningful for the case at hand. A scattering of undigested PCA data in a text does not improve it. 2. The danger of overemphasizing the importance of statistical control of analysis figures can lead to the neglect of other pitfalls in interpretation and understanding of pollen data.

No useful purpose is served by reducing the purely statistical noise to a level lower than that created by non-statistical noise in the material (numerical methods in themselves don't differentiate). The important matter is to know at which level of confidence one is working, not to press this level unnecessarily. Statistical parameters never give absolute data: confidence intervals can be narrowed, but never to zero.

Numerical methods deal with numbers only. To distinguish, in a biological registration, data on a purely mathematical, numerical basis, gives a mathematical result, which may be mathematically meaningful, but that is no guarantee that it is also biologically so. Unless the result of a statistical analysis can be translated back into terms of vegetation it is, pollen-analytically, an empty exercise.

In many, perhaps most, cases pollen-analytic experience renders complicated mathematical methods less important. After all, *vegetational* changes can be evaluated only by a mental integration of the whole data set: the diagram curves are not variables to be discussed as such, they are registrations of vegetation and no mathematical analysis, however refined, can replace botanical common sense in the final phase of interpretation. That much said, it must also be clearly stated and understood that the numerical analysis of the same data can be a very important and sometimes powerful help, especially in doubtful or complex cases.

The problem is aptly summarized in von Post's words (in a slightly different context) in what may have been one of the first major international discussions of the statistics of a pollen analysis (in *Geologklubben*, 1947: 239, orig. Swedish): that one 'is right in warning seriously against the statistical error devil. He may pop up at any time. . . . Even against this potentate common sense is a good protection'.

Computer programs suitable for pollen data have been produced at various laboratories and for various purposes. Some of them are commercially available. Others may be had at a more or less nominal price from the laboratory of origin.

8. How to use the information. Applications of pollen analysis

Motto: We have the tools: let's do the job.

8.1. Direct information: vegetation and flora

The immediate information of a pollen-analytic investigation comprises the pollen spectrum and the pollen diagram, which are discussed in Chapter 6. These are in themselves very valuable documents, indicating the composition of the vegetation of a certain area around the sampling point at a certain time. Depending on sampling conditions the area from which vegetation is registered may be large or small, and the degree of resolution of the vegetational record correspondingly general or specific. Not infrequently it is both, the record containing pollen representing the general vegetation as well as a smaller complement of strictly local origin. It is part of the primary understanding of a pollen spectrum to keep these two components apart, which can never be done 100%. The interpretation of the pollen record is dealt with in Chapter 7.

Another factor which greatly influences the specificity of the pollen record is the taxonomic level down to which the pollen can be identified (cf. the keys). If local pollen can be identified down to species level the botanical analysis can become very accurate. Especially if the pollen flora of the deposits analyzed is largely of very local origin and includes pollen of entomophilous plants, there is a possibility of a phytosociological analysis of former plant communities (cf. Smit and Janssen, 1983). Pollen from species indicative of certain plant communities of various rank ('character species' and/or 'differential species' cf. Rybničková and Rybniček, 1971, 1985) may occur in these deposits, though sporadically. Linked with the indications deriving from the quantitative relationship existing between major pollen producers (R_{abs} values) these may contribute to establish the sociologic character of specific plant communities. For a further analysis of conditions of formation of the pollen spectrum such narrowly defined plant communities are much sharper instruments than the always vague general pollen record, and also more indicative than the record of individual species.

However, for a numerical analysis it is safer to base inferences on the wider ecological demands of individual species rather than on the more specific ones of plant communities. The former only presumes that the species in question is found within its whole potential area and that it has not changed its ecological characteristics in the meantime. The latter in addition involves so many qualifications that the possibility for a biased selection is a real danger (cf. Birks, 1981); the lack of a modern analogue which can act as a calibration is also a drawback.

A third point which is directly registered in the pollen record is the *immigration* of the flora (cf. Section 7.2).

The three fields mentioned represent the only information given *directly* in the pollen record. All other information that can be derived from the record is indirect and must be based upon inferences about the interdependence between plants and ecologic factors, and can never be better than are those inferences.

8.2. Indirect information

The causes of vegetation changes registered in the pollen diagram may be general or local, or usually general ones modified by local conditions. General causes are major climatic changes and major geologic events: isostatic effects, glaciation, etc. Local effects are, above all, human, but also minor geologic events (landslides, volcanoes), soil development and 'endogenous' effects of plant successions and competition, windthrow, etc. Effects of general causes are (sub-)synchronous, but not necessarily identical. Effects of local causes are metachronous, but often—within the same vegetation region—identical.

In drawing non-vegetational inferences from a pollen-analytic record it is imperative to realize that diagram changes are not necessarily, as a matter of fact are probably not, the result of a single change in ecological factors. The arrival of exigent plants in northern Europe after the Ice Age was for many years interpreted as a one-to-one reaction of plants on climate. Only later has the concept gained acceptance that the distance from refugia and factors of dispersal ecology may have had a decisive influence (Birks, 1986a). Similarly, but more complicated, the *Ulmus* decline in European vegetation at about 5000 BP has been attributed variously to climatic deterioration, human influence, and disease, all factors discussed individually as *the* cause. Possible synergistic effects have been neglected, and so have additional, but less obvious, causes like increased virulence (mutations!) of pathogens, dispersal of pathogens and/or vectors.

The statistical observation of synchroneity between the elm decline and the first appearance of agricultural indicators suggests (but does not prove) a possible causal relation—direct or indirect. However, elm declines observed in regions where no trace of agriculture, or even of human occupation, are known do indicate that causes other than human influence are also involved, but do not preclude the possibility of human impact. It only shows that this explanation does not cover all cases.

A conceptual model is a great help in clarifying a case like this. Take as a starting point the truism that in a (semi-)stable vegetation there must exist an equilibrium between an organism and its parasites, including pathogens. If not, one or the other becomes too strong, the opposite part is exterminated, and if that is the host, the parasite also dies out as a consequence. Let us presume that somewhere there existed such an equilibrium between elms, and an elm pathogen (and its vector). If the pathogen is moved out of this area into another where the elms have not achieved the same degree of immunity the condition is there for a major epidemic. Under natural conditions this spread will go hand-in-hand with a spread of immunity. If spreading is accelerated, e.g. by human influence, the pathogen may outrace the immunity. This may happen if there is human transport of the pathogen (or the vector) out of its endemic area. Such a human-induced dispersal would conceivably be possible with migration of whole human populations, or simply with the spread of an agricultural technique, notwithstanding who practices it. The Neolithic introduction of agriculture is just such a case.

Once arrived outside the area in which the elms

possessed a relative immunity the pathogen could spread very fast through the local elm population without any more human help. The result would be two-fold: (1) momentary killing of all susceptible elms, a drastic reduction in number in a short time; (2) establishment of a new equilibrium between the remaining (semi-)immune biotypes and the pathogen. Experience shows that such equilibria are often much lower than before the epidemic, as immunity also confers unfavourable genetic traits. Once dispersed to a new area the pathogen (whether a new species or a more virulent type) is able to look after itself and spread also outside the immediate human domain.

This model, so far, does not presume any direct human action against the elms, but that does not mean that it excludes such an action. As a matter of fact, utilization of elms for fodder or differential destruction of elm forest because of clearing of the better soil would not only diminish the number of trees, it would also with great probability render the rest more disease-prone, since the whole population would be weakened.

Does climate come in? Not necessarily, but, again, this does not preclude climatic influence. Climate is never irrevelant. But what climatic effect: on the elms? on the pathogen? on the vector? These possibilities must be left open until further evidence —or an analysis of the earlier data—elucidates them.

The effect of dispersal of pathogens outside the area of (semi-)immunity has been observed many times in the history of humankind: the Black Death, measles epidemics in the South Seas, *Phylloxera* on European wine cultures, the chestnut blight, the Dutch elm(!) disease. In the interpretation of the past we should not forget the present. The elm decline is a good example of a simple reaction upon a possibly very complicated series of causes, and is discussed here as such, not as a suggestion of the final solution of a vexatious problem.

An ecological factor which is frequently overlooked in drawing conclusions from changes in the diagram is competition. Vegetation changes are usually caused by a change of *competition* relations caused by some external factor, e.g. climatic change, but competition may also change without major non-vegetational changes, e.g. by the immigration of another (sub-)dominant species. Welten's diagram from Simmental (Fig. 7.8) gives an impressive picture of how the immigration of *Abies* completely changes the vegetation picture, out of all proportion to the changes of climate as inferred from other independent evidence. On the other hand, the difficulty of entering a climax or near-climax vegetation may prevent a species from establishing itself inside an area which would be climatically suitable.

A pollen diagram in which nothing happens (curves run parallel to the base lines) is not particularly exciting. However, it is important because it suggests that, within the time span recorded, vegetation was in equilibrium with habitat conditions: climate, soil, human impact. Further deductions as to the absolute value and relative importance of the various parameters resemble the solving of a single equation with three unknown —the major technique for which in our case is elimination.

On the other hand a diagram in which curves change—as is usually the case—testifies to a vegetation not in equilibrium with external conditions, a situation that exposes it to a stress. The result is a succession, which may be an immediate, catastrophic one, but is usually an event over a long time, decennia, perhaps centuries, during which external factors change continuously. Plants and plant communities are sensitive ecologic indicators, but their reactions follow their own laws, those of plant successions, not those of climate or any other parameter. Absolute data for such parameters are difficult to come by in a changing diagram, but the direction of the change is usually clear. What does vegetation response to climatic (ecologic) changes really mean? Climatic change, especially, is much faster than plant successions, which therefore integrate short-period climatic variations. The effect of changes in ecological parameters is less dependent on the numerical magnitude of the change than on its duration (Fægri, 1950a). A small change lasting a century may have profound effects; great changes lasting a year of two may have next to no effect. After all, the change between seasons in temperate regions today is probably greater than any long-term climatic change the Earth has seen since the Cambrian, but the general pollen-analytic, i.e. successional, effect is nil. It is integrated away. On the time scale of a century a temperature change of a few degrees would mean a climatic revolution.

Pollen analysis shows only one, albeit fairly important, aspect of plant life: flowering, which gives a perfect description of the occurrence of annuals. It is a very blunt instrument for the analysis of momentary composition of a forest, the components of which, i.e. the individual trees, may persist in a semi-sterile state for very long periods, especially in the case of an alternative vegetative reproduction. Such biological complications as these make the predictive value of simple differential equations (like Webb, 1986: 79) rather weak.

As mentioned above, an exception must be made for catastrophic events which, however, are usually followed by a relatively rapid recuperation. If they are not, this may be an indication of a pre-existing disequilibrium which was hidden, e.g. by the homeostasis of the plant cover, or it may have been upset by factors causing or concomitant with the catastrophe. An example would be an immigrant species which got a chance to establish itself on the ground laid bare by the catastrophe.

A major catastrophic event, although a benign one, was the recession of the Ice Age glaciers, exposing areas of virgin ground. Since the reaction of major glaciers to climatic change is slow, the climate at the glacier's snout was probably considerably more favourable than the proximity of the ice would indicate. Plants from many different climatic zones could establish themselves provided they could live in a soil with little biological activity. As the buffer capacity of such fresh soils is very low, calcicolous species could grow in a soil derived from rather acid rocks, and vice-versa.

Since neither climate nor soil exerted any stress on plants, and there was ample space, the first plant assemblages establishing themselves after the recession of the ice were quite chaotic: plants occurred together which would under ordinary circumstances never do so. On the other hand the more or less fortuitous delay of immigration of the various species potentially able to grow there gives the impression of a poorer plant community than the potential one (Iversen's (1958) 'protocratic phase').

After some time, competition and, later, soil development establish ecological stresses which lead to the formation of more definite communities. The time necessary for this process to take place varies greatly, above all with climate. In the pollen record this is expressed by an initial mixed flora of disparate elements, being replaced by the more uniform composition characterizing one major plant community.

The reaction of vegetation to changes in ecological parameters varies both with the parameter in question and the type, not least the direction, of change. There is usually a distinct time-lag in the response to a positive ecological change (Fig. 8.1), whereas the response to a negative change may be almost instantaneous, especially for flowering (pollen production), but may be slower for vegetative development, especially in long-lived organisms. Because of this time-lag, vegetation practically never reaches its potential maximum development and is only momentarily in equilibrium with climatic parameters (cf. discussion in Davis, 1986). Pedologic changes are slower than climatic ones, and vegetation may adjust to them with shorter time-lag, but the problem of how long a soil type and its vegetation can persist against unfavourable climatic conditions is not easy to solve. The effect of human impact varies enormously, from immediate destruction under landnam conditions to slow, indirect reactions, as exemplified in our days by the effects of acid precipitation. Similarly, the effect of immigrant species is variable depending upon the

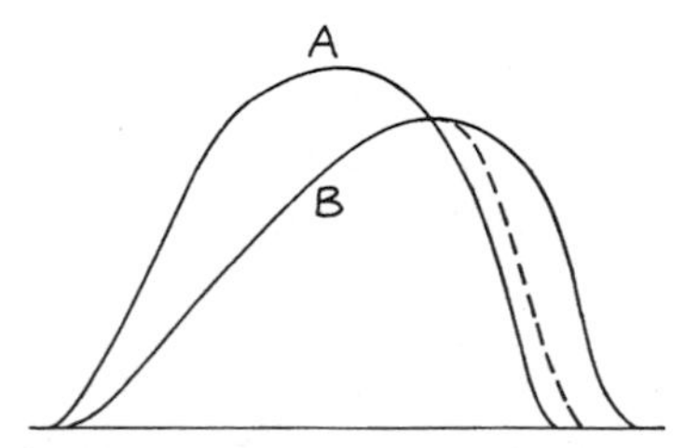

Fig. 8.1. Reaction lag of vegetation (effect, B) in relation to climatic change (cause, A). The dotted line may represent flowering, the continuous one vegetative persistence.

ability of the immigrant to penetrate into a (para)-climax vegetation and to establish another, very fast in some cases (*Abies* in Fig. 7.8), much slower in others (*Picea* in Scandinavia—Moe, 1970; Hafsten, 1985).

Under changing external conditions vegetation is in equilibrium with changing external conditions only for very short periods (cf. Fig. 8.1). Successions may even be positive after the general trend has changed into the negative.

8.2.1. *History of climate*

The classical non-vegetarian problem is that of *climate* history. Vegetation is not a meteorological universal instrument, and the climatic factors which chiefly affect vegetation are only part of the climatological complex, viz. those climatic elements that are at minimum then and there. For instance, in the moist and cool climate of northwestern Europe, moisture is generally adequate, but temperature is too low for optimal development of more exigent vegetation types. Accordingly, vegetation will react sharply to changes of temperature: with increasing temperature the warmth-demanding species will spread from their local stations in favourable positions and on favourable soil, and conquer larger areas. With decreasing temperature they will again recede. The climatic fluctuations registered in a pollen diagram from such a region are therefore chiefly those of temperature. Temperature and precipitation do not vary independently of each other; however, the laws of this interdependence are not so well understood that we can infer from the one to the other. If we want information about precipitation in an area like northwestern Europe where there has always been plenty, we must resort to phenomena which are not, or at any rate are less, conspicuously represented in the diagram. On the other hand, a pollen diagram from a semi-arid region will, in the varying dominance of forest and steppe elements, give a representation of variation of humidity (i.e. precipitation less evaporation), whereas the influence of temperature will be less conspicuous, and less direct (see Fig. 8.3).

However, secondary phenomena may give additional information about the fluctuations of climatic factors not conspicuous in the regional diagram. The changing character of the deposit itself may give evidence about hydrologic changes not always registered in the pollen diagram, which in the main registers upland vegetation. Effects of hydrologic changes have been used as arguments for and against climatic change since the days of Blytt (1876). In addition to water-level changes in more or less closed basins, ombrogenous bogs are especially sensitive to hydrologic changes and have been extensively studied (Granlund, 1932; Aaby, 1974). Changing bog sediments are usually without noticeable pollen-analytic effects, whereas water-level changes may—or may not—be registered by strong differences, occasionally amounting to distortions of the diagram, because of local pollen production by aquatic, beach or dry-land vegetation. By comparison with a regional diagram hydrologic registrations usually date themselves pollen-analytically, i.e. are properly placed in the time scale of vegetational history.

Whereas *climatic change* (towards warmer, colder, drier, more humid, etc., climate) is relatively easily deduced from pollen-analytic information, absolute *climatic data* present difficulties. The basic method is to define the ecologic niche(s) of the relevant taxon (taxa), and by a system of overlapping niches restrict the originally wide-ranging individual taxa down to a single defining value. This was done long before pollen analysis came into being. Andersson (1902) used the value for the summer temperature at the present northern limit of *Corylus* to deduce the summer temperature at the limit of the same species during the Holocene hypsithermal and deduced a drop of summer temperature of some 2–3°C since then, which was supposed to have caused the retreat from the limit of maximum distribution.

Samuelsson (1915) improved Andersson's method by using two (semi-)independent climatic parameters, introducing the duration of summer as the second. Such binary correlation diagrams, which define the ecologic niche in terms of two independent parameters, were used by Iversen (1944) to give more precise information about the conditions at the limit of various distribution areas

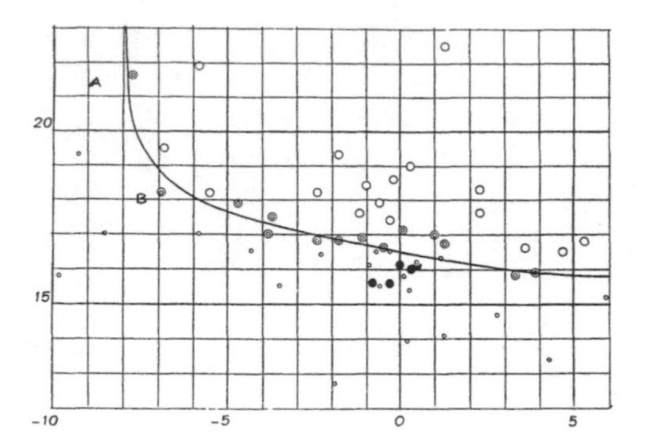

Fig. 8.2. Diagram illustrating the climatic demands of mistletoe (*Viscum album*). Ordinate: mean temperature of the warmest month; abscissa: mean temperature of the coldest month. Rings indicate the position of meteorological stations in the coordinate system. Small rings: stations outside the area of distribution of the species; large rings: stations inside that area; double rings: stations on the present distribution limit. The dots indicate some finds of fossil mistletoe outside the present area. The limiting temperatures seem to be *ca.* −8°C in winter and 15.5°C in summer. Samuelsson's rectilinear regression evidently represents the part of the curve between *ca.* −5°C and +5°C winter temperature. Redrawn after Iversen, 1944.

of thermophilous indicator species (cf. Fig. 8.2). When using climatic data inferred from the present-day distribution of plants one should keep in mind that the assumption of equilibrium between ecologic demands and existing climatic conditions is in any case unproven and sometimes overtly fallacious (e.g. *Picea* or *Fagus* in Scandinavia). However, it is the best estimate we can get, and we must use it as a benchmark.

The introduction of pollen analysis has made possible refinements by introducing quantitative surface pollen registrations instead of the simple presence/absence data used previously. Webb (1985) has published scatter diagrams of the correlation between *Picea* and *Quercus* pollen registrations and temperature data, between prairie forbs and precipitation data. The data indicate a maximum value of *ca.* 4°C for *Picea* and a minimum value of the same level for *Quercus* (Fig. 8.3). Further, they indicate that a surface pollen registration of *ca.* 5% indicates the limit of local occurence of oak and spruce. It should be emphasized that such figures must be treated with the utmost care, and not generalized without careful scrutiny of all circumstances.

The prairie forb curve differs in presenting both an upper and a lower limit (which obviously exist also for other taxa, but are not represented in the diagram). Both for *Picea* and even more for prairie forbs there is reason to believe that the limits may be caused more by competition than by direct effects of climate.

Further improvements were introduced by Bartlein *et al.* (1986), who presented the pollen percentages in a temperature/humidity gradient. The *Picea* diagram (Fig. 8.4) shows an upper gradient correlated with temperature data and a lower (left-hand) one correlated with precipitation. *Quercus* shows a clear optimum with gradients to lower pollen values towards lower and higher values of both temperature and precipitation. The prairie forbs data indicate a very strong combined influence which may suggest that evapotranspiration would be the parameter showing the best correlation.

Apart from the complications caused by possible vicariant taxa with different ecologic demands, such distribution vs. climate diagrams are subject to severe constraints.

1. They are valid only within a climatically uniform area. In another area with a different

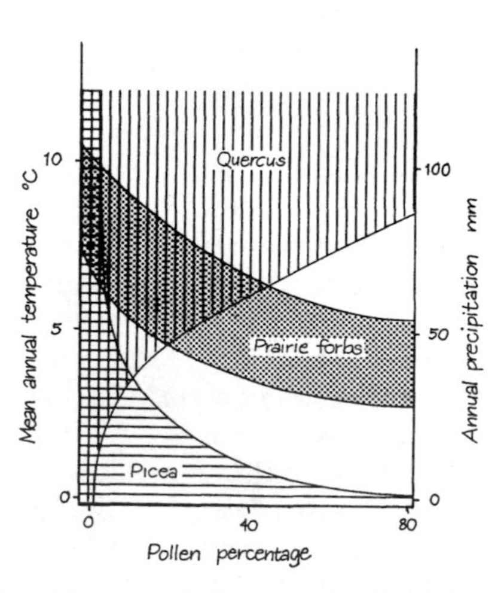

Fig. 8.3. Diagram of the climatic demands of *Picea*, *Quercus* and prairie forbs in relation to temperature or humidity. Data from Webb, 1985.

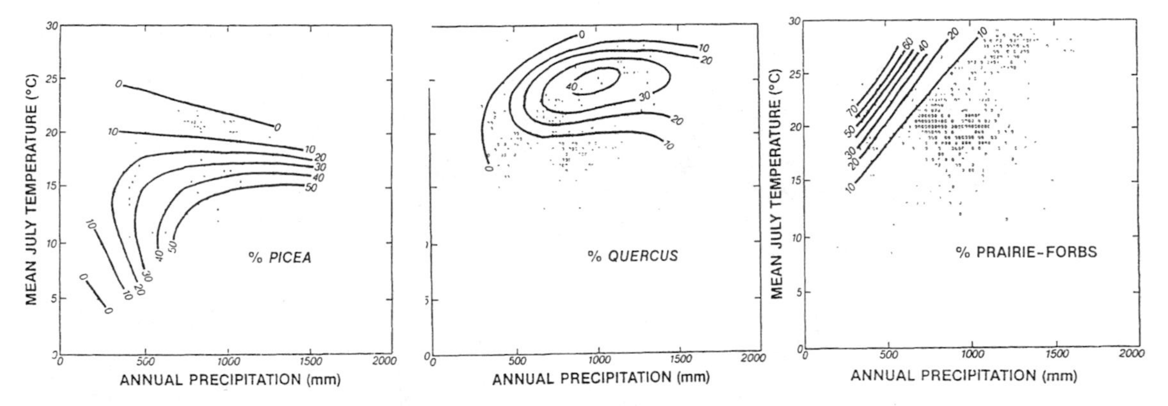

Fig. 8.4. Quantified scatter diagrams of the distribution of *Picea*, *Quercus* and prairie forbs in relation to annual precipitation and July temperature. From Bartlein *et al.*, 1986; redrawn. Reproduced by permission of the copyright-holder: Blackwell Scientific Publishers, Oxford.

qualitative or quantitative temperature or precipitation seasonality their validity cannot be taken for granted. This is important when drawing inference about climates of the past.

2. Webb's data are percentages, which means that they are influenced by the presence of other strong pollen producers: a reduction of prairie forbs occurrence from 100% to 50% may reflect in a reduction of pollen percentage from 100% to 16% if the competing taxon produces 5 times as much pollen per area unit. This might be obviated by using absolute pollen registrations. Enough such data for diagram construction are not available.
3. It should be pointed out again that the deduction of climatic parameters from the distribution of taxa presumes (a) that the taxa are in equilibrium with climate, i.e. are distributed within the whole of their climatically potential area of distribution, and not outside the same; (b) that registration does not necessarily mean that the taxon grew there, nor does absence of registration inevitably mean that it did not. Pollen registers flowering only, not the presence of individuals in a vegetative state.
4. The usefulness of the climate correlation diagram depends on the independence of the parameters used. In none of the approaches dealt with here is it very good, and especially in the data from Bartlein *et al.* there is a very strong autocorrelation between them, which, combined with the absence of data from the area to the right of the diagonal, makes especially the *Picea* diagram fairly incomplete.

Generally it is not possible to give a conclusive answer to the question of climatic inference on the basis of distribution data. Owing to the interchangeability of ecological factors, the answer will always have to be given as alternatives. Pollen analysis comes in as an important method owing to the many distributional facts which may be inferred from diagram and analysis, but it must never be forgotten that even the best correlation *is not a causal explanation*. It indicates that certain factors correlated with our data and in the diagram represented by them are conditional for the occurrence of these taxa under the circumstances given. What these factors are is an ecological, not a climatic problem. In particular, the diagrams do not indicate whether a given distribution is due to measurable physiological causes or to competition.

By multivariate analysis more factors could be brought into the picture, but one would very soon run into difficulties with autocorrelations, and the immediate perspicuity would be lost. Also, plants may react to parameters not measured by routine meteorological observations and their distribution may in addition be influenced by non-meteorological factors.

Factors delimiting the distribution of a taxon towards colder or drier areas are relatively simple to envisage. Factors that would delimit an area of distribution towards a warmer region are largely unknown, but it is reasonable to assume that competition is one of them. E. Dahl (1951) has advocated the view that high summer temperatures are in themselves limiting, but has not given a plausible physiological explanation (water stress because of inefficient water transport system? cf. Fægri, 1989).

An alternative to the use of simple climatic parameters is to base the analysis on air masses, as done by Webb and Bryson (1972) and others. The problem is to find sufficient criteria for defining air masses in the pollen-analytic data. Skre (1982) could not classify to air mass more than 40% of the meteorological situations studied. Such a low hit figure with prevailing mixed types detracts from the usefulness of the method.

8.2.2. *The human impact*

In addition to climate, *cultivation* is an important factor, influencing the pollen diagrams directly. As soon as a tract of land is cleared for agricultural or other purposes the quantitative composition of the flora changes materially. A number of species previously unknown in that region, or of very limited distribution, become (dominant) weeds, and their pollen subsequently becomes a regular constituent of the pollen rain. Similarly, crop plants may provide part of the pollen rain, though some of the most important of them (viz. most of our cereals) are self-pollinating and release very little pollen. Even the pollen of those weeds and crop plants that are very sparsely represented in the pollen rain may yet be of great value as indicators; in addition to the direct information they give, their appearance may explain changes in the diagrams, that would otherwise be inexplicable. The appearance of such plants marks not only the beginning of agriculture but at the same time the introduction of a powerful non-climatic factor in vegetational development. Cultivation diagrams are of secondary importance as documents of climatic change, but they often give surprisingly intimate glimpses of (rural) life in the past. These matters are treated in Chapter 9.

8.2.3. *Pollen analysis for dating*

As conceived by von Post, pollen analysis was primarily an instrument for dating of peat and sediment sequences (cf. von Post, 1950). When radiocarbon dating was introduced this role became less important. For purely chronologic information ^{14}C dating is usually faster and cheaper, but pollen analysis can date material which is not dateable by the other method, and also often gives additional information about the ecological setting as an extra bonus. Whereas radiocarbon dating is absolute, pollen-analytic dating is relative: diagrams need chronologic checkpoints from other sources.

Radiocarbon *dates the object itself and nothing else*. Inference about the age of the surrounding deposit or of accompanying objects is indirect, and subject to all the usual uncertainties due to irregularities of sedimentation, etc. Whereas dating of a discrete object, e.g. a piece of wood, is usually unequivocal as far as the object goes (cf. however, Kaland, 1976), the dating of sediment or peat is more dangerous. Both can be heavily metachronous due to incorporation of younger carbon from deep-lying roots of much younger vegetation penetrating the deposit (Kaland *et al.*, 1983). As long as roots and rootlets are living or recently dead they can be picked out mechanically, but the time inevitably arrives when they become indistinguishable from the surrounding, amorphous and much older matrix: the more so the older the deposit. Radiocarbon dates, especially in peat, will therefore tend to give too young dates unless individual constituents, e.g. *Sphagnum* stems, can be washed free from adhering amorphous matrix before dating (cf. technique in Vorren, 1979). This may be very cumbersome for classical dating, but with the small samples needed for activation analysis it is a more workable proposition.

Roots also penetrate into sediments, but in this case there is another influence as well, which may contribute to misleading ^{14}C dates, viz. the presence, in the basin, of water of high age and corresponding low ^{14}C activity. This water enters into the metabolism of organisms and is later incorporated in, or even forms, the sediment. Fossil

$CaCO_3$ in alkaline lakes has the same effect, and sea water has a radiocarbon age of several hundred years.

Although usually less precise, pollen-analytic dating is independent of some of the problems that beset ^{14}C datings. Unless secondarily disturbed, the pollen flora gives a consistent, synchronous date of the matrix, as the vertical displacement of pollen can, on the whole, be discounted. Also, radiocarbon dating is restricted to organic material, whereas pollen analysis can also be carried out in minerogenic sediments (the concentrating of organic material for activation analysis is, under such circumstances, extremely unsafe because of possible incorporation of 'secondary', redeposited material). Some non-organic deposits can be dated—absolutely or relatively—by other physical methods, but in many cases a pollen-analytic dating is the best, sometimes the only, alternative. It also carries with it the serendipity of ecological information obtained at the same time. On the other hand, pollen diagrams are often very local, and zoning systems may be metachronous from one basin to the next, so that data from one diagram cannot without reservation be carried over into another.

Ideally, one should use only datings of high quality, whether by ^{14}C or pollen analysis. Datings by different methods on the same material give a mutual check upon the reliability, as the sources of errors, if any, will be different. In cases of discrepancy it is important to go deeper into the case and not assume automatically that there is one method that is always right. For a discussion of pitfalls of radiocarbon dating refer to Olsson (1986).

The general procedure for the classical pollen-analytical dating of events and objects that have left their marks in the development of a deposit is as follows. One or more diagrams are prepared from the finding-place of the material in question, from sections as near the site as possible, preferably including the find. If the find is an object that has been removed from the deposit, its location is ascertained by measurement and controlled pollen-analytically. Like recent bog surfaces, those of older date were not necessarily completely flat, and we have no guarantee that any synchronous level is a horizontal plane in the deposit. Once the object has been removed and its impression in the deposit has disappeared, even the most exact indications of the depth are of limited value. Besides, small objects frequently drop out of the peat cutting and are recovered from the bottom of the trench—or where the peat is dried. In such cases no direct information is available. However, with the exception of highly polished stone and metal objects, most of the objects found in pollen-bearing deposits will contain in some crevice or cavity material for analysis.

The reliability of information from such finds varies. The danger of contamination is slight if samples are secured in the field by or under supervision of a specialist at the time of recovery, or if taken from interior surfaces of cavities (not from outer surfaces!) shortly after. Contamination is more likely if the object is investigated some time after recovery, or if metachrony between find and stratigraphy (sinking down during embedding) cannot be excluded. Material that has been wilfully buried in the bog, has been redeposited or probably contaminated by recent pollen, should never be used for dating purposes.

If the possibility of contamination can be ruled out, the pollen spectrum, or preferably spectra, from the object may be fitted into its proper place in the diagram from the deposit. This may be done simply by drawing the spectrum close to the edge of a slip of paper, and then sliding this along the reference diagram until the point is found at which the two spectra coincide best. Owing to statistical uncertainty it is not possible to define the place exactly; a better understanding of the statistical possibilities is achieved if confidence intervals are indicated by shading along the curves. More sophisticated statistical methods are available but, again, the apparent greater rigour of methods is not always commensurate with the quality of the original data.

In this way the position of the object in the local sequence and the local diagram may be established. In many cases this local diagram will be badly distorted by overrepresentation, selective destruction, etc., and the next problem is the fitting of the local diagram into a general diagram which carries independent dates. Only after this question has been solved—and here again no fixed rules can be laid down to cover all cases—will it be possible to date our object in relation to the vegetational history of the region. On the other hand, if the age of the object is known, the same technique may be reversed to insert an absolute date in the regional pollen diagram and vegetation history.

In an area from which no adequate pollen-analytical data are previously known a dating by pollen analysis will entail much work. If the regional diagram type has been worked out previously, the work will be greatly reduced, but even in such cases the dating of an object may involve work covering weeks or months. It may prove impossible to date objects found in bogs, the pollen contents of which are poorly preserved. The special case of pollen-analytic dating in archaeology will be dealt with separately in a later chapter.

8.2.4. Pollen data in geology

As mentioned above, the main geological interest in palynology is for dating purposes. Together with other microfossils, pollen permits dating also where only a small quantity of material is present, e.g. in oil drilling. Because of their omnipresence and indestructibility pollen grains and spores are most important instruments for this purpose. But even dating apart, pollen analysis finds many uses in geologic investigations, and we cannot in this book deal with the subject exhaustively.

The deglaciation of former glaciated areas is most directly approached by sedimentologic studies, but pollen analysis comes in later, alone or together with ^{14}C dating. The definition and identification of an unstable, usually rich (cf. above) pioneer vegetation is sometimes possible, narrowing the possible hiatus between deglaciation and sediment accumulation.

Outside deglaciated areas microfossil analyses are important for the ecologic recognition of sediments: salt-marsh, aquatic, etc. The constant, or at any rate slowly changing, total pollen rain also affords a yardstick against which the productivity of other microfossils may be evaluated.

Like other growing sediments, growing *glacier ice* incorporates pollen that can be retrieved, in this case simply by melting the ice and, if necessary, treating the samples in the ordinary ways. Vareschi (1952) used this method for an analysis of glacier stratification: summer layers contained much pollen whereas winter layers were pollen-free. Using this as a basis he was able to analyze details of glacier movement.

Ambach *et al.* (1966) agreed that there were great differences in quantity of pollen between late summer horizons (up to 7000 grains per litre ice) and layers formed at other seasons with 100 grains or less. This provides a means of recognizing 'false' ice layers in the firn, i.e. layers not formed during the summer melt-off: they are distinguished by low pollen counts.

Bortenschlager (1969) showed that glacier ice is also an effective trap for long-distance pollen, *Ephedra* being the most indicative. Sufficiently high up, the relative composition of the pollen rain varies very little with the seasons, except for locally produced pollen types. The significance of this observation is not yet clear.

Other events manifest themselves in the sequence and form part of it. Hydrologic fluctuations and their climatic interpretation have been referred to above. Another example of such events is the change of sediment types due to *marine transgressions and regressions* which result in alternation between lacustrine and marine deposits, in most cases with transitional brackish-water deposits. The age of contacts between sediments is determined by pollen analysis in the same manner as that of objects; first, the place in the local sequence must be established, then the change is dated in relation to vegetational history (cf. Chapter 10).

If a number of basins at different altitudes within a restricted area can be investigated, the established age of various marine–lacustrine contacts will provide data for a curve of displacement of the shore-line (see Fig. 10.2). Such curves are easier to construct in areas of predominant regression than in areas of predominant transgression, in which most deposits will be sub-marine. It should be noted that if a basin extends across the general isobase direction for some distance, isostatic movements will result in unilateral transgressions and regressions in the parts of the basin opposite the outlet. The same applies, *mutatis mutandis*, to basins with no outlet, but in this case the phenomenon may be complicated by multilateral changes of lake level, due to climatic fluctuations, being superimposed upon the unilateral evidence of isostatic shortline displacement. In all cases, however, pollen analysis provides a much firmer factual foundation for shoreline investigations than can be obtained by other methods.

8.3. Recent pollen as markers

The techniques dealt with so far are based on the recovery and analysis of sub-fossil pollen grains. In contrast, other techniques use the living grain, immediately after release from the stamen, as an instrument. Since the deposit is in many of these cases very rich in pollen, preparation techniques are often correspondingly simple. On the other hand, quantitative data are more often desirable. A very important field is that of *pollen allergy* studies (Wodehouse, 1945). Though other airborne substances may also lead to allergic reactions, the toxins of pollen grains are of major importance. In many countries a continual check is kept upon the pollen content of the air, partly in order to study the general rules of pollen incidence, finding areas and periods which are comparatively safe for allergic persons; partly in order to give forecasts of the pollen incidence for the following day or days.

If, and to what extent, pollen grains entering the organism through persorption (cf. Jorde and Linskens, 1974) cause allergic reaction is unknown, but the main reactions seem to be due to pollen grains entering the respiratory organs with the intake of air. The decisive factor is therefore the quantity of pollen floating in the air. Sampling for the purpose of registering such pollen incidence accordingly constitutes a problem different from that of sampling for sedimentation studies. As pollen is just one type of contaminant in air pollution, the usual filter methods (electrostatic filters, vacuum cleaner-type samplers, etc.) may be used with advantage (Raynor *et al.*, 1961). However, relatively simple devices consisting of sticky, vertical cylindrical surfaces ('impact sampler'—Mirams, 1953), which are aerodynamically more suitable than flat samplers (Gregory, 1951), are sufficient to give a first approximation. Pollen grains are usually investigated as they are, with no further treatment except, perhaps, staining. The stain may be incorporated in the surface coating of the sampler, giving stained grains without any more preparatory work.

Data collected for pollen allergy studies are often used predictively: warnings being given that the air will be strongly pollen-contaminated in the immediate future. However, they can also be used for other predictions. Cour and van Campo (1980) have demonstrated a correlation between pollen content of the air in the spring and the subsequent harvests of wine, olives and grain.

Honey pollen analysis (mellittopalynology or mellissopalynology) is an important technique in the analysis of the work and working conditions of honeybees; also as a honey contamination control (Zander, 1935: 37). Contaminants originating from exotic honey are easily detected by a qualitative analysis: the presence of certain pollen types gives away the contamination, since they cannot derive from local vegetation. Quantitative analyses of the composition of the pollen flora are even more revealing. Mellittopalynology is also concerned with the methods of measuring absolute pollen frequency. Sediments vary from fractions of a mm^3 per 10 g honey to 50 mm^3 or more. Special, graduated centrifuge tubes (Trommsdorff tubes as used in dairy research) are used for this purpose. Such tubes should be centrifuged floating in water-filled casings in order to take the strain off the graduated bottom part. Checks may also be carried out to test if honeys conform to the designations under which they are marketed. The quantity of pollen present in a usual honey sample is in most cases so large that it can be expressed by volume. The ordinary APF technique (exotic marker) may be redundant. Untreated honey samples also include a varying admixture of non-pollen particulate material, which disappears if the sample is acetolyzed. Because of differential characteristics of the flower and of the bees' behaviour when collecting, some honeys are very poor in pollen, e.g. *Tilia* honey if it is pure—which it rarely is. Among honeys with large concentrations of pollen are the heath honeys; even greater numbers are found in *Myosotis* and *Eucryphia* honeys. For evaluation of the actual quantities of nectar contained in mixed honeys it is necessary to use a conversion table of the specific pollen contents of various honeys (Demianowicz, 1961). It is, indeed, easy to see what would be found, e.g. in a 50–50% mixture of genuine *Tilia* honey, with *ca.* 200, and *Myosotis* with *ca.* 15 million pollen grains per gram. Unfortunately, so far very few such data are available; besides, figures vary greatly with external circumstances (Maurizio, 1949). Correcting for the volume of the pollen grains would to a certain extent reduce the pollen figure for species with many very small pollen grains like *Myosotis* or, more important for the trade, *Eucryphia.* Analysis of the contents of corbiculae gives information about the pollen-collecting activity of bees (Maurizio, 1953; Louveaux, 1955). Exines remaining in the honeycomb after newly hatched larvae have emerged may serve as indicators of the relative importance of plant species as larvae fodder.

The methods of mellittopalynology have been standardized by the International Commission of Bee Botany (Louveaux *et al.*, 1970; cf. Maurizio and Louveaux, 1967). There is also an official French Codex for honey analysis (Louveaux, 1970).

The investigation of *pollination problems* is closely related to mellittopalynology (Høeg, 1924). Pollen adhering to blossom-visiting animals gives evidence of the species visited. Analysis of pollen in corbiculae or other pollen-carrying devices shows which types of blossoms have been visited for pollen collecting. Similarly, pollen contents in honey stomachs and—less reliably—in honey cells of social insects give evidence of which plants have been visited for nectar (Fægri, 1961). Pollen tests—not always quantitative analyses—are now a regular part of pollination studies. For the analysis of diffuse pollen loads on pollinators the animals may either be washed and the washing-water centrifuged, or, if that is contraindicated, the presence of pollen can be studied by autofluorescence microscopy.

A similar field is the study of *animal feeding*. This is frequently done in modern archaeological investigations: the pollen found in animal (and human) faeces gives information about the food intake. However, the same technique is successfully applied also to recent animal dung, e.g. in game management or selectivity of grazing studies. Mostly, the pollen record from such studies registers ingested flowering or near-flowering plants only, but fruits or seeds may also carry great quantities of pollen with them (cf. Greig, 1982). With the same constraint pollen studies of coproliths give information about the feeding of animals in former periods, including extinct ones: mammoth (Kupriyanova, 1957) and giant sloths (Martin *et al.*, 1961).

For obvious reasons little has been published regarding examination of pollen for *criminal investigation* (cf. Erdtman, 1962:225); cases may be found hidden in police archives. Soil particles frequently have characteristic pollen spectra, and soil or dust found in the clothes of a person who has lain on the ground (dead or alive), or merely walked through a field, may have a pollen composition indicative of the piece of ground involved, and thus be useful in the reconstruction of events. It is obvious that a negative conclusion is more likely to be decisive than a positive one. A case in which straw, etc., had been fraudulently substituted for an expensive commodity may also be mentioned. The weed composition and pollen flora of the straw gave definite indications about where and when the substitution had taken place.

9. Archaeopalynology. Pollen analysis of the human environment

Motto: With man, sin came into the world.

Originally, the effect of humans on the development of vegetation was regarded as being of minor importance in the interpretation of pollen-analytical data because no pollen-analytical technique was then known with which to assess such effects. Other environmental factors (climate, soil, plant migration) were thought to have dominated the postglacial vegetation development, at any rate in northern Europe. During the 1930s Firbas demonstrated methods for tracing human influence in the pollen diagrams, and in 1941 Iversen presented the definitive demonstration that the activity of prehistoric humans could cause major vegetation changes. This opened the possibilities for interdisciplinary cooperation between pollen analysis and archaeology, history and ethnology to elucidate the human impact on the vegetation. The study of anthropogenic vegetation has become a major field in pollen analysis, linked to problems of environmental adaptation in the past as well as the present.

It is now generally accepted that humans have caused major vegetation changes, such as deforestation over major parts of the world's surface, and even induced desertification. Human effects on vegetation are on a scale similar to, or even exceeding, that of other major environmental factors. Pollen-analytical studies of the development of human-influenced vegetation types have documented the actual successions from undisturbed to anthropogenically influenced, indicated the time relations of these changes and improved the understanding of causal relations. By comparison of regional ('natural') and local ('disturbed') pollen diagrams it is possible to reconstruct in time and space the historical transition from natural to cultivation landscapes. This is important also because it demonstrates the interaction between humans and landscape (vegetation). The pollen-analytical record demonstrates in historical detail the development both of settlement and of land-use techniques.

Recent research has also proved the possibility of documenting short-term (decades) effects of submodern and modern activities leading to changes in natural conditions, e.g. the presumed effects of acid precipitation on freshwater resources and forest death. Integrated in the environment research team, pollen analysts and diatomists have documented the start of acidification, of changes in water chemistry and changes in land-use techniques within the catchment areas. The work on these problems has demanded new standards of resolution in pollen analysis (cf. Report of the surface water acidification programme, Mid-term review conference, The Royal Society, etc., July 1987).

The increased interest in interdisciplinary work has also necessitated research on anthropogenic soils, which cover much more extensive areas than hitherto realized: arable fields, middens, culture layers of habitation sites and redeposited anthropogenic sediments in lakes and on the sea floor. Increased understanding of the processes of formation of such sediment types is of paramount importance for the interpretation of pollen-analytical data. The results are also in themselves important for archaeology, providing data on a wide range of human activities registered in the sediments of the habitation sites: resource utilization, function of excavated structures, composition of food (latrines), extent of commerce (exotic pollen).

Since this approach has developed independently in various countries, terminology is not quite consistent. *Environmental archaeology* (UK) refers in general to the use of the methods of natural science in the interpretation of archaeological sites and their environment. *Paläoethnobotanik* and *Archäobotanik* (Continent) are biased towards macrofossil (seeds, etc.) analysis in the sites themselves. In French one would speak more specifically about 'palynologie archéologique' or 'archéo-palynologie' (Helly, in Renault-Miskovski *et al.*, (1985). The latter term is easily translated into English, and will be used here.

9.1. The problem: human activity

What is the objective of pollen-analytical investigation of an archaeological site? On the one side the results may be of importance to the pollen analyst in his/her own work. On the other side data from pollen analysis may serve to elucidate and explain archaeological problems by adding more evidence, coming from another angle.

As to the first, it should be remembered that pollen analysis is a technique only. This means that pollen-analytical investigation of human activity throws light on other problems that the pollen analyst is studying by the same technique at the same time, e.g. the history of vegetation of the area in general. Obviously, the existence and activities of a local human population is an important ecological parameter.

On the other hand, most human activity has some palynological effect, changing the natural vegetation, i.e. the composition of the pollen rain, by intentionally or unintentionally introducing polleniferous material into the site, by causing erosion and redeposition of old material, etc. Such changes may lead to conclusions which support, corroborate or contradict results obtained by other techniques, including the classical ones of archaeology itself. Some questions can be answered with greater precision by pollen analysis (and related techniques) than by purely archaeological (sensu stricto) methods. Food remains—whether digested or not—may be completely disintegrated and amorphous, but nevertheless contain well-preserved pollen indicating where the food came from and what it existed of. Analyzing an archaeological site, the pollen analyst is not so much interested in the vegetation of the site itself (which is generally destroyed) as in the conditions of productivity of the area, leading to the necessity of investigating a wider area than just the site itself. It is the surrounding area that determines the total productivity at the disposal of the population at the site.

As a consequence, archaeological field investigations have changed towards multidisciplinary teamwork where pollen analysis, botanical macrofossil analysis, osteology and soil chemistry play important parts in the interpretation of data (notwithstanding occasional specimens of arrogant snobbishness like Welinder, 1985: 96). More specifically, the pollen analyst brings into the picture data from pollen-bearing deposits outside the habitation site, which may provide a data set on human interference with vegetation, a set independent of the archaeological source material, which is concentrated inside the site.

The cultural landscape has been created by the action and interaction of diverse forces: the physical effect of the settlement, the vegetational effect of utilization of the ground, the effect of selective utilization of native plant species, etc. These effects depend both on land-use strategies and on the size (number of individuals) of the local population. Consequently, the pollen-analytical study of the cultural landscape presumes input from and interaction with archaeology, history, geography, economics, etc., and the pollen analyst must be able to obtain relevant information from the other sources. The pollen analyst must be prepared to use a wide spectrum of methods and rely on data not usually considered or used in pollen analysis, e.g. resulting from investigation of disturbed anthropogenic deposits which were formerly avoided in pollen analysis, if possible.

As always, pollen-analytical procedures must adapt to the types of deposits also in the study of human-influenced vegetation. If they have been laid down at some distance from the actual settlement, conditions are (semi-)natural and ordinary procedures may be followed with proper caution for pollen accidentally dispersed from the site. Methods and data do not in principle differ from the usual. The situation is illustrated by the left-hand side of the flowchart (Fig. 9.1), which has been discussed in earlier chapters.

Human activities may result in the classical deposition model being inapplicable, both with regard to pollen transport and deposition and with regard to deposit stratigraphy.

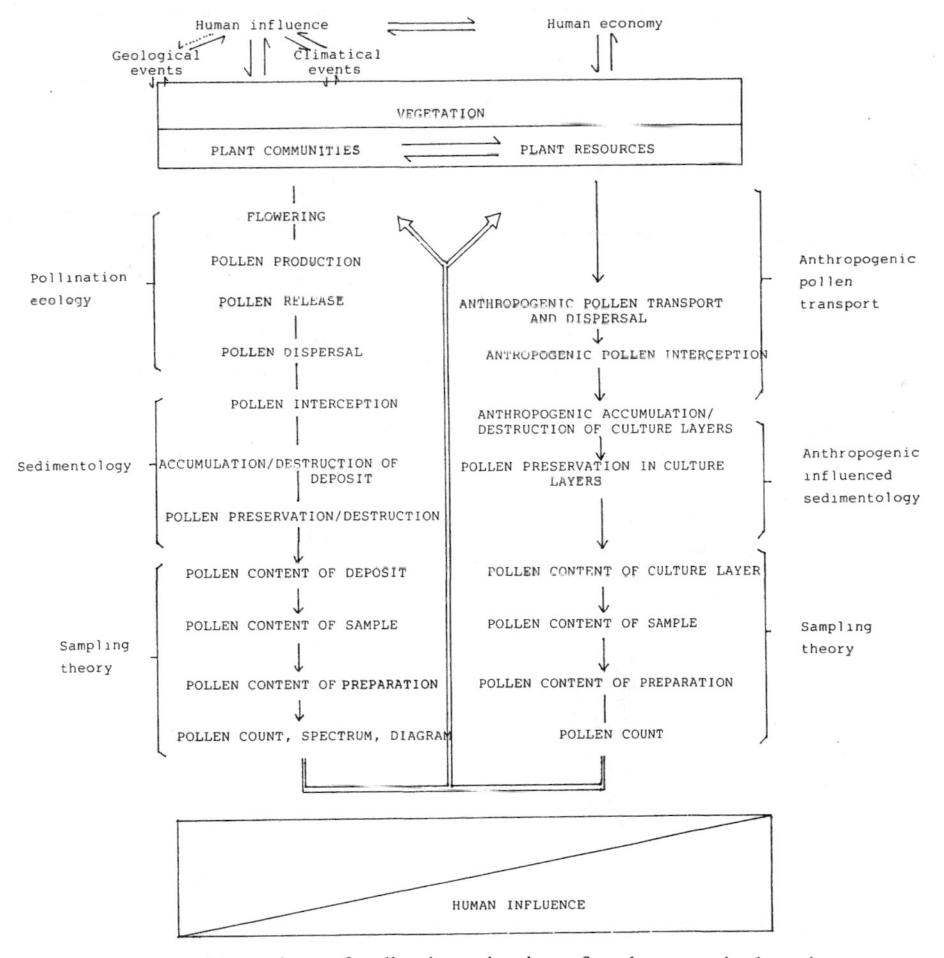

Fig. 9.1. Flowchart of pollen investigation of anthropogenic deposits.

Under such conditions the pollen flora, in addition to the regular components, contains a variable quantity of pollen that has been transported in bulk to the site, e.g. as part of human collection and use of plants for food, or with weeds associated with the food and discarded *in toto*. This component, although giving a statistically skewed picture of the pollen composition of the air, is contemporaneous with the deposition process, and gives information about the actual conditions. Older pollen may also have been brought to the site, directly or indirectly, e.g. with sods used for roof cover. This part of the pollen deposit is metachronous and disturbs the contemporary record.

Sedimentation is influenced by human use of the soil, by ploughing, digging, trampling, by excavating for construction, by burying objects, by carrying soil from one place to another, e.g. for improvement or maintenance of arable fields, etc. The strategy of sampling from such deposits varies with the aim of the investigation; approaches and techniques will more often be similar to those used in macrofossil studies. However, in all cases the pollen analyst must have the ultimate control of the sampling. Samples brought in without proper control of deposition conditions are not suitable for pollen analysis, which under such circumstances mostly represents a waste of time.

The interpretation of pollen-analytical evidence from anthropogenic, potentially disturbed, soil requires a different way of thinking from the interpretation of an ordinary pollen record. In this context ordinary diagrams are meaningless, as both the statistical and stratigraphical basis for diagram presentation are absent. It may be possible to reconstruct former conditions from the (qualitative) pollen record. The spectrum does not in this case carry any meaning. When it comes to indoor sediments with a pollen flora brought together by the activity carried out on the premises (cf. Krzywinski and Soltvedt, 1988), any resemblance to any plant community, if present, is fortuitous. On the other hand such data can give valuable information about utilization of plant resources, in a similar way to plant macrofossils. Often the data from the two approaches differ and supplement each other (Fægri, 1985).

The habitation site represents the local centre of human activity, where the composition of the deposited pollen flora and the deposition mode are most deeply influenced by 'non-natural' processes (cf. Fig. 9.2). With increasing distance from the site the ever-present, constant flora of wind- and waterborne pollen becomes more important in relation to the decreasing quantities of pollen brought in by humans. However, the composition of the surrounding vegetation (delivering the windborne pollen) is far from being undisturbed. A network of pollen diagrams at increasing distances from the site will illustrate the gradients of the pollen flora

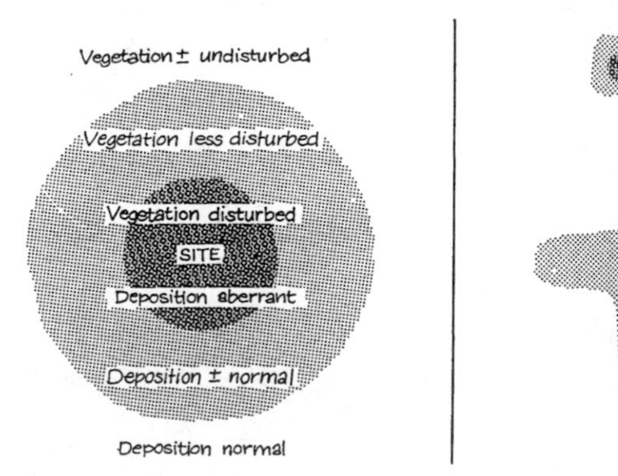

Fig. 9.2. Schematic representation of conditions at a single site (left) and in a culture landscape with several sites and communication lines (right). Within the darkest areas the normal *deposition* model is invalid. Within the light-shaded area the model is (conditionally) valid, but *vegetational* development is changed by humans. Deposition at communication lines is often aberrant.

between that of the site and that of the 'uncontaminated' surroundings. With a caveat for possible stratigraphic disturbances, diagrams from outside the site can be interpreted in accordance with the ordinary deposition model. The activity at the site usually creates changes in the pollen flora. At archaeological sites it is important to establish vegetation gradients towards less disturbed vegetation, which serves as a standard against which changes in space and time can be elucidated.

Within the cultural landscape, human activities (field cultivation, grazing, burning, peat cutting) start increased leaching and soil erosion. Nutrients and pollen included in the old soil are liberated and may be redeposited. In lakes, redeposited material both contaminates the sediment and the pollen flora. This introduces errors in dating (Johansen, 1978; Tutin, 1969; Pennington, 1975) and creates metachronous pollen spectra. If, however, such contaminants can be identified, dated and/or isolated, their occurrence can be used for information about the onset and development of the human influence.

The following discussion will first consider the interpretation of pollen-analytical data when the normal deposition model may be applied. Secondly, we shall consider problems in sampling and interpretation when this is not so.

9.1.1. *Cultural landscape development*

We define the cultural landscape as a mosaic of plant communities directly or indirectly influenced by the activities of man (cf. Fig. 9.2). In a material sense it expresses itself in a mosaic of spontaneous vegetation, cultivated and waste areas, habitations and communication systems.

This landscape has existed as long as man has actively influenced his surroundings. Finds of hazelnuts in refuse layers of Mesolithic sites show that man utilized the nuts for food. The rapid spread of *Corylus* in northwestern Europe in the early Holocene may indicate that man played an active part in its dispersal (Firbas, 1935, 1949: 149) and possibly also, intentionally or not, changed the vegetation near the sites in favour of the hazel (Smith, 1970). As long as habitations were transient the effect of this activity was necessarily limited. Only the introduction of farming changed this, as farming activities favoured (seasonal or all-year) permanent sites, which is the precondition for the establishment of a true cultural landscape.

The effect on the vegetation was *selective*, because the farmers cleared what was, for them, fertile and easily arable soil. In such clearings new and completely artificial communities were created, consisting of cultivated plants (favoured) and weeds (disfavoured, but persistent). Some of the weeds were newcomers, brought in with the exotic cultivated plants. Others (archaeophytes) occurred in the area already, but now got a new chance to spread. Some of them may have belonged to the open vegetation on recently exposed ground in Late Glacial times, and may have survived *in situ* as relicts until man's opening up of the landscape gave them a second chance.

Plant communities of arable fields are directly created by and completely dependent on human activities. Those of the surrounding area were less disturbed and less dependent. Here man adjusted an existing, natural vegetation to satisfy the demand for pasture, fodder collection, communication, etc. This vegetation, albeit transient, possesses a certain persistence, in contrast to that of arable fields, which immediately after abandonment undergoes a rapid succession to more static types.

The recognition of the historical cultivation influences and the resulting cultural landscapes must be based on the knowledge of present-day or historically documented conditions, especially old cultivation techniques, many of which are on the way towards extinction together with the landscapes they created. By backwards extrapolation the recent and sub-recent pollen deposition of such landscapes can serve as a guide to the vegetational interpretation of pollen diagrams. For the explanation of causal relations, interdisciplinary cooperation is often essential.

Man influences vegetation in three ways: directly through plant cultivation, indirectly through animal husbandry, and also through technical utilization of resources. Each of these produces a characteristic vegetation pattern, leaving a more or less distinct and recognizable imprint on the pollen registration. Geographical, ecological and ethnological differences create different patterns, but across these patterns, certain common features can be discussed.

Arable fields usually carry the following characteristics:

1. Trees and shrubs were removed to allow light to penetrate to the light-demanding crop plants.
2. Certain native plant communities were preferred for the establishment of arable fields. The constituent species lost their natural habitat and were more or less exterminated unless they found refuges in newly created habitats, persisting as ruderals or weeds, if hemerophilous.
3. The original soil profile was changed and the availability of plant nutrients was depleted, especially as the cultivated species were gross feeders. Good farming practices could maintain soil fertility. Manure technology being relatively advanced, primitive cultivation mostly relied on some types of shifting slash-and-burn technique for restoring soil fertility. The deep-rooting trees could reach plant nutrients not available to shallow-rooted cultivated species. The first step in the restoration of fertility after soil exhaustion was by abandonment, thus allowing forest regeneration, which replaced the nutrients from deeper soil layers. Self-manuring (grazing) could slow down the soil depletion, even to equilibrium, imitating natural conditions (paraclimax).
4. The open soil permitted the growth of weeds which, because of their gross feeding and vulnerability to competition, could not persist in natural vegetation. These weeds initiate the successions following abandonment.

Weed communities depend on cultivation methods. In mixed agricultural economies with long-term rotational systems, perennial weeds are favoured. Under continuous, short-term cultivation, annual species are better suited because the more extensive root systems of perennials are sensitive to the yearly cultivation activity. The type of tool used is also important: light tools such as the ard and the mattock do not disturb the root system of many perennials. Heavier tools such as the spade or the plough do, and therefore favour annual weeds. This may reflect in the pollen record and give important information about production techniques (cf. Fig. 9.3).

Animal husbandry

If the ground is cleared especially for animal husbandry, or the cattle graze the abandoned fields, the regeneration of forest is slowed down or prevented by methods of hay-making or grazing. The effect depends not only on the intensity of utilization, but also on the periodicity (winter grazing). Another point is the absence of open ground: the vegetation of grazed areas is formed by competition-resistant

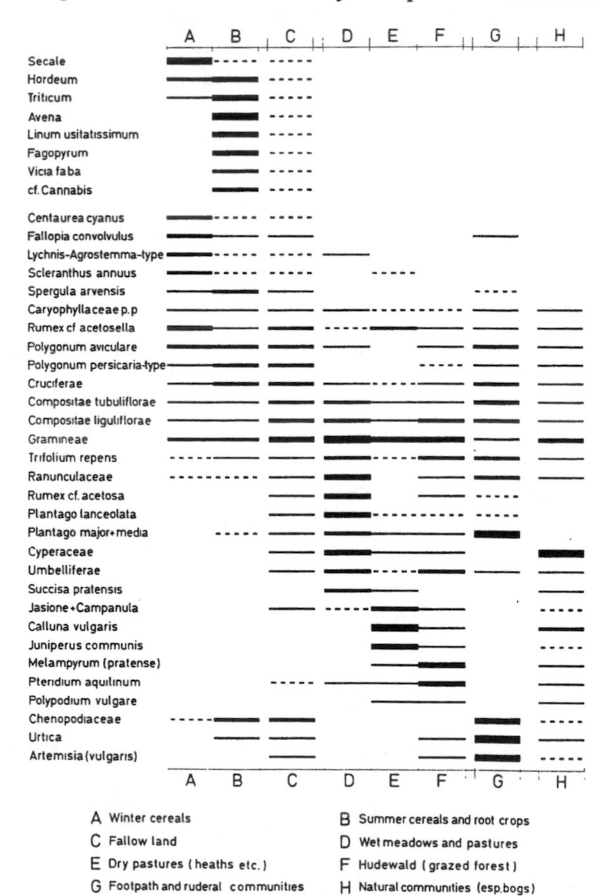

Fig. 9.3. Anthropogenic indicators in transalpine diagrams (Behre, 1971).

species. The weed flora of such areas is poorer and usually less indicative than that of arable fields. Like people, domesticated animals require a *continuous* supply of fodder and water, as well as protection against extreme climatic conditions (e.g. cold, dryness) and against predators. The need for fodder can be met in two ways:

1. By utilizing large grazing areas where the animals can find fodder all year round or, as the case may be, by moving the herd between grazing areas when the quality of the pasture is seasonally impoverished (transhumance, nomadism).
2. If climatic conditions preclude all-year grazing, and the mobility between areas is restricted, fodder must be collected and stored for the lean season(s). If intensified, fodder collection turns into fodder production, approaching the effects of the field cultivation syndrome above.

Techniques of fodder collection and storage depend on the technological level. Pastures must be maintained by some technique. Cutting sufficient quantities of grass for winter fodder needs an iron scythe or at any rate a sickle. Before such tools were available, fodder collection must have concentrated on easier-to-collect material, especially twigs and leaves of trees. Obviously, this has very specific pollen-analytical consequences, as the pollen is produced from just the same (flowering) twigs. In areas with a cold climate the feeding season might extend to several months each year. For protection the animals also had to be stabled (although with regard to winter climate this was certainly less important with the primitive cattle races of early husbandry).

In addition to what has been stated above on maintenance of soil fertility, maintenance of pastures could be effected by proper treatment of the vegetation, e.g. by burning non-palatable plants (old heath), leaving room for fresh, more nutritious species. The often very complicated (sub-)modern systems of double-use of the ground (forest pastures, etc.) certainly have very old forerunners.

Animal husbandry is a less intensive and more flexible technique than plant cultivation, and the effects on vegetation can vary from almost none to very strong influence by grazing. The composition and stability of the natural vegetation determines its carrying capacity: the number of heads per unit area and the composition of the herd that can be grazed without permanently changing the plant cover. Activities to improve the pasture by burning and fertilizing may delay depletion of soil nutrients. The effects of fodder collection or production differ with the method used (pollarding, shredding, grass production, etc.).

Natural resources

The earliest one to be utilized was probably timber for houses and boats, firewood for heating and, at a much later time, for metal production (reduction of ores to metal). In pre-industrial days forestry practices were extensive and had a limited effect on the vegetation, except in marginal areas. In local pollen diagrams forestry can be traced as temporary depressions of AP curves, indicating clearing and, later, successions during re-establishment of climax forest. Again, the demand for wood might have been so great as to lead to permanent deforestation in marginal areas.

Communication systems

A habitation site is an area of intensive intercommunication: trampling, transport of material, etc. From this nucleus area communication spreads as linear structures (cattle passes, footpaths, trails) into and through less influenced surroundings, often crossing or ending in more or less 'natural' vegetation (cf. Fig. 9.2). The palynological effect is usually very localized, restricted to a narrow zone with a different flora which, under fortunate circumstances, may leave a record. Communication also means material transport, material which often (intentionally or not) is polleniferouos. Cattle or animals of transport having grazed in one vegetation region may excrete pollen from that region into a completely different vegetation. Alternatively, they may bring seeds of plants which may flower and deposit pollen for a season or two, even if they cannot maintain themselves permanently. Such finds are scarce and extremely scanty, but they do occur (Moe, 1974a). They may cause interpretation problems, e.g. when cultivation indicators, which are usually scarce, are recorded from areas in which no cultivation is otherwise recorded. Likewise, the narrow (permanent) strips of aberrant vegetation accompanying communication lines may show up, sometimes rather forcefully, in pollen records (Moe, personal communication).

The cultural landscape is usually the result of a combination of these types of land-use. Pollen-analytically it may be very difficult to find suitable sites where the individual components in a diversified land-use economy can be identified.

9.2. How to solve the problems in the field and laboratory

In contrast to the regular laminar growth of deposits presumed in the stratigraphic model of pollen analysis, the deposit at a habitation site is irregular: material is dumped in heaps, holes are dug into material already deposited and the material is heaped up elsewhere on top of younger material, etc. (cf. Fig. 9.12). Material may be pollen-free (ashes, sand) or it may contain great quantities of human-transported pollen (faeces, waste material). Leaf fodder or building material shed concentrations of single-species pollen into a deposit within a limited space. Certainly, the activity also tends to even out the original variation over time, but the normal deposition model is obviously useless. This has repercussions both for sampling and for interpretation. The basis for usual pollen analysis, viz. the basic pollen sum, is invalid under the circumstances prevailing in habitation sites, so the quantitative approach must be replaced by a more qualitative one, where the main emphasis is on the occurrence of individual taxa or taxon combinations.

9.2.1. Pollen as cultivation indicators

In pollen analysis, land-use is most commonly identified by *indicator species (taxa) pollen* (Iversen, 1941), based on preference for, or avoidance of, anthropogenic communities. The narrower and the more well-defined the ecological tolerance of a taxon towards human influence, the more indicative it is in a pollen diagram. The registration of a stenoecious taxon or, better, a group of taxa, indicates the former existence of a specific plant community and the land-use technique that created it. This suggests the existence also of species that have not been registered, if they usually form part of the same community. Obviously, the crop plants themselves are the best cultivation indicators, together with weeds and ruderals such as (in transalpine Europe) *Plantago lanceolata*, *P. major*, *Rumex* spp., *Artemisia* and Chenopodiaceae.

As the indicator species method is based on the occurrence of a limited number of pollen taxa, the interpretation is relatively simple, even for investigators not familiar with the methods of pollen analysis or with ecological problems generally. After Iversen's first publication the range of anthropogenic indicator pollen types has been extended and critically evaluated, e.g. by Troels-Smith (1954) and Behre (1981) and Behre and Kučan (1986). Figure 9.3 shows the most important anthropogenic indicators in transalpine pollen diagrams. Exclusive indicators are rare: most of them overlap several cultivation communities. Since the totality of indicators of various communities is specific, interpretation is greatly improved by using this totality instead of basing the interpretation on one or two indicators only (cf. Fig. 9.4). The possibility should be kept in mind, though, that ancient anthropogenic communities, even with the same main crop plant, may have differed not only in the quantity, but also in the specific composition, of the

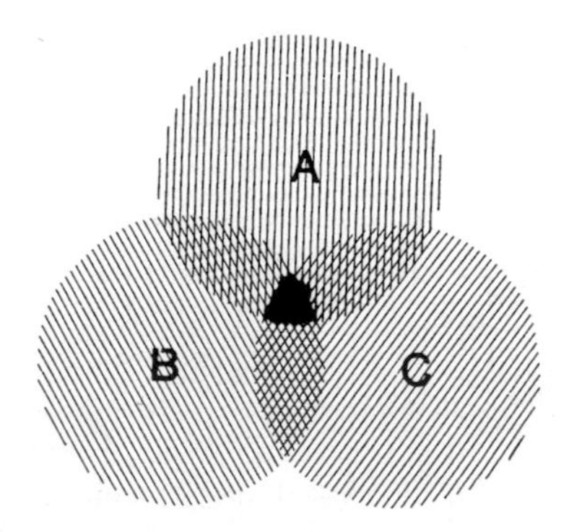

Fig. 9.4. Niche overlap. By the use of more than one indicator the niche is narrowed to a more informative size (dark area).

weed flora. It is enough to think of some of the former very serious weeds, which have today disappeared (*Chrysanthemum segetum*, *Convolvulus arvensis*, *Agrostemma githago*). To a certain extent this has been taken into account in the establishment of Fig. 9.3.

Indicators may be positive: occurrence means human impact, but there are also negative indicators. These are plants which, for one reason or another, are disfavoured by, or intolerant of, man's presence (hemerophobous species). The most important negative indicators are forest trees. Whether they have been actively harvested or merely cleared away, their pollen frequencies decrease with the arrival of agricultural man. The so-called *Ulmus* fall (Fig. 9.5) recorded 7000–5000 BP is a classical case in point. It is a good example of the reaction of a taxon to human manipulation of the natural vegetation. This interpretation is locally corroborated by the simultaneous reaction in curves of positive anthropogenic indicators (Cerealia, *Plantago lanceolata*, *P. major* and *Rumex acetosella*), curves which are not statistically influenced by the changes in the *Ulmus* curve. Various causal explanations have been offered, among them the potential influence of selective fodder collection (cf. also Section 8.2). Usually, pollen diagrams show both positive and negative indicators, and interpretation should be based upon as many as possible.

A non-pollen microfossil of great (positive) indicator value is microscopic *charcoal dust* derived from burning of wood and other plant material (Tolonen, 1986). The heat of the fire effectively carries the dust fraction up into the atmosphere, in which it is dispersed like pollen. Autochthonous charcoal is usually characterized by great variation in particle size, whereas allochthonous charcoal deposition is usually more uniform owing to differential fall-out during dispersal. However, it is hardly worth the effort to record particle sizes, especially as charcoal particles are (uncontrollably) crushed during preparation for microscopic investigation. Peaks in the charcoal dust curve may represent accidental forest fires, spontaneous or human-induced. Regular habitation and farming activities produce a more uniform, continuous curve (cf. Figs 9.7 and 9.8).

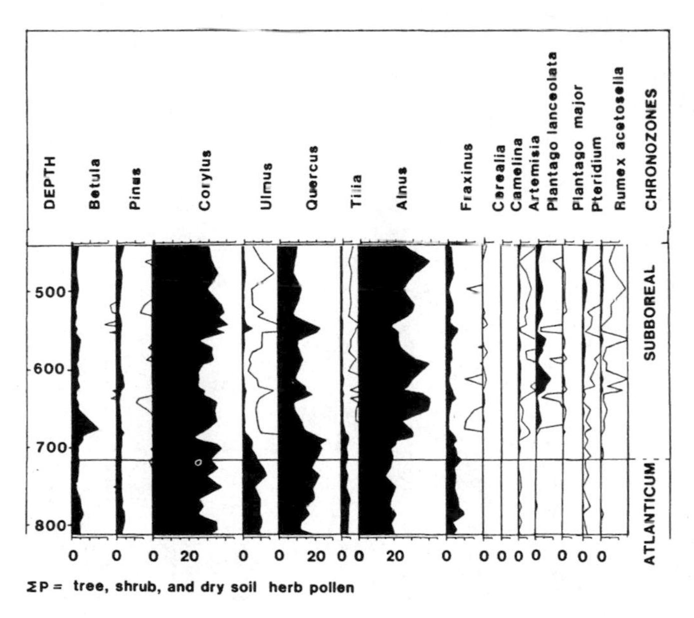

Fig. 9.5. Pollen diagram showing the *Ulmus* fall shortly after the Atlantic–Subboreal transition with simultaneous increase of anthropogenous indicators (Cerealia, *Camelina*, *Artemisia* etc.; from Kaland, 1981). Reprinted from: Behre, K. E. (ed), Anthropogenic indicators in pollen diagrams, 1986. A. A. Balkema, P.O. Box 1675, Rotterdam, Netherlands.

The actual indicator species vary from one region to another, but the general trend of pollen curves—opening of forest (negative indicators) accompanied by increase of plants favoured by farming activities (positive indicators)—are in principle the same in all parts of the world where forests have been cleared for agriculture, cf. Figs 9.6 and 9.7). The extensive opening up of North America for European-style cultivation in the nineteenth century is of great methodological interest since other indicator taxa were involved than in Europe. The process is well documented historically and by archaeological study. The clearance is also documented in many pollen diagrams (McAndrews, 1967; Davis *et al.*, 1971, and others). In a pollen diagram from a lake in Ontario, Burden *et al.* (1986; cf. Fig. 9.6) could demonstrate successive clearings by Huron Indians (until 1650) and Europeans (after 1876). The Indian clearing (X–X) is characterized by a slight decrease of forest components (*Fagus*, cf. also *Acer saccharum*) and an increase of light-demanding herbs and ferns (*Artemisia*, *Pteridium*; *Zea* would have been the best indicator of Indian farming but its pollen is rarely found in sediments). The pollen diagram shows that the cleared area was reforested after the Hurons had left (increase of *Betula*, *Quercus*, total AP). The European clearing (Y–Y) is indicated by the first appearance of introduced taxa (*Plantago*, possibly also *Rumex*) together with abundant *Ambrosia*) and grasses, and a sharp decrease of AP.

A similar development can be demonstrated by early clearings in southwestern Japan (Tsukada *et al.* 1986; cf. Fig. 9.7) 6500 years BP, although the species are different. The clearance is indicated in the pollen diagram by a sharp decline of forest trees; a sharp increase of charcoal indicates fire as the agent. Cultivation is indicated by pollen of *Fagopyrum esculentum*, introduced from the Asiatic mainland, and by grass pollen of cereal type. The diagram indicates decrease of cultivation and partial reforestation *ca.* 3100 BP with a distinct decrease of charcoal. At *ca.* 2000 BP another phase starts with buckwheat as well as rice pollen.

The indicator species method suffers from several limitations, and must be used with circumspection. First, there is a clear limitation in specific identification due to the collective character of many pollen taxa. The pollen of many good anthropogenic indicators cannot be identified to species level, and cannot be distinguished from that of species not associated with human activity. Among the major cereals only *Secale* and *Zea* are specifically identifiable. *Triticum*, *Hordeum* and *Avena* and most tropical/subtropical species can be distinguished only to the level of types including several closely related species, not all of which are or have been cultivated. Likewise, *Vicia faba* and

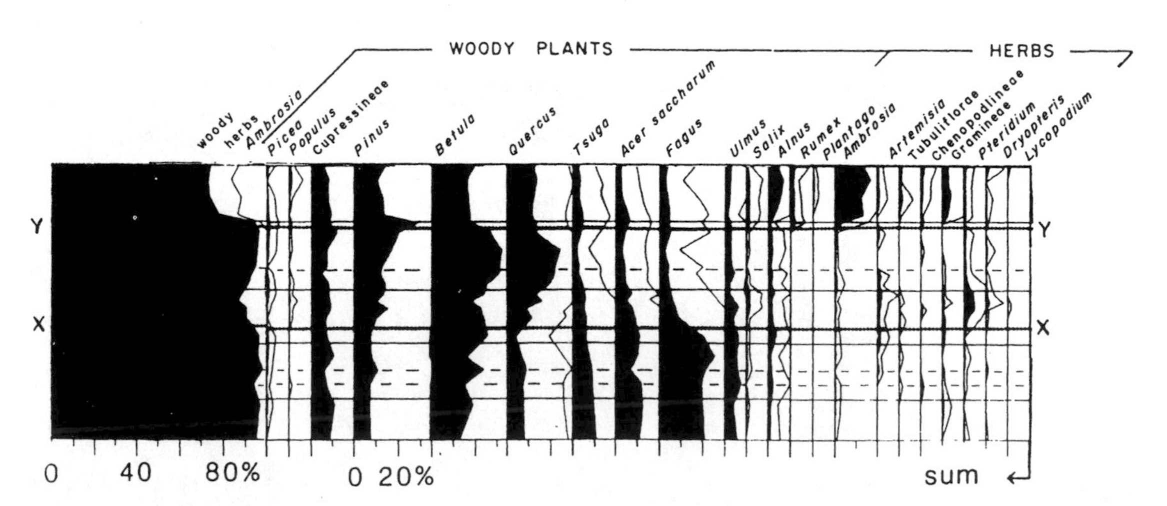

Fig. 9.6. Anthropogenic influence in the pollen record from Second Lake, Ontario, (Burden *et al.*, 1986). For explanation, cf. text. Reproduced by permission of the copyright holder: National Research Council Canada.

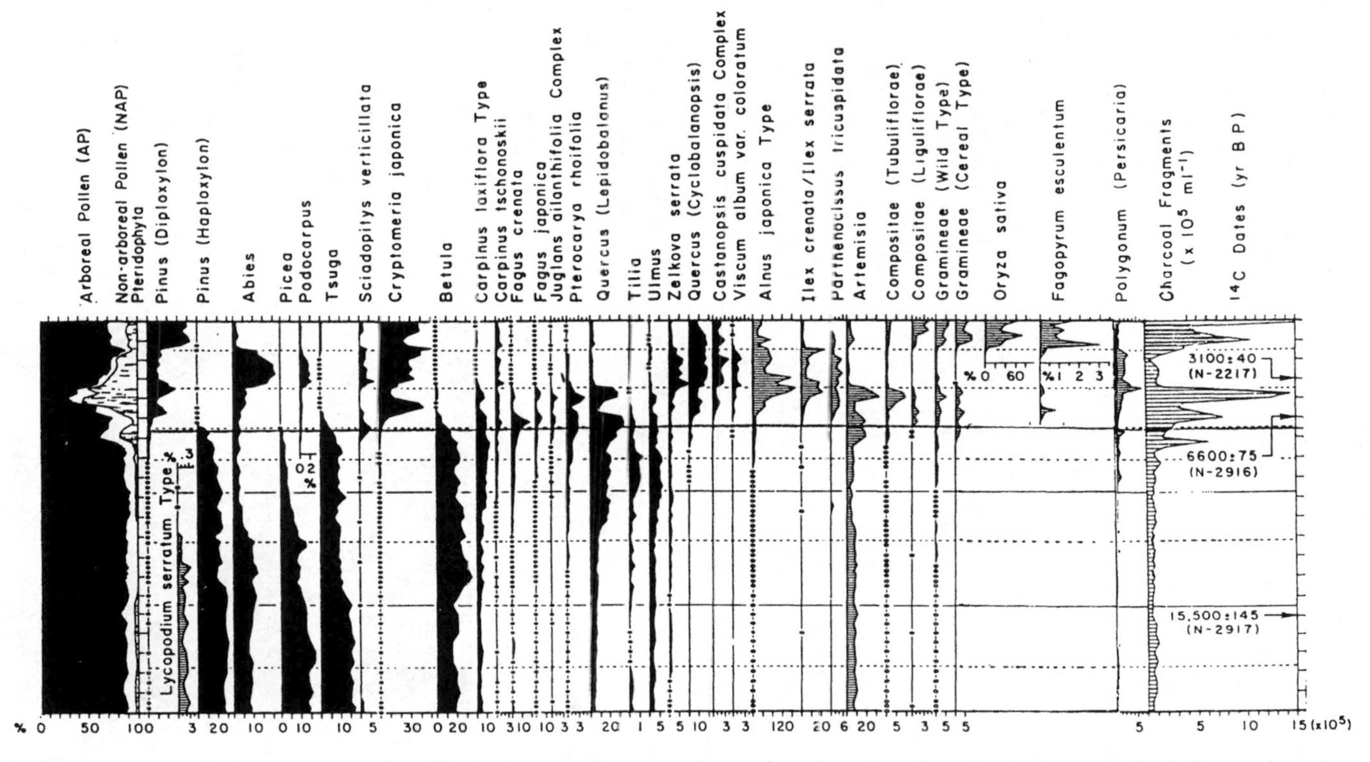

Fig. 9.7. Anthropogenic influence recorded in Ubuka bog, southwestern Japan, based on data from Tsukada *et al.*, 1986. For explanation, cf. text. Reproduced by permission of the copyright holder: Macmillan Magazines Ltd.

Pisum can be identified, but important crops such as *Camelina sativa*, *Isatis tinctoria* and most other crucifers and legumes cannot be specifically identified at present. The same pertains to many important weeds and ruderal species. However, the rise from a former insignificant level of a curve potentially representing both native species and weeds, e.g. crucifers, is in itself a strong indication of disturbance. Whether this disturbance has 'natural' causes or is caused by cultivation must be evaluated from coincidental evidence.

Second, many important crop plants and weeds are insect-pollinated or autogamous and disperse very little pollen in the air. They are therefore only accidentally registered in the pollen count and *it is not possible to exclude their occurrence on negative evidence*. The first find does not necessarily indicate the first occurrence, and the percentages in themselves are not indicative of quantitative relations. On the other hand, pollen may also have been deposited by agents other than the regular ones. Vuorela (1973) has shown that more pollen of autogamous cereals (*Hordeum* and *Avena*) is deposited during the threshing season, when pollen is beaten out from between glumes, etc. (cf. Greig, 1982) than during the flowering period. Similarly, pollen of cultivated plants is often combined in garbage, and is accumulated in waste heaps and other anthropogenic deposits with no relation to any pollen-rain deposition model.

The third limitation is the immigration history of cultivation indicators. If they were introduced together with the knowledge of the new type of land-use, their indicative value is much higher than if they are archaeophytes, which grew (in small numbers) in the area before. In northern Europe *Plantago major* is a case in point: was it introduced with farming and spread to the beaches subsequently, or did it grow on the beach before and spread to farmland when the latter came into being? Mostly the question is of academic interest only, but it is important for the interpretation of those very few pollen grains coming before the advent of more massive evidence for cultivation.

Finally, there may exist no modern or historically known equivalent to past land-uses and their concomitant plant communities. The use of indicator species may therefore mislead by indicating a modern plant community instead of the extinct one in which the indicator originally grew. Pollen-analytical interpretation of past cultivation techniques by the indicator method is therefore full o pitfalls, especially in the interpretation of the oldes farming techniques from the very beginning of agri cultural economy.

Some of these difficulties are obviated by con sidering *total pollen spectra* from modern cultura landscapes instead of the individual species. If foss pollen spectra resemble those from modern cultura landscapes they should indicate a similar type (vegetation. The difference from the indicator spe cies method is that, in this case, one bases inte pretation on a definite *constellation* of species, (which each alone would have been of limited i dicator value or none. This spectrum method h primarily been used in the study of 'natural' veget tion (Birks and Birks, 1980, and others), but only a very limited degree for cultural landscape studie where it should also have potentialities.

Since the emphasis is on combinations of ta and their quantitative relations, the method r quires a comprehensive collection of modern poll spectra from as many anthropogenic plant co munities as possible. This first step can be realiz by establishing the relationship between curre land-use and pollen deposition, e.g. in moss cu ions. The method includes a greater number parameters than the indicator method, and therefore more difficult to handle. Handling large data sets has become more effective with use of multivariate statistics (cf. Chapter 7).

We shall first present an example from a fa simple anthropogenous landscape, viz. the Atla coastal heath in Norway (Fig. 9.8). The traditio land-use of this area was based on all-year outd grazing of stock, made possible by the mild wi climate. The evergreen *Calluna* and other er ceous species provided winter fodder, while gra and herbs constituted the summer grazing. Reg burning of the heath regulated the balance betw summer and winter pasture and, together with ter grazing, prevented the re-establishment forest. The vegetation is rather monotonous includes very few indicative taxa (Gimingh

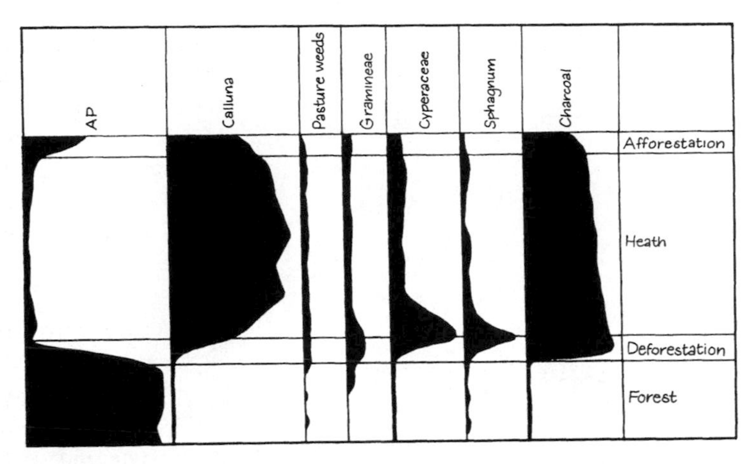

Fig. 9.8. Cultivation recorded in a diagram from the Atlantic heath region in western Norway (Kaland, 1976). For explanation cf. text.

1972; Øvstredal, 1985). With the discontinuation of the traditional land-use, the heath at present develops into other vegetation types, including the reversion to forest. Figure 9.8 (Kaland, 1986) shows the development of the heath and the beginning of the return to (planted) forest, so that the sub-recent pollen spectrum differs in a characteristic way from the previous ones. With this exception all pollen curves are parallel, from the sub-recent down to the deforestation level. The same plant community has prevailed all the time. In heaths such as this one the spectrum method has the great advantage that the deposit is uniform: changes in pollen records are not influenced by statigraphic development, which is often the case in lake sediments or peat. In studies of this kind all pollen taxa are of (equal) importance, consequently allochthonous pollen rain may exert a disturbing influence.

The second example is a pilot study performed in South Sweden (Berglund *et al.*, 1986). Sixteen specific anthropogenic modern plant communities were classified for land-use, vegetation composition and pollen composition. The three classifications were then compared by cross-classification tables to test if there was any relationship between land-use and vegetation, vegetation and pollen assemblage, and pollen assemblage and land-use. PCA and correspondence analysis were thereafter applied. Figure 9.9 shows the results of the correspondence analysis, which divides the material into three groups of localities and pollen characteristic of burned vegetation, high and low tree cover.

Like the indicator species method, the spectrum method suffers from various pitfalls. One is that—in contrast to the example in Fig. 9.8—spectra and recent comparison material established in different sediment types are strictly not comparable. Moss cushion spectra differ from those obtained from lake mud. Also, the collection of modern spectra may have missed the vegetation type in question even if it might still be extant. Especially for strict mathematical treatment this is a serious shortcoming which may cause difficulties, as such methods do not distinguish between taxa according to their known ecological significance. A good example is that of *Fagus* in Fig. 9.9, which has ended up in the category 'Low tree cover'. For the material at hand this is undoubtedly so, simply because beech forests were not included among recent vegetation type studies, being irrelevant to the problem under consideration.

The low pollen productivity and poor dispersal of many species of anthropogenic communities may cause difficulties in separating between different types of land-use. Even with multivariate statistics, Berglund *et al.* (cf. Fig. 9.9) had difficulties in establishing criteria differentiating between registrations of differences in land-use.

A fourth limitation of the method is that the

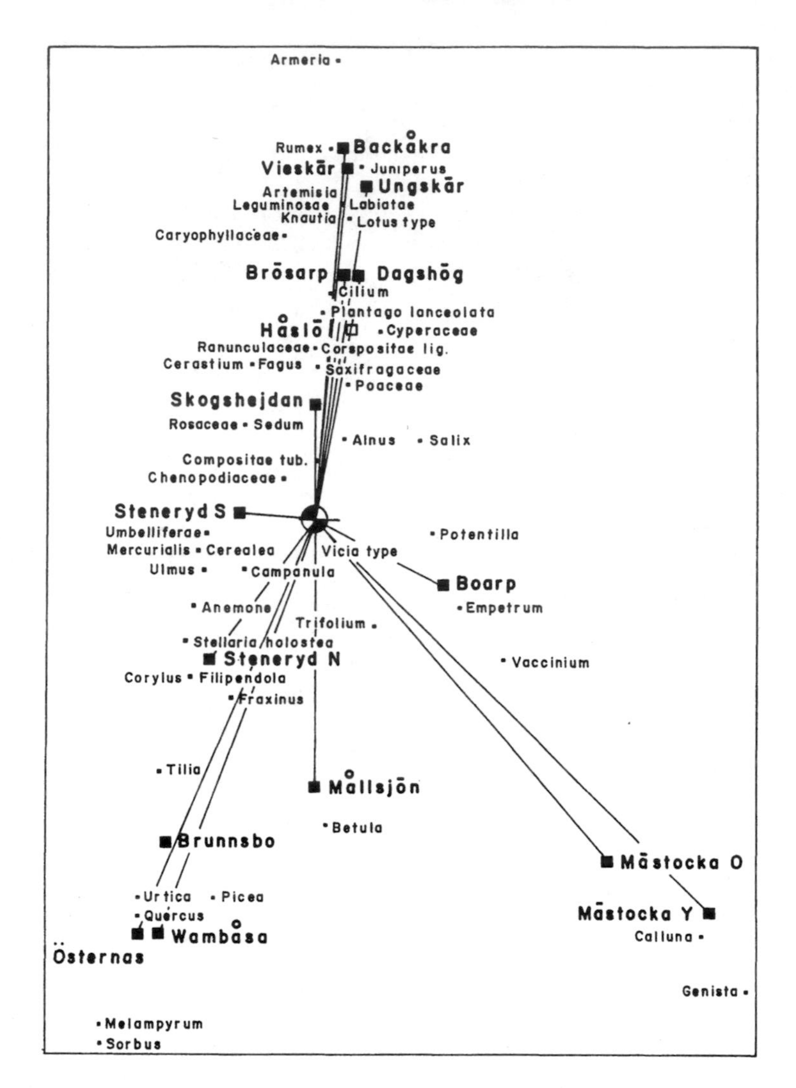

Fig. 9.9. Result of correspondence analysis of pollen deposition in sixteen different plant communities in southwestern Sweden (Berglund *et al.*, 1986). Reprinted from: Behre, K. E. (ed), Anthropogenic indicators in pollen diagrams. 1986. A. A. Balkema, P.O. Box 1675, Rotterdam, Netherlands.

same land-use technique may formerly have been applied in different vegetation regions, but is now known from a restricted area only. A good example is the slash-and-burn technique, which has survived in Finland until recently and the pollen-analytical impact of which has been studied by Vuorela (1973). The results describe the effects of this technique within the vegetation of Finland. However, the same technique has been used all over Neolithic Europe in much richer vegetation types, and conclusions from Finland may not be immediately applicable in richer areas.

The spectrum method has its strongest potential in the investigation of large-scale, rather homogeneous cultural landscapes where the ecologically important taxa have a relatively high pollen output, e.g. the coastal heath (Fig. 9.8), where *Calluna* is a strong pollen producer, and it is possible to correlate its presence and the presence of lesser producers with the high frequencies of the charcoal dust indicating fire. The indicator taxon method is more flexible, allowing the botanist to weight the ecologically important components of the vegetation. This method is therefore not so sensitive for separation of pollen from different plant communities. Gradually the distance between the methods has become shorter. The trend within the indicator method goes from the recognition of exclusive specific indicators towards groups in order to obtain a better delimitation of the plant community. An opposite trend is traceable for the spectrum method when using a multivariate method like correspondence analysis. This method points to which pollen types are most important for the different anthropogenic vegetation types (cf. Fig. 9.9). The two methods do not exclude each other, but rather give the researcher greater freedom when designing a project and in the interpretation of the results.

Both indicator species and spectrum methods are based on modern analogues of former anthropogenic plant communities. Artificial fertilizers, heavy mechanization and rationalization of farming methods and extensive forest planting on marginal land have changed the vegetation of the cultural landscape. In intensively cultivated landscapes it is today often impossible to find modern analogues to yesterday's anthropogenic communities and, even more so, to find examples of the primitive cultivation techniques which initiated the agricultural epoch. In these circumstances both methods may be misleading: their use requires great care.

If no modern analogue is known, the pollen analyst is forced to base interpretations exclusively on the fossil material at hand, leading to *ex analogia* conclusions that are certainly not as reliable as those based on the methods discussed above, and which may lead to circular reasoning. The usual method is to compare pollen spectra before and during the (assumed) cultivation period. The empirical base for this may be improved by producing more than one pollen diagram from the area. The consistency of the pollen curves can be checked between the sites.

In some cases validity of interpretation can be checked experimentally. A classical example is the experiment of Iversen and Troels-Smith in 1954 to reconstruct early farming methods in Denmark (cf. Iversen, 1973). The first cultivation phase in Danish pollen diagrams had been interpreted as indicating a slash-and-burn technique, but there were no historical sources to corroborate this. An experiment was conducted in which a slash-and-burn clearing was made in a deciduous forest presumed to be very similar to those of 5000 years ago. Approximately one hectare was burnt and cultivated in accordance with the traditions of Finnish practice. The ensuing vegetation successions have been monitored since then; correspondence with the development inferred from pollen diagrams has been good (however, cf. Göransson, 1982, 1984).

The more the pollen-analytical investigations of a site can benefit from the cooperation in an interdisciplinary research team, the more pitfalls can be identified and avoided, both in planning and executing the field-work, and in the interpretation of the results.

In northwestern Germany Behre and Kučan (1986) have made a detailed study of the effect of a Neolithic clearing in an area which has been extensively investigated by archaeologists. Seven pollen diagrams were made from kettle-hole bogs and a raised bog within the area, measuring the effect of the clearing along a gradient between the site (Swienskuhle) and a point 1.5 km away. Figure 9.10 shows the effect of the settlement for *Quercus* and *Tilia* pollen and for some positive anthropogenic indicators. *Quercus* and *Tilia* both react negatively. Pollen of cereals and weeds is very poorly dispersed away from the fields. Very low quantities of cereal pollen were recorded in the nearest bogs only, whereas pollen of some weeds (*Rumex*, Compositae Tubuliflorae and especially *Plantago lanceolata*) are better dispersed. Grass pollen incidence is very high at the site, but is fairly great also at the other stations.

The authors also illustrate the dispersal of anthropogenic indicator pollen from an archaeologically investigated mediaeval site in the same area (Fig. 9.11). There is a distinct inverse correlation between distance from the site and the possibility of identifying indicators of anthropogenic influence. Outside the field area, pollen of cereals,

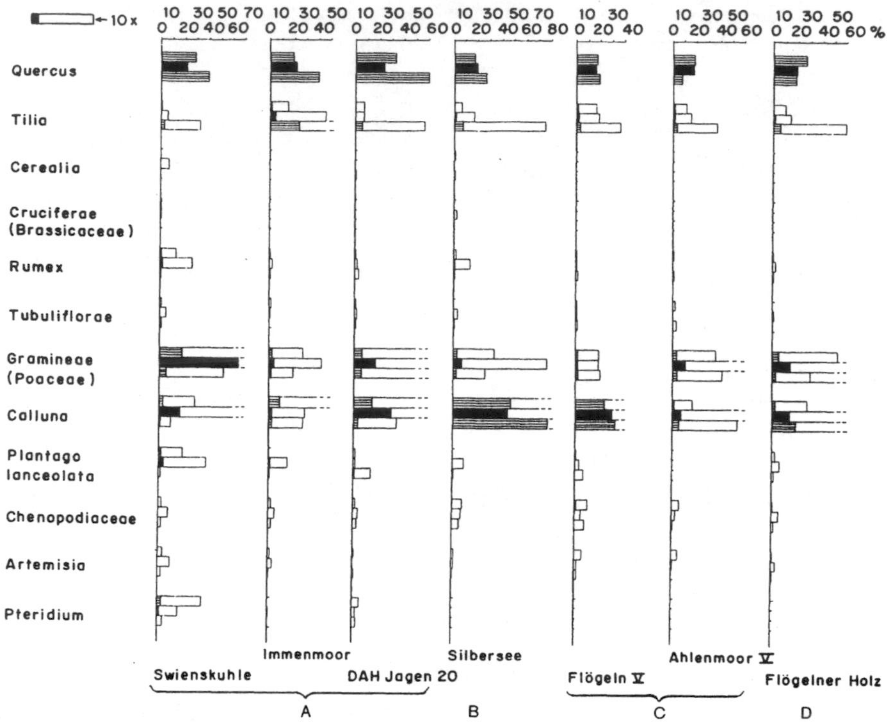

Fig. 9.10. Pollen record of the Neolithic clearing of an area in northwestern Germany according to Behre and Kučan, 1986. The three columns at each entry indicate, respectively, conditions before, during and after the cultivation phase. The letters indicate a grouping of sites, but no gradient in relation to the centre of cultivation. Reprinted from: Behre, K. E. (ed), Anthropogenic indicators in pollen diagrams. 1986. A. A. Balkema, P.O. Box 1675, Rotterdam, Netherlands.

even of the wind-pollinated *Secale*, is only found in very small quantities. With the exception of *Rumex (acetosella)*, weeds of the fields also disperse their pollen very poorly. Up to 0.4 km from the village, indicators of cereal production and cattle grazing might be discerned in the pollen diagram. Three kilometres away it would be very difficult to detect even such a large village as this.

9.2.2. Sampling

In the work at the site itself the archaeologists usually have the leading edge, and as much as possible of the pollen-analytical fieldwork should adapt to their working pattern, not only in using their reference grid but also, if possible, their stratigraphy. On the other hand, this must not become a straitjacket: there is always the possibility that details of minor interest to the archaeologist may be important for a pollen-analytical evaluation. The amorphous matrix, which is of marginal, if any, interest to the archaeologist, may be full of pollen. Another point is that the action area of a pollen analyst at a site is invariably and necessarily larger than the archaeological one, simply because pollen grains have a dispersal completely different from the dispersal of the archaeological objects. Much of the relevant information is therefore hidden in the gradient between the conditions at the site and those of the 'natural' vegetation of the wider surroundings. Likewise, the pollen analyst will usually want to penetrate deeper into the stratigraphy, beyond the obvious culture layers. The population of the site has arrived from somewhere else and their history, if any, is documented there. The people, and the flora they brought in, do not represent any endemic development; they arrived ready-made. The vegetation which existed there before had de-

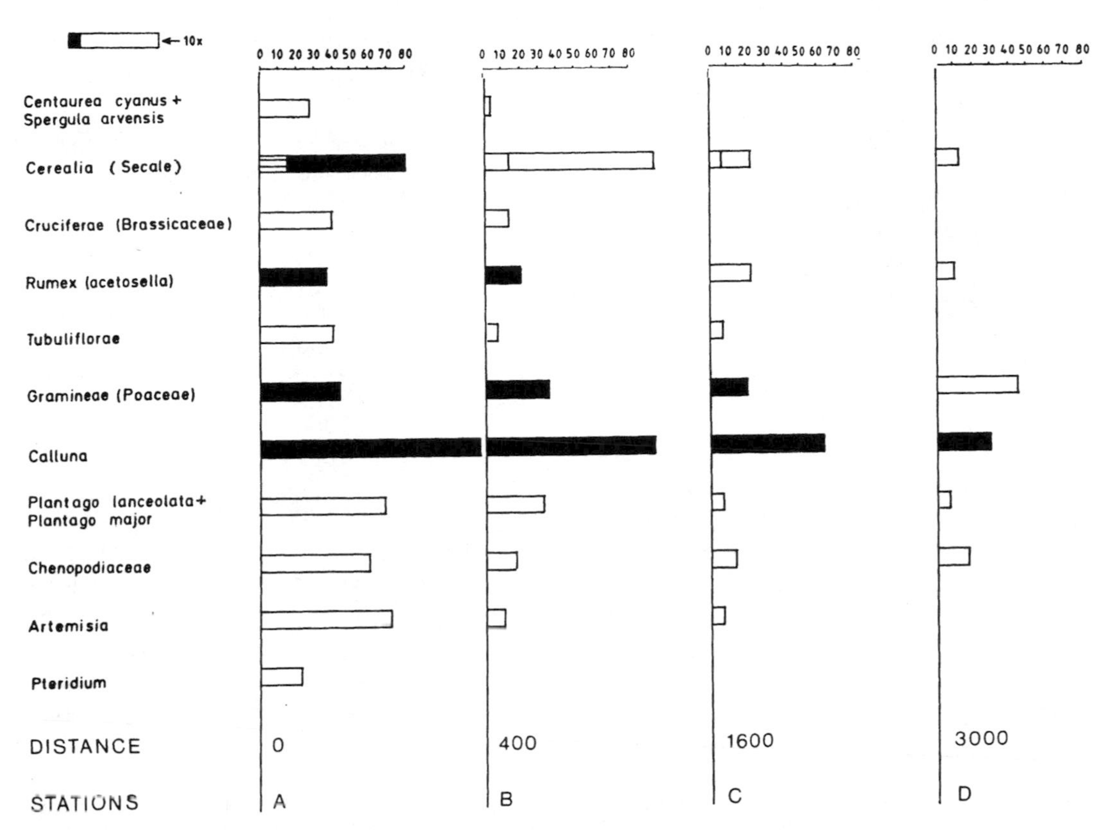

Fig. 9.11. Pollen dispersal from a mediaeval field in northwestern Germany according to Behre and Kučan, 1986. Distance from the field is indicated (metres). The letters indicate the (groups of) station from Fig. 9.10. The outer parts of the Cerealia columns are *Secale*. Redrawn. Reprinted from: Behre, K. E. (ed), Anthropogenic indicators in pollen diagrams. 1986. A. A. Balkema, P.O. Box 1675, Rotterdam, Netherlands.

veloped *in situ*; changes caused by cultivation are also a function of what was there before. The development under cultivation influence can only be understood and measured against the background of conditions before their arrival. These points may need some elaboration before being properly appreciated by the archaeological part of the working team.

In archaeological excavations sampling is easy inasmuch as the material is usually immediately available at the surface. The difficulties are: (1) the often very confused stratigraphy and (2) the impermanence of the sections which are continually being destroyed by further excavations. The latter means that there will be no chance later to supplement the material. It is therefore better to take too many samples and too much material than to take too few and too little.

Archaeological excavations are often carried out horizontally and give no immediate profile through the deposit. The stratigraphy in such cases has to be reconstructed at the drawing board. If the palynologist attends the excavation from the beginning, a series of test pits can give some idea of the stratigraphy of the deposit. If the palynologist is only invited after horizontal excavation has gone on for some time, the upper members of the stratigraphy are irrevocably lost. It is, for this and other reasons, usually impossible to obtain a complete vertical series of samples at one point, representing the whole stratigraphy of the site. In addition, the character and fossil contents of the individual members may change radically over very short horizontal distances, so that each stratigraphic unit should be represented by samples from various parts of its areal distribution, in addition to an adequate representation in the profiles. Deposits are frequently so coarse that monoliths cannot be taken out, even if an exposed face is available. In taking discrete samples one must take into account that the accretion of material may, at times, have been very slow, and that samples must therefore be taken at close intervals. Often, a minor part of the sequence embodies more abrupt change than the whole of the remainder.

Sampling from individual finds of extraordinary interest should follow the triple principle: below, above and within, giving terms *prae* and *post quem* and conditions during the time when the artifact was deposited. That pollen contents of vessels, etc., can give information about food and drink is too well known to need elaboration.

If an artifact is of some size, several samples should be taken from all sides, if necessary, during the excavation of the find. If sampling can be undertaken while the actual excavation is being carried out, no cleaning should be necessary: samples should be taken directly, before any contamination could have taken place. If, on the other hand, the artifact has—as is usually the case—been removed from its original place, or has been exposed for some time, great care is necessary to ensure that the samples are meaningful and uncontaminated.

The reliability of pollen-analytical information referring to archaeological or other finds may be classified as follows (cf. Krog, in Degerbøl and Krog, 1951: 71, 105):

A. Contamination improbable.
- A 1. Samples collected in the field by a specialist at the time of recovery.
- A 2. Samples collected by a specialist in recently recovered material. In this case samples must be taken from interior cavities, etc., not from the outside of the object, which may have been contaminated during handling by a non-specialist.
- A 3. Like the preceding, but sample collected by a non-specialist following careful instructions.

B. Contamination not improbable.
- B 1. As A 2–3, but the material has been recovered some time before, and has in the meantime not been adequately protected.
- B 2. Metachrony possible between object and sediment (sinking down during embedding).

C. Contamination probable. The material has been (1) recovered from surface layers (contamination by recent pollen); (2) buried in a hole or ditch in the deposit or has sunk through its layers; (3) redeposited; (4) found long ago and has been kept in a collection, etc.

Pollen analysis of objects in the C category must be regarded as a waste of time, since no reliable information can be gathered from such material. The B type may give positive results, but care must be taken in assessing the reliability. Before undertaking any pollen-analytical investigation the investigator should form an opinion about the reliability of information pertaining to the circumstances of recovery and about the chances of contamination. Pollen analysis is expensive both in manpower and in time, and should not be wasted on unsuitable material.

Since archaeological site excavations are usually open ditches or larger areas, sampling procedures are, in the main, those indicated in Section 4.3.1. As long as excavations are not too deep it even pays to make an open section (if possible) to obtain samples outside the excavation area, e.g. to complete a horizontal grid. As most matrix material is removed during an archaeological investigation, the constant presence of a pollen analyst or another specialist is necessary to ensure that information is not unnecessarily lost that way. It is not always easy to convince an archaeologist that the matrix may contain more positive information than the most glorious object!

The usual precautions against contamination of samples in the field must obviously be taken, only they are much more pertinent under the conditions of an excavation. The great open areas expose metachronous deposits which dry out and produce enormous quantities of dust blowing around during the work. Waste heaps are especially important sources of such contamination. The large number of people usually active at the excavation whirl up dust in dry weather and splash mud in wet, both dangerous sources of contamination. Equally great care should be taken to avoid trampling and deposition of waste material on top of any section under investigation.

In ordinary pollen analysis the sampling site is selected on many criteria, one of the most important of which is the—expected—preservation of the pollen in the deposit. A good pollen spectrum requires a good pollen preservation. In archaeopalynology the choice—if any—is strictly limited by other considerations, and the pollen analyst often has to contend with deposits in which the pollen is in a very poor state of preservation. As most sites are on dry land, pollen floras may be strongly distorted or obliterated by oxidation after deposition. The validity of the pollen record suffers correspondingly, and the interpretation of the record must take this into account.

No sampling starts without presuppositions—about the representativity of the samples to be obtained, about the relevance of the site, etc. This is also the case in a habitation site, only one must take into account the horizontal variation in addition to—and usually more than—the vertical one. In the normal model a synchronous level is presupposed to possess—within limits—uniform pollen flora; in a site this is not so. The horizontal argument is important for the placing of sampling points. In the same way as a vertical sampling tests subsamples in a definite pattern (distance) along the vertical, horizontal sampling tests subsamples in a definite horizontal pattern (grid) on a *synchronous surface*. Usually, the grid of the archaeological excavation can serve; however, just as samples are intercalated at critical levels in a vertical series, extra samples may have to be introduced into the horizontal grid.

Because irregularities of deposition are always suspected in archaeopalynology it is important to correlate pollen records with adequate standards. The record from inside a structure should be correlated with that of the matrix surrounding the structure. On a larger scale the record from a site should, if possible, be compared to that of a stratigraphically less contaminated site immediately outside. The best situation is found where the site is situated at the edge of, or very close to, a lake.

A lake registers the vegetation of the whole catchment area, even if the more distant parts contribute less than the proximal ones. As the catchment area is gradually transformed into a cultural landscape, it is necessary to take into account the effect of agriculture, etc., even if the site of activities is not immediately adjacent to the lake itself.

The position of the sampling point in relation to the site of activities is fundamental for the understanding of the pollen record. The nearer the sampling point is to the site, the greater is the danger of soil erosion and the possibility of redeposition of sediment, which introduces a term of metachrony in the deposit and the record (Fig. 9.12), leading, in the end, to a complete collapse of the basic principle of pollen analysis. If this is overlooked the situation may become dangerous.

9.2.3. *Interpretation*

In dealing with analyses from sites of human habitation the idea of the uniform dispersal of pollen grains is invalid. This problem must be considered more closely for an understanding of what really happens in the analysis of such deposits. Figure 9.13 shows three (pollen) deposition types. The only strictly uniform dispersion is that shown by A: the distance between all units is the same. Apart from beehives, such patterns (overdispersion) hardly occur in nature. The usual pattern of uniform dispersal is the pattern in B: units are stochastically dispersed, and distances between them form a Gaussian curve. This is the reality behind the terms 'even' or 'uniform' dispersion as used in pollen analysis. Figure 9.13C shows underdispersion: units are irregularly clustered and distances between them tend towards a Poisson distribution.

In pollen dispersal the time unit is the deposition year. An event lasting for a deposition year (season) or less is, pollen-analytically speaking,

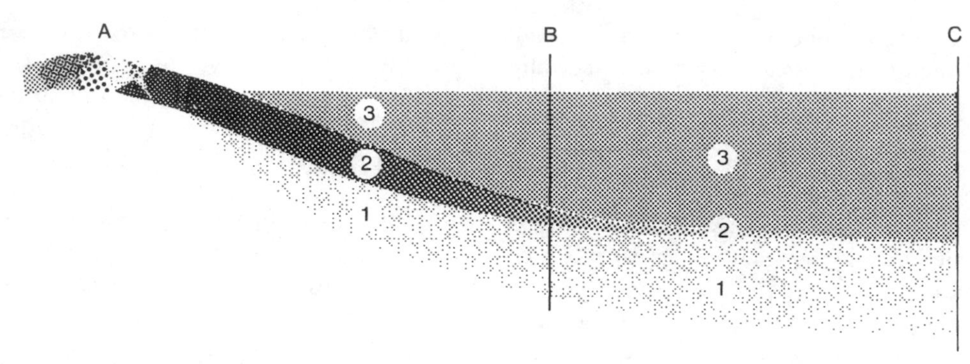

Fig. 9.12. Deposition model for a habitat site near a lake. Within the site area, A, stratification is disrupted. Stratigraphic elements are dislodged, mixed and secondarily deposited in the lake. At point B the stratigraphy is formed by three layers: 1 is pre-cultural; the normal model is fully valid. The upper and lower limits of layer 2 represent synchronous niveaus, but inside the layer there is chaotic (re)deposition where quantitative relations are fortuitous, becoming more regular distally. In layer 3 deposition is more or less normal again, but disturbances must be taken into account in interpretation. At point C layer 2 is not lithologically distinct but forms a chronological part of the sequence to be interpreted like layer 3.

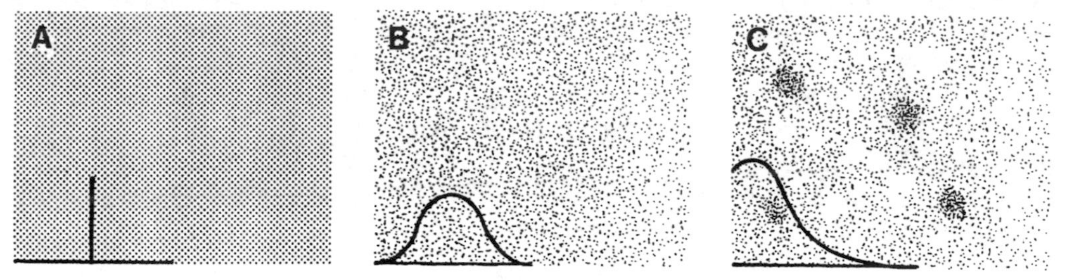

Fig. 9.13. Dispersion models. A: overdispersion, B: normal dispersion, C: underdispersion. The curves indicate the variation in distance between neighbouring dispersed units. Cf. text.

momentary. Momentary underdispersion is probably a very frequent occurrence in ordinary pollen deposition and has been recorded repeatedly (cf. Lüdi, 1947; A. Danielsen, 1970). However, it is rarely observed because, being momentary, it is 'integrated away', as a sample usually represents a sequence of time units (seasons). In the deposit from a habitation this does not apply: the pollen record from a soil sample may easily represent a singular event and, because of the irregular sediment formation, may not be obliterated by subsequent pollen accretion.

On the other hand, if in our underdispersed matrix the concentrations themselves are taken as units, their dispersion may be normal. The question of normality is a question of scale—in time or space. In a normal pollen analysis any (momentary) underdispersed pattern is integrated over time: if a spot received an excess of one pollen type it may not receive the same again. In pollen analysis of an archaeological site deposits are geologically momentary, but can be integrated over a *synchronous* surface, or even a volume of sediment, as long as the deposit is considered synchronous. Such an integrated record may give an overall registration of the activity at the site.

In other cases it is often more desirable to study the deposition pattern in more detail to obtain information about short-term changes, e.g. at the establishment of the settlement, or activity at a definite spot, indicating a specific function. In that case we need the individual, non-integrated records.

In samples from a habitation site both pollen and sediment accretion vary uncontrollably according to local activity, and do not provide generally applicable quantitative data. Since both pollen and sediment are extremely underdispersed, both pollen percentages and influx data from individual samples are meaningless, and can only be used qualitatively. Even so, this primary *pollen record* is a very important document. The find of, e.g., *Vicia faba* pollen in a soil sample indicates beyond doubt that the piece of deposit which gave the record came from a cultivated field where *V. faba* was grown, or from the immediate neighbourhood of such a field (with reservation for commercial import of beans). But absence of *V. faba* pollen does not imply that the species was not grown there, nor does the number of finds (which can be increased *ad infinitum* by counting more material) give any quantitative data, as long as there is no yardstick to measure the data against, neither pollen universe nor sediment thickness.

The immediately available constant, against which other data can be measured, is the *influx of allochtonous AP*. As shown by Middeldorp (1982), this can be considered constant, at any rate within a limited time span. With this is a basis one can calculate the *relative influx* (i.e. relative in relation to ΣAP) of other pollen types and produce meaningful statistical data (cf. Fig. 9.14).

These data suffer from the same shortcomings as the qualitative data of *V. faba* above: no conclusions can be drawn on negative premises, but we obtain a quantitative measure of the incidence of that pollen in that sample (only!).

Such data can easily be transformed into percentages, but they would have no meaning as long as the pollen universe is accidental. It is therefore better to keep them as (relative) influx data, giving the influx of pollen type x as a unique event. Figures may be calculated as %Total, %(ΣAP + x), or as n.ΣAP. All calculations give more or less the same results, but since, in this case, ΣAP is the yardstick, the third form is theoretically the best, giving the influx data. Numbers vary by one or two magnitudes and are better represented in a semi-logarithmic grid, except for very small fractions, which are best represented as regular numbers.

The use of ΣAP as a constant is invalidated if the activities on the site introduce additional AP, e.g. by building or leaf-foddering, leaving nests of high AP concentrations. Usually samples from such deposits give themselves away by a skewed percentage composition for one taxon (or a few, ecologically related ones). Such overrepresented taxa can be reduced to lower values by the following operation schedule.

1. Prepare percentage AP diagrams at all sampling points. Calculate average values.
2. If a sample shows great differences indicating macroscopic pollen accretion, correct the percentage of the relevant taxon to the average value for all samples (cf. Section 7.7.2).
3. Correct ΣAP and the value for the overrepresented taxon accordingly.*
4. Calculate relative influx of other pollen types in relation to the corrected AP sum (as n.AP) (cf. Fig. 9.14).

It is important to keep in mind that the basic assumption is that of a constant influx. As long as the rate of growth of the sediment is unknown, the influx cannot be expressed quantitatively, but that of other taxa may be expressed as n.ΣAP, i.e. in multiples of the AP influx in order to avoid confusion with percentage data.

The figures obtained by this procedure are mathematically meaningful, relating one influx to another, better-known one. However, conditions of formation of the soil sample studied may render the basic assumption invalid. The calculation assumes that the matrix of the sample contains at least one season's pollen deposition, and that the AP and NAP registered is deposited more or less simul-

* If there are two taxa, A and B, and the sum of all taxa is S, and, further, the percentage value of B is presumed corrected ($\%B_{corr}$), the corrected percentage value for other taxa and corrected numbers for the total (S_{corr}) and B (B_{corr}) are given by the three formulae:

$$\%A_{corr} = A \times \frac{100 - \%B_{corr}}{S - B}$$

$$S_{corr} = \frac{100\,A}{\%A_{corr}}$$

$$B_{corr} = \frac{\%B_{corr} \times A}{\%A_{corr}}$$

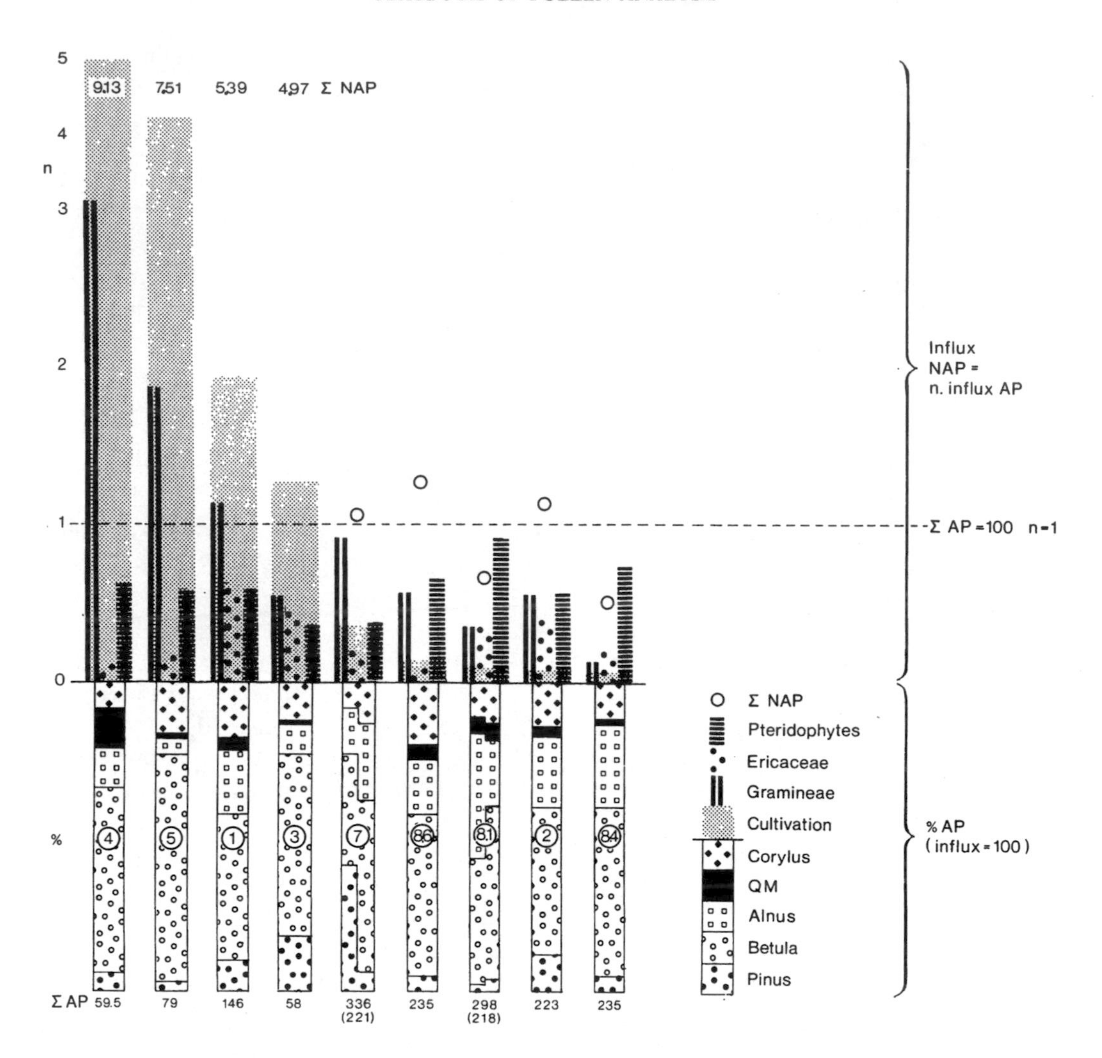

Fig. 9.14. The analyses from Bergenhus published by Krzywinski and Fægri (1974, Fig. 1), recalculated on the basic assumption of constant AP influx. All samples represent bottom and/or refuse layers in the ancient castle ground. Below the line: *Percentage* AP spectra demonstrating the constancy of the composition of the regional AP flora. Samples 7 and 8.1 are also presented after having been corrected, respectively, for assumed excess *Pinus* and *Alnus*. The correction does not materially alter the picture. ΣAP indicated at the bottom of the figure, sample numbers as in Krzywinski and Faegri (1974, Fig. 1) in circles in the columns. Above the line: *n*.AP diagram showing (shaded) 'Cultivation spectra': cerealia, *Centaurea cyanus*, *Vicia faba*, *Plantago*, *Rumex*. Signatures indicate Ericaceae and pteridophyte spores: hemerophobous, and grasses: indifferent. Note change of scale above $n = 3$. The diagram shows two phases: one representing a pre-existing farm, Holmen; the other representing the discontinuance of farming and building of the castle when the local vegetation was eradicated. The NAP values indicate that sample no. 3 still belongs to the farming period. Incidentally, the existence of the farm was disclosed by these analyses.

taneously. If not, the attenuation of the pollen deposit by differential accretion of matrix will give non-comparable results. Another source of invalidity is the possibility of instantaneous (matrix) deposition, i.e. deposition taking less than a full pollination season and therefore not receiving the full pollen complement at deposition. If it is then closed off, e.g. by an overburden, the pollen picture will be skewed. If the matrix is a polleniferous deposit secondarily deposited it will contain a full complement of pollen plus a skewed representation of those taxa flowering at the moment of deposition. If the matrix was sterile (e.g. ashes), the final deposit will contain nothing but the skewed record. Both bases will give non-meaningful data, the latter the worse. In such cases qualitative records are the only possibility.

The above discussion presumes that the sampled soil has, on the whole, been undisturbed since its incorporation in the archaeological deposit. This is usually the case in a habitation site after it has been left by the inhabitants, but in arable fields, for example, conditions are different. Ploughing disturbs the stratigraphy, but is fairly innocuous, the main effect being to mix the pollen flora of different layers, thus blurring the record.

Manuring is more destructive to the record, as animal (and human) manure contains great and uncontrollable quantities of ingested pollen. Especially, if straw or similar matter has been used for feeding the animals, the quantities of cereal pollen coming into the ground that way widely exceed those coming directly (Greig, 1982). However, since the plants fed to the animals were (a selection of) those that grew in the surrounding field and pasture, there still is the fairly simple combination of 'pollen rain' material, symbolized by the AP, and the human-induced pollen production and transport. It is a situation not very different from that already discussed, although complications may arise if animals are fed on exotic plant material (e.g. brewers' grains).

A more complicated picture develops when manure is not used directly, but is absorbed into polleniferous bedding material, which is later spread on the fields, as in the northwestern European Plaggenwirtschaft (Behre, 1976; Bakels, 1988). If the material used in this process is distinctly older than and/or comes from a plant community different from that of the site and its surroundings, the ultimate pollen flora may consist of disparate and metachronous elements: (1) the old 'pollen rain' flora of the original soil, (2) human-induced pollen present in that soil, (3) 'pollen-rain' corresponding to the time of using the manure, (4) human-transported pollen from the same time. It may prove impossible to distinguish between (1) and (3) or between (2) and (4) unless distinct floristic differences can be observed or postulated. Figure 9.15 demonstrates such a case from the deforested heath region of western Norway (Kvamme, 1988), in which the fields were manured with soil dating back to a pre-deforestation and pre-cultivation period. Consequently, the pollen record from the fields shows an excess of AP (component (1)) in relation to diagrams from outside the field. A smaller excess of the same pollen types in the diagram from immediately outside the field may be due to secondary redeposition (dust, trampling, etc.) from the field. Again the seriousness of such contamination depends on the floristic difference between the old and the new components, but for a closer analysis of the individual field it may be serious enough. Due to the great differences between components (1) + (2), and (3) + (4), in the above case, the contamination was sufficiently dis-

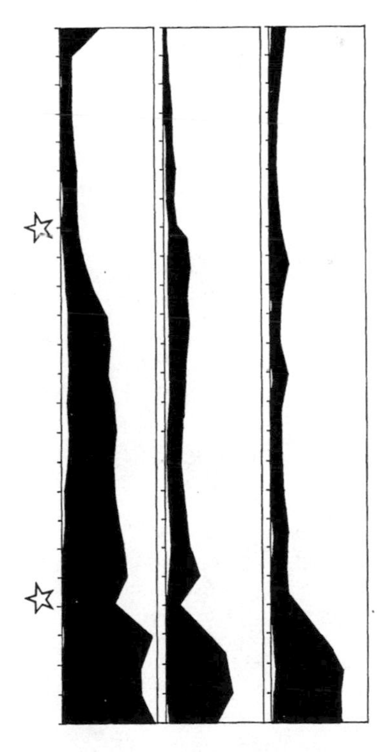

Fig. 9.15. AP curves in total diagrams from inside (left), immediately outside and at some distance from (right) a cultivated Iron Age field in the western Norwegian heath region. The stars indicate beginning and end of the cultivation period. Based upon data from Kvamme, 1988.

tinct to stand out. Usually it is more subtle and difficult to detect.

Complications like those caused by this type of manuring are also met with when soil, and especially turves, have been used for construction purposes (grave mounds, walls, etc.).

The ever-present possibility of stratigraphic turnover by the soil fauna (earthworms, etc.) is another factor which should be considered in the interpretation of, especially, analyses carried out in aerated dry-land habitation deposits. Also, the selective oxidation in older strata (Fægri, 1971) may complicate the picture by removing the organic component of the matrix, i.e. the old manure.

9.3. **Summary** (see Table 2.5)

In dealing with anthropogeneous deposits (Fig. 9.16) one must establish three relationships:

1. The stratigraphy. Has the deposit grown by more or less uniform accretion across relatively extensive surfaces or has material been added in lumps, irregularly distributed? In the first instance a normal stratigraphic procedure can be used; in the second it is necessary to treat individual lumps separately, relating them to some standard not influenced by the same mode of deposition.
2. Has the pollen accumulated independently of the deposit, or was pollen incorporated into the material before its final deposition? In the latter case there are chronological problems.
3. Is the pollen of local origin or was it transported to the site in bulk? If the latter, the pollen flora does not give adequate information about contemporaneous local vegetation.

The primary consideration in pollen analysis of

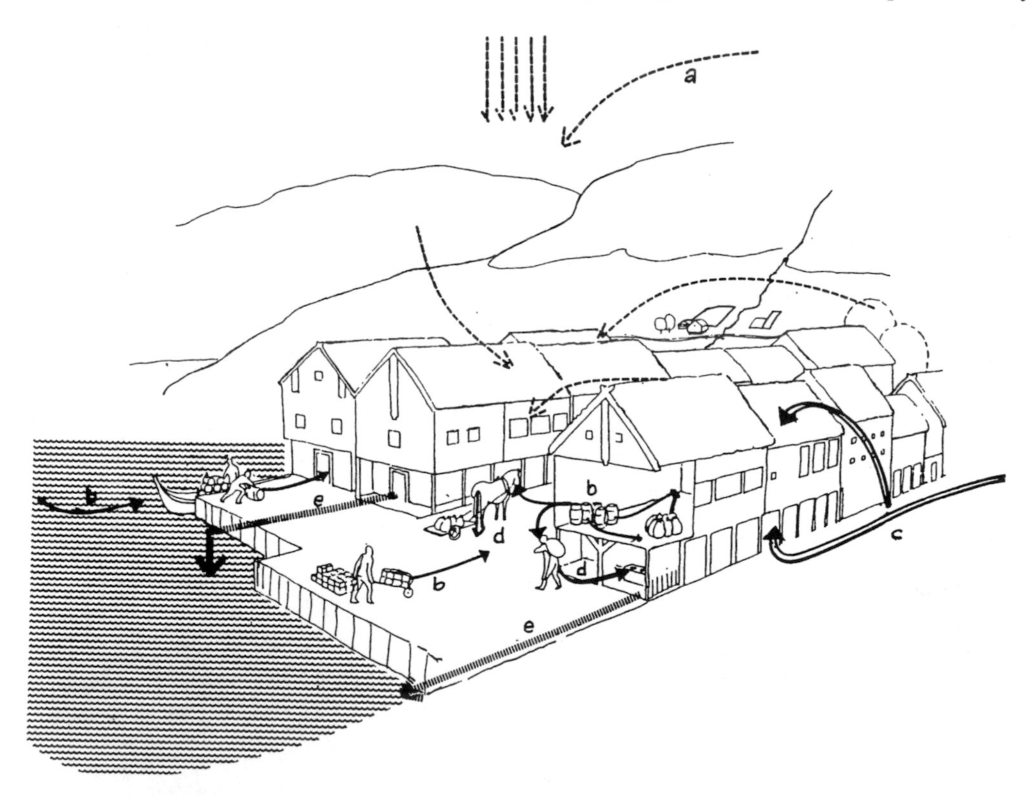

Fig. 9.16. Pollen transport model for an urban shore settlement. a: Air transport, 'pollen rain'; b: with goods; c: brought in with soil on floors and turf on roofs; d: excreta from inhabitants and domestic animals; e: redeposition of refuse. From Krzywinski *et al.*, 1983; redrawn.

anthropogeneous deposits is to collect data and interpret them as they present themselves without, in advance, trying to fit them into a preconceived system, archaeological or palynological. Many of the (unnecessary) discussions and misinterpretations between archaeologists and palynologists are due to straitjacketing data for the sake of established systems and chronologies. If they do fit in, well and good; if they do not, one will have to develop and utilize them as far as they go, and leave the rest to the future. *Eo ipso* it is not certain that one system is, under given circumstances, superior to another. Probably both need modifications in the light of each other.

10. Some other microfossils

10.1. Microfossils

A pollen-analytical preparation contains much material in addition to pollen grains; the quantity depends on the nature of the deposit and the treatment of the sample. Acetolysis removes many potentially useful fossils. In some publications there has been a tendency to include all and sundry discrete microscopical objects under the term 'palynology', notwithstanding the origin of that term. This practice should not be encouraged. Micropalaeontology is still a valid word.

10.1.1. Spores

In acetolysed samples the quantity of extra material is frequently negligible; it chiefly consists of spore walls, etc., of a chemical composition comparable with that of pollen exines. The chemical resistance of spore walls varies; whereas the spores of *Sphagnum* are easily corroded and in part disappear when acetolysed, those of *Lycopodium* and *Selaginella* are extremely resistant, even more so than pine pollen.

Obviously, there is no great difference in principle between spores and pollen grains, and spores should be counted during analysis. Spore-producing plants may represent a separate area and the spores thus may be important for the interpretation of vegetation. In other cases, however, spores will be of strictly local origin. Owing to the position of the spore-producing organs (generally under a leaf, a pileus, etc.) cryptogam spores do not, on the whole, seem to be so well dispersed in and with the air as pollen of wind-pollinated trees, or even herbs. Many spore-producing plants are members of the bottom layer in forests—a very unfavourable position when it comes to wind dispersal. Water transport may be a more effective alternative (cf. van Campo, 1975: 48). Spore curves are therefore more local in character; frequently they are very irregular and inconclusive. Important exceptions do, however, exist, e.g. the curve of *Pteridium* in post-Glacial deposits in western Europe, which shows distinct relations to land clearance (cultivation). Some difference seems to exist between spores that are spread by explosion (Polypodiaceae) and those that are spread more passively (*Lycopodium*). Because of their extremely small dimensions, fungus spores are very effectively dispersed once they have become airborne. Also, in some tropical vegetation types tree ferns form important constituents; in this case there is no reason at all to assign spores to a subordinate position. As an example we may mention *Hemitelia* (Cyatheaceae), with a highly characteristic spore, which has also been found as a fossil (Kuyl *et al.*, 1955). When fern spores are dispersed from higher strata of vegetation, some of the difficulties resulting from the low-level origin of pteridophyte spores in more temperate areas disappear.

As we go back in geological history, pteridophytes become increasingly important; their spores are indispensable parts of the microfossil record, and pollen analysis automatically becomes pollen and spore analysis.

Illustrations of spores of most European, and a number of exotic, *ferns* have been published, chiefly in some of the publication previously mentioned (cf. Moe, 1974b; Sorsa, 1964) but here, too, illustrations are of rather limited value. Some of the most indicative and easily recognized spores are habitually counted. In some vegetational types, e.g. as found in the Late-Glacial of western Europe, fern spores (*Gymnocarpium dryopteris, Botrychium*, etc.) are frequently produced in profusion and ferns certainly represent an important part of the vegetation. There is no doubt that fern spores may in many cases provide the pollen analyst with additional information; in some cases they may even be included in the 'pollen' sum for total diagrams. In some areas, e.g. the US Pacific coast, very high spore counts, probably of local origin, are consistent features of the diagrams (cf. Heusser, 1960).

Unfortunately, in most Polypodiaceae (sensu latior) the outer layer of the exosporium, which provides diagnostically important characters, is very thin and only loosely connected to the inner and heavier layer. This outer layer is frequently lost, and with few exceptions (e.g. *Gymnocarpium dryopteris*) the resulting 'bald' spore cannot be specifically identified.

Spores of *Lycopodium* (some of which may be identified to species) and the easily recognized microspores of *Selaginella* should also be counted; they may be used for conclusions as to the history of vegetation. The enormous spore production of *Lycopodium* must be taken into account. In which way, if any, *Lycopodium* spores should be entered in the pollen diagram, depends on circumstances.

Equisetum spores are frequently encountered, especially in hydroseres in shallow lakes. They are spherical and resemble *Larix* pollen grains, but lack the annular thickening and usually have an outer, crumpled and colourless 'envelope'. *Isoetes* microspores with their characteristic double wall are well known from limnic deposits.

While the spores of vascular cryptogams are to a certain extent known, those of the lower plants have, with few exceptions, been ignored by pollen analysts. *Sphagnum* spores are easily recognized as such, and other moss spores may also be identified (Boros and Járai-Komlódi 1975), but their importance, if any, for the analysis of former vegetation is largely untested.

The characteristic double teleuto spores of *Puccinia* are old acquaintances of most pollen analysts, and so are the small spores produced in parasitized *Sphagnum* sporangia and variously called *Tilletia sphagni, Helotium schimperi* and *Bryophytomyces sphagni*, but the correct name of which seems to be *Hymenoscyphus schimperi* (Naw.) Eckblad (Eckblad, 1975). Other fungus spores—and there are many others, though in most cases numerically less frequent—are generally neglected. Some of them may provide important information (Graham, 1962), and several recent pollen-analytic investigations take them more seriously, even if few of them can be specifically or even generically identified. Identification often has to stop at higher taxa or as 'types' of unknown affinity. No comprehensive treatment has been published, but identifications of various types are found, for example, in van Geel *et al.* (1976–83) which also contain material referring to the other fossil groups mentioned below. Owing to their allergenic effect and very frequent occurrence some fungus spores play an important role in medical aerobiology.

Some of these spores are explosively released during the night, and dispersal problems are therefore different from those concerning pollen grains. Some algae produce spores with resistant and identifiable walls, e.g. Zygnemataceae (cf. van Geel and van der Hammen, 1978). The usefulness of such fossils is commensurate with our knowledge of the ecological niches they represent.

10.1.2. Tissue fragments

Plant *tissue fragments* are particularly important in peats, in which they constitute most of the deposit. Very few of them survive acetolysis treatment, the object of which is, of course, to remove them. Easily recognized fragments like the epidermis of *Carex* or *Phragmites* rootlets, *Sphagnum* leaves, etc., serve to determine the character of the mother formation. A special KOH preparation may be made for analysis of perishable constituents to identify the original composition of the deposit.

Hyphae of certain soil fungi are resistant against both KOH and acetolysis, and are often found in disturbing quantities in terrestric peats where, together with pollen grains, they may constitute an appreciable part of the acetolysed deposit. A 'hypha index' serves to illustrate this. It may be calculated as number of hypha fragments/number of pollen grains, or more quantitatively by taking into account the variable size of fragments, e.g. by measurement in a cross-rule field.

Van Geel (1976–83) has published illustrations of tissue fragment types. One important tissue fragment type is charcoal which, when plentiful, indicates local fire—natural or human-made. A quantitative assessment is difficult; probably many of the fragments were crushed during the preparation of the sample. Nevertheless, the only practicable method is to count number of charcoal fragments/number of pollen grains, neglecting pieces below a certain size (2 or 5 μm), cf. Figs. 9.7 and 9.8. Quantitative measurement by microscope is very time-consuming and the results do not deviate much from those obtained by counting.

Among the tissue fragments that may be recognized are stomata cells of Pinaceae; how far specific determination can be undertaken in an area with many species is not known. Other cuticular fragments may certainly be determined, and it is not improbable that a cuticular analysis similar to that evolved by palaeobotanists could be used advantageously in Quaternary microfossil analysis, too. However, little work has been done along these lines (Katts and Katts, 1933: Istomina *et al.*, 1938). Phytolith analyses have been published.

Plant *trichomes* are in many cases well preserved, and are easy to determine. Among those important in Europe are the easily recognized peltate hairs of *Hippophaë rhamnoides*. As the pollen representation of a species may be due to long-distance transport, the occurrence of such fossils as trichomes or stomata cells is of great value, since they are most probably of local origin.

10.1.3. Algae, diatoms

Various *algae* may form important constituents of the deposit. *Scenedesmus* gyttja, and, above all, *Cyanophyceae* gyttja, are well-known types. *Botryococcus* is rarely absent from limnic deposits, etc.

More important than these are, however, *diatoms*. They occur practically everywhere in open water or very moist places. Their great importance is due to this fact, as well as to the facts that diatoms (1) have siliceous 'shells' (frustules) very resistant to decay, though they are broken by mechanical action in sandy deposits; and (2) are very sensitive to the chemical composition of the medium in which they live, e.g. pH and trophic condition (cf. Meriläinen, 1967). Above all, however, diatoms are sensitive indicators of the salinity of the mother formation. Various species are more or less exclusively bound to a specific degree of salinity. Diatom 'halobion spectra' are indispensable in investigations of former positions of the shoreline.

Diatom spectra originated in studies of the history of the Baltic, the brackish waters of which house many stenohaline species with very narrow tolerances. Under oceanic conditions, changes between a stenohaline sea-water flora and an equally stenohaline freshwater flora are often very abrupt, with euryhaline species bridging the gap, e.g. at the isolation of a basin from the sea. There is no niche for the sensitive brackish-water specialist species (cf. Fig. 10.1).

Complete diatom analyses can be undertaken only by specialists. The average analyst will have to rely on the occurrence and distribution of a small number of easily recognizable indicator species. Their presence may be expressed as percentages of a certain basic sum, in the same way as pollen grains. Because some diatoms, especially the epiphytic ones, are highly gregarious, the basic sum

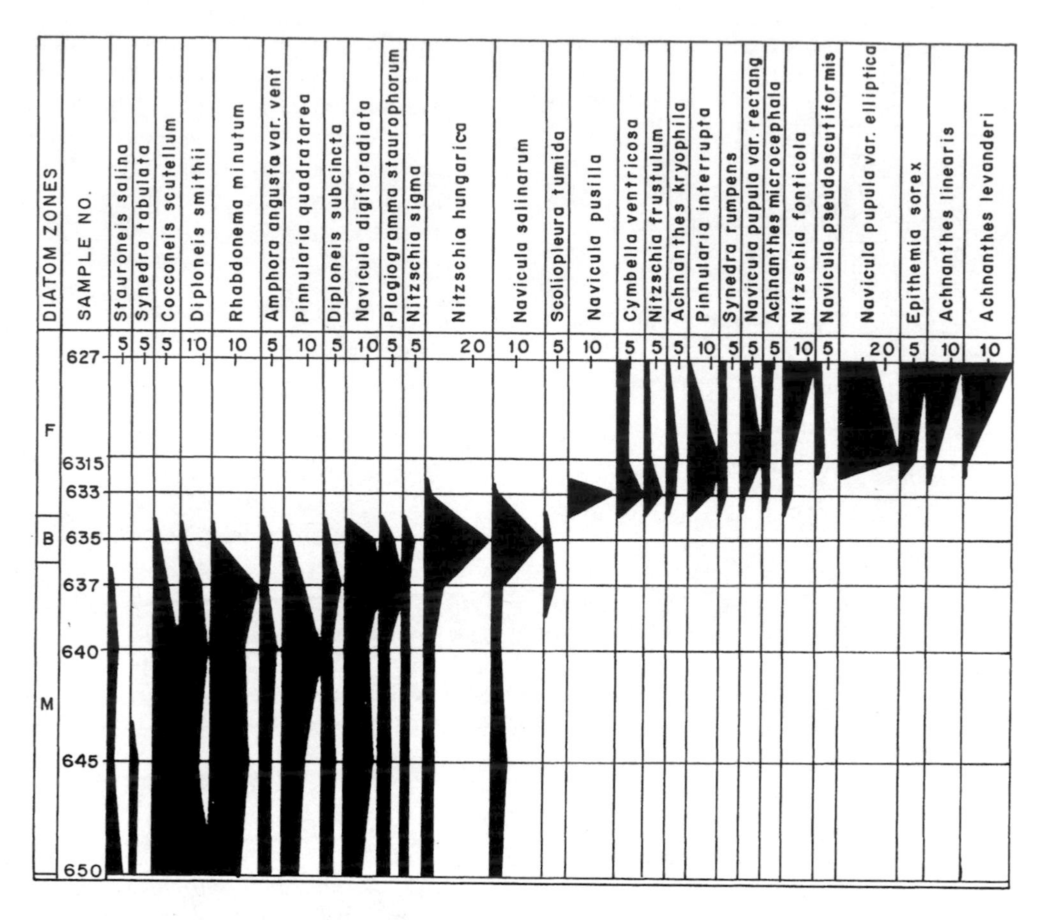

Fig. 10.1. Isolation of a basin as displayed by diatom analysis. Samples 650–640 are marine, followed by a short brackish phase at 635 and then freshwater, becoming gradually more oligotrophic, in 633–627. Redrawn from Stabell in Lie *et al.*, 1983.

must be made up with some care. One good basic sum is formed by the naviculoid diatoms, which are easily recognized, present in all salinity types, and show no marked tendency to mass occurrence. The number of frustules of indicator species are then expresed in percentages in relation to the total number of naviculoid diatoms of the sample. As most diatom species are found in very small numbers only, much work may be saved, without sacrificing much information, by neglecting rare species, the identification of which consumes most of the time spent on a complete diatom analysis.

Preparation methods for diatoms vary according to the ultimate object of the analysis. For such simple analysis as will be carried out by most pollen analysts the following method will be adequate.

1. A small quantity of substance is boiled (bleached) with perhydrol in a beaker until the organic matter has been removed.
2. If coarse sand and particles are present, they may be concentrated in the middle of the beaker bottom by carefully rotating the beaker by hand. The remaining material is then drained off in suspension and centrifuged. Alternatively, a combination of sieving and filtration may remove coarse and very fine siliceous material.
3. After being washed with water in the centrifuge, the diatom suspension is transferred to cover-slips and left to evaporate at ordinary room temperature. This will give more uniform preparations than evaporation over a flame.

The addition of a trace of detergent to the suspension will cause it to spread more evenly over the whole of the cover-slip. The proper concentration of the suspension must be learned by practice.

4. To be clearly visible, diatom shells must be embedded in a matrix with sufficiently high refractive index, of which there are several proprietory brands on the market. They are best applied as a solution of resin in xylene or some similar fluid, which is afterwards evaporated over very low heat. When the resin is sufficiently tacky, the cover-slip with the adherent diatoms is placed on a slide and is ready for examination. Scanning electron micrographs are easily prepared and are diagnostically very useful.

Taxonomic diatom literature is widely scattered, and there is no modern comprehensive flora. For European workers Hustedt's two works (1930 and 1930–62) are the most convenient. For northern Europe Cleve-Euler's manual (1951–5) should be consulted.

10.1.4. Animal remains

Animal remains are very frequently met with in pollen-analytic prepartions. Particularly, chitinous skeletal remains are often well preserved; cf. the papers by van Geel *et al.* quoted above.

Rhizopods. Monothalamine forminifera live in fresh water, and their easily recognized remains are characteristic components of *Sphagnum* peat, in particular. They have been dealt with by Hoogenraad (1935), and are of great importance when dealing with recurrence surfaces (cf. Section 3.1.2.1). The calciferous shells of the marine polythalamine species usually disappear by fossilization, but in some cases an inner *Discorbina*-like chitinous shell remains and gives evidence of the marine origin of the deposit. Theoretically, remains of other rhizopods (heliozoa, radiolaria) should be found as well, but so far they have not been of great importance in the analysis of Quaternary deposits.

Remains of *sponges* (spicules), *tardigrades*, *entomostraca* (Frey, 1960), *chironomids*, etc., are frequently encountered, but very few attempts have been made to evolve a systematic treatment of these microfossils, which are generally neglected in pollen analysis. The so-called *Hystricosphaeridae* ('Hystrix'), marine cysts, chiefly resting stages of marine dinoflagellates, are in some cases important as indicators of the presence of redeposited inter- or pre-Glacial material in fresh-water deposits (cf. Iversen, 1936). In marine deposits these fossils may be primary.

Hystricosphaerids are now commonly treated

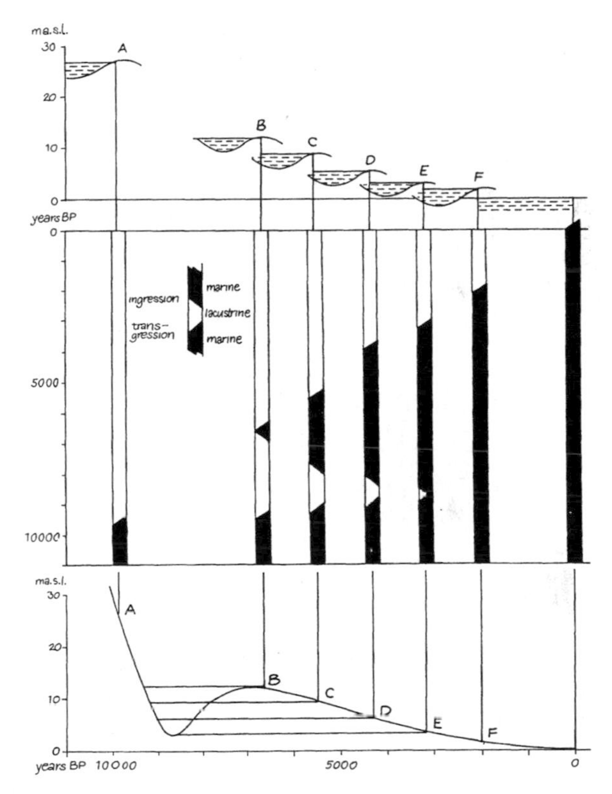

Fig. 10.2. A 'pollen staircase' showing a Holocene regression at the Norwegian west coast. Top: basins in different levels; middle: diatom registrations in the bottom sediments of the lakes; bottom: regression curve indicating when basins were isolated from the sea and, as the case might be, re-inundated. Basins A and F were isolated once only, the others inundated by a secondary transgression, which just reached the level of basin B. On the other hand, in basin E there was an ephemeral brackish phase at the first maximum regression; cf. Fig. 7.19. After Kaland *et al.*, 1983; redrawn. Note that the time scale is vertical in the middle figure, horizontal in the bottom one.

under the general concept of *acritarchs*, which may include cysts of various groups of microorganisms (Traverse, 1988). Together with other (artificial?) groups of identifiable palynomorphs, *scolecodonts*, etc., of more or less uncertain origin, acritarchs are important in pre-Quaternary stratigraphy (oil geology, etc.), but are of more limited importance in Quaternary microfossil analysis. It is beyond the scope of this work to deal with such objects.

Whereas mineral particles in pollen-analytic samples are generally a nuisance, the occurence of *volcanic dust* is of great importance, providing means of very exact dating and correlation of deposits—cf. investigations in Iceland by Thorarinsson (1944), in Fuegopatagonia by Auer (1956–70), in the USA by several investigators, especially in the northwest, and in the fossil explosion craters of Middle Europe (Straka, 1985).

HF treatment removes volcanic dust, which can therefore not be observed in a regular preparation. Such particles will be present, e.g., in a diatom preparation, or they may be observed in a deflocculated sample, if necessary after sieving. Volcanic dust (brimstone) particles are identified by their glassy appearance and sharp edges, in contrast to the blunter edges of ordinary mineral particles. A test by polarized light (crossed Nichols) indicates if they are actually amorphous. Even so, they may derive from local rocks if there are lava rocks in the area. Apart from the heavy tephra deposits near the

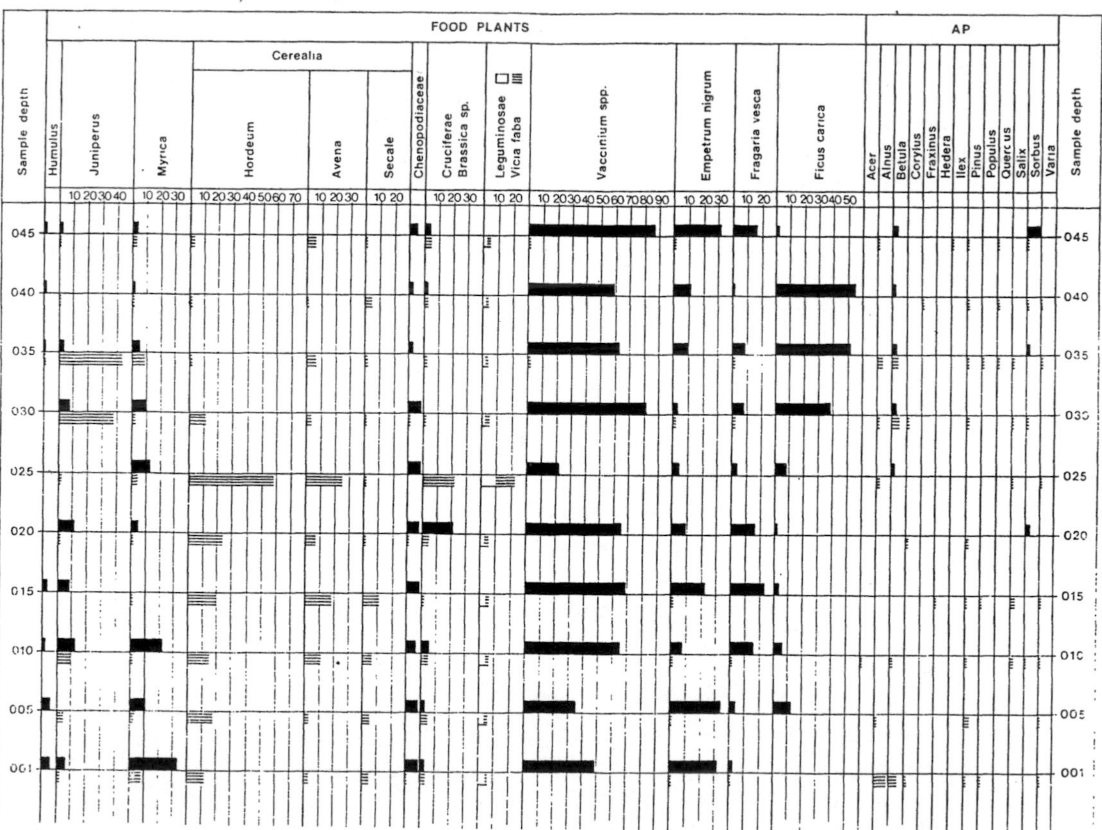

Fig. 10.3. Pollen and macrofossil analyses from the same deposit. Pollen percentages (lined) under the line, macrofossil concentrations (black) above the line. From Krzywinski *et al.*, 1983; redrawn.

site of the volcanic event, thin tephra layers are found, often hundred of kilometres downwind as a narrow fan, often irregularly deposited because of wash-down with precipitation.

10.2. Macrofossils

There is only an arbitrary limit, some 500–1000 μm, between micro- and macrofossils. Small macrofossils are often observed in samples taken for and prepared for pollen analysis, and can be picked out by an attentive investigator: macrospores of *Selaginella*, Characeae spores, small seeds, etc. Most macrofossils—more seeds, beetle elytrae, impressions of leaves etc.—are found by special investigation carried out for that purpose. Apart from really big specimens, tree trunks, large animal bones, etc., samples for the study of macrofossils are usually of the magnitude of 1–5 kg, and are prepared specially. The aim of the preparation is the same as for microfossils: deflocculation and removal of non-informative material. Owing to the size and brittleness of such fossils methods must be less vigorous than those used in pollen analysis, which means lower concentrations of chemicals, lower temperatures, and longer time. Deflocculation can be achieved by immersion in water, better by cold dilute KOH or nitric acid in which the sample is left for some time. After deflocculation the material is sieved in a set of decreasing hole diameters to withhold on one side large pieces, often of little information value, and on the other to get rid of the fine detritus. The rest is divided by sieving into fractions, subsamples of which are picked through, usually in water. The finer fractions are investigated under the stereomicroscope and individual fossils, mostly seeds, are picked out by brush, fine forceps or capillary tube. Preservation is by freeze-drying or wet in Hantzsch's fluid: equal parts of glycerol, water and 96% alcohol with some disinfectant—a thymol or phenol crystal.

The presentation of macrofossil data is more complicated than presenting pollen data, since there is no natural sum on which to calculate percentages. A 'seed sum' is possible, and may be extended to include some non-seeds as well, but there are many other fossils that fall outside such a sum. Concentration data are in most cases the most natural ones to use. Macrofossil diagrams can give completely different information from a pollen diagram from the same deposit and thus give important additional information (cf. Fig. 10.3). Air transport of macrofossils is practically non-existent except over very short distances. Of natural agents water is the most important, but animals also transport seeds some distance. 'Pack-rat middens' is a well-known type of deposit. Transport as a consequence of human activity is extremely important for macrofossils. Natural macrofossil assemblages are very local and give information about the vegetation of the nearest neighbourhood. Assemblages due to human activity tell very little about local vegetation, but a great deal about the activity of the people concerned: travel, trade, etc.

11. What pollen is and does: Form and function of the pollen grain

In principle, the pollen grain corresponds to the microspore of lower plants, representing therefore the *asexual* generation. Whereas the general character of pollen grains corresponds to that of spores, it is not *a priori* certain that pollen grains of gymnosperms, even less cryptogam spores, are built the same way as angiosperm pollen. Apparent inconsistencies in the interpretation of the sub-microstructure of spores and pollen may be caused by taxonomic differences between the taxa studied, which may even represent separate major lines of evolution.

In Quaternary (and pre-Quaternary) pollen analysis grains are (sub)fossils: all non-resistant parts have disappeared. For this purpose we therefore mainly take into account morphological characters that, directly or indirectly, pertain to fossil grains. Characters that disappear during fossilization (colour and structure of cell contents, intine features, colour of external substances adhering to the grain, etc.) are, on the whole, neglected. When dealing with recent material for comparison one should prepare grains so as to resemble, as much as possible, the fossil ones. As the presence of cell contents, extraexinous oil, etc. obscures the diagnostically important exine characters, all such superfluous matter should be removed.

In some other fields of research where pollen grains are used (mellittopalynology, allergology, etc.) grains are living, and other morphological characters may be of importance for taxonomic identification beyond the level achievable by use of fossil or fossilized grains alone. Such characters will not be discussed here; they may be consulted in the relevant literature.

11.1. Pollen morphology in relation to function

The 'proper' function of a pollen grain is to germinate on a stigma and give rise to a pollen tube, which takes the sperm nucleus down to the ovarial complex. This presumes transport across a shorter or longer distance from the anther to the stigma. The transport vector may be abiotic, usually wind, exceptionally water, or biotic, predominantly insects, but also various other animals, flying or not. Wind dispersal and wind pollination are the most important in relation to pollen analysis.

A major, but often overlooked, constraint in wind pollination is the extreme restriction of the possible germination bed: unless a pollen grain lands exactly on a compatible stigma, a surface that is usually measured in square millimetres, it is a failure. This constitutes an enormous difference as compared with most cryptogam (micro) spores which, however specialized, have a potential germination area several magnitudes larger. This causes demands with regard to the quantity of pollination units, dispersion, transport, etc., as well known in pollination ecology (cf. Fægri and van der Pijl, 1979).

A last difficulty in wind pollination arises when the grain is to land on the stigma. The grain is carried with and in the air stream, almost as part of it, as its independent movement (gravity fall) is much less than the speed and, especially, the turbulence of the wind. The smaller the grain, the better it is integrated in the air stream, and the longer it is (potentially) transported. When the wind abuts against an obstacle it bends around it, eddying in various ways. If the pollen grain follows the wind exactly, it will also bend away, in this case from the stigma, and not be deposited. By having a slightly higher inertia, the pollen grain will continue its straight trajectory out of the bending stream, and hit the stigma the wind avoids (cf. Fig. 2.6). For this to operate, the pollen grain must have a not too small mass, i.e. it cannot be smaller than a certain value. It is therefore seen that wind-dispersed pollen grains, in spite of the necessity to be light, are fairly big, magnitudes of 30–40 μm being quite common. Many insect-dispersed pollen grains are much smaller.

Another set of conflicting constraints refer to surface configurations. A strong sculpturing would increase the surface of the grain and, presumably, its buoyancy by increasing the friction against the air. On the other hand, a complicated surface would cause grains to adhere to each other, building up pollination units too heavy for successful transport. As the sculpturing elements have no function in relation to pollenkitt, which is usually missing in such grains, sculpturing is usually absent or very reduced in typical wind-pollinated grains. —cf. the pollen of *Ambrosia, Artemisia* or *Xanthium* in relation to ordinary Compositae pollen (fig. 2.2).

The very large and heavy pollen grains of some conifers may be seen in connection with the strong eddies estabishing themselves around a female flower (Niklas, 1981; Niklas and Buchman, 1986). It takes a great deal of extra inertia to tear a pollen out of an air stream at an obstacle of that magnitude compared to eddies at a single stigma of a grass, or a moving catkin of a birch.

In insect-pollinated plants much of this is reversed. Pollen grain size is of little significance in relation to transport problems and may be regulated by different factors. The sculpturing functions as a receptacle for pollenkitt, etc. But there is one complication that has recently been realized: wind-dispersed pollen is part of the air stream and its movements in relation to the air is negligible. Insects move with their pollen at a different speed (and direction) from the surrounding air; there is a strong friction, which establishes a static electric potential. When the pollen-loaded insect approaches a grounded conductor the opposite potential establishes itself there, and pollen grains jump (a distance of, perhaps, the magnitude of 100 μm) across to the stigma which, of all parts of the flower, takes up the strongest opposite potential (Corbet *et al.*, 1982). Chaloner (1986) has suggested that the strong, sometimes very strong, sculpturing of insect-pollinated grains may serve to isolate the grain from the stigma, so that it does not lose its potential immediately, but will remain electrically attracted until further processes cement it more securely.

11.2. Terminology and its use

Morphologic terminology in palynology has been, and still to a great extent is, a controversial and even chaotic subject (cf. Kremp, 1965). One reason for this state of affairs has been the fact that pollen grain characters are so small as to be at the physical limit of ordinary light visibility, i.e. *ca.*0.3 μm. Many of the features named in the early literature were at that time more of an educated guess than a concrete observation, and the terminology correspondingly tentative. The use of electron optics with a much higher resolution has, on the whole, ended such controversies. It is possible that computer-assisted image analysis, which has till now not been tried in the evaluation of pollen micrographs (LM and SEM), may add another dimension to our understanding of the morphology of the grain.

However, there is another point that is still unsettled. Should our terminology be strictly descriptive or influenced by the concept of function as well? We may compare with the terminology necessary to describe the gearbox of a car. A purely descriptive account will enumerate wheels and cogs and transmissions, etc.; a general account which —provided the observations are correct—is unassailable now and for ever. If we perceive the gearbox as an instrument for transferring variable power from engine to wheels, the individual parts obtain a new meaning, and another terminology may be developed for the same parts, indicating the use in the function of the box, as perceived. This terminology is also unassailable; it gives more information, but it can only be used under the special conditions evisaged by our conception of function. It is not of general usefulness.

In contrast to the gearbox the function of various parts of the pollen grain is still more or less conjectural. The aperture, as an opening in the exine, is a well-defined feature, the morphology of which may be described and analysed, but what is its function: harmomegathy and/or exitus of pollen tube and/or what? A functionally influenced terminology is therefore not advisable in palynology.

To describe a pollen grain as exineless is to state a fact. If we call it omniaperturate we have introduced a functional aspect, indicating our assumption that the whole surface has the same function as the individual apertures in a normal grain, whatever that may be. Since an aperture by definition is an opening in an otherwise closed envelope the term omniaperturate is meaningless: an opening presumes something delimiting it.*

Whereas we agree that an interpretive functional terminology is acceptable for its own purpose we shall in this book endeavour to maintain a strictly descriptive terminology and present interpretations and the corresponding terminology independently.

It follows that if a term is based on a fallacious observation or assumption it has no right of recognition, it is a *terminus nudus*. Otherwise the principle of priority should be observed, if not so narrowly as in the nomenclature codes.

11.3. Techniques in pollen morphology

11.3.1. Preparation techniques

Pollen grains for reference samples can be taken from dried (herbarium) material or from fresh, living specimens. For general use the origin of the material is of no importance, but voucher specimens should always be available in a public herbarium and cross-referenced against the sheet and the preparation. Also, it is important to update if the herbarium sheet has been re-identifed.

Preparation techniques for recent pollen for morphological studies are the same as those dealt with in Chapter 5, with minor changes due to the quantity and concentration of the original material. The procedure is summarized in the flowchart (Fig. 11.1). Steps 3 and 4 are redundant if the material is already dry.

For recent pollen preparations it is recommended that a quantity of a standard pollen of known

* Omniapert, derived from Latin *apertus*, uncovered, would be linguistically defendable, but morphologically redundant.

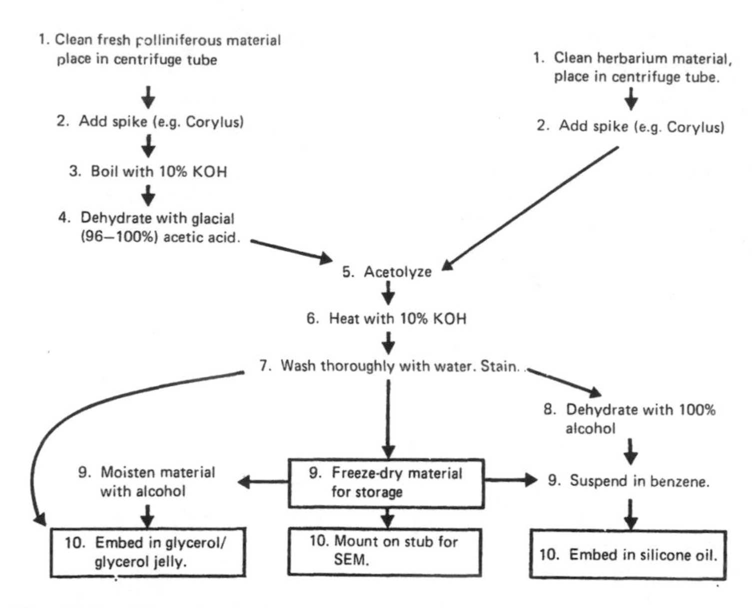

Fig. 11.1. Flowchart of preparation of recent pollen for comparisons.

behaviour and size be added to the preparation. The standard pollen (*Corylus avellana* in our case) should be collected at the outset, a sufficient supply to last for years. It should be added to the pollenife-rous material before the preparation procedure starts.

The advantages of the *Corylus avellana* pollen is that (in northern Europe) it seems to be constant, there is no other species with which it can be con-fused, and its pollen wall characteristics are identi-cal with the characteristics of most of the other pollen species in the analysis material. In areas where *C. avellana* pollen is not easily available one may use some other indicator or get some *Corylus* pollen from a colleague more fortunately situated. It is a very small job to collect a large portion of it during flowering time. Beware of commercial pol-len preparations! We have no experience with other spikes, but are sceptical with regard to very thin-walled grains, e.g. *Eucalyptus* spp.—thin-walled grains seem to swell in a different way from those of thicker walls (cf. Fig. 11.14). When the be-haviour of spike pollen under stress is known, its state in the individual samples also gives indi-cations of the 'health' of the other pollen types present. If and when the structure of *Corylus* grains in a mount is no longer distinguisable, the mount is due for scrapping.

The other function of the spike is size control. If the spike pollen grains have changed size, the other grains have done so as well, and probably proportionally. We operate with a 'standard *Cory-lus*'. If a check shows a change to size *x*, a corre-sponding correction factor gives a presumably more correct value for the other grains, too.

The quantity of the spike should be adjusted to that of the pollen type for which the mount is made, so that neither type is swamped by the other. How-ever, sometimes the number of desired pollen grains is very small, and a greater quantity of spike facilitates the manipulation of material, saving it at the expense of a heavier search.

For the preparation of recent pollen samples Tatzreiter (1985) has developed an alternative method, which is complicated but gives superb results as can be seen from his publication:

1. Prepare solutions:
   A: 84 ml HCL 5%
   16 g $MgCl_2 \cdot 6H_2O$, 98% purity
   8 g $NgSO_4 \cdot 7H_2O$, 99.5% purity
   12 g $AlCl_3$, non-aqueous, 97% purity.

1 g Al phosphate, basic, Merck.
B: 60 g KCNS, 98.5% purity or
50 g NaCNS, 98.5% purity.
Dissolve in aq. dest to 100 ml.

2. Place 0.5 g dry plant material in centrifuge tube with 5 ml solution A and 1 ml solution B. Heat in double boiler, stirring with a glass rod *ca.* 20 min. Hood! Poisonous fumes develop!
3. Sieve on ceramic filter 0.5 mm mesh. Wash down with 5% $MgSO_4$ solution from squirt bottle. Centrifuge.
4. Wash with 96% alcohol—fill centrifuge tube. Centrifuge.
5. Add three to five drops lactic acid *ca.* 90 conc. and 1–2 ml H_2SO_4 conc. Heat in double boiler *ca.* 15 min. until dark brown-blackish. Stir.
6. Cool, add alcohol and shake. Centrifuge. Repeat three or four times.
7. Squirt 3% $K_2Cr_2O_7$ solution into, or stir up, bottom sediment. Heat in double boiler some minutes.
8. Heat with aq. dest. in double boiler three or four times.
9. Embed or prepare for SEM.

The process takes some 3–4 hours and is impractical for individual samples. One should prepare batches of about a dozen samples at the same time.

Which-ever preparation technique is used, there is always a great deal of local know-how implied, varying from one laboratory to the next. A major problem is often scarcity of material, for which reason it is important to guard against losses when decanting supernatants from centrifuge tubes. They should be slowly tipped to a horizontal position, but not beyond. The drop at the lip of the tube is removed with absorbent tissue, preventing it from running back and disturbing the deposit. This also makes the washing more effective.

After the last washing it is not necessary to pour off the supernatant as thoroughly as in the previous operations. An adequate piece of glycerol jelly is dropped into the tube and melted. The solution is then poured out onto a watchglass and left to evaporate—usually overnight, but not too much: dry glycerol jelly is rather unmanageable. If there is very little material, one may reserve a little more (*ca.* 1 ml) of the supernatant and use a very small piece of glycerol jelly. The material can then be concentrated by slight rotation of the watch-glass. After drying only a thin film of jelly is left on the watch-glass. The polleniferous part can be cut out under a dissecting microscope and mounted with some more glycerol jelly. The technique gives useful mounts from a single, small anther, e.g. in *Primula* (Wendelbo, 1961). The spike may function as a carrier, but care must be taken not to overload the mount.

Sometimes the material is even more scarce and may consist of single pollen grains picked up individually from a herbarium sheet, e.g. by 'fishing' with glycerol jelly under the dissecting microscope. In such cases one may use the method introduced by Avetisyan (1950) for acetolysis on the slide. Whereas it is said not to give too good results in its original form (Brown, 1960), a modification by Punt (1962) is more useful. The material is placed in the pit of a hollow-ground slide; a few drops of acetolysis mixture is poured into the hollow, and the slide gently heated, e.g. on a metal strip that is heated at one end. Evaporation of the mixture is compensated for by adding a few more drops. The result of the process may be followed in the dissecting, or even better an inverted, microscope. We would recommend neutralizing with KOH, careful washing with water, once or twice (taking the water away with a piece of blotting paper), and staining, if necessary. In this way it is possible to treat pollen grains more or less individually.

The method is also useful if time is a consideration, but mounts made this way are difficult to wash completely free of reagents, and may be short-lived.

The intended effect of acetolysis is to clear away disturbing constituents not present in the fossil grain. The same can be achieved by prolonged boiling in 10% KOH on a reflux cooler. Some exines resist many hours' boiling, others are dissolved after a relatively short boiling time (cf. Fægri and Deuse, 1960).

Pollen walls are usually translucent and the

inner structure is visible in a whole mount; optical sections in various planes give the necessary details. Some very thick and complicated walls are not immediately understandable and should be sectioned for a closer investigation.

Sectioning individual fossil grains needs a great deal of technique and experience, but mass sections of recent material are easily obtained. Various methods have been proposed (cf. Christensen, 1949); those involving dehydration have a tendency to turn exines rather brittle and difficult to cut. For optical microscopy it is better to use an embedding medium that does not require dehydration. The simplest procedure is to embed in glycerol jelly or in gum arabic to which has been added some glycerol to prevent it from becoming brittle. Leave the pollen grains, after acetolysis if needed, in a watch-glass with a thin solution of glycerol jelly in water and cut after the jelly has become sufficiently firm, if possible on a freezing microtome. Blackmore and Dickinson (1981) have described the method in greater detail. By freezing in liquid nitrogen and cutting in a cryostat at −20°C they also could achieve sections good enough for SEM. Cooled-down material can also be fractured with a razor-blade and a hammer (Blackmore and Burnes, 1986).

At ordinary temperature the jelly is resilient, and cuts vary greatly in thickness, but among the great number of sections obtained in a mass preparation there will be some that can be used. Water-miscible embedding waxes would appear to be an answer to that, but the use hardly pays. When the upper half of a grain has been cut the lower part frequently tears out of the medium, so there are many uncut halves in the mount, but again, there will be a sufficient number of sections from the upper, cut part. With some luck, sections can be cut down to 2 μm without freezing. Pollen-rich deposits may also be sectioned after proper treatment (cf. the polyethyleneglycol process, Section 5.2) for the study of grains *in situ*.

For SEM studies sections through the exine can also be obtained by ultrasonic fracturing. The method presumes the availability of special apparatus and does not give results superior to those achieved by ordinary sectioning.

Simple methods of fracturing exines for SEM observation are used in our laboratory: placing the pollen material—acetolysed or not—between two pieces of stongly adhesive tape, which are forcibly pulled apart. A number of grains are pulled apart as well. Alternatively, grains may be frozen between two SEM stubs, which are broken apart, taking grain fragments with them.

The ordinary methods of preparation for SEM and TEM studies can also be used for pollen grains, but are too specialized to be detailed here.

11.3.2. The reference collection—the pollen herbarium

During analysis the principal problem is that of recognizing pollen grains. It is chiefly a question of practice, and the beginner cannot learn very much about this part of the work from books. We can only recommend that he/she start work in the laboratory of some more experienced colleague. A well-stocked 'pollen herbarium' containing all the more important types is a necessity. The preparation of such a collection for personal use is excellent practice for a beginner. Even the best illustration can hardly serve as more than a reminder to more experienced workers. The beginner will have great difficulties in interpreting the morphology of a pollen grain by means of illustrations alone, even SEM photographs.

The slides of the pollen herbarium should correspond as closely as possible to the grains met with in actual analysis work, usually showing exines only. If a special method is standardized in a laboratory, preparation and mounting of the 'pollen herbarium' should (if possible) follow the same procedure (including embedding medium) to facilitate comparisons. No matter whether fresh or herbarium material is used, peparations are cleaner if as much as possible of extraneous matter (the rest of the flower, etc.) is removed beforehand.

Some few pollen types explode in contact with water (e.g. *Mahonia*, cf. Roland-Heydacker, 1979), and must be concentrated by picking the anthers out separately. Alternatively, one may grind the material carefully in a mortar with some drops of chloroform or alcohol, sieving afterwards under a jet of alcohol. A very careful acetolysis would be indicated—no KOH treatment in advance.

In other taxa the exine is very delicate or non-existant and in extreme cases such grains must be embedded without any kind of preliminary treatment (Najadaceae, Juncaceae, Musaceae). Such grains are usually without any practical importance in pollen analysis, even if they may be retrieved under extremely favourable circumstances (Juncaceae: Fredskild, in Degerbøl and Fredskild, 1970: 185). In Greenland material *Juncus/Luzula* pollen grains are even said to be frequent (Fredskild, 1973: 27).

When preparing recent pollen one should be careful to avoid contamination. The pollen of some anemophilous blossoms (e.g. of grasses) is dispersed almost instantly when the anthers open, and must therefore be collected in the field when flowering occurs, best immediately before the anthers open. Herbarium material of such plants is quite useless for this purpose. It also happens that the pollen of the blossom has already been dispersed, and that foreign pollen has been deposited by a later visitor/pollinator. As long as the pollinator is constant (in the sense of pollination ecology) there is little risk of contamination, but not all pollinators are, and results may be misleading, e.g. a case of an *Asclepias verticillata* flower that contained nothing but *Ambrosia* pollen (personal observation).

All work with recent pollen should be kept strictly separate from that of fossil analysis, to prevent contamination. Different sets of utensils should be used, and treatment carried out in different rooms, not intercommunicating with those where preparation of fossil material and pollen analysis itself take place.

Since results differ according to treatment, it may be useful to prepare recent material in more than one way. For example, thick-walled grains may be very dark after acetolysis. A short oxidation will clear the exines, but at the same time swell the grains; such slides are of no use for any purpose other than qualitative morphological analysis. Pollen prepared in different ways may be combined in one mount, but is better kept as separate mounts on the same slide. Recent preparations may be stained or not, depending on laboratory practice. If preparations for analysis are stained, those of the pollen herbarium should also be. The durability of some stains is unreliable. Stained and unstained grains must be mounted separately. For more discriminating work, particularly size statistics, care should be taken that treatments of fossil samples and of recent comparison material are as identical as possible.

Even more so than in reference slides from an investigation (cf. Chapter 5) those of the pollen herbarium should be prepared and protected so as to be useful as long as possible. In practice today there are only two media of importance: silicone oil and glycerol/glycerol jelly. Both have advantages and drawbacks.

Silicone oil is the simpler: once samples have been dehydrated and embedded they are long-lived. The size of the pollen grains does not change after embedding. The medium is non-volatile and does not dry out. Mounts may be left unsealed, especially if stored horizontally. The disadvantages are, in addition to the more complicated embedding procedure, the smaller size of pollen grains in this medium, which may complicate the observation of the very finest details at the border of visibility, and the low refractivity that causes too strong contrasts, especially disturbing Becke lines. Another drawback, more rarely making itself felt, is that a pollen grain that has been embedded in silicone is almost impossible to clean if it should later be needed for SEM.

The advantages of glycerol and glycerol jelly preparations are above all the much simpler embedding procedure, especially for preparations meant to be short-lived. Further, and more important, is the better distinction of finest details with less vigorous contrasts.

The drawbacks of glycerol and glycerol jelly preparations have been known and discussed for practically the whole time pollen analysis has existed, and a great deal is known about them, and can be controlled. The main drawback is that glycerol and glycerol jelly preparations are labile in relation to water vapour: taking up water or drying out. They must be sealed to prevent leakage.

Mounts made with glycerol jelly fix the individual grains so that they cannot be turned for inspection of other sides. In reference preparations this does not matter, as there is generally so much material present that is not necessary to turn any individual grain. Mounts can also be made between two cover-slips, and then be viewed from both sides. For inspection, such mounts are easily fixed to a slide with adhesive tape.

The often-observed impermanence of glycerol jelly preparations has been quoted against the use of that medium, certainly with some justification. On the other hand, we possess in our pollen reference collection glycerol jelly preparations that are 30 or 40 years old, and are still perfectly good. Also, if the procedure as given above is followed, it is such a simple task to make another preparation that it fully compensates for the more demanding silicone procedure.

Some immersion fluids are apt to dissolve sealing substances. The best plan is to put the seal under the cover-slip, if possible. Water-miscible media can be very easily sealed with paraffin simply by melting a small quantity of paraffin (melting point above the expected maximum air temperature) with the mount. When a cover-slip is added, the paraffin spreads around the medium and seals it off.

The simplest and most elegant technique is in advance to coat the edges of cover-slips with paraffin (or any other sealing wax) by dipping them into the molten wax. Place a small drop or piece of pollen suspension on the cover-slip (upper side) and place a (not too) hot slide on top. The melting wax seals the drop off. The operation should be carried out on a piece of absorbent tissue to drain off the wax from the other side of the cover-slip. The drop should not be too large because of the risk that it will break through the seal. This technique can also be used for sealing preparations in fluid mounts, which may be difficult by other techniques.

Paraffin does not adhere to glass that has been soiled with glycerol and glycerol jelly. They must be thoroughly washed first.

Latex paint ('Spread' poster paint, etc.) is oil-proof and can be applied as a protective coating on top of wax and lacquer seals.

In addition to the necessity for sealing (glycerol or) glycerol jelly preparations have three other drawbacks: Fresh glycerol jelly mounts are likely to melt in hot climates. With increasing age this tendency ceases, and with appropriate sealing (high melting point of the paraffin), preparations are safe even if the medium melts.

Sometimes—far from always—crystals form in the glycerol jelly. The reason in unknown, but most probably it is the presence of impurities (salts) which have not been washed out.

The most serious disadvantage of glycerol jelly is the hitherto unexplained tendency of pollen grains to swell and lose their characeristics in this medium. This may partly be due to compression of the mount because the medium dries out, as described by Cushing (1961). This effect is most serious in the best samples that contain little but pollen grains, and also more serious for large grains than for small ones (Whitehead and Sheehan, 1971). Apparently, however, there are other effects entering into the picture, too, causing swelling. Insufficient washing of the grains before embedding may be of importance. The composition of the jelly itself does not seem to be responsible (Fægri and Deuse, 1960). The 'Cushing effect' may be counteracted by avoiding too thin mounts, e.g. by adding small cover-slip splinters to the preparation, and by a paraffin seal under the cover-slip (cf. above). The latter will prevent both drying-out and crushing. Owing to possible swelling effects, size measurements of pollen in glycerol jelly mounts must be considered suspect unless carried out immediately after embedding.

Many authors have discussed the reasons for bad preservation of samples in glycerol jelly, and various suggestion have been put forward: acidity, water contents, the phenol added to keep the jelly sterile, etc. However, none of these seems to be valid in all cases. Admittedly, there may be unsuspected variations in the gelatine used for preparation of the glycerol jelly.

The most reasonable explanation seems to be that samples have not been properly washed after chemical treatment. In spite of the obvious shortcoming of glycerol jelly there is for the time being no generally accepted alternative.

Praglowski (1970) found that the degree of

collapse of pollen grains in various media depends (1) on the species: large, thin-walled grains collapse more easily, grains with many apertures are more resistant: (2) on the treatment; grains collapse more easily after KOH treatment alone than after acetolysis; and (3) on the embedding medium: greater percentages of collapse in very viscous and rapidly solidifying media. The last can be prevented by the use of diluted glycerol jelly which is allowed to set by evaporation as described above.

Preparations in glycerol jelly or other solidifying media should be left inverted until the medium has set, leaving the grains close to the cover-slip, the thickness of the mount is in itself relatively unimportant.

As for the practical arrangement of the pollen herbarium, it is important that new mounts can be easily intercalated in the right, systematic order. Ordinary slide boxes, whether for vertical or flat storage, are impractical. We use slide holders of approximate file-card size (*ca.* 9 × 1 cm, e.g. Fisher scientific no. 12–588–25) which can be stored in ordinary file cabinets. The holders take four slides each, and intercalation is very simple. This is more important, the larger the collection grows (our present collection has *ca.* 5000 slides).

Surplus material should be stored for future use. Storage in glycerol or glycerol jelly is simple, but the unpredictable effects of long-time storage in these media advise against their use for this purpose. The best seems to be freeze-drying. The prepared pollen grains must be frozen in water suspension in advance and kept frozen until dry to prevent collapse and clogging of grains. Sufficient material for another preparation is taken out on the tip of a needle.

11.3.3. Microscopic technique

The microscopic techniques and equipment used in pollen morphology are, perforce, very similar to the ones used in practical analysis work, and reference is made to Chapter 5. However, even in a non-specialized laboratory more special techniques than those used in analysis can be used for morphological studies, including the recognition of pollen species. In addition to light microscopy (LM), scanning electron microscopy (SEM) and transmission electron microscopy (TEM) are both important for the understanding of pollen morphology even if they cannot (yet?) be used in routine analysis. Light optics are and will probably always remain the usual technique in analysis, but the additional morphological knowledge achieved by the use of other techniques helps in understanding features which can be seen, but not properly studied, by light microscopy.

In light microscopy ordinary light, phase contrast and interference contrast are all useful. Dark-field is less useful because of its lower resolution.

Phase contrast microscopy shows density differences in the field. As specific density differences within the objects are small, the main difference shown is that between embedding medium and pollen wall, which means that the phase picture is, above all, an expression of the thickness of the wall. A thicker part (a spine, columella, etc.) absorbs more light than a thinner one and comes out darker. Phase contrast is very easy to use and, in our laboratory, is also used all the time in routine analysis work. If desirable, it is very easy to shift between ordinary light and phase contrast. In focusing techniques (cf. below), phase light is as adaptable as direct light.

Interference contrast microscopy has not been much used in practical analysis, but it should also be fairly easy to use. It needs special optical equipment and a very strong source of light. Interference contrast differentiates surface features, and is also superb for making optical sections by focusing down into the grain. No special preparation is necessary: the technique can be used for ordinary pollen analysis preparations once the technical equipment is at hands. It is not suitable for routine analysis.

In addition to ordinary magnifying, direct light can be used for two techniques, both connected with focusing. Since it is practically impossible to section pollen grains for study and identification during analysis, optical sections must be used instead, by focusing to different levels, preferentially from the surface of the grain towards the centre. Because of the sphericity of the grain, only small (annular) parts of each level are visible at the same time. Sculpturing elements in the periphery are skewed in the microscopical picture; only in the centre line are the elements seen head-on. In focusing it is important not to have a deep focus: open diaphragms! By focusing up and down one can form a picture of the morphology of the various levels of the wall, (cf. Fig. 11.2—which neglects sphericity).

Another focusing technique bases itself on the so-called Becke line—a bright line delimiting two morphological elements of different refractivity. If the focus is raised a little from the exact sharp position, the Becke line wanders into the component having the highest refractivity, usually the pollen wall. A knob or spine on the surface thus gives the opposite reaction from that of a hole in the exine. This is the basis for Erdtman's so-called LO and OL analysis.

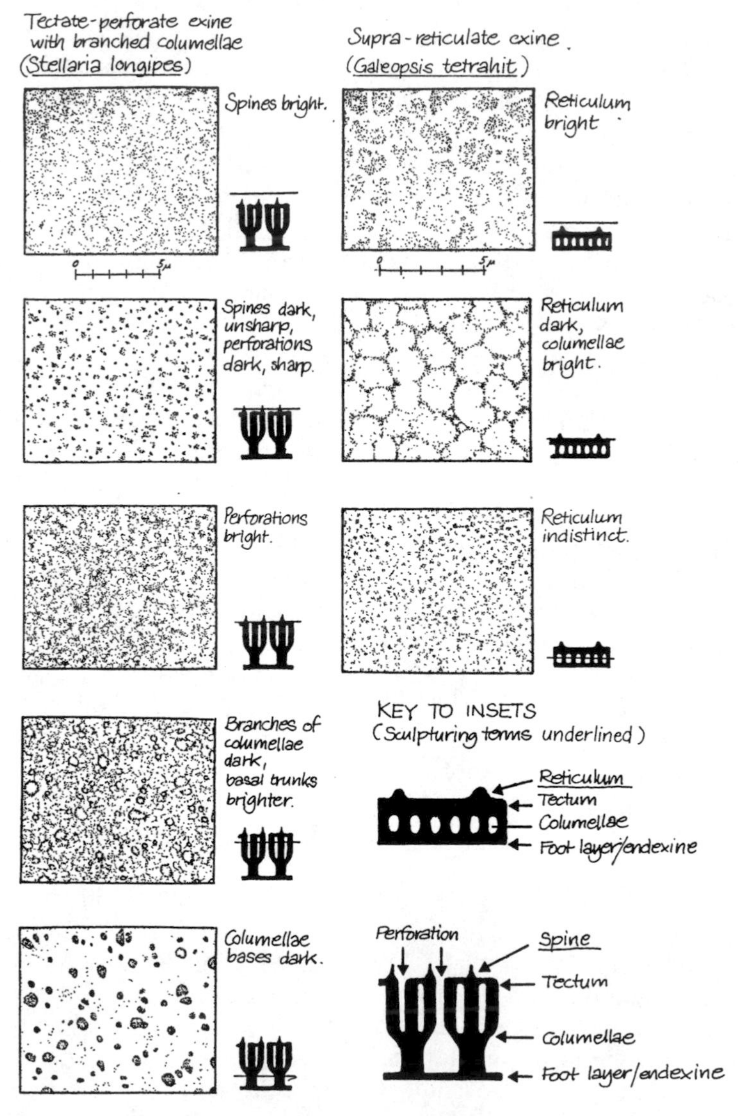

Fig. 11.2. Structure and sculpturing in varying focus as indicated by the insets. The horizontal line indicates the level of the focal plane. Drawings by B. Brorson Christensen from Iversen and Troels-Smith, 1950; redrawn.

It will be seen that the three techniques: staining, phase contrast and LO analysis, complement and enhance each other and can—should—be used together.

To obtain the optimum microscopical picture one should use an embedding medium of proper refractivity. If the difference between medium and pollen wall is too small, contours are indistinct. If it is too great, the Becke line may be so strong as to be disturbing. Various embedding media have been proposed and tested. The ones in general use are, in addition to glycerol, glycerol jelly and silicone oil. In favour of the two former speaks their greater ease of use; in favour of the latter, the greater permanence of mounts.

Because of the greater difference in refraction the silicone oil preparations exhibit Becke lines that are so strong as to be disturbing when it comes to the observation of the finest details. Pure glycerol or glycerol jelly gives greater resolution, with weaker contrasts.

11.4. General pollen morphology

The subject of pollen grain ontogeny has sprouted a very rich literature which is, on the whole, of marginal interest for practical pollen analysis. In the following we shall give a very condensed account of the earlier stages and the development of those parts of the grain that are not fossilized, concentrating on the exine, which is the part left in deposits.

11.4.1. The tetrad stage: gross form

With few exceptions each pollen mother cell gives rise to four pollen grains initially forming a tetrad. In most plants they are ultimately free from each other: monads. In some genera they do not separate and are dispersed as tetrads or other rarer types of composite grains (dyads, polyads; cf. list in J. W. Walker and Doyle, 1975: 686). In Cyperaceae (Carniel, 1972; Dunbar and Strandhede, 1973) and Epacridaceae (Ford, 1972) three nuclei out of each tetrad degenerate, and only one grain develops from the tetrad.

Even if the grains are eventually isolated from each other, the tetrad stage constitutes an important step in their development, and the surface pattern is related to the orientation of the grain in the tetrad (Wodehouse, 1935: 159). The part of the pollen grain which is nearest to the centre of the tetrad is called *proximal* (Wodehouse, 1935) and the line between the proximal and distal pole is also usually the axis of symmetry. The plane perpendicular to this axis through the middle of the grain is the *equatorial plane*. Exceptions exist, e.g. the strongly distorted tetrads inside orchid pollinia. In isolated pollen grains the tetrad relations cannot be observed directly, but with the exception of the regularly globular grains one is rarely left in doubt as to the position of the axis, which is usually the axis of symmetry of the grain. Generally, pollen grains can be regarded as more or less regular rotation ellipsoids, the polar axis also being the rotation axis.

Problems arise when the rotation axis does not coincide with the ontogenetic polar axis, e.g. in *Ephedra* or in grasses (Fægri, 1978). In agreement with our general terminological principles we shall also use this term in a strictly descriptive sense, i.e. in relation to the observable symmetry of the isolated grain, which can always be defined (except in isometric grains where the distinction does not apply). We shall use the term polar axis in a strict morphological sense: the unique axis of maximum symmetry of the monad.

Positions on the surface of the grain can be identified by their *latitude*, corresponding to the latitude on a regular sphere. Similarly, surface features perpendicular to the equatorial plane are called *meridional*. The terms *exterior* and *interior* refer to the distance from the centre of the individual grain; alternatively, when dealing with details of exine structure, the terms *above/beneath* may be used for features of greater/smaller distance from the centre of the grain.

The form of the rotation ellipsoid is frequently an important characteristic of the pollen grain, e.g. the *oblate* types of Betulaceae or the *prolate* of most Umbelliferae. Erdtman (1943a: 45) has proposed a series of shape classes, based upon the relation between the length of the polar axis (A) and of the equatorial diameter (E) of the rotation ellipsoid corresponding to the pollen grain, ranging from perprolatae (A/E > 2) to peroblate (A/E < 1/2). As the shape of pollen grains, especially that of the thin-walled ones, depends very much on the previous treatment, the embedding medium, etc. (Christensen, 1946: 12), a too elaborate subdivision in shape classes serves no real purpose. Many of the differences between form taxa of *sporae dispersae* lie within the shape variability of recent pollen from one species, or even one specimen.

All references are to grains without cell contents. Dry, living grains are frequently subprismatic owing to contraction of the exine, whereas the empty exine assumes the ellipsoidal shape. Some grains that are globular when moistened, or in a fossil state, may be saucer-shaped when dry. The latter shape has aerodynamic advantages in wind pollination, and thus contributes to the long-distance dispersal effect.

Departures from the ellipsoid form are frequently met with (cf. Fig. 11.3). In pollen grains with equidistant equatorial apertures the equatorial plane is frequently distorted to assume the shape of the corresponding polygon (the semi-triangular outline of *Corylus* or the sub-quadratic of four-pored *Alnus*, etc.); in many cases this distortion is increased by the addition of strong pore rims to the basic form (*Epilobium*). The extreme form, the *lobate* grain, is found both in fossil material and in recent plants (cf. Feuer and Kujit, 1980).

It is well known that the (etectate) pollen of some hyphydrophilous, predominantly marine angiosperms is thread-like and extremely long, reaching 5000 μm (Ducker *et al.*, 1978), but also in terrestric angiosperms one may find very long, thin (tectate) grains (520 μm in *Crossandra*: Brummitt *et al.*, 1980). In other cases the shape of the grain is that of a triangular prism rather than that of an ellipsoid (e.g. *Centaurea cyanus*). This state is brought about by the unequal thickness of the exine in different parts of the grain (much thickened between the furrows). In a few cases the three axes of the ellipsoid are of different size (*Myriophyllum alterniflorum*). The triaxial pollen described from living material of other species (Risch, 1940) does not always seem to keep that pattern on fossilization. Finally there are a few types that are completely irregular, e.g. the pear-shaped pollen of

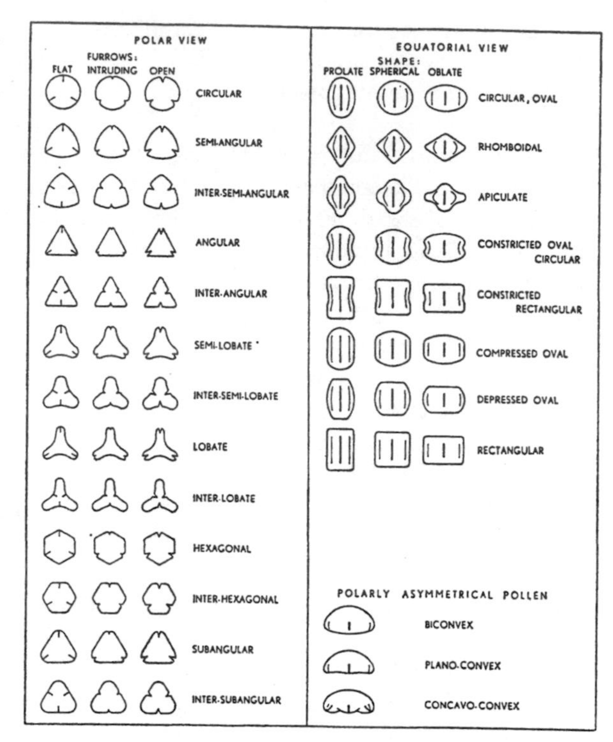

Fig. 11.3. Various possible shapes of a tricolpate, radially symmetrical pollen grain in polar and equatorial view, showing the equatorial (left) and meridional (right) limb. Any type in the one main column may combine with any one in the other. The top line represents the regular ellipsoid. From Kuyl *et al.*, 1955.

Cyperaceae. Grains which deviate from the ellipsoid gross form are rather frequent in pre-Quaternary material. The shapes are not always easy to interpret.

A special non-ellipsoid gross form is represented by the boat- or bean-shaped grain with a strongly elliptical equatorial outline and a single, polar furrow. This probably represents a phylogenetically primitive stage in the development of the angiosperm pollen. In recent plants it is found chiefly in monocotyledons, but also in some primitive dicotyledons (J. W. Walker and Doyle, 1975).

11.4.2. Cytoplast and intine

The angiosperm pollen grain is built up of three main concentric layers. The central part is the living cell, which germinates on the stigma.

The middle layer is the *intine*. It is present in all pollen grains, and envelops the whole of the grain in an apparently uniform sheath. It has been relatively little studied, and the information about its chemical composition is rather conflicting. It is frequently stated to consist of cellulose, but there is great disagreement as to the percentage. Some authors indicate figures down to a few per cent (Sitte, 1960: 23), whereas others seem to indicate that cellulose is a major constituent (Brooks and Shaw, 1968b: 552). Other classes of chemical substances that have been indicated as constituents of the intine are pectic substances and callose, also other polysaccharides, proteins, and enzymes (Knox and Heslop-Harrison, 1969). Of these, pectic substances apparently may form structural parts of the intine.

No part of the intine is known to be fossilized, and those grains of which the intine forms the outer sheath, or in which the exine is extremely delicate (Juncaceae, Musaceae, *Zostera*), are not found as fossils, or found only under extremely favourable conditions.

If we define the intine as those extra-cytoplasmic parts of the pollen grain that are not fossilized, we should also include the so-called *Zwischenkörper* (Fritzsche, 1837, onci: Hyde, 1955): thickenings immediately underlying the apertures and apparently having a somewhat different chemical composition from the rest of the intine, possibly pectic (Heslop-Harrison and Heslop-Harrison, 1980). Onci may play a part in nourishing the pollen tube by dissolving, when the pollen grain germinates; at any rate this is so with a corresponding body occurring in conifers (Martens and Waterkeyn, 1962).

11.4.3. Exine, sporopollenin

If a pollen grain fails to reach its proper destination it soon perishes, and both the cytoplasmatic interior and the intine substances are easily destroyed and disappear, leaving, for a shorter or longer time, the third layer, the *exine*.

The density of the (acetolysed) exine seems to vary a little between species, but the values obtained by Flenley (1971) fall between 1.4 and 1.5. Juvigne (1973) finds that the density increases with the age of the deposit, irrespective of taxonomic relationship, from 1.4 in recent pollen to 2.1 in tertiary. All these values are remarkably high for organic substances.

The exine contains one of the most extraordinarily resistant materials known in the organic world. Apparently unchanged spore walls (consisting of the same or a closely related substance) are found in Paleozoic and even older rocks, where all other organic remains have been carbonized and distorted. In an oxygen-poor atmosphere, recent pollen grains can for a short time be heated to almost 300°C (Zetzsche, 1929: xxix) or be treated with concentrated acids or bases with very little effect on the exine. Zetzsche (1929) and Vicari (1936) showed that the exine was susceptible to degradation by oxidation. Jentys-Szaferowa (1928) demonstrated layers of different resistance to oxidation by chromic acid in the exines of species of Betulaceae. This line has not been followed up. Bailey (1960) made the observation that exines are easily dissolved in hot (*ca.* 100°C) monoethanolamine. The reaction is instantaneous when the right temperature is achieved. In that way one may remove ektexines and leave the rest of the grain intact (except for effects of the heating). According to Kress (1986) some of the disagreements about the effect of ethanolamine on exines may be due to misunderstood nomenclature and consequent use of the wrong substance: (mono)ethanolamine, $H_2NCH_2CH_2OH$ and diethanolamine, $NH(CH_2CH_2OH)_2$ are both effective.

The chemical substances responsible for the resistance of exines against chemical attack were called sporopollenins, but their nature was unknown until Brooks and Shaw (1968b) showed that sporopollenin is formed by oxidative polymerization of carotenes and carotene esters. Incidentally, it was shown at the same time that *Lycopodium* exospores conformed to the same general pattern, but exhibited different proportions between constituents. It is possible that the different resistivity to corrosion exhibited by various exines is due not only to differences in wall thickness and percentage of sporopollenin in the wall substance, but also to different qualities of sporopollenin, i.e. various proportions of different monomers and various degrees of polymerization.

There are still a number of unsolved problems in the formation of sporopollenin and the nature of its precursors, chemical as well as topographical, within the loculus. There are indications that the fully mature sporopollenin does not form until the pollen is quite ripe (Rowley, in lecture). Also, the protocolumellae (see below) may consist of quite different substances.

Sporopollenin has been synthetically polymerized from the monomers (Brooks and Shaw, 1973b). Based on an arbitrary number of 90 carbon atoms, the unit contains 140–160 hydrogen, and 25—45 oxygen. According to Brooks and Shaw (1973a), the sporopollenin content of exines varies between 24% (*Pinus*) and 3.5% (*Phleum*). Cryptogam spores go down to 1.8% (*Equisetum*) with 14% cellulose. In angiosperms the cellulose content of exines varies between 10% and 20%. There is a (weak) positive correlation between the contents of the two substances, indicating that they do not substitute for each other (Brooks and Shaw, 1973b).

As carotenes and carotenoid esters are abundantly present in anthers—at any rate after exines have been formed (Heslop-Harrison, 1968b)—the idea that they have a function in the synthesis of pollen substances is a very natural one, and the Brooks–Shaw explanation of the composition of sporopollenins also explains the occurrence of the so-called *Ubisch bodies*: small, more or less spherical, bodies of a composition apparently corresponding to that of the exine, but occurring freely in the anther loculus. The Ubisch bodies are produced by the tapetum; even if they may sometimes exhibit a structure reminiscent of that of the exine, their occurrence hardly means anything more than that the tapetum cells also have the capacity to catalyse the polymerization process.

Unstructured sporopollenin has also been found in pre-Cambrian rocks and in carbonaceous meteorites (Brooks and Shaw, 1968c, 1969). How this can be fitted into the general picture seems more controversial. The latter observation can hardly be taken as unequivocal evidence of extraterrestrial life, as long as the mechanism of polymerization is not known.

The occurrence of acetolysis-resistant (= sporopollenin?) membranes also in the more peripheral layers of the anther may not represent any radical departure from the pattern already given above (cf. Heslop-Harrison, 1969).

The strata of pollen exines are usually periclinally dispersed upon each other, but in many cases one stratum may have 'roots' penetrating into the one(s) below, even into the intine.

When pollen is dispersed in tetrads or larger *pollen-units* (J. W. Walker and Doyle, 1975: 685) individual grains may be bound by a continuous exinous membrane (*calymmate* units *sensu* van Campo and Guinet, 1961) or invested in individual exinous envelopes (*acalymmate*), usually reduced at the interior face (Guinet and Thomas, 1973).

Sporopollenin is also the substance forming the so-called *viscin* threads that bind together grains and tetrads in certain families (especially Onagraceae and Ericaceae—cf. Hesse, 1981 and Waha, 1984). Remains of viscin threads may be found in fossilized material, whereas analogous organs in other materials disappear, especially after acetolysis. According to Hesse (1984) viscin threads can be considered to be special columellae which continue growth out of and beyond the tectum.

The terminology of exine stratification has been the subject of prolonged controversy, and has changed a great deal over the years. Some of the changes represent necessary adjustment because of new knowledge; some have been due to other reasons. It is unfortunate that the terms do not always cover the same concepts (cf. Fig. 11.4). The strata of the exine seem to be traversed by usually submicroscopical channels, *microtubuli* (Rowley and Dunbar, 1967; Rowley and Flynn, 1969) which may form pathways of exchange of substances between the pollen cytoplast and the surrounding matrix. Later possibly they may also play a role in the recognition processes on the stigma.

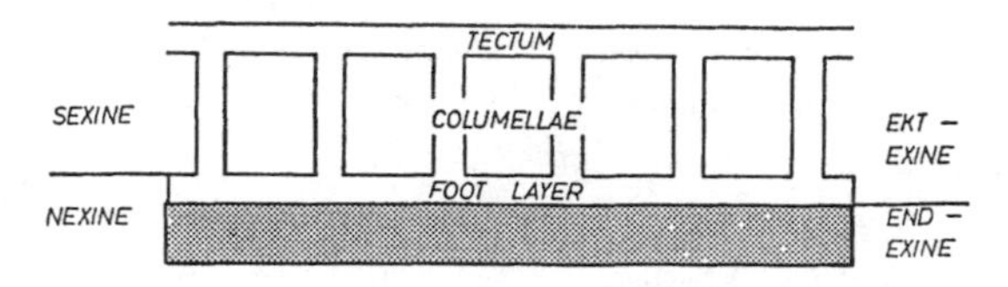

Fig. 11.4. Main stratification of the exine, diagrammatic.

11.4.4. Endexine

In a typical exine it is possible to distinguish between two main layers, discovered by Fritsche as early as 1837. The inner one he called *intexine*. Later, Erdtman (1943: 43) coined the terms *endexine*, and *ektexine*, which we shall use here. In sections the two layers may be distinguished optically by their staining reaction: the endexine appears faintly pink after treatment with fuchsine B when the ektexine is already dark red. A similar reaction is observed by staining with auramine O (Kress, 1986). Also, the resistance to chemical action differs in different layers (Fischer, 1890: 15, Jentys-Szaferova, 1928). Kress (1986) has pointed out that endexines are not dissolved in ethanolamine, as are ektexines. The chemical basis for this is unknown. The two layers are very distinct in the transmission electron microscope. In *Borago officinalis*, Saad-Liman and Nabli (1984) found a distinct discontinuity (a 'white line' in TEM) between the ekt- and endexine even in the mature grain.

The inner layer forms a microscopically homogeneous, continuous membrane with few morphologic developments except for those connected with apertures (cf. below). Some observations of pores or cracks in the endexine may be due to artifacts, but there can hardly be any doubt that rather wide pores are a genuine feature of the endexine of, at any rate, many species. Electron micrographs frequently show them. In the pollen grains of Malvaceae there are wide channels running from the interior of the grain into the endexine. Some scanning electron micrographs suggest sculpturing of the inner face of the endexine (van Campo, 1971, 1978). The occurrence of 'endo-cracks' in the endexine has been described by various authors (cf. Southworth, 1983) and even been used diagnostically (Oldfield, 1959). If they are not artifacts, their function is at any rate unknown.

The following discussion is mainly based upon the angiosperm pollen grains. Whereas it has been maintained (van Campo, 1971) that there is a difference in principle between the morphology and genesis of gymnosperm and angiosperm ektexines, the endexines are apparently formed in the same manner in both taxa.

The endexine seems to form on usually periclinal membranes of non-sporopollenin nature—cf. above (Rowley and Southworth, 1967). In electron micrographs these membranes, or their remains, show up as lighter streaks in a more opaque medium. In the mature grain they are usually obliterated.

Thomas and Lugardon (1972) have also demonstrated the occurrence, within Annonaceae, of some endexine types in which the lamellar structure is well represented, but the lamellae form a very loose tridimensional net instead of the usual compact layer. Also the structure of the lamellae is preserved, not obliterated as usual in ripe grains.

In the ontogeny of the exine the development of the endexine starts later than that of the ektexine. Consequently, it has been proposed (Godwin *et al.*, 1967) that the former be distinguished as secondary exine in contrast to the primary exine (= ektexine). Since the terminology would depart from the principle of purely descriptive terminology, and introduce an ontogenetic perspective, we see no justification for change. Phylogenetically, the laminar type, corresponding to the endexine, seems to be older, and to be the only one in lower plants (liverworts: Rowley and Southworth, 1967; *Lycopodium*: Afzelius *et al.*, 1954, and Brooks and Elsik, 1975). Thus, there does not seem to be any reason to abandon the traditional terminology. Van Campo and Lugardon (1973: 188) maintain that in the pollen of *Nerium oleander* there is no endexine. This seems to be a rare exception.

11.4.5. Exine ontogeny

Whereas the formation of the endexine seems to follow a fairly consistent pattern in the species studied, that of the ektexine is still controversial. Mostly, ektexines are characterized by a more or less pronounced radial structure, as opposed to the periclinal one of the inner layer. However, an equally consistent system of membranes of deposition has not yet been found.

After the pollen mother cell has differentiated, a thick callose wall develops inside the PMC wall. Only after this has developed does meiosis take place, and the tetrad is then formed. After the four sporocytes have formed, each of them is also surrounded by a similar callose wall, within which the development of the exine takes place. It is not quite clear to what extent the process is influenced by the surroundings outside the callose wall, which perhaps does not seal off the sporocyte completely. Since fragmentary cells with dying nuclei go through many of the same processes, they seem to be, at any rate partly, independent of the cytoplast itself. Apparently there are various control mechanisms at work at different stages (cf. Rowley and Flynn, 1969), and some systems may be independent of the presence of a living cell.

The other important structure in the anther, in addition to the archespor and its derivatives, is the *tapetum*, the cells of which at this stage go through a pronounced degenerative phase, in extreme cases dislodging themselves in the thecal fluid. Even so, the actual role of the tapetum in the metabolism of the anther—and it must be a crucial one—is not fully understood (cf. Echlin, 1973). The formation of sporopollenin globules at the tapetum cells even before the pollen grains are liberated indicates the presence of enzyme systems in (or on) the tapetal cells similar to those in the pollen grains, but absence of the morphogenetic principles (cf. also the Ubisch bodies).

Whereas many details are still controversial, there is a general agreement on the main phases in the ontogeny of the ektexine, originally formulated by Heslop-Harrison in 1962. Apparently, vesicles of the endoplasmatic reticulum migrate centrifugally and place themselves in close apposition to the plasmalemma surrounding the cytoplast. The places where this happens become the future apertures, and will represent voids in the template pattern. Apart from these voids the exine starts development via a cellulosic *primexine* surrounding the grain uniformly. In the primexine there appear radiating strands of granules, the *protocolumellae* (probacula). These are initially not acetolysis-resistant, but become so even before the pollen cells are released from the callose walls. Thus, one may conclude that the first sporopollenin (the '*protosporopollenin*') is laid down while the cell is still under complete control of the haploid nucleus or its derivatives.

The location of the protocolumellae on the plasma membrane also determines the later development of the pattern of structure and sculpturing. The determining mechanism is still not understood. The especially frequent reticulate pattern of mature grains, also very primitive ones, may be an expression of a stress pattern in the developing exine.

The protocolumellae are the focal structures for the (production and?) deposition of sporopollenin: the two other strata of the exine develop from the columellae. In some ways the protosporopollenin differs from the mature substance by being more reactive. Its composition is not known; some of its reactions resemble common tests for lignin and may be responsible for repeated—unverified—indications of the occurrence of lignin in exines (Heslop-Harrison, 1968b).

At present, various forms of the primary exine unit seem to exist. The most usual form seems rod-like, but may actually be, or at any rate be derived from, lamellar structures, which persist in mature *Cycas* pollen (Audran and Masure, 1976). In other pollen types the first sporopollenin seems to come as very fine granules (*exine grenue, sensu* van Campo and Lugardon, 1973: 172) which may coalesce at later stages, but may also be distinguishable in mature grains. Rowley *et al.* (1981) describe 'tufts' of primexine material. The alveolar structure of the ektexine of Coniferae and Cycadaceae does not seem to correspond to the usual radial pattern.

At a much later stage of development columellae may break off and leave extensive cavities between tectum and foot layer/endexine: *cavate* pollen (cf. Skvarla and Larson, 1966b).

After the grains have been released from the callose wall the primexine degenerates and the cavities between ektexine elements are filled with thecal fluid, i.e. mainly with tapetum derivatives which are most probably the matrix delivering monomers for sporopollenin polymerization and deposition (Godwin, 1968).

If, as it seems, some polymerization takes place on the cytoplast before breakdown of the callose wall, it may be that monomers (and polymerization processes) of tapetal and cytoplasmic origin may differ, and also that there is a difference between young polymers and mature sporopollenin (cf. Dickinson, 1976).

Which factors are active in bringing the polymerization and accretion process to a stop, so that a consistent pattern develops, is unknown, nor is it known whether control is exerted from within (via microtubuli) or from without.

Within the interstices of the fully mature pollen grain and on its surface, electron micrographs show a variety of substances, the origin of which is not always easily explained. Apart from possible artifacts, the assemblage is known as 'tryphine' and seems to consist of remains of primexine—if any, of tapetal substances, and of others that might be present, all of which in the final stage consolidate and adhere to the outside of the pollen grains. The main constituents are lipids, waxes, etc. In *Lilium* the tapetal cells go through an almost explosive development during the period immediately before the final maturing of the anther, and only the U-bisch bodies are left to show where the tapetum used to be (Heslop-Harrison, 1968b). The so-called *pollenkitt*, which is very heavy on *Lilium* pollen, is synthesized during this period. On the surface of the much drier *Tradescantia* pollen, dehydrated tapetum material is deposited during the last 12 hours before 'anthesis' (=dehiscence? Mepham and Lane, 1968).

Whereas the exospore of many lower plants seems to be composed of sporopollenin-like substances, the ontogeny of these exospores is largely unknown. In pteridophytes tangential layers are found, similar to those of the endexine of angiosperms. The ontogeny and structure of the remarkable excrescences on the exospore of some mosses (e.g. Miyoshi, 1969) are not yet properly understood.

11.5. The ektexine

In contrast to the usually homogeneous or tangentially orientated inner layer, the ektexine in most angiosperm pollen grains comprises small, radial, rod-like elements (*Körnchen*, Fritzsche; *granules*, Wodehouse), the development and distribution of which underlie the extreme morphologic variability of the exine.

The different staining and electron density of endexine and ektexine indicate that either their chemical composition or their physical strucuture must differ in some way which is not yet fully understood.

A 'complete' ektexine may be regarded as a three-layered structure in which the granules form small columns, *columellae* (or *bacula**) separating an outer and an inner stratum, called, repectively, *tectum* and *foot layer*. The latter strata may be more or less fragmentary or even absent. Geometrically, tectum and foot layer may be considered fused distal (or proximal), parts of the columellae.

This complicated, three-strata exine is found in the large majority of angiosperms, but in taxa, which are also for other reasons considered primitive, J. W. Walker and Skvarla (1975) have demonstrated a system of simpler exines which they consider phylogenetically older (Fig. 11.5). The simplest type, which they call *atectate*, has a homogeneous exine with no internal structure and usually no sculpturing. Walker's study (1976) strongly indicates that the simple condition is also the most primitive, and that radial elements belong to a later stage in the development of the angiosperm pollen, even antedated by angiospermy itself

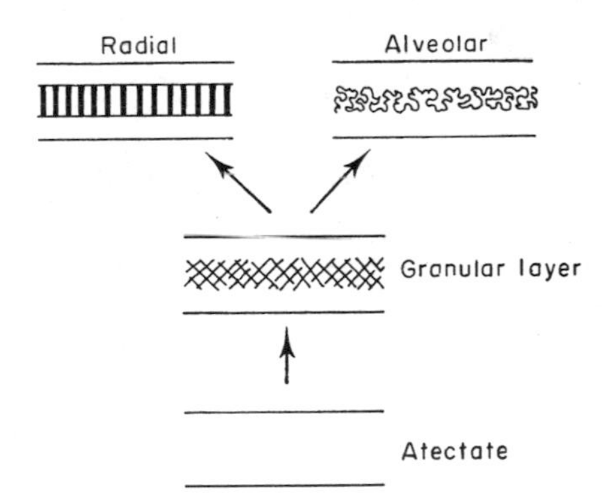

Fig. 11.5. Phylogenetic development of exine structure according to Walker and Skvarla, 1975.

* In former editions the term baculum was used for radial rods which were not topped by a tectum whereas those supporting a tectum were called columellae. Two names for what is esentially the same structure seems an unnecessary complication, and we therefore in this book consistently use the term columella. A columna is a column, whatever is on top of it.

(Fægri, 1979). The next stage recognized is the development of a middle, granular layer, foreshadowing the tripartite ektexine. Both the usual radiate type and the alveolar exines can be derived from this. On the other hand, a columella-like structure is observed in pre-angiosperms even back in the Carboniferous (Meyer-Berthaud, 1986) but not necessarily in the immediate ancestry of the present angiosperms.

If the tectum (fused distal parts of columellae) cover all or most (*ca.* 75%), of the surface of the grain, the latter is called *tectate*. Grains which have no tectum, in which no columellae—if present—are fused are referred to as *intectate*. If there is some fusion, but less than by definition in tectate grains, the transitional types are called *semitectate*. Within the columellate exine type a typological reduction series may be established, ranging from the 100% tectate–imperforate to the 100% uncovered–intectate (cf. Fig. 11.6). In a few taxa, mainly aquatic, the tectum and possibly the whole exine seems to be very reduced or vestigial. After acetolysis no structure is left. J. W. Walker (1976) has called such grains *etectate*. Pettitt and Jermy (1975) describe vestigial exine membranes, said to be more or less structureless, in etectate grains, e.g. of *Halodule* (their Fig. 15). However, a faint lamellation seems indicated, which would mean that the exine remains represent the endexine (cf. also Hesse and Waha, 1983).

The absence, practically speaking, of exine in hyphydrophilous plants is easily understood. An apparently identical organization in Zingiberales is more difficult to understand ecologically (cf. Kress *et al.*, 1978). Also in some Lauraceae there are extremely thin ektexines overlaying thick intines (Hesse and Kubitzki, 1983). Whether the phylogenetic evolution always follows the sequence in Fig. 11.6, or if there are also reversals, as indicated by Walker (1976: 279) is an open question.

The tectum may be thicker or thinner in relation to the total thickness of the ektexine. In some pollen the impression is that it is composed of the outer ends of individual columellae, which may be contorted in various ways. In other cases the interior morphology of the tectum is less obvious. Examples of thick tecta on top of extremely short

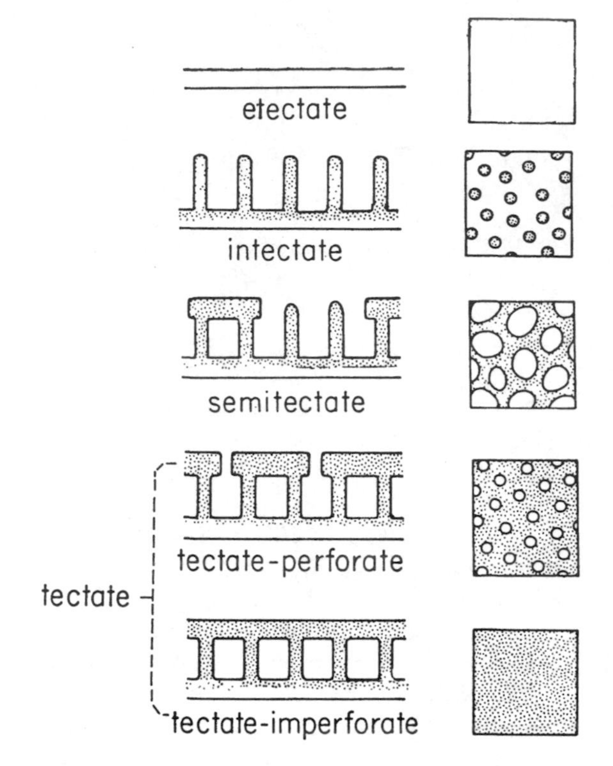

Fig. 11.6. Typological development (reduction) of the ektexine in angiosperms. Left in section, right as seen from the outside. The composition of the etectate exine is conjectural. Redrawn with permission from data in J. W. Walker, 1974; cf. 1976. Ektexine stippled.

free columellae are found in Betulaceae. Columellae are usually simple, but may be branched (Figs. 11.2 and 11.7). In Compositae columellae are divided into two—a lower, coarse part with few branches and an upper part that usually forms an extremely intricate, anastomosing net. The two parts are separated from each other by a membrane; another thin one seals off the network. In *Geranium* there is a similar organization, but much coarser.

In semitectate grains columellae are generally found under the fragmentary tectum, but there may also exist free granules that do not support a tectum, e.g. in the lumina of some *Salix* pollen grains.

In intectate grains all columellae are free, and may be more or less crowded (e.g. *Populus tremula*), or somewhat scattered (*Callitriche*). They may be evenly dispersed or form patterns, and their shape

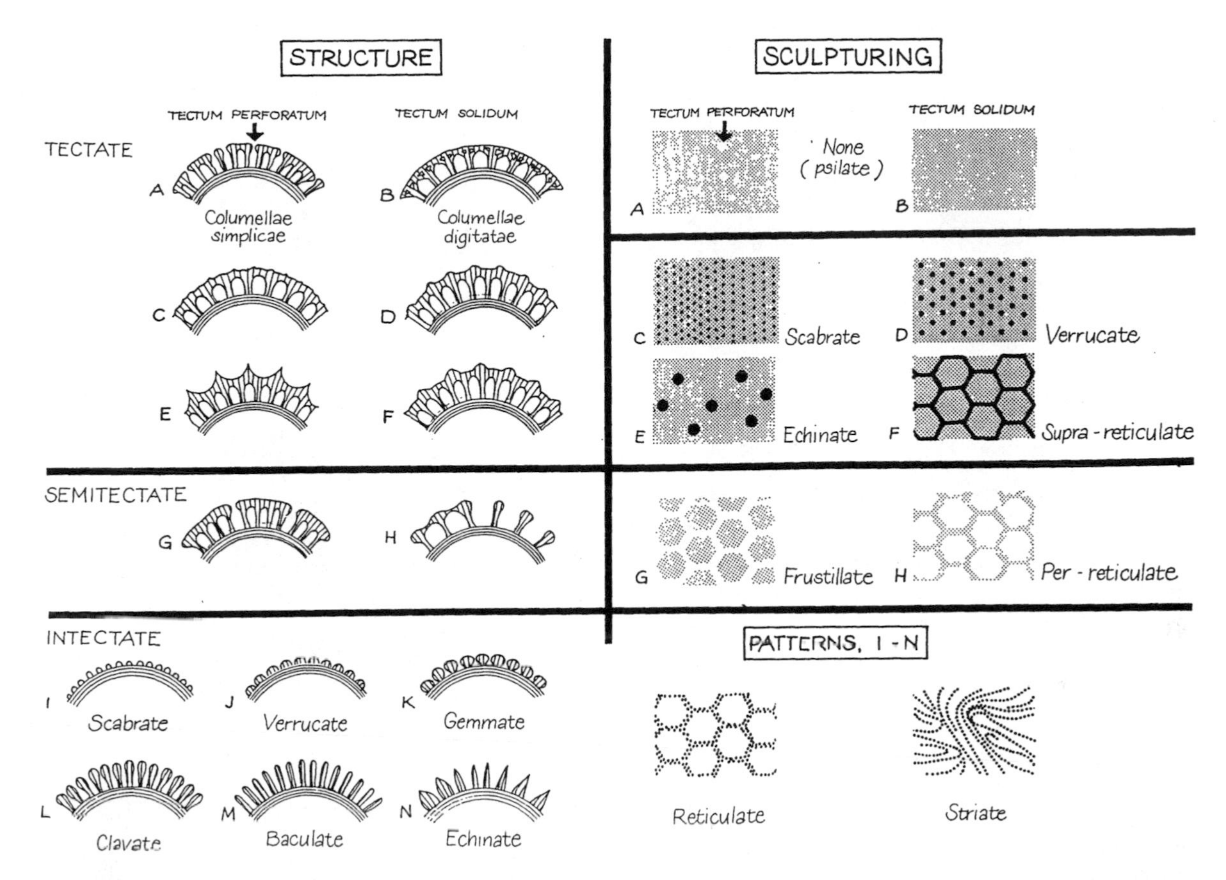

Fig. 11.7. Main types of structure (left) and sculpturing (right). Endexine/foot layer: concentric signature in section, white in surface view. Columellae black in pattern figures I–N. Columellae not indicated in surface view of tectate exines. Tectum shaded; black: sculpturing features. From B. Brorson Christensen, in Iversen and Troels-Smith 1950; redrawn.

varies a great deal, but the club-shaped type (as in *Ilex aquifolium*) predominates. Columellae are generally simple.

In the Liguliflorae the tectum is broken by a limited number of rather large openings or extremely thin areas (*lacunae*, Wodehouse, 1928) arranged in a symmetric pattern. We class it as a (semi-)tectate pollen of a special type (*fenestrate*, i.e. equipped with windows). The lacunae occupy an area of the total surface of the grain greater than the distinctly tectate parts of the exine. Similar types are found in other taxa as well.

11.5.1. Structure and sculpturing

In accordance with Potonié (1934) we distinguish between the *structure* (*texture*) and the *sculpturing* of the exine. The term structure comprises all those characters which are due to the form and arrangement of the exine elements *inside* the tectum. It also comprises the form and arrangement of the individual elements in intectate types. Accordingly, all the ektexine characters that have been dealt with above belong to the structural features.

The term sculpturing, on the other hand, comprises the *external* features without reference to their internal construction (cf. Figs. 11.7 and 11.8). In practice, sculpturing is what the SEM picture reveals. Thus a spine, a sculpturing feature, may consist of a single columella, e.g. in *Geranium*, or as in *Malva* of a few columellae fused in their upper part (forming in reality an extremely thick frus-

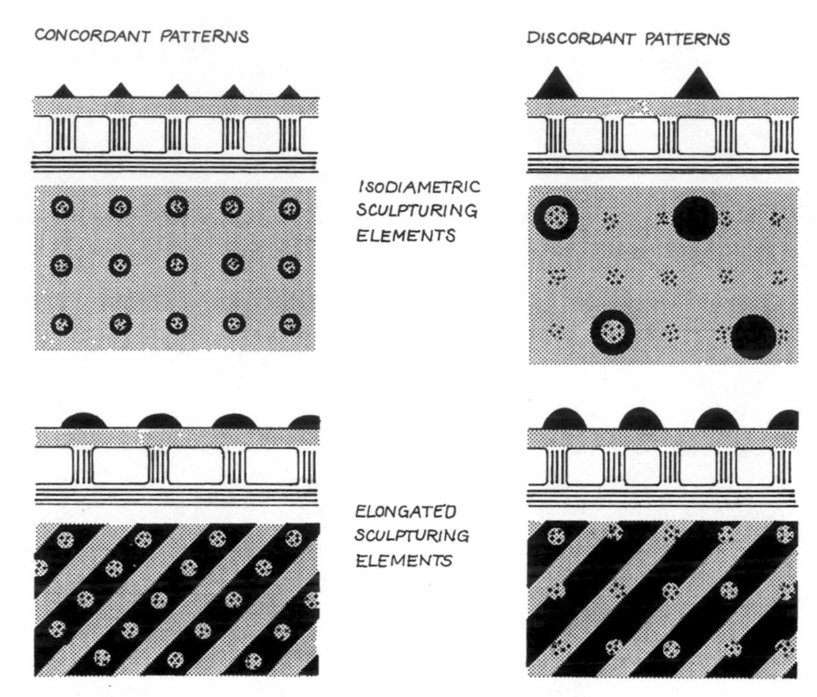

Fig. 11.8. Concordant and discordant patterns in tectate grains. Signatures: endexine/foot layer: concentric; columellae: radial in section, dotted in surface view; tectum: shaded; sculpturing features: black.

tillum). The spine may also be a highly complex structure comprising a number of subdivided columellae, as in Compositae.

Both conceptually and observationally, it is not always easy in the light microscope to distinguish between structure and sculpturing. In intectate grains they are identical; the pattern of columellae also forms the surface pattern. In semitectate grains (e.g. *Lilium*) they are usually concordant: the tectum follows the columella pattern. Such exines are referred to as per-reticulate, etc. In some tectate grains, e.g. *Galeopsis tetrahit* (cf. Fig. 11.2), the sculpturing, i.e. the surface configuration, is reticulate, whereas the underlying columellae are evenly distributed without any distinct pattern: supra-reticulate.

On the other hand it is self-evident that sculpturing is in many cases due to the arrangement and form of structure elements, e.g. the striations of the pollen grains of most Rosaceae are just as much a structural feature, i.e. a linear arrangement of columellae, as a sculpturing type, i.e. ridges on the tectum.

Whereas some pollen grains are smooth and thus possess no sculpturing (under the light microscope!), all types with an exine seem to possess a structure, even if it is sometimes difficult to observe. For instance, in *Pseudotsuga* it is practically impossible optically to discern the structure except under phase illumination. Even when the individual columellae cannot be discerned, they confer a certain graininess on exines, which distinguishes them from the plastic-like clear appearance of most non-pollen met with during actual analysis. Cryptogam spore coats usually also appear nonstructured.

Since the possibility of discerning small morphological features varies with equipment and technique, it is for practical reasons necessary to introduce an arbitrary limit for 'visibility'. Features which can be seen only in SEM are not useful in practical analysis. Unless otherwise stated, the limit of 'visibility' in our keys and descriptions lies at *ca.* 1 μm. Features are easily seen even if smaller, but it is difficult to observe and describe their detailed morphology. Thus, in Table 12.1, exines with sculpturing elements smaller than *ca.* 1 μm are classified as smooth or scabrate (the latter an unfortunate term as there is no corresponding noun: scabra is a linguistically meaningless construction by Beug, 1961). Likewise pits (negative sculpturing elements) are neglected if smaller than 1 μm (psilate).

Electron scanning micrographs show very intricate and probably also specific submicroscopic sculpturing of pollen grain surfaces. However, as actual analysis perforce must be carried out with the light microscope, this knowledge is not always helpful even if some of these features may also be seen in the optical microscope under optical conditions, and may serve in the distinction between, e.g. *Corylus* and *Myrica* pollen grains (cf. the key). Also, the better understanding provided by the greater resolution of SEM may be useful in day-to-day work.

The sculpturing of the pollen grain is generally a fairly constant character and is in most cases an excellent means of recognition. It is, however, frequently difficult to describe in exact terms, and consequently somewhat unsatisfactory for analytical description. We distinguish between the types indicated in Table 12.1 and in Figs 11.7, 11.9 and 11.10, which are based upon the form and arrangement of *sculpturing elements*, i.e. those elements which project beyond the (imaginary) even surface, usually of the rotational ellipsoid, either the endexine in intectate pollen or an imaginary surface corresponding to the tectum in tectate ones. In

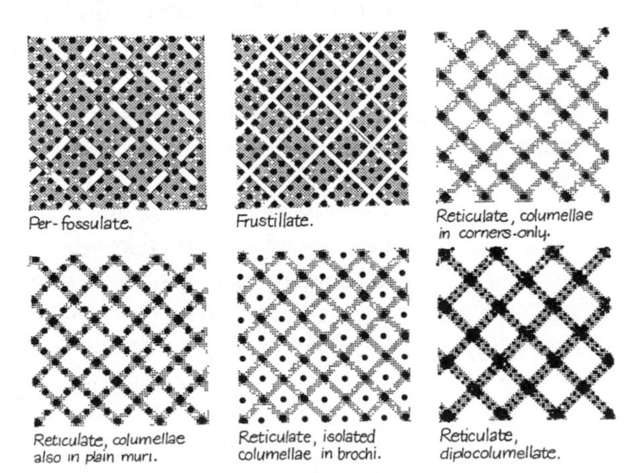

Fig. 11.9. Exine types. Columellae indicated in black.

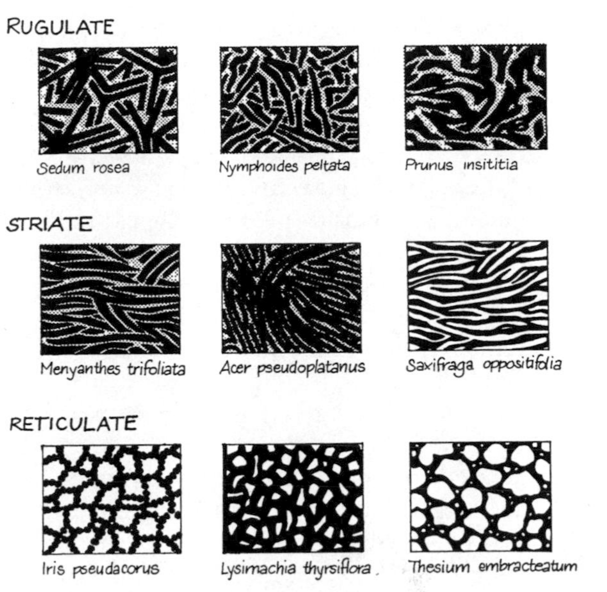

Fig. 11.10. Sculpturing types. From B. Brorson Christensen, in Iversen and Troels-Smith, 1950; redrawn.

cryptogam spores other types of sculptural elements occur (cf. Puttock and Quinn, 1980). They will not be described in this book.

Sculpturing in tectate grains may be given the prefix supra-. Thus, *supra-reticulate* refers to a reticulation on the outside of the tectum (many Papilionaceae), not necessarily accompanied by one beneath, whereas *intra-reticulate* refers to columellae forming a reticulate *pattern* beneath the tectum notwithstanding what is present above. Usually, the two types of reticulation are combined in a (supra)-reticulate grain, but not always. Finally, *per-reticulate* refers to the reticulation of a semitectate grain which is essentially intra-reticulate, but where the tectum forms a top reticulation.

The specific constancy of the sculpturing type may be less than has hitherto been assumed. Under experimental conditions Hebda and Lott (1974) could vary the exine pattern with varying conditions of temperature and humidity.

A survey of sculpturing types is given in tabular form in Chapter 12.

11.5.2. Apertures

Most pollen grains possess openings in the ektexine, collectively referred to as *apertures* (cf.

Blackmore and Crane, in preparation). In non-angiosperm pollen and spores there may be thin parts of the exine taking over some of the supposed functions of the apertures. These will be dealt with separately. According to their shape, apertures are called, respectively, *pores* and *furrows* (*colpi*). Apertures manifest themselves both in the ekt- and endexine, although to a varying degree. In some grains the ekto- and endoapertures are concordant, in others not (cf. Fig. 11.11). The inner part of the aperture is often filled with non-exinous material, including the onci.

Apertures are thought to serve several functions. They form weak parts of the pollen wall through which the pollen tube almost universally emerges at germination: *exitus* effect. In most pollen grains possessing them furrows also function as *harmomegathi* (Wodehouse, 1935: mechanisms accommodating change in volume of the semi-rigid exine). There are other types of volume regulation as well. Which role apertures play for the exchange of substances between the grain and the surroundings, especially the stigma, is not quite clear, but such an effect seems very probable.

The difference between furrows and pores seems to be purely morphological and a matter of definition: furrows are elongate, boat-shaped and the ends are more or less acute. Pores are generally isodiametric, and if they are elongate the ends are

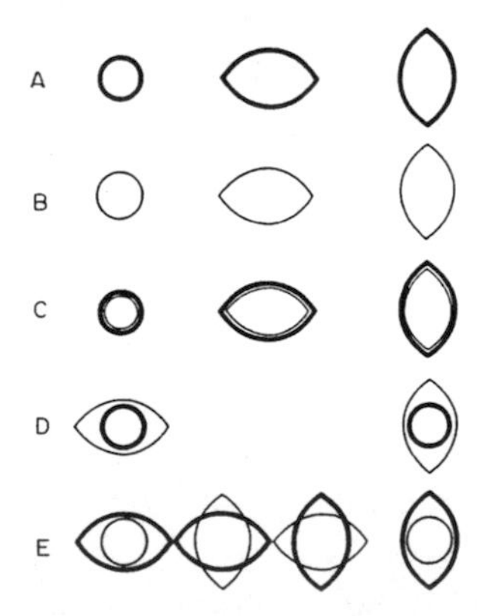

Fig. 11.11. Aperture types. A: Pore, transverse and meridional furrow in the ektexine; B: same, in the endexine; C: concordant apertures; D: discordant apertures, pore in ektexine, furrow in endexine; E: the same, furrow in ektexine, pore or furrow in endexine.

rounded. For practical reasons the limit between pore and furrow may be defined by a length/breadth ratio of 2/1, but it is in many cases irrelevant whether an aperture is regarded as a short furrow or an elongated pore. Transitional forms occur. Phylogenetically, furrows are apparently the primitive form; pores developed later by contraction.

In accordance with our general policy we shall use a restricted descriptive aperture terminology. Some investigators, especially Erdtman, have proposed a number of terms (Kremp, 1965, refers to 92 terms under the general heading colpate) indicating whether the aperture is meridional or polar, distal or proximal, etc. In pollen-analytic practice this is quite immaterial, and in sporae dispersae the differences cannot be observed, only inferred.

Aperture-like organs are found also in some cryptogam spores, where they form either a single longitudinal slit or a triradiate scar. These aperturoids form where sporocytes contact each other in the tetrad stage. A subtetraedric spore with a triradiate scar ('trichotomosulcate') is related to a tetrahedral tetrad arrangement, whereas a boat-shaped spore with a longitudinal scar is produced in a tetragonal or other plane arrangement of sporocytes in the tetrad. In both cases the scar is polar in relation to the tetrad axis and proximal. Anatomically, such scars are quite different from the ektexinous apertures of angiosperms.

In pollen grains, triradiate scars are very scarce, but are found e.g. in Palmae (Thanikaimoni, 1970) and some Liliaceae. On the other hand, slit-shaped polar apertures are not infrequent among primitive angiosperms, possibly indicating a similar origin to those of cryptogams, in spite of anatomical differences. Even if we accept the connection between the pollen of primitive angiosperms and similar organs of primitive pteridophyte-like ancestors, the development from the polar, monoaperturate arrangement to the equatorial (or isometric) arrangement of a multiple (usually $3n$) of apertures in the great majority of angiosperms is still an unsolved problem in spite of intensive work. Indubitably transitional forms have not been observed, and the usually triaperturate state characterizing most angiosperms may conceivably have arisen *de novo* from an inaperturate exine.

Also in evolved angiosperm grains the surface configuration, including the sites of apertures, shows a geometric relationship to the position of the grain in the tetrad, but the mechanism behind this is certainly more subtle than envisaged by Wodehouse (1935: 177 *et seq.*). It also seems that the position of the pollen grain in relation to the loculus wall is of importance in this respect.

The volume of the desiccated, living grain is relatively small, the exine is contracted and the furrows appear as narrower or broader slits, according to species. When the grain is moistened it expands, and the thin membranes of the furrows bulge out (this is the state generally depicted by Wodehouse (1935), and explains his choice of terms). When the cell contents have been removed —by fossilization or by chemical action—the furrows appear as grooves in the exine. It is obvious that, as soon as the pollen grain is wetted and its contents swell, the very thin membrane of the aperture is in danger of being ruptured. Some grains (e.g. *Crocus* spp., *Berberis* spp.) are very susceptible to wetting, which causes an almost explosive rupturing of the whole exine. Such grains are consequently not found as fossils.

In the ektexine furrows run meridionally or form a polyhedric (*isometric*) pattern. In some types furrows also run equatorially in the endexine, either forming one continuous equatorial attenuation (*colpus aequatorialis*), or forming discrete ones (*colpi transversales*).

Thus, in a non-isometric pollen grain, three different types of attenuations may occur: pores, meridional furrows or transverse furrows. These attenuations may combine in various manners as shown in Fig. 11.11. In isometric pollen grains the distinction between meridional and equatorial furrows becomes meaningless.

Pollen grains are called *colporate* (Erdtman, 1945a) if a furrow and a pore are combined in the same aperture. The term *pororate* has been used for porate grains in which the ektexine aperture is not of the same shape or size as the endexine aperture. For practical reasons no account of this is taken in the keys. It goes with the porate type. Generally there is one pore in each furrow, but grains in which pores occur only in some (generally in a half or one-third of the furrows (*heterocolpate*) are known. The remaining furrows then probably have a harmomegathic function, and their homology with other furrows may be questioned. They are often referred to as *pseudocolpi*.

In some rare cases there are more than one pore per furrow, which are then referred to as diploporate, triploporate in Buxaceae (Köhler and Brückner, 1982). If pores are formed, phylogenetically, by lengthwise contraction of furrows, such diploporate furrows might be expected to produce pores sitting two and two together. This actually seems to be the case, (cf. also Feuer and Kujit, 1980: Figs. 29 and 31).

The existence of an acetolysis-resistant aperture membrane can be inferred from the occurrence of isolated ektexine elements within the aperture region in many pollen grains (cf. Fig. 11.12). They must be held in place even if the membrane that does so is invisible. In some pollen types, aperture membranes carry an isolated piece of exine, structurally more or less comparable to the exine of the rest of the grain. Such heavy aperture coverings, separated from the rest of the exine by the surrounding thin membrane, are called *opercula*. In many cases, e.g. in grasses, they are easily lost, being thrown off when the grain expands on being moistened. In a pore the operculum is usually a central disc, in a furrow it forms a median strip. Beautiful opercula are found in the furrows of many Rosaceae (*Potentilla* type). In *Sanguisorba officinalis* the operculum separates the two sides of the primary furrow, the ends of which may even coalesce with those of the adjacent ones at the poles. In some grains, e.g. *Aesculus*, there is no continuous operculum, but isolated ektexine elements on the furrow membrane.

Electron micrographs indicate that the aperture membrane is not always homologous in comparable grains. According to Roland (1966, 1968) the membrane varies between in all-but-indistinguishable endexine skin to a rather heavy participation of the ektexine (also outside the operculum). The heavier type forms a transition towards those pollen types in which the aperture is

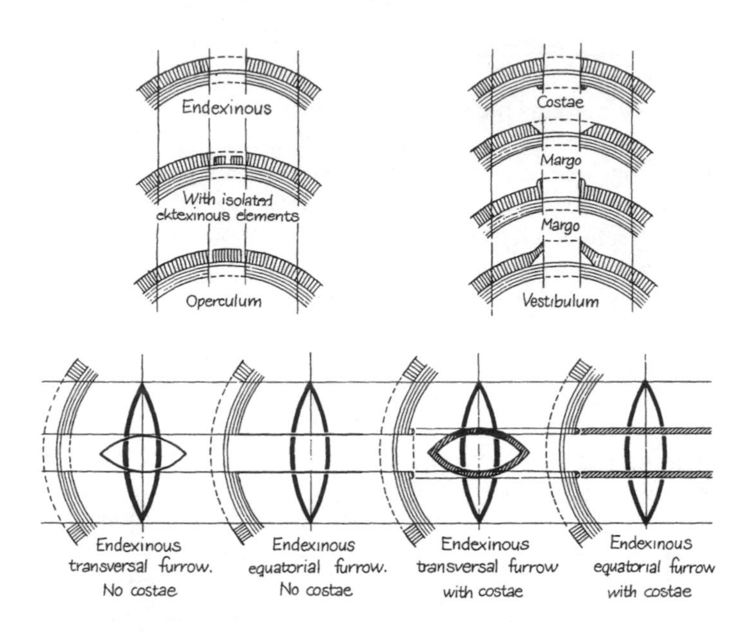

Fig. 11.12. Apertures. Top left: aperture membranes; top right: aperture edges; bottom: colporate apertures in section (left) and straight on (right). Meridional ektexine furrow. Endexine: concentric signature. Ektexine/columellae: radial signature. Costae: oblique hatching.

represented only by a thinning of the exine (cf. also Rowley, 1964).

Edges of apertures may be like the rest of the exine (non-bordered) or apertures may be *bordered.* In intectate grains bordering can take place only by changing the size and/or density of granulae, in tectate and semitectate grains the edges of apertures are frequently thinner or thicker than the rest of the exine. The terminological usage is not quite consistent. Edging in the ektexine around a pore is generally called an *anulus*, that around a furrow a *margo* (cf. Fig. 11.12). The typical anulus, as found in grasses, forms a thickened ring around the pore, but anuli that are thinner than the rest of the ektexine are also known. On the other hand, thinning of the ektexine is a regular feature of the margo, but since the ektexine elements are at the same time also generally more densely packed, these thinner margins hardly represent weaker parts of the exine. In some genera (very distinct in *Artemisia*) exines are gradually thinner towards the furrows, without any other morphological change, appearing crescentic in polar projection. Such furrows are not regarded as marginate.

Thickenings in the endexine accompanying apertures are termed *costae* (*c. colpi, c. pori*). costae colpi may also be transverse (*c. transversales*) or, if they are continuous around the equator in uninterrupted, equatorial rings: *c. aequatoriales*. Whether costae and anuli are simple thickenings of the respective layers, bulges filled with some other material, or more complicated conduplications (cf. van Campo and Hallé, 1959: 814) are questions of great theoretical interest. Becaue possible differences are probably very difficult to observe, they are of secondary importance in pollen analysis.

The effect of edging of apertures is easily seen in colpate grains in which furrows frequently collapse, unless reinforced by margins or costae.

In some families (Betulaceae, Onagraceae) pores are rather complicated, forming a separate outer chamber that communicates with the interior of the grain (*vestibulate* grains) (Fig. 11.12). Other aberrrant types occur as well, but are of marginal practical interest in European pollen analysis (cf., however, *Borago officinalis*).

The number and arrangement of normal furrows and pores varies between species. In most cases three pores and/or meridional furrows are arranged equidistantly from each other along the equator of the grain. If more than three are present they may be arranged in the same way (*stephanocolpate/stephanoporate*) or be evenly distributed over the whole surface of the grain (*isometric* distribution Wodehouse, 1935, *pericolpate/periporate* syn. *pancolpate*, etc.). A special type is the single furrow or pore of the great majority of the monocotyledons.

The number of apertures varies from one to thirty, forty, or even more. The most frequent number in dicotyledons, viz. three, corresponds to the three contact surfaces with the other grains of the tetrad (Wodehouse, 1935: 182). Next in frequency to the equidistant equatorial arrangement comes the isometric (with increasing frequency the higher the number of apertures) and then the completely irregular arrangement. The polar arrangement is practically exclusive to the monocotyledons, in which it is dominant. In *syncolpate* grains two or more furrows are combined into rings or a spiral (*spiraperturate*) surrounding the whole or parts of the grain.

Furrows extend to a higher or lower latitude on the surface of the pollen grain. As *polar area, apocolpium*, (cf. Fig. 11.13) we define that part of the pollen grain which is situated at higher latitudes than all apertures, anuli or margins. The relative size of this area is of considerable diagnostic value and may be expressed by the angles formed by the

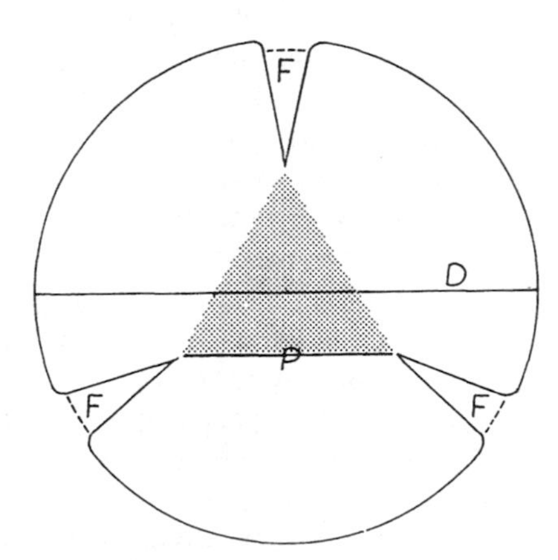

Fig. 11.13. Polar area: shaded. Polar area index: p/D (F = furrows).

latitudes. For practical reasons we have, however, measured the greatest distance between the ends of two furrows (or in the case of a well-delimited margo, between the ends of the margins) and express the '*polar area index*' as the ratio between this measure and the greatest breadth (equatorial diameter) of the pollen grain.

The aperture pattern of polyploids, hybrids, etc., is frequently aberrant or disturbed.

The number and arrangement of apertures are easily observed and can be described exactly in comparatively simple terms. These characters are of great diagnostic value and are useful for practical classification even if they are less constant than some other characters (e.g. in *Ranunculus*, which fluctuates between tricolpate, isometric hexacolpate, etc.).

11.5.3. Size

The smallest regular pollen grains measure about 5 μm, the largest more than 200 μm. However, size is even more dependent on the history of the grain than is shape. The living pollen grain has neither definite form nor definite size, expanding and contracting according to the quantity of moisture available. The empty exine—as found in sediments under good conditions of preservation—seems to have a rather constant size, which is one of the characteristics of the species and is not appreciably affected by some customary procedures, e.g. deflocculation with 10% KOH and transfer to glycerol. Other treatments, e.g. HF treatment or acetolysis, and other embedding media seem to influence size more or less substantially. Conversion factors between the results of one method and the other are more or less constant for the pollen of a given species, but may differ from one species to the next (cf. Fig. 11.14).

Reitsma (1969) has published a study on the size modifications of pollen due to various treatments. Unfortunately he does not include the statistical parameters necessary to evaluate the data, nor has he carried out the last KOH treatment after acetolysis. The following points seem evident: dry material increases in size, notwithstanding whether it is wetted by KOH treatment or by means of a detergent or wetting agent. This is a purely physical

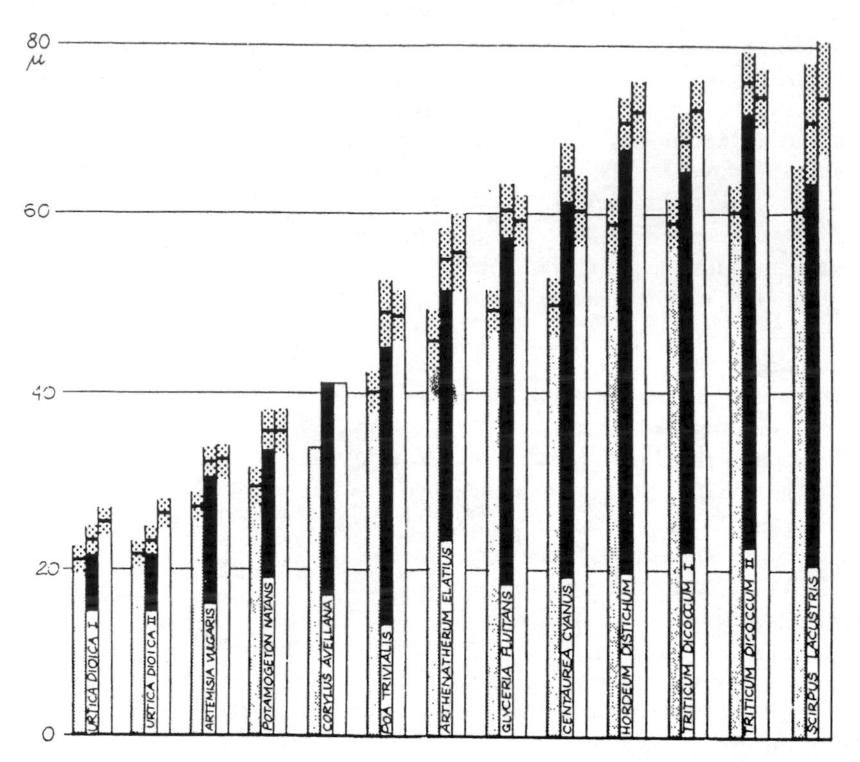

Fig. 11.14. Pollen sizes in silicone oil AK 2000 (left) and glycerol (middle, black). Right-hand column: glycerol size calculated from silicone data, presuming the same increase as for *Corylus* (factor 1.22). Shaded area on top indicates ± 1 sigma.

process, due to uptake of water. When a sample is subjected to the acetolysis mixture the size of the grains increases suddenly; with further treatment it decreases again and reaches a plateau after some 3 min.

These changes must be due to chemical reactions within the exine. According to Reitsma's results the preparation methods described in the preceding should give consistent results. However, it is important to maintain strict consistency in preparation procedures.

The increase in size of pollen exines with acetolysis treatment may be related to an observation of Afzelius *et al.* (1954) that the inter-lamellar spaces open up during treatment. Yamanoi (1969) has tested various combinations of the usual treatments—KOH, HF, acetolysis—and has shown that as long as acetolysis is one of the treatments, variation curves are reasonably similar. Without acetolysis, pollen exines are on the whole smaller.

Some of the size variations of pollen grains seem to be caused by the cell content of the grain, and disappear when the cytoplasm is removed (Wagenitz, 1955). However, there are other species the pollen grains of which exhibit a geographical size variation. In *Pinus echinata*, Cain and Cain (1948) found a geographical cline 'with the grains of the north-eastern half of the area of the species having consistently larger mean values than those of the southern half of the area of the species' (p. 330). Usinger (1980) insists that there are geographical clines in northern European *Betula* species, but the variations—if realistic al all—are fairly insignificant. In other taxa there are apparently subspecies which differ in pollen size and/or sculpturing. Birks (1978) maintains that on the basis of pollen morphology, including size, the European spruce can be divided into a number of distinct morphological types with different area distribution.

Attempts to relate variations in pollen size to definite environmental factors have not been successful (cf. Mikkelsen, 1949). How much of this can be generalized remains to be seen (cf. Prentice *et al.*, 1984). There is no reason why size should not occasionally be variable, like other morphological characteristics.

By means of the morphological exine characters mentioned above, one may distinguish 'morphological pollen species', which are of greatly varying taxonomic value, sometimes comprising part or all of a large family (Chenopodiaceae, Gramineae), sometimes a single or a few closely related species only (e.g. *Polygonum amphibium*). They may even serve to distinguish sub-specific taxa (*Lepisanthes*, Muller, 1970).

If it is necessary to subdivide the pollen further, this can sometimes be done by statistical treatment of quantitative characters which in themselves are insufficient for identification. Thus Eneroth (1951) developed the statistical treatment of pollen grain size for the distinction between Scandinavian *Betula* species. As (1) the size of a pollen grain is a rather indefinable quantity, as (2) the method is always extremely laborious, and as (3) the reliability of the results is generally somewhat dubious, the method should be used only when no other is available (for a discussion of the method, cf. Fægri, 1945: 103; Praglowski, 1966; Robertson, 1973; Usinger, 1980 and earlier).

Only when the state of preservation is almost perfect—which seems to be the case in acid sediments only—can size-statistical determinations be carried out with some hope of success. It is obvious that the treatment of both samples and comparison material should be carried out with the utmost care. Christensen (1946) has suggested that one should use the dimensions of an indicator pollen type of known 'real' size in the same preparation (recent or fossil) as a factor of correction. Thus, in many northwest European samples the size of *Corylus* pollen grains will show whether the pollen grains of a given preparation are of 'normal' size or not. Further, an index of expansion/diminution of the *Corylus* grains of the preparation may be used as a conversion coefficient, since pollen grains of not too aberrant types have been proved to react more uniformly than might have been expected (cf. Fig. 11.14). Deterioration of pollen grains in the mounts is not taken into account in this statement. *Corylus* pollen has been considered uniform. Usinger (1980) hints that there may be a size cline in *Corylus* pollen, but gives no data. Also, his preparation method includes HF treatment, which is known to change the size of pollen grains.

Multivariate techniques may make it possibe to distinguish more closely between pollen taxa than can be achieved by simpler statistical methods (cf. Birks and Peglar, 1980). So far the technique has not been widely applied, and it also demands a great deal of the material.

A common characteristic of all statistical methods is that whereas the individual grain cannot be classified, the percentages of various species within the total may be calculated. Borderline cases are those in which the variation curves of the components are almost completely separate, e.g. most cereals and ordinary grasses in Middle Europe (cf. Firbas, 1937), or *Isoëtes* microspores (Andersen, 1961: 31).

If it may be presumed that size changes do not influence the shape of a grain, relative measurements ('indices'), e.g. size of apertures in relation to size of the grain, may be more useful than simple size measurements. Also, numerical data—number of elements per area or length unit, etc.—can be used for identification purposes instead of size measurements (cf. Iversen and Troels-Smith, 1950). An example of elaborate use of linear and angular measurements and indices is given by Ting (1966), cf. also Dyakowska (1964). With access to modern numerical treatment, such data can be more efficiently utilized.

On the whole the variability of pollen grains (and spores) should not be underrated. Both with regard to size and to morphological details pollen grains from the same species may exhibit great variations, and may transgrede into the sphere of variation of pollen of related species. When making pollen preparations for general comparative use one should take care to procure pollen from several individuals. Another question is the pollen dimorphism associated with heterostyly and related phenomena.

11.6. Gymnosperm pollen and cryptogam spores

As conifers are generally wind-pollinated, and their pollen plays a great part in pollen rains, gymnosperms have been studied very intensively from a palynological point of view (Erdtman, 1957; Ueno, 1960), also with regard to phylogeny (van Campo-Duplan, 1950 and later). Even so, we do not yet have a full understanding of the structure of the wall. This may partly be due to the fact that gymnosperms constitute no phylogenetic unit, and, consequently, the pollen walls of various groups may be differently built.

In Cycadaceae there is a special modification of the lamellate exine type inasmuch as the columellae are here represented by vertical lamellae which form an alveolar system of radial pipes. The pronouced lamellate structure in *Ephedra* and *Welwitschia* is different from the structure of the angiosperm exine (Afzelius, 1956). In Coniferae there is great variation; the alveolar structure of Pinaceae is unique, and the very simple exines of, e.g., *Larix* or *Pseudotsuga* may be considered reduced.

The breaking-up of the connection between tectum and endexine that causes various types of air bladders to form on the grains in some families is an interesting ecological adaptation, but is of less morphological importance than the alveolar structure of the bladders (van Campo and Sivak, 1972). The presence of sporopollenins seems to be comparable to that of angiosperm exines.

Apertures of gymnosperm pollen grains differ from those of angiosperms in being less distinct, and also by their position in relation to the tetrad. In many grains the aperture is a rather indistinct, thinner part of the exine (Pinaceae) and the grain regularly bursts through this thinner part (Cupressaceae). In some genera (*Sequoia, Taxodium*) the aperture is represented by a protruding papilla.

When it comes to cryptogam spores one would prefer to speak about *exospore* instead of exine. Exospores are chemically closely related to exines, being equally resistant, staining in the same or a similar way, etc. However, the morphology is completely different, and in recent cryptogams there is no columella layer.

In some ferns the outer layer appears periclinal, the inner apparently massive. It is possible, under favourable circumstances, to make electron micrographs of fossil pollen and spores (Pettitt, 1966; Meyer-Berthaud, 1986), both SEM and TEM of sectioned grains.

The coats of megaspores are essentially different from those of microspores.

12. Which plant? Identification keys for the northwest European pollen flora

Motto: Even the best key is inferior to a preparation

Partly in series (*World pollen and spore flora, The NW European pollen flora*), partly in monographs, an increasing number of palynological identification keys with an increasing degree of sophistication are being published. With few exceptions (e.g. Kapp, 1969; Moore and Webb, 1978; cf. also the original approach of Ikuse, 1956) such keys deal with individual families or similar taxa. To be able to use them one must possess an advance knowledge of the taxonomic relationship of the species.

In practical analysis of fossil material this knowledge does not exist: one must be able to enter the key on a purely morphological basis. On the other hand it is not necessary in a key of this kind to pursue the matter to the finest breakdown. Its main purpose is to serve as an entry to the palynological keys to families and lower taxa. A key attempting to achieve the ultimate in taxonomic breakdown on purely morphological characteristics would assume unmanageable proportions.

Also, in practice much work can be saved by exclusion of species on ecological grounds, taxa the occurrence of which at that site and that time is unlikely—with proper caution for the danger of circular reasoning.

The aim of the key presented here is to be a practical help in the identification of the pollen grains more or less likely to be encountered in analysis work with special reference to northwest Europe. It should primarily serve as an entry to the ever-expanding body of pollen morphological data in literature, for which we must refer to current bibliographies (cf. Section 1.2.5). The key does not pretend to give complete morphological descriptions, and above all *any identification must be considered tentative until tested against an actual specimen* in the pollen herbarium.

In addition to taxa occurring naturally or in cultivation in the region today, some pre-Pleistocene taxa have been included, which are sometimes observed as secondary components of pollen assemblages. Furthermore, a few taxa have been included the pollen of which is probably always long-distance transported from outside the area (*Ephedra*) or exotics so frequently planted and such great pollen-producers that they may occur in the recent pollen rain (*Eucalyptus*).

As it is meant for practical pollen-analytic purposes, the key is based upon light microscopy. Electron microscopy in pollen analysis may become a practical possibility some time in the future. It is not now. On the other hand, modern pollen analysis presumes microscopie equipment that reveals anything that can be discerned by light, which also implies phase contrast, high-resolution objectives, properly trimmed microscopes etc. Characters given as 'fine' or 'minuscule' invariably presume phase contrast and/or high resolution.

Only in a few cases is there a reference to SEM, viz. when it is felt that knowledge of the SEM picture is essential for understanding what the light microscope shows. However, the whole key has been revised at the hand of SEM data.

The taxonomic breakdown in the key varies, depending on several conditions: (1) the number of taxa included in the unit; (2) the possibility of discerning between taxa within a unit, e.g. a family or a genus; (3) the importance of such discrimination for the (ecological) understanding of the pollen assemblage; (4) the availability and quality of previously published detailed treatment of the pollen morphology within the unit. As a consequence, we have incorporated or added special keys going further than the general breakdown. The breakdown in the key presented here also depends on the use we have made of it in this laboratory. Taxa that have proved to be of importance for the interpretation of ecological or archaeological problems, and which have therefore been subject of closer study, are treated more thoroughly. Others, which may in other areas or other contexts be equally important, have been treated more lightly, even if pollen-morphological monographs are available. In some cases we have found that existing monographs—published or not—did not work satisfactorily for keying out unknown pollen types, and we have therefore not made use of them, even if our own researches have been inconclusive.

The units of the key are understood to be the taxa occurring in northwest Europe. Thus '*Acer*' or '*Ulmus*' refer to northwest European species only. The pollen morphology of species not occurring within the region has not been taken into account. Also, reference to a comprehensive taxon, e.g. a family, does not necessarily mean that all northwest European species would key out at that place. In many cases identification can be carried out beyond the units of the key. For this we must refer to the ever-increasing number of palynological studies with identification keys.

Being, on the whole, easy to observe, specific and fairly constant, the number, shape and arrangement of apertures are major entries in the key. However, it should be kept in mind that the ar-

Table 12.1. Sculpturing types (cf. Figs. 12.1 and 11.7, 11.9 and 11.10)

Key	Description	Type
A_2	No distinct sculpturing elements sensu stricto present.	
B_3	Surface pitted, diameter of pits less than 1 μm	psilate.
B_2	Surface pitted, diameter of pits equal to or more than 1 μm	foveolate.
B_1	Surface with grooves	fossulate.
A_1	With sculpturing elements sensu stricto	
C_2	Radial projection of sculpturing elements $\pm$ isodiametric.	
D_2	No dimension equal to or larger than 1 μm	scabrate.
D_1	At least one dimension equal to or larger than 1 μm.	
E_2	Sculpturing elements not pointed.	
F_2	Lower part of elements constricted.	
G_2	Greatest diameter of radial projection equal to or greater than height of element	gemmate.
G_1	Height of element greater than greatest diameter of projection	clavate.
F_1	Lower part of element not constricted.	
H_2	Greatest diameter of radial projection equal to or greater than height of element	verrucate.
H_1	Height of element greater than greatest diameter of projection	baculate.
E_1	Sculpturing elements pointed	echinate.
C_1	Radial projections of sculpturing elements elongated (length at least twice the breadth).	
I_3	Elements irregularly distributed	rugulate.
I_2	Elements $\pm$ parallel	striate.
I_1	Elements form a reticular pattern	reticulate.

rangement is not absolutely constant, even if the percentage of deviating grains is in most taxa fairly low. Triaperaturate grains may occur in a periaperturate, usually hexa- or tetra-aperturate state. The opposite transition is less frequent, and so are deviations in the stephanoaperturate pattern. Especially if a grain cannot be identified from the triaperturate keys, a glance at the periaperturate ones may help.

Exine structure and sculpturing come next to apertures in importance for the identifcation of pollen grains. The main types are summarized in Table 12.1 and Fig. 12.1. The features are often quite tricky to interpret, and the keys often leave doubts in deciding between alternatives. The illustrations may help to visualize what is meant by the various terms, but in actual practice features may vary a great deal, even if characterized by the same term as is seen by a comparison of the realistic Fig. 11.10 with the schematic Fig. 11.9.

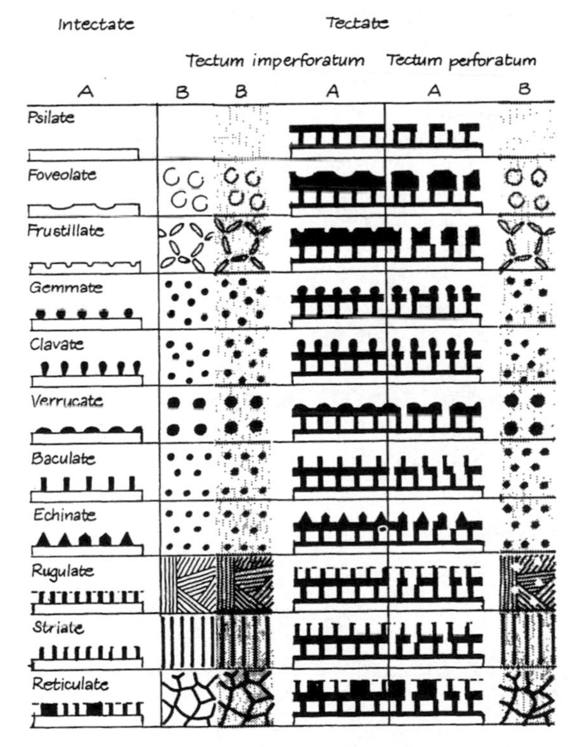

Fig. 12.1. Summary of terms in exine morphology. White: endexine; black: ektexine (footlayer, columellae and tectum) in section; shaded: tectum in surface view. Columns: A: sections; B: surface views.

Fig. 12.2. Diagrammatic section through a discordant striate exine. Where the (optical) section cuts the lower part of the exine (middle of figure) the evenly dispersed columellae are seen. The outer (peripheral) parts show the striate surface pattern.

Except for occasional free columellae semitectate patterns are concordant (Fig. 11.8), whereas tectate exines may also be discordant (Fig. 12.2), i.e. the pattern formed by structural elements is different from that formed by sculptural elements. In tectate grains sculpturing elements sit on the tectum with or without infratectal correspondence.

The distinction between supra- and per-reticulate patterns is not always easy to make, especially not in grains with a fine sculpturing. In case of doubt, both keys should be consulted.

Measurements, the third group of general identifying features, are the least satisfactory. They depend on many partly uncontrollable factors. They are cumbersome to make, and in many cases give just a point on overlapping curves of variation. Nevertheless, they are necessary. Reticulate patterns, polar areas, etc., represent special problems in measuring (cf. Figs. 12.3 and 11.13).

It is often difficult to establish the number of isometrically distributed sculpturing elements (pores, spines, etc.) on the grain by visual inspection alone, especially if the number is great and/or the exine is thick. McAndrews and Swanson (1967) developed a formula for calculating the number from the relation betwen the (average) length of the chord between the centres of two adjacent elements

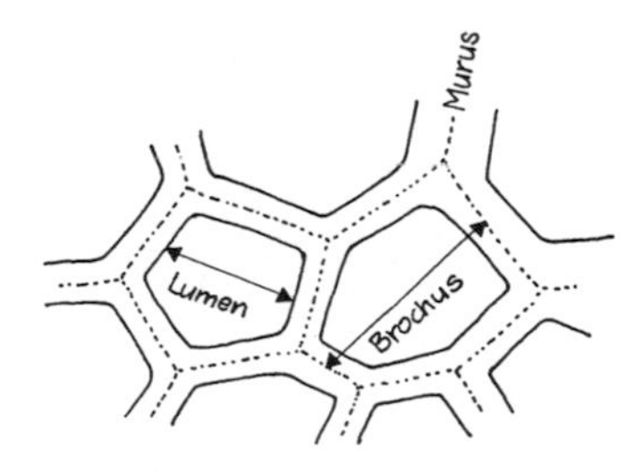

Fig. 12.3. Dimensions in a reticulate pattern.

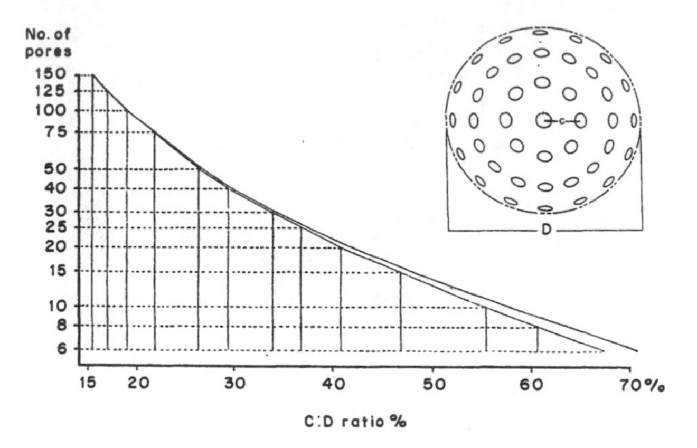

Fig. 12.4. Number of isometric elements on the pollen surface as function of the chord/diameter ratio c/D. The space between the curves of the lower figure gives the mathematical correction by Christensen.

and the diameter of the grain. Mathematically, the formula has been corrected by Christensen (1986); in practical analysis the difference is immaterial, cf. the original papers for details. Figure 12.4 summarizes the data for identification purposes.

In principle the keys are binary, but more than two entries at the same level have been used when it has been considered practical, e.g. differentiating between several sculpturing types. The way in which the keys are constructed may seem unfamiliar. It has been chosen (1) to indicate clearly and at the outset if the division is binary or not, and (2) it gives a simple, unique designation of the place of each entry, each taxon, in the key. Thus D: B_1 is the unique reference to *Myriophyllum alterniflorum* and F 11,5: G_2 that to *Hedera*.

The right-hand column gives the ultimate breakdown to taxa as far as the key goes. The term 'type' refers to pollen types common for several taxa, either a number of—but not all—species in a genus or species belonging to related genera. Taxa included in the 'type' are indicated by 'Incl.' in the description part of the key. If a major taxon is identified in the key, and subsequently broken down into minor units, its name is inserted in the description part of the key after a colon, e.g. *Arctostaphylos* in Tables B1–B3: H_3. References to taxa which do not really belong in the relevant key are sometimes given in parentheses. Under no circumstance should the mention of a name be interpreted exclusively: even within the northwest European flora other taxa, not mentioned, may belong to the same category.

The term 'grain' is used for characters pertaining to the whole sporomorph. Size measurements referring to 'grain' are the largest dimension unless expressly excepted. P/E is the ratio between length of polar axis and equatorial diameter.

The small, diagrammatic figures in the margin are not intended as naturalistic illustrations of pollen grains, but only serve to illustrate a single feature indicated (italicized) in the key. The use of conventions follows the same principles as Fig. 12.1 (except of silhouettes). With the exception of a few, which are indicated, the figures are not to scale. Figures showing parts (details) of grains are framed in a circle.

To identify unknown pollen grains from a key is never easy. Interpretation of the features of the microscopical picture requires acute observation and long practice, especially to interpret in three-dimensional term the essentially two-dimensional picture seen. The 'Special keys' included in this book are definitely not for beginners. They presume experience in evaluation of pollen morphological data.

Master key

A_2 Grains with air-sacks . . . . Table A

A_1 Grains without air-sacks.

B_2 Composite grains . . . . Table B

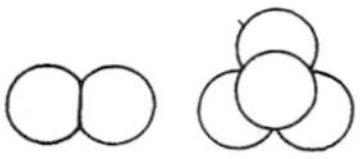

B_1 Grains single.

C_3 Grains polyedric. . . . . Table C

C_2 Grains bilateral . . . . Table D

C_1 Grains isodiametric or one axis longer/shorter, rarely heteropolar . .

D_2 Grains subtetrahedral . . . . Table E

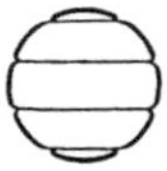

D_1 Grains rotational . . . . Table F

Table A. Vesiculate (Pinaceae and Podocarpaceae)

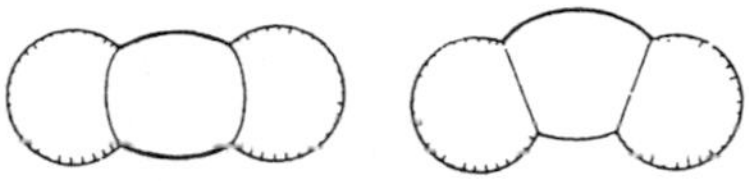

(A_3 *Three meridional*, airsack-like *crests* . . . . **Trapa**)

A_2 *One marginal*, anular, ± rudimentary *air-sack* . . . . **Tsuga spp**

A_1 Distinct, subglobular or pouch-shaped air-sacks with internal, 3-dimensional reticulum, usually two.

B_2 Proximal part of the exine *('crest') more than 5 μm thick* . . . . . .

C_2 Body of the grain (excl. air-sacks) 80–100 μm. Air-sacks more than semiglobular . . . . **Abies**

C_1 Body of the grain ca. 50 μm. Air-sack semiglobular. . . . . **Cedrus**

B_1 Proximal part of exine less than 5 μm thick.

D_2 Proximal entry angle of air-sack very blunt; *no obvious constriction* between air-sack and body of the grain . . . .

E_2 Gradual transition between structure of the air-sack and that of the wall of the body of the grain. Colpus membrane smooth . . **Picea**

E_1 Abrupt transition between structure of air-sack and that of the wall of the body of the grain. Colpus membrane dotted . . . . **Pinus haploxylon t.**

D_1 Proximal entry angle of air-sack sharp; distinct constriction between air-sack and body of the grain.

F_2 Body of the grain *ca.* 50 μm.

G_2 Colpus membrane smooth . . . . **Pinus diploxylon t.**

G_1 Colpus membrane dotted . . . . **Tsuga mertensiana**

F_1 Body of the grain less than 40 μm and/or grain with *more than two air-sacks* **Podocarpaceae**

Table B. Composite grains

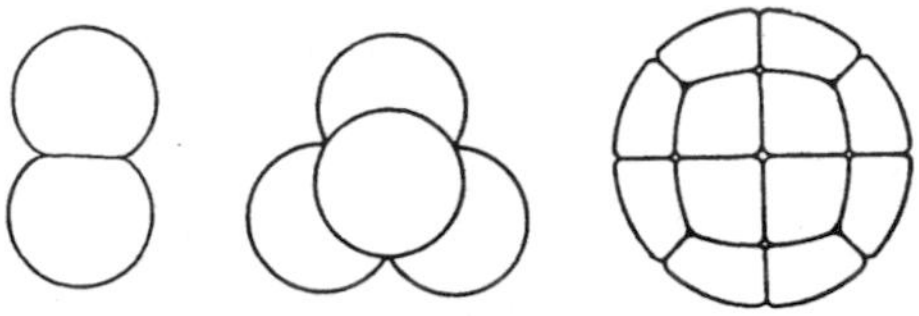

Owing to the great variability and mutual similarity of many Ericales pollen types the key does not invariably catch all grains from the taxa indicated. Atypical grains usually end up in one of the collective taxa, especially the *Vaccinium* type. The 'normal' position of a tetrahedral tetrad is the one in which the polar axis of one grain is parallel with the optical axis of the microscope (cf. the illustrations).

Table B1. Dyads . . . . **Scheuchzeria**

Table B2. Tetrads

A_2 Tetrads linear, *flat* or irregular. Grains reticulate, each with two apertures. .

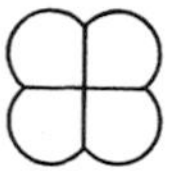

B_2 Tetrad *linear* Grain with one pore, *ca.* 3 μm wide . . . . . **Typha spp.**

B_1 Tetrad *irregular*. Grain with single pore 6 μm wide (*Epipactis palustris*) or two apertures . . . . **Orchidaceae**

A_1 *Tetrads tetrahedral*, sometimes irregularly so . . . .

C_3 Grains echinate.

D_2 Tricolporate, furrows broad, not concealed . . . . . . **Aldrovanda**

D_1 Stephanoporate, apertures concealed . . . . **Droseraceae**

C_2 Grains psilate, inaperturate. Exine extremely thin, normally not preserved/recognized in deposits . . . . **Juncaceae**

C_1 Grains scabrate or verrucate.

E_2 Triporate, pores with vestibulum. Grains loosely connected. Cf. Table F15: Triporate. . . . .**Onagraceae**

E_1 Tricolpate/tricolporate. Pyrolaceae, Ericaceae (cf. Oldfield, 1959)

F_2 *Tetrads open*, usually with a gap in the middle; grains united by a peripolar circular region . . . . **Chimaphila**

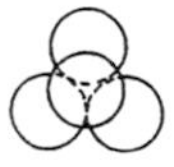

F_1 *Tetrads not open*; grains united by their proximal faces . . . . . .

G_2 Tetrads *loose*, often irregular or flattened. Edge of furrows irregular, costae absent or indistinct. Frustillate–verrucate **Calluna**

G_1 Tetrads compact, Furrows regular, costae usually distinct.

H_3 *Tetrads globular. Arctostaphylos* . . . .

I_2 Interior walls densely perforate. . . . . . . **A. alpina**

I_1 Interior walls with few perforations . . . . **A. uva-ursi**

H_3 Tetrads *subtriangular*. Colpi long with distinct costae. Tetrad less than 35 μm . . . . **Cassiope**

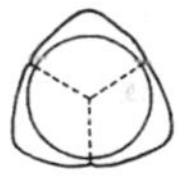

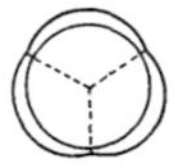

H_1 Tetrads *subglobular–subtriangular* . . . .

J_2 Grains usually in normal position. Interior walls thick and straight, *Y figure rectilinear* . . . .

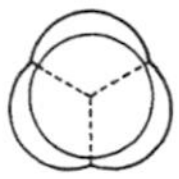

K_2 *Interior walls distinctly thicker than the exterior ones.* Costae indistinct. Colpi with transverse cracks . . **Empetrum nigrum**

K_1 Interior and exterior walls equally thick. Costae heavy. Colpi without traverse cracks . . . **Ledum**

J_1 Grains usually not in a strictly normal position. *Y figure of inner walls not rectilinear* . . . .

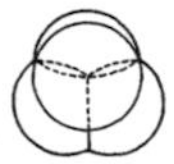

L_2 *Colpi very short.* Costae indistinct. . . . **Moneses**

L_1 *Colpi long*, usually with margo. *Vaccinium* group . . . .

M_2 Interior walls densely perforate . . . . **Andromeda**

M_1 Interior walls with few or no perforations.

N_2 Coarsely verrucate. Incl. *Phyllodoce* . . . . **Erica tetralix t.**

N_1 Psilate to irregularly scrabrate–verrucate.

O_2 Costae absent/indistinct.

P_2 Polar index large. Columellae evenly distributed. Exine thick **Rhododendron lapponicum t.**

P_1 Polar index smaller. Exine thinner. Incl. *Pyrola* spp. . . . **Empetrum hermaphroditum**

O_1 Costae distinct.

Q_2 Interior walls thicker than the exterior ones. . **Loiseleuria**

Q_1 Interior walls not thicker than the exterior ones

R_2 Tetrads subglobular. Walls thick . . **Chamaedaphne**

R_1 Tetrads subglobular. Incl. *Andromeda, Chamaedaphne, Empetrum hermaphroditum, Erica* spp., *Phyllodoce, Oxycoccus* . . . . **Vaccinium t.**

Table B3. Polyads

(A_2 *Regular polyads* of 8–16–32 grains. . . . . **Acacia**)

A_1 Polyads irregular; tetrads usually indistinguishable.

B_2 Loose aggregations. **Chimaphila**

B_1 Tetrads densely crowded in massulae, usually irregular.

C_2 Monoporate . . . . **Orchidaceae**

C_1 Triaperturate . . . . **Asclepiadaceae**

Table C. Polyedric grains

Most recent polyedric grains are easily reduced to the corresponding rotational shape by abstracting prominent apertures, ridges, etc. The empty exines of a few pericolpate grains are polyedric. The stephanocolporate grains of the *Viola tricolor* group appear polyedric, due to flattening of the polar areas. In pre-Quaternary material more or less distinctly polyedric sporomorphs are found. Some recent cyptogam spores, cysts, etc., would also come in here.

Table D. Bilateral grains

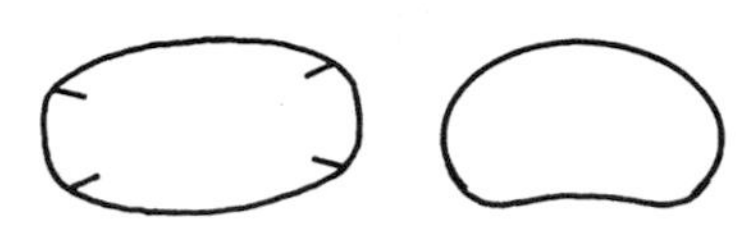

Some monocolpate and biporate grains are faintly bilateral or pear-shaped (heteropolar). They are found in Table F.

A_2 With more than one aperture.

B_2 *Tetra-aperturate*, reticulate . . . . **Impatiens**

B_1 *Bi-tetra-aperturate*—pores aggregated. Psilate **Myriophyllum alterniflorum**

A_1 With one or no aperture.

C_2 With columellae. Grain reticulate.

D_2 *Grain slightly bent.* Intectate at both ends. Columellae distinctly separate . . . . **Ruppia**

D_1 Grain not bent. Heterobrochate, semitectate, columellae forming muri. Liliaceae (cf. Table F4).

C_1 No columellae. Pteridophyte spores.

E_2 Perine loose, crested or folded, not echinate but in some spores spinulose.

F_2 *Perine saccate* . . . . **Dryopteris t.**

F_1 Perine lophate–cristate.

G_2 *Crests few, 1–4*, long, sometmes as long as the spore . . . **Matteuccia t.**

G_1 Usually more than 4 crests, shorter.

H_2 Crests with few or no spinules. . . . . . **Woodsia t.**

H_1 *Crests densely spinulose*. . . . . **Asplenium t.**

E_1 No loose outer perine or, if present, not folded or crested.

I_4 *Verrucate*, sculpturing elements coarse, more than 3 μm high **Polypodium**

I_3 Scabrate.

J_2 *No perine* visible . . . . **Athyrium fillix-femina**

J_1 Perine thin, *slightly wrinked* . . . . **Blechnum spicant**

I_2 Psilate.

K_2 Wall double with *loose outer cover* . . . . **Isoëtes**

K_1 Wall one-layered or outer cover not loose.

L_2 Wall faintly undulate . . . . **Gymnocarpium dryopteris t.**

L_1 Wall smooth, structureless. Unidentifiable fern spores that have lost their perispore.

I_1 Echinate.

M_2 Spines scattered, slender, *uniform*. . . . . **Cystopteris filix-fragilis**

M_1 With spines and (blunt) papillae.

N_2 Mainly papillae . . . . **Cystopteris spp.**

N_1 Sculpturing elements *variable* . . . . **Thelypteris palustris**

Table E. Subtetrahedral grains (spores)

A_2 With an outer, loose, folded perine, cf. Table D . . . . **Matteuccia**

A_1 No loose perine. Cf. *Thesium*, Table F19.

B_3 *Grain subglobular*, exospore thin . . . .

C_2 Spore echinate(–foveolate) . . . . **Hymenophyllum**

C_1 Spore baculate–clavate . . . . **Osmunda**

B_2 Spore triangular, semi-lobate.

D_2 *Corners flattened* . . . . **Lycopodium selago**

D_1 *Corners rounded*. . . . .

E_2 Verrucate . . . . **Cryptogramma**

E_1 Scabrate–fimbriate, usually smooth by loss of ornamentation. . **Pteridium**

B_1 *Equatorial limb triangular–subcircular*. Polar axis shorter than equatorial diameter . . . .

F_5 Echinate . . . . **Selaginella**

F_4 Foveolate. . . . . **Ophioglossum**

F_3 Reticulate, muri sharp. *Lycopodium* s.l.

G_2 Heterobrochate.

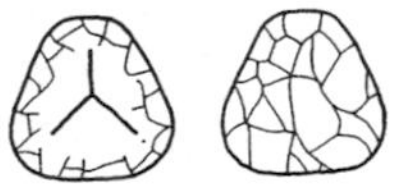

H_2 *5–6 brochi in median of distal face; reticulum sparsely developed on proximal face* . . . . **L. annotinum**

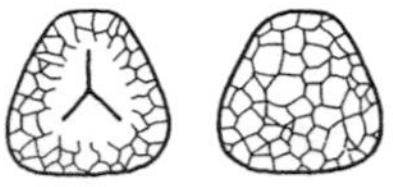

H_1 *Reticulum developed on proximal face*, rarely covering more than halfway in to centre . . . . **L. complanatum t.**

G_1 Homobrochate, brochi present also on proximal face.

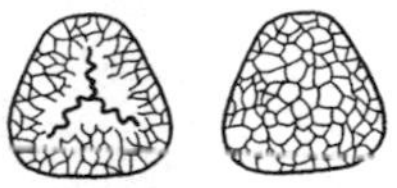

I_2 *Laesura arms undulating; 10–12 brochi in median* . **L. clavatum.**

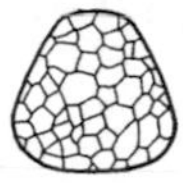

I_1 Laesura arms straight; *7–9 brochi in median* . . . . **L. alpinum**

F_2 *Rugulate* . . . .

J_2 Valla narrow, on the distal side only . . . . **L. inundatum**

J_1 Valla broad, indistinct pattern also on proximal face. . . . .
Botrychium multifidum

F_1 Verrucate.

K_3 Verrucae indistinct, exine stains weakly with fuchsin. *Polar axis short.*
Sphagnum

K_2 *Verrucae baculoid–echinate* . . . . **Ophioglossum**

K_1 *Verrucae rounded*, sometimes spinulate. *Botrychium* . . . .

L_3 Spore more than 54 μm diameter. Sculpturing on proximal face less distinct . . . . **B. boreale t.**

L_2 Spore 43–58 μm. Sculpturing on proximal face as distinct as on the distal . . . . **B. lunaria t.**

L_1 Spore smaller than 43 μm . . . . **B. virginianum**

Table F. Rotational ellipsoidic (or ovate) grains: main key

The left-hand figure gives the polar view, the right-hand one the equatorial, except for B_1, H_1, J_1 and L_1, which are apolar. The two figures at E_1 represent two different types (ring and spiral).

A_3 No distinct aperture.

B_2 With meridional groves and ridges . . . .**F1 Polyplicate**

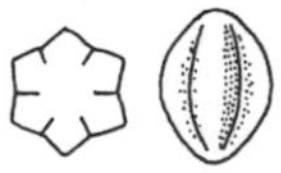

B_1 No meridional groves or ridges . . . . **F2 Inaperturate**

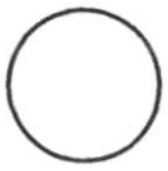

A_2 With one aperture.

C_3 Aperture three-silt . . . . **F3 Trichotomocolpate**

C_2 Aperture elongate . . . . **F4 Monocolpate**

C_1 Aperture circular . . . . **F5 Monoporate**

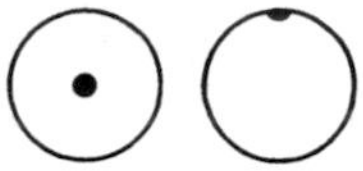

A_1 More than one aperture.

D_2 No lacunae or, if present, not in a fixed geometric pattern.

E_2 Apertures not fused.

F_3 With furrows, no pores or transverse endexinous furrows (colpate)

G_3 Two furrows . . . . **F6 Dicolpate**

G_2 Three furrows . . . . **F7 Tricolpate**

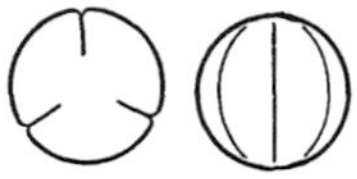

G_1 More than three furrows.

H_2 All furrows meridional. . . . . . **F8 Stephanocolpate**

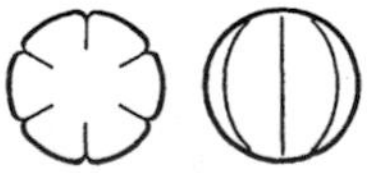

H_2 Some or all furrows not meridional . **F9 Pericolpate**

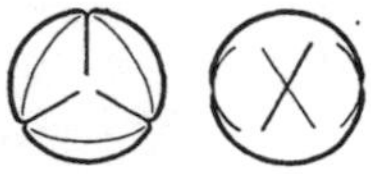

F_2 Both furrows and distinct pores or transverse furrows present (colporate). Usually one pore per furrow, occasionally missing; in some taxa more than one pore per furrow (not in our material).

I_3 Two furrows . . . . **F10 Dicolporate**

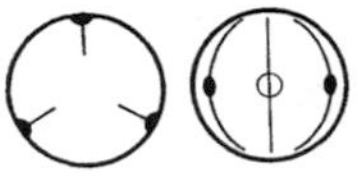

I_2 Three furrows . . . .**F11 Tricolporate**

I_1 More than three furrows.

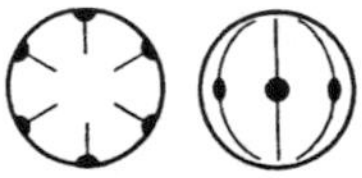

J_2 All furrows meridional. . . . . . . **F12 Stephanocolporate**

J_1 Some or all furrows not meridional . . . **F13 Pericolporate**

F_1 Free pores present, no furrows (porate)

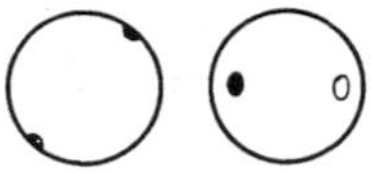

K_3 Two pores . . . . **F14 Diporate**

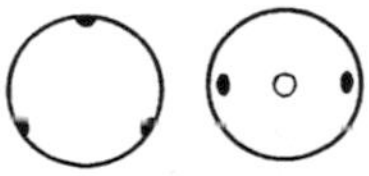

K_2 Three pores . . . .**F15 Triporate**

K_1 More than three pores.

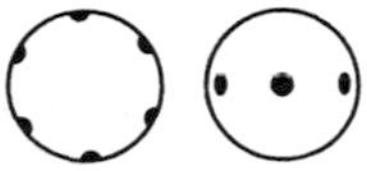

L_2 Pores confined to a circular (sub-)equatorial belt . . . . . **F16 Stephanoporate**

L_1 Pores ± uniformly distributed on the surface of the grain **F17 Periporate**

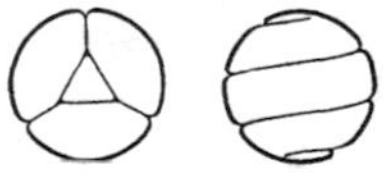

E_1 Apertures fused to rings or spirals . . . . **F18 Syncolpate**

D_1 With lacunae in a fixed geometric pattern

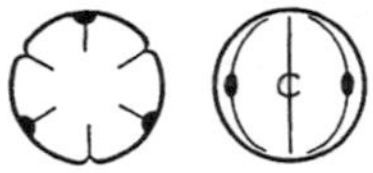

M_2 Lacunae elongate, meridional (pseudocolpi). . . . . **F19 Heterocolpate**

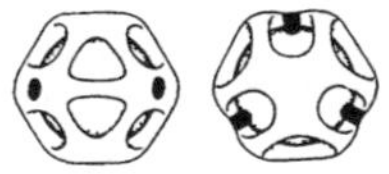

M_1 Lacunae not elongate . . . . **F20 Fenestrate**

*Table F1. Polyplicate (*Ephedra*)*

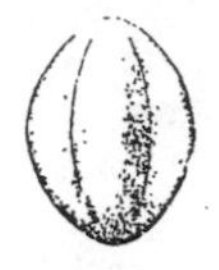

A_2 Groves *branched* . . . . **E. distachya t.**

A_1 Groves simple.

B_2 More than 10 groves.

C_2 Meridional ridges low, indistinct in polar view . . . **E. strobilacea t**

C_1 Meridional ridges high, distinct in polar view . . . . **E. fragilis t.**

B_1 Fewer than 10 groves . . . . **E. fragilis t.**

Table F2. Inaperturate

Many cryptogram spores, cysts, etc., are also inaperturate. They are, on the whole, easily distinguished from pollen grains by the (absence of) structure of the wall. They are not included in the key.

A_4 Cavate (verrucate) or echinate.

B_3 *Cavate.* Grain more than 50 μm . . . . **Tsuga**

B_2 Echinate. Cf. also Lemnaceae, *Nymphaea*: Table F5, monoporate, and *Nuphar*, Stratiotes: Table F4, monocolpate . . . . **Hydrocharis**

B_1 Verrucate, cf. *Plantago*, Table F17.

A_3 Scabrate or frustillate.

C_2 *Grain heteropolar, with thin areas* (lacunae). Cf. special key. . **Cyperaceae**

C_1 No lacunae or perforations. *Populus*.

D_2 Scabrate, with minute dark dots. . . . . **P. tremula t.**

D_1 Frustillate, exine easily broken . . . .**P. balsamifera t.**

A_2 Psilate/gemmate.

E_2 Psilate.

F_2 Grain more than 50 μm. Exine not perforate

G_2 Grain *ca.* 90 μm . . . . **Pseudotsuga**

G_1 Grain 60–85 μm. **Larix**

F_1 Grain less than 50 μm. Exine perforate (some areas sometimes not).

H_2 Grain *ca.* 25–30 μm. Sculpturing apparently vermiculate. One or two ruptured areas . . . . **Calla**

H_1 Grain *ca.* 15 μm, smooth . . . .**Acorus**

E_1 Gemmate

I_2 *With a well-marked exitus papilla* . . . . **Taxodiaceae**

I_1 No exitus papilla.

J_2 *Fossil grains usually two-split.* Gemmae scattered, deciduous, rounded, with constricted base, mostly lost in fossil material. Cf. *Pedicularis*, Table F16 . . . . **Juniperus**

J_1 Gemmae rather crowded, size variable, not deciduous. SEM: microbaculate . . . . **Taxus**

A_1 Clavate–baculate or reticulate.

K_2 *Grain bilateral, slightly bent.* Intectate at ends; columellae distinctly separate **Ruppia**

K_1 Grain rotational.

L_2 *Isolated clavae/baculae in reticulate pattern* . . . . . . **Callitriche**

L_1 Reticulum distinct.

M_2 Columellae distinct.

N_2 Diameter of lumina subequal to breadth of muri **Daphne**

N_1 Diameter of lumina several times the breadth of muri. *Potamogeton.*

O_2 Columellae widely spaced, *mostly confined to corners of brochi* . . . . **Coleogeton t.**

O_1 Columellae denser, also *between corners of brochi*; cf. also *Triglochin maritimum* . . . . . **Potamogeton t.**

M_1 Columellae indistinct.

P_2 Diameter of lumina subequal to breadth of muri. *Muri (indistinctly) duplicolumellate* . . . . **Daphne**

P_1 Muri much narrower than width of lumina, simplcolumellate. . . . . **Triglochin**

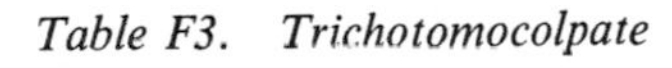

Table F3. Trichotomocolpate

Triradiate scars are very common in cryptogam spores (cf. Table E). In angiosperm pollen they are found in some taxa, e.g. in Palmae and Liliaceae, or Annonaceae. In the area none is known, apart from occasional faint traces in some grains of *Trapa natans*.

Table F4. Monocolpate

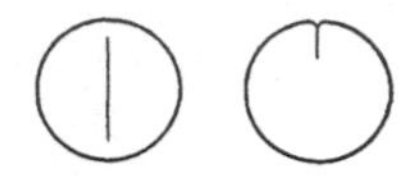

Compare also Table F5, Monoporate, the pore of which may become colpus-like by collapse of the grain

A_4 Verrucate.

B_2 *Verrucae uniform*, hollow, *multicolumellate*. Aperture broad, irregular . . **Sciadopitys**

B_1 Verrucal dimorphic, smaller on the proximal side. . . **Gagea pratensis t.**

A_3 Echinate–baculate with *sculpturing elements projecting above a bottom of minute, crowded projections* . . . .

C_2 Grains usually prolate. Echinate; spines more than 5 μm long . **Nuphar**

C_1 Grains subspherical. Baculate–echinate, large projections less than 5 μm long. Aperture sometimes indistinct . . . . **Stratiotes**

A_2 Reticulate (brochi equal to or more than 1 μm).

D_2 Brochi more than 3 μm diameter.

E_2 *Heterobrochate*. Margo very broad . . . . **Anthericum**

E_1 Homobrochate or brochi not falling into discrete size groups.

F_2 *Muri consisting of transverse elements*. No free columellae in lumina. Margo present . . . . **Lilium**

F_1 Columellae approximately cylindric. *Free columellae* in some brochi. No margo . . . . **Iris pseudacorus t.**

D_1 Brochi 1–3 μm diameter.

G_2 *Furrow extends to the proximal side of the grain* . . . . . . **Muscari**

G_1 *Furrow restricted to the distal side of the grain* . . . .

H_2 Heterobrochate. Grains long, narrow. Few, large brochi, mostly on the proximal side of the grain. Incl. *Ornithogalum, Polygonatum verticillatum*. . . . . **Scilla t.**

H_1 Homobrochate.

I_3 Grain less than 30 μm. Brochi *ca.* 1 μm; *reticulum with gradient towards colpus* . . . . **Narthecium ossifragum**

I_2 Grain 30–50 μm. *Margo present*. Muri narrow. Incl. *Butomus, Veratrum* . . . . **Narcissus**

I_1 Grain more than 50 μm.

J_2 Pollen often ruptured with ragged colpus edges. No margo . . . . **Iris pseudacorus t.**

J_1 Margo present. Columellae distinct . . . . . **Fritillaria**

A_1 Microreticulate (brochi less than 1 μm), psilate or scabrate.

K_2 *Furrow extends to the proximal side of the grain. Allium porrum, A. scorodoprasum, A. vineale*

K_1 Furrow restricted to the distal side of the grain.

L_2 Grain less than 30 μm.

M_2 Columellae imperceptible. Brochi *ca.* 1 μm . . . . **Narthecium ossifragum**

M_1 Columellae distinct. Brochi less than 1 μm . . . . **Asparagus**

L_1 Grain more than 30 μm.

N_3 *Grain shaped like an orange segment. Allium.* . . . .

O_2 Psilate with minute perforations and columellae. Incl. *Allium angulosum, A. montanum, A. paradoxum* . . . **A. ursinum t.**

O_1 Foveolate with *distinct perforations and columellae*. Incl. *A. schoenoprasum* . . . . . . **A. oleraceum t.**

N_2 *Grain curved with thickened ends* . . . . . **Sisyrhinchium**

N_1 Grain neither segment-shaped nor curved with thickened ends.

P_2 Reticulate, muri broader than width of lumina.

Q_2 Homobrochate . . . . **Tulipa silvestris**

Q_1 Heterobrochate . . . **Ornithogalum umbellatum, Smilacina stellata**

P_1 Psilate–scabrate, perforate.

R_2 Grain more than 60 μm. Columellae evenly distributed. Colpus edges ragged, almost marginate **Brasenia schreberi**

R_1 Grain less than 50 μm. Columellae unevenly distributed.

S_2 *Columellae distributed in relation to major perforations*. Colpus edges ragged . . . . . . **Gagea lutea t.**

S_1 *Columellae irregularly distributed.* Colpus edges straight. . . . **Majanthemum bifolium**

Table F5. Monoporate

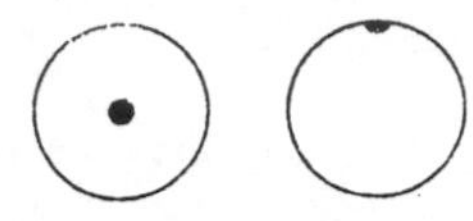

A_2 Pore large, diameter more than half that of the grain.

B_2 *With operculum.* Grain verrucate, gemmate, clavate, baculate and/or echinate. Sculptural elements unicolumellate. Grain oblate–subsphaeroidal **Nymphaea**

B_1 No operculum. Grain verrucate, *projections multicolumellate.* Grain subsphaeroidal . . . . **Sciadopitys**

A_1 Pore smaller, diameter usually much less than half that of the grain.

C_2 With distinct anulus. Grain tectate, psilate, scabrate, verrucate. Cf. special key . . . . **Gramineae**

C_1 No anulus. Cf. *Juniperus*, Table F2.

D_3 Grains psilate, densely perforate *with lacunae*, one of which forms a rudimentary pore at broad end of usually pear-shaped grain. Cf. special key . . . . **Cyperaceae**

D_2 Grain reticulate, heterobrochate. Pore not always distinct . . . . . **Sparganium, Typha spp.**

D_1 Grain echinate, spines *ca.* 1.5 μm with broad conical base. Pore may be indistinct. Cf. cysts and other microfossils of unknown origin **Lemnaceae**

Table F6. Dicolpate

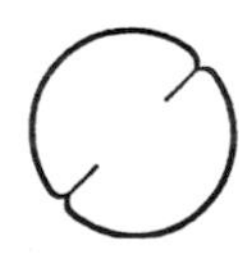
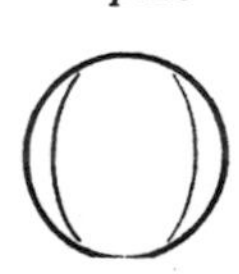

A_2 Columellae distinct. Per-reticulate. Margo present . . . . **Tamus**

A_1 Columellae indistinct. No margo.

B_3 Per-reticulate. Grain prolate. . . . . **Tofieldia**

B_2 Psilate–rugulate and perforate or occasionally ± reticulate. Shape irregularly sphaeroidal. . . . . **Calla**

(B_1 Grain usually two-split, edges ragged. Structure and sculpturing very faint or nil. Cf. Table F18. . . . . **Pedicularis**

Table F7. Tricolpate: Main key to subsections

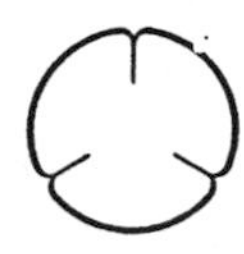
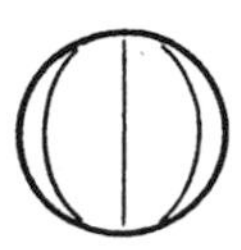

A_2 Tectate

- B_5 Psilate . . . . Table F7.1
- B_4 Scabrate . . . . Table F7.2
- B_3 Echinate–verrucate. . . . . Table F7.3
- B_2 Rugulate–striate. . . . . Table F7.4
- B_1 Supra-reticulate . . . . Table F7.5

A_1 Semitectate or intectate

- C_2 Per-reticulate . . . . Table F7.6
- C_1 Intectate . . . . Table F7.7

The crested grains of *Trapa* are not included in the keys; cf. Table A.

Table F7.1 Tricolpate, psilate

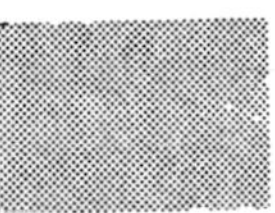
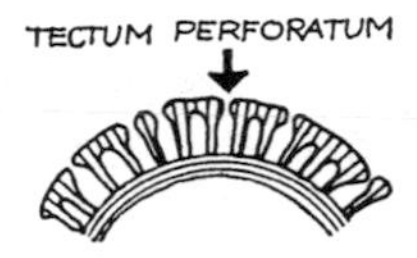

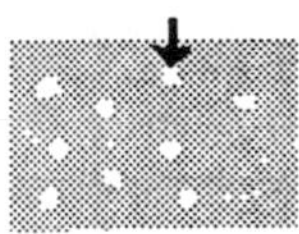

A_2 *Furrow without equatorial constriction*. . . . .

B_3 Exine equal to or thicker than 3 μm. Columellae shorter and exine thinner next to furrow (*crescentic*). Convolvulaceae . . . .

C_2 Exine *ca*. 5 μm thick. Columellae branched. Grain 70–100 μm. . . **Convolvulus pp.**

C_1 Exine *ca*. 3 μm thick. Columellae simple. Grain *ca*. 30 μm . **Cuscuta**

B_2 *Exine thinner in middle of intercolpium*. Furrow narrow, straight, edges sharp. Grain *ca*. 20 μm . . . . **Melampyrum**

B_1 Exine equal to or thinner than 2 μm. *Edge of furrow ragged* . . . . .

D_2 Exine thinner next to furrow, micro-echinate. Columellae distinct and/ or irregularly distributed.

E_2 Grain globular–prolate. Exine thicker in polar area, without perforations. Columellae present in furrow, grouped . . . **Veronica**

E_1 Grain globular–oblate. No polar thickening of exine. Tectum with perforations. Columellae in furrow, if present, scattered. *Spergula* s.l.

F_2 Distinctly micro-echinate. *Number of perforations and columellae approximately equal. Columellae in a dense, regular pattern* . . . . **S. arvense t.**

F_1 Not echinate. *Perforations fewer than columellae. Columellae irregularly districted* . . . . **Spergularia t.**

D_1 *Exines same thickness throughout, ca.* 1 μm. Surface smooth . . . . .

G_2 Grains oblate. Polar area small, index below 0.3

H_2 Grain *ca.* 30 μm. Structure imperceptible . . . . . **Myricaria**

H_1 Grain 40–50 μm. Structure distinct. . . . . **Euphrasia**

G_1 Grains subglobular/prolate, 30–40 μm. Surface uniform. Incl. *Bartsia. Odontites* . . . . **Rhinanthus t.**

A_1 *Furrow with equatorial constriction.* Cf. Table 11.1 . . . .

I_2 Furrow with prominent, undulating edges. Special key . . . . . **Rosaceae**

I_1 Furrow edges flat, straight.

J_2 Microreticulate. Columellae imperceptible . . . . **Saxifraga nivalis t.**

J_1 Exine not reticulate.

K_2 Columellae imperceptible. Exine thin . . . . . **Viola palustris t.**

K_1 Columellae distinct.

L_2 Columellae regularly dispersed. Perforations minute, dense . **Lobelia**

L_1 Columellae irregularly dispersed, minute. Perforations scattered, irregular . . . . **Myricaria**

Table F7.2. Tricolpate, scabrate

A_2 *Furrow with equatorial constriction* . . . .

B_2 Equatorial limb triangular. *Costae colpi distinct. Cornus* . . . .

C_2 Grain more than 50 μm, scabrate. Tectum perforate. . **C. sanguineus**

C_1 Grain less than 25 μm, with scattered microverrucae. Tectum imperforate . . . . **C. suecica**

B_1 *No distinct costae colpi.* Columellae crowded, minute . . . . **Myricaria**

A_1 Furrow without equatorial constriction . . . .

D_3 Endexine very thin. Columellae scattered, their mutual distance much greater than their diameter. Furrow more or less operculate . . . . . . **Saxifraga hirculus t.**

D_2 Endexine distinct. Exine thick, at any rate in the polar area.

E_2 Exine more than 3 μm thick, *crescentic*. Endexine heavy. Columellae distinct, their mutual distance 1–2 times diameter. *Furrow membrane with columella-like granules* . . . . **Valerianella**

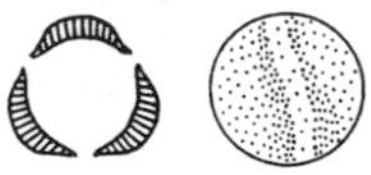

E_1 Exine thinner except in polar area.

F_2 Furrow edge ragged, *granules at membrane much coarser than columellae* . . . . **Aconitum**

F_1 Furrow sharp, membrane without or with very few granules . . **Teucrium**

D_1 Not so.

G_2 *Low ridges at base of spinules form a reticuloid pattern.* Exine with multiple perforations. *Papaver* . . . .

H_2 Spinules (scabrae) regularly distributed . . . . . . **P. rhoeas t.**

H_1 Spinules distinct, often sub-verrucate. . . . . **P. radicatum t.**

G_1 No ridges between bases of spinules.

I_2 Structure and sculpturing very fine. Grain oblate . . **Myricaria**

I_1 Structure and sculpturing distinct.

J_2 Spinules of uniform size, regularly distributed.

K_2 Spinules distinct. *Tectum with numerous, dense perforations* . . . . **Anemone t.**

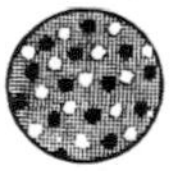

K_1 Spinules minute. Tectum with few, distinct perforations. *Columellae fine, in vermiculoid pattern.* Grain less than 25 μm . . . . **Caltha t.**

J_1 Spinules irregularly distributed, size variable.

L_2 Tectum without perforations. Columellae distinct, uniform, rather crowded. Furrow narrow . . . . **Quercus**

L_1 Tectum with perforations, (sub-)verrucate–echinate. Columellae distinct, partly absent below verrucae. Furrow open, membrane granulate. Grains often pericolpate, cf. Table F9. *Ranunculus.*

M_1 Columellae in open pattern, *forming a bright area ('halo') around each spinule.* Perforations follow the same pattern . . . . **R. acris t.**

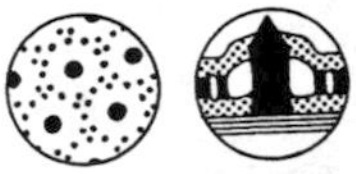

M_1 Columellae in a denser pattern, *haloes less distinct.* Perforations more evenly distributed **R. flammula t.**

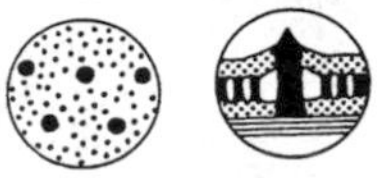

Table F7.3. Tricolpate, echinate

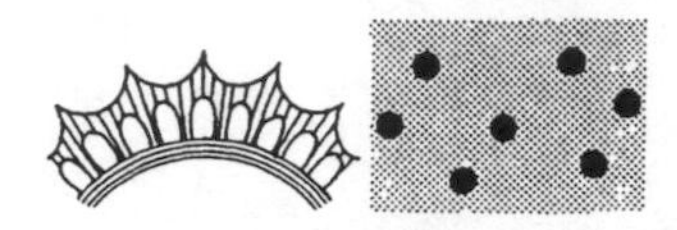

A_2 Grain large, 50 μm or more.

B_2 Spines prominent, long, widely separated.

C_2 Polar area very small. *Long spines on shield-shaped verrucae* . . . . **Valeriana**

C_1 Polar area large. Equatorial limb subtriangular. Spines coarse, not on shield-shaped verrucae. Incl. *Lonicera caprifolium, L. coerulea* . . . **Lonicera xylosteum t.**

B_1 Spines smaller.

D_2 Exine 5–10 μm. Columellae branched. *Spines of different size classes.* Furrow short, broad, Cf. Table F15, Triporate. . . . **Dipsacaceae.**

E_2 Structural and sculpturing elements regularly distributed. *Exine thicker at poles* . . . . **Scabiosa**

E_1 Not so. . . . . **Succisa**

D_1 Exine less than 3 μm. Spines of uniform size.

F_2 Exine *ca.* 1 μm. Spines widely separated. *Colpus edges protruding* **Lonicera periclymenum**

F_1 Exine 2–3 μm thick. Spines crowded . . . . **Linnaea**

A_1 Grains smaller.

G_3 *Spines constricted at base, frequently flaccid* (bent over at base) . . . . . **Rubus chamaemorus**

G_2 Spines conical, low. Tectum thick. Colpi short, indistinct (Fig. 2.3) . . . **Ambrosia, Xanthium**

(G_1 Verrucate/scabrate, not really echinate, Cf. Table F7.2.

H_2 Tectum perforate, colpus membrane with coarse granules . . . . . **Ranunculus pp.**

H_1 Tectum imperforate . . . . **Quercus)**

Table F7.4. Tricolpate, rugulate–striate

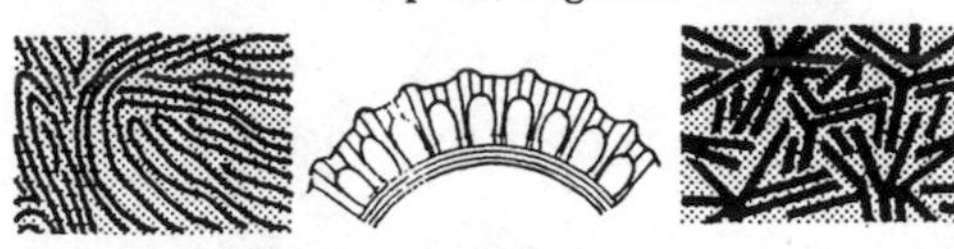

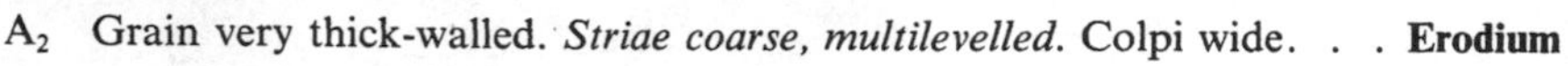

A_2 Grain very thick-walled. *Striae coarse, multilevelled.* Colpi wide. . . **Erodium**

A_1 Exine thickness and striation normal–fine.

B_2 Exine with or without perforations. Edges of furrows undulating, often prominent. Rosaceae, cf. Special key

B_1 Exine perforated. Edge of furrows straight.

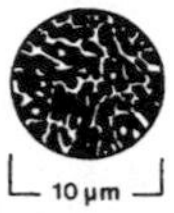

C_2 *Rugulate-scabrate. Vallae short*, in reticuloid pattern. . **Oxalis acetosella**

10 μm

C_1 Distinctly striate.

D_3 Sculpturing fine, *striae predominantly meridional*. Grain 30 μm or more. Cf. *Menyanthes*, Table F11.3 . . . .**Acer**

D_2 Direction of striae variable . . . .

E_2 Grain smaller than 25 μm. Sculpturing fine . . . . . **Trollius**

E_1 Grain more than 25 μm. Sculpturing coarse. Incl. *Saxifraga adscendens, S. aizoides, S. cotyledon, S. osloensis, S. paniculata, S. tridactylites, S. tricuspidata* . . . . . **Saxifraga oppositifolia t.**

D_1 Striation diffuse. Vallae in very low relief. Incl. *S. rivularis* . . . . . **Saxifraga cernua t.**

Table F7.5. Tricolpate, supra-reticulate

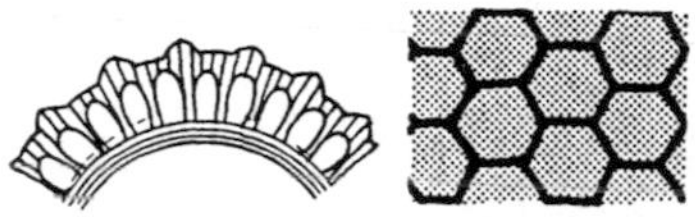

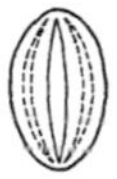

A_2 *Furrow without equatorial constriction* or poroid area. No costae . . . . . .

B_2 Meridional limb compressed oval, grain subcylindric, usually prolate-perprolate. Furrow narrow, not boat-shaped. *Onobrychis* type.

C_2 All brochi large, *ca.* 1.5 μm . . . . **Onobrychis**

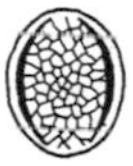

C_1 Brochi small, *ca.* 0.5 μm, or *a median field of small brochi* in each intercolpium . . . . **Hedysarum**

B_1 Meridional limb circular, oval or rhomboidal, not compressed. Furrow boat-shaped, often ruptured. Sect. Stachydeae. Incl. *Ajuga, Scutellaria* . . **Stachys t.**

A_1 *Equatorial part of furrow constricted or ruptured.* Costae interrupted at equator

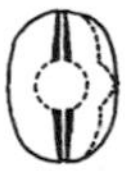

D_2 *Poroid area large*, usually ruptured. *Branches of costae colpi short.* Incl. *Ulex, Cytisus* . . . . **Genista t.**

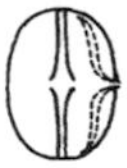

D_1 *Poroid area small. Branches of costae colpi long. Hypericum* . . . . . .

E_2 Grain less than 30 μm. Columellae in most cases not observable. . **Hypericum spp.**

E_1 Grain more than 30 μm. Muri duplicolumellate. . . . . .**H. elodes**

Table F7.6. Tricolpate, per-reticulate

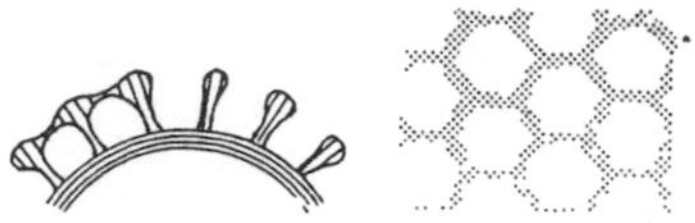

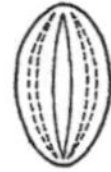

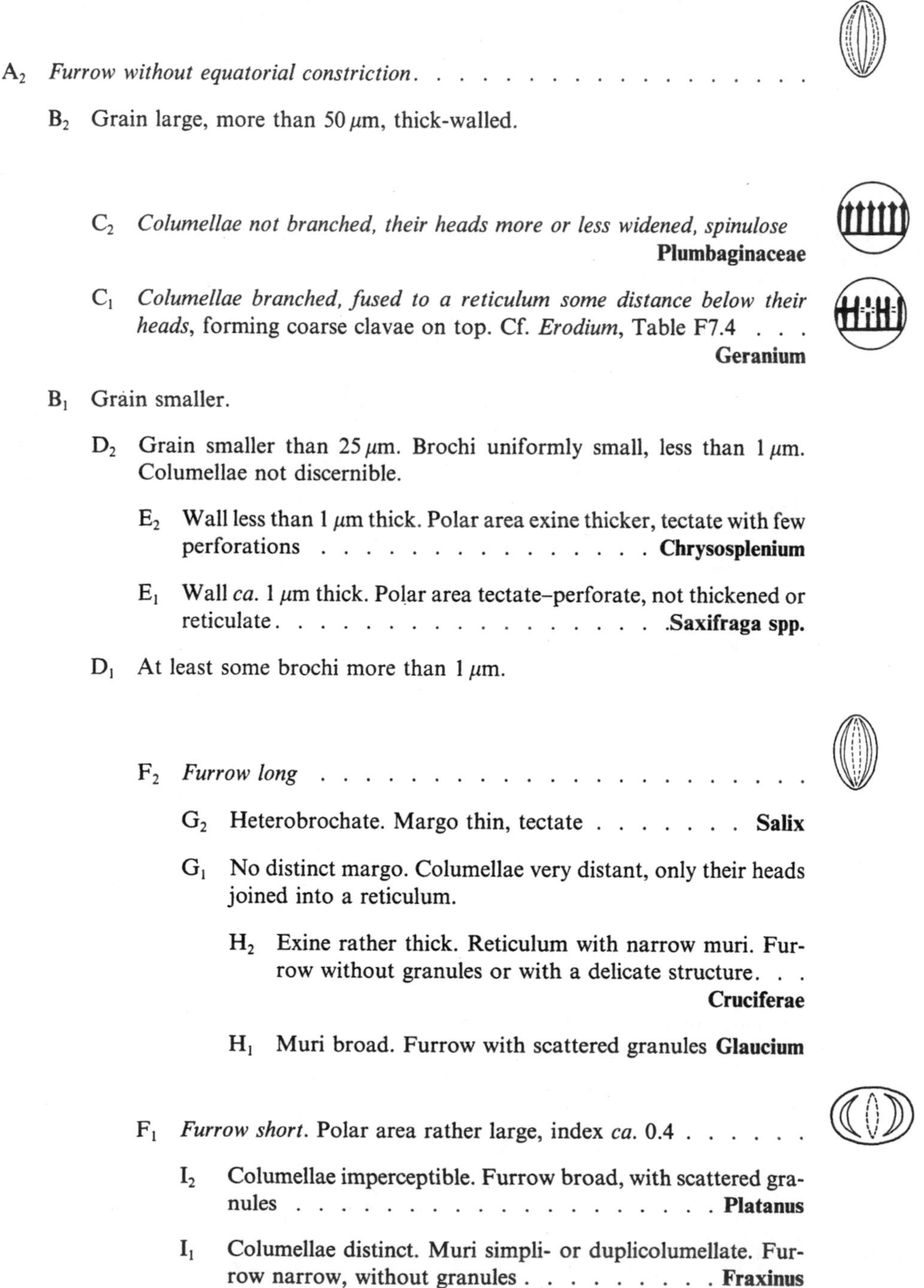

A_2 *Furrow without equatorial constriction* . . . .

B_2 Grain large, more than 50 μm, thick-walled.

C_2 *Columellae not branched, their heads more or less widened, spinulose* **Plumbaginaceae**

C_1 *Columellae branched, fused to a reticulum some distance below their heads*, forming coarse clavae on top. Cf. *Erodium*, Table F7.4 . . . **Geranium**

B_1 Grain smaller.

D_2 Grain smaller than 25 μm. Brochi uniformly small, less than 1 μm. Columellae not discernible.

E_2 Wall less than 1 μm thick. Polar area exine thicker, tectate with few perforations . . . . **Chrysosplenium**

E_1 Wall *ca.* 1 μm thick. Polar area tectate–perforate, not thickened or reticulate . . . . **.Saxifraga spp.**

D_1 At least some brochi more than 1 μm.

F_2 *Furrow long* . . . .

G_2 Heterobrochate. Margo thin, tectate . . . . **Salix**

G_1 No distinct margo. Columellae very distant, only their heads joined into a reticulum.

H_2 Exine rather thick. Reticulum with narrow muri. Furrow without granules or with a delicate structure. . . **Cruciferae**

H_1 Muri broad. Furrow with scattered granules **Glaucium**

F_1 *Furrow short*. Polar area rather large, index *ca.* 0.4 . . . . . .

I_2 Columellae imperceptible. Furrow broad, with scattered granules . . . . **Platanus**

I_1 Columellae distinct. Muri simpli- or duplicolumellate. Furrow narrow, without granules . . . . **Fraxinus**

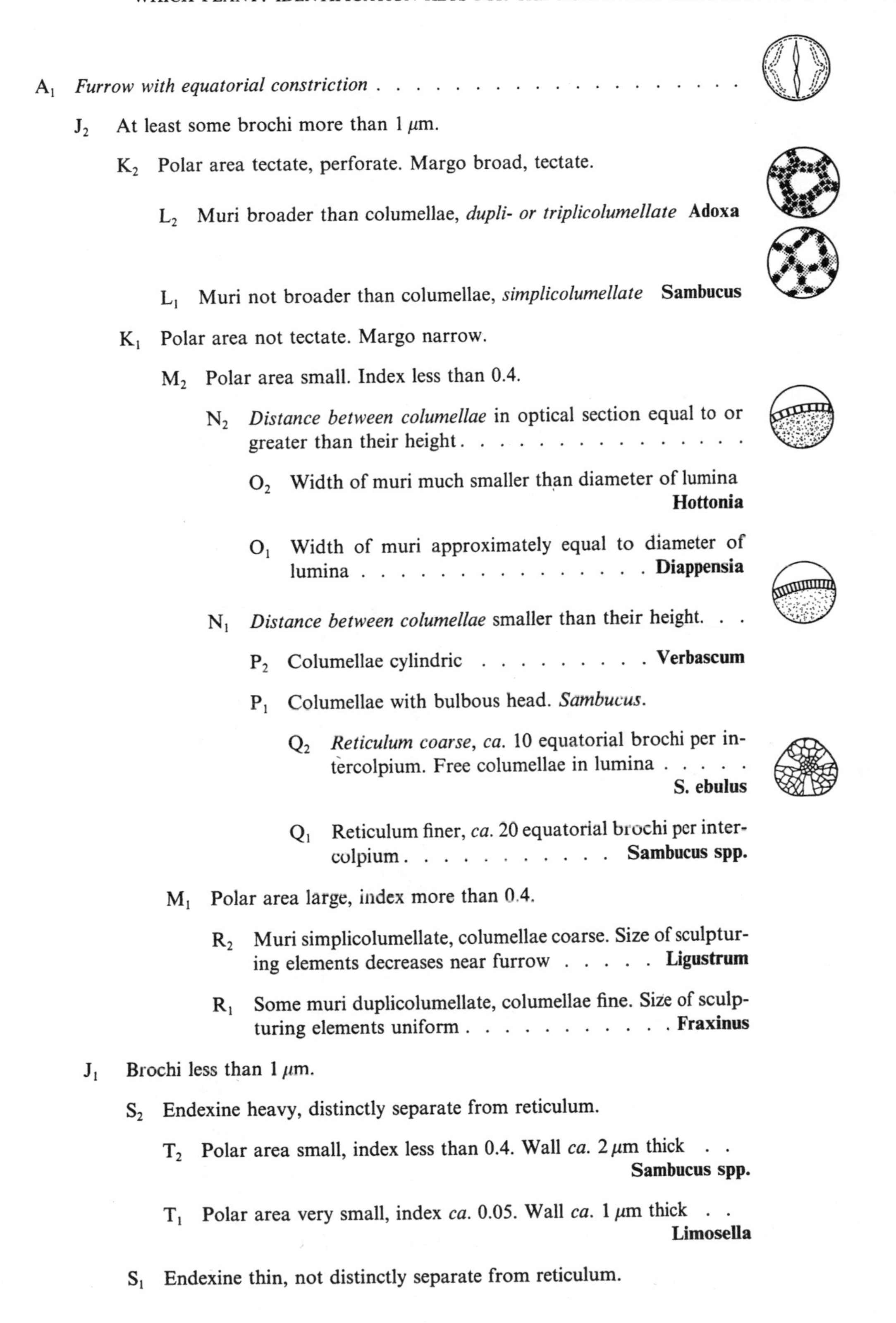

A_1 *Furrow with equatorial constriction* . . . .

J_2 At least some brochi more than 1 μm.

K_2 Polar area tectate, perforate. Margo broad, tectate.

L_2 Muri broader than columellae, *dupli- or triplicolumellate* **Adoxa**

L_1 Muri not broader than columellae, *simplicolumellate* **Sambucus**

K_1 Polar area not tectate. Margo narrow.

M_2 Polar area small. Index less than 0.4.

N_2 *Distance between columellae* in optical section equal to or greater than their height. . . . .

O_2 Width of muri much smaller than diameter of lumina **Hottonia**

O_1 Width of muri approximately equal to diameter of lumina . . . . **Diappensia**

N_1 *Distance between columellae* smaller than their height. . .

P_2 Columellae cylindric . . . . **Verbascum**

P_1 Columellae with bulbous head. *Sambucus*.

Q_2 *Reticulum coarse*, *ca.* 10 equatorial brochi per intercolpium. Free columellae in lumina . . . . . **S. ebulus**

Q_1 Reticulum finer, *ca.* 20 equatorial brochi per intercolpium . . . . **Sambucus spp.**

M_1 Polar area large, index more than 0.4.

R_2 Muri simplicolumellate, columellae coarse. Size of sculpturing elements decreases near furrow . . . . . **Ligustrum**

R_1 Some muri duplicolumellate, columellae fine. Size of sculpturing elements uniform . . . . **Fraxinus**

J_1 Brochi less than 1 μm.

S_2 Endexine heavy, distinctly separate from reticulum.

T_2 Polar area small, index less than 0.4. Wall *ca.* 2 μm thick . . **Sambucus spp.**

T_1 Polar area very small, index *ca.* 0.05. Wall *ca.* 1 μm thick . . **Limosella**

S_1 Endexine thin, not distinctly separate from reticulum.

U_2 *Costae colpi heavy, usually turning outwards at the equator*, Diameter of lumina subequal to breadth of muri. Cf. Table 7.5 . . . . **Hypericum**

U_1 *No costae colpi* . . . .

V_2 Margo distinct, tectate . . . . **Linaria**

V_1 Margo indistinct or absent . . . . **Hottonia, Samolus**

Table F7.7. Tricolpate, intectate (clavate)

A_2 Size of clavae variable.

B_2 Intercolpium and polar area with *scattered, large, free clavae* in addition to many small ones, more or less fused. Furrow not distinct . . . **Viscum**

B_1 All clavae free. *Small clavae scattered beneath the large ones.* Near the furrow there are only small clavae, forming a margo. Cf. *Rubus chamaemorus*, Table F7.3 . . . . **Ilex**

A_1 Clavae dimorphic or uniform. Linaceae

C_2 *Clavae dense, almost contiguous*, fine, 1–1.5 μm, dimorphic, the smaller ones dominant. Incl. *L. angustifolium* . . . . **L. usitatissimum t.**

10 μm

C_1 *Clavae distinctly separate, coarse*, longer than 1.5 μm, uniform or dimorphic. If dimorphic, both size classes subequally represented. Spinules on top of clavae (SEM), cf. Fig. 13.3 . . . . **L. catharticum t.**

10 μm

Table F8. Stephanocolpate

A_2 Psilate–scabrate.

B_2 Grain large. 4–6 broad furrows . . . . **Hippuris**

B_1 Grain small. 6–10 narrow furrows . . . .**Galium t.**

A_1 Reticulate.

C_2 *Polar area large*, index higher than 0.6 . . . .

D_2 *4 furrows. Grain oblate, subrectangular.* Supra-reticulate; no columellae discernible in muri. . . . .**Impatiens**

D_1 Per-reticulate. Columellae distinct in muri.

E_2 5–10 furrows . . . . **Primula vulgaris**

E_1 4 furrows. Muri simpli-duplicolumellate. . . . . . . **Fraxinus**

C_1 *Polar area medium–small*, index lower than 0.5. Labiatae pp. . . . .

F_2 Supra-reticulate with free columellae in lumina . . . . **Prunella t.**

F_1 Per-reticulate. Columellae restricted to muri.

G_2 Muri *beaded* . . . . **Salvia**

G_1 Muri not beaded. *Mentha* t.

H_2 Lumina small, *thickness of columellae more than width of muri* **Satureja**

H_1 Lumina large, *thickness of columellae* more or less *equal to width of muri*. Incl. *Origanum, Thymus* . . . . **Lycopus subt.**

Table F9. Pericolpate

A_3 Reticulate.

B_2 *Lumina deep, with free, coarse clavae. 30 short furrows in dodecahedral arrangement* . . . . **Polygonum amphibium**

B_1 No free elements in lumina. 4–6 furrows. Cf. Tricolpate, Tables F7.5 and F7.6.

A_2 Echinate.

C_2 *Spines slender*, regularly distributed. *Furrows short, numerous* . . **Koenigia**

C_1 Spines small, irregularly distributed. Tectum perforate. *Central area of intercolpium with coarse columellae* in reticuloid pattern. 12 furrows . . . **Montia**

A_1 Psilate, microechinate, scabrate, verrucate.

D_3 Psilate with minute, evenly distributed spinules (SEM or careful focusing in ph!). Minute perforations visible in SEM. Some grains with internal transverse furrows. Cf. Table F13 . . . . **Polygonum oxyspermum t.**

D_2 Irregularly distributed spinules, verrucae or scabrae of varying size. Tectum with perforations (ph, SEM): *Papaver*, Ranunculaceae. Cf. Table F2.

D_1 Psilate–scabrate with uniform minute spinules and perforations. . . . . **Spergula arvensis**

Table F10. Dicolporate

Not known from the area.

Table F11. Tricolporate: main key to subsections

 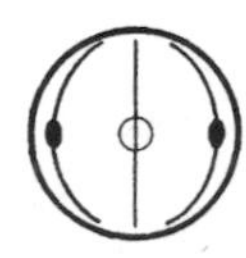

Together with tricolpate grains, the tricolporate are the most numerous and also the most troublesome of the list. The key presumes very close observation, and even then a certain subjectivity in the perception of morphological features is unavoidable. Some users may disagree with our treatment of individual taxa. In cases of doubt we have tried to cover conflicting opinions by entering the same pollen taxon in different places in the key. Especially, the transition colpate–colpate with equatorial constriction–colporate is not easy to define in categorical terms. Consultation of Table F7 is recommended if Table F11 does not lead to a result.

A_2 Tectate.

- B_4 Psilate or scabrate . . . . Table F11.1
- B_3 Echinate . . . . Table F11.2
- B_2 Striate or rugulate . . . . Table F11.3
- B_1 Foveolate or supra-reticulate . . . . Table F11.4

A_1 Semitectate. Per-reticulate . . . . Table F11.5

Table F11.1. Tricolporate. Psilate–scabrate

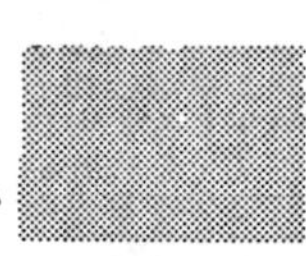 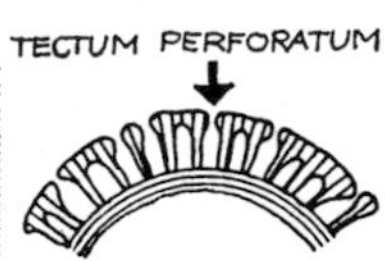 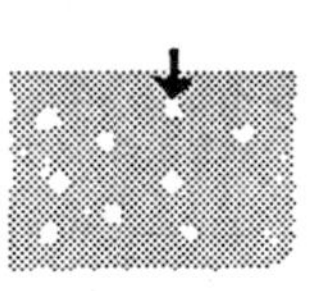

A_2 *With costae aequatoriales and transverse furrows* . . . .

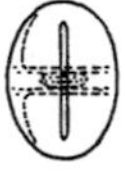

B_2 Equatorial limb intersemiangular. *Exine more than 4 µm thick, crescentic* Columellae coarse, branched. *Centaurea* sect. *Cyanus* . . . .

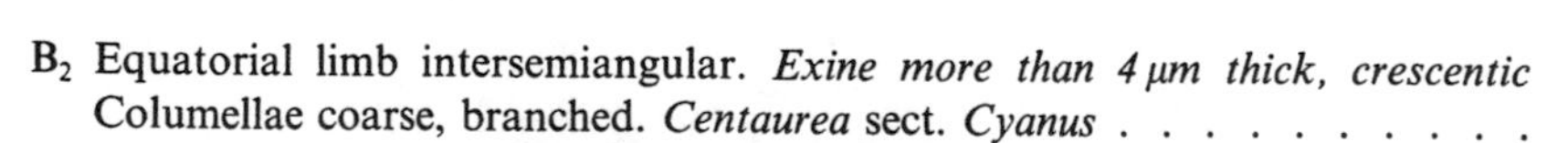 

C_2 *Meridional limb compressed oval. Costae aequatoriales sharply projecting* in meridional optical section . . . . **C. cyanus t.**

 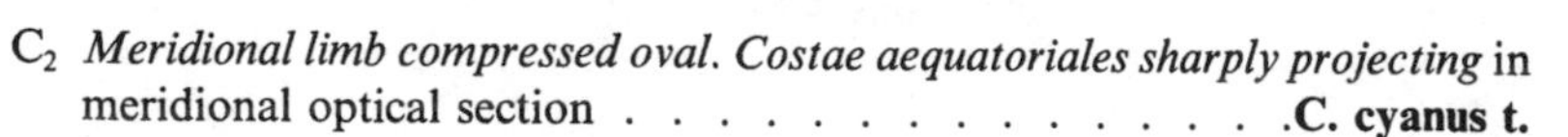 

C_1 *Meridional limb oval.* Costae aequatoriales less prominent **C. montana t.**

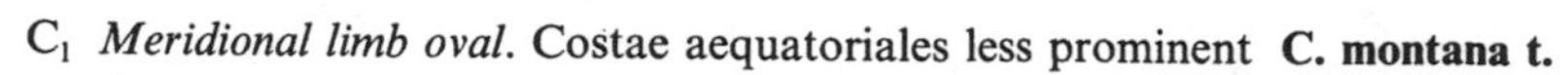 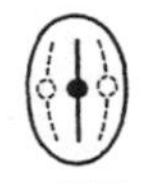

B_1 *Equatorial limb subcircular.* Exine thinner, not crescentic. Columellae finer. *Costae aequatoriales less prominent* . . . . **Polygonum aviculare t.**

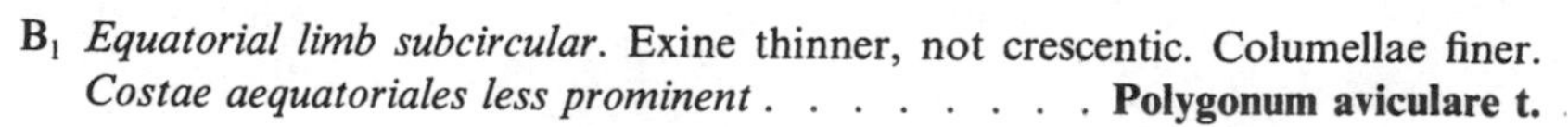  

A_1 *No costae aequatoriales* . . . .

D_2 Meridional limb *compressed oval*. . . . .

E_2 Grain distinctly *prolate* (cf. Cerceau, 1959) . . . . **Umbelliferae**

E_1 Not so distinctly prolate.

F_2 Exine more than 3 μm thick. Grain more than 50 μm, dimorphic, one type with denser structure. Tectum with scattered perforations. *Columellae* branched, *coarser at poles* . . . . . . **Polygonum bistorta**

F_1 Exine thin.

G_2 Grain more than 30 μm. *Pore edges distinctly protruding*. Scabrate, polar area suprareticulate . . . . **Anthyllis**

G_1 Grain less than 25 μm, psilate . . . . **Lotus**

D_1 Meridional limb oval or circular.

H_2 *Intercolpium flattened* or concave *in polar view* . . . .

I_2 Exine thicker and columellae coarser at poles. Furrow narrow, slit-shaped.

J_2 *Prolate–apiculate. Pore rectangular* . . . . . . **Pleurospermum**

J_1 Subglobular. . . . . **Hydrocotyle**

I_1 Thickness of exine uniform. Furrow not slit-shaped.

K_2 *With transverse furrow* . . . .

L_2 Scabrate. Tectum perforate. Grain 20–35 μm **Rhus coriaria t.**

L_1 Psilate. Tectum not perforate. *Solanum*.

M_2 Grain more than 20 μm . . . . **S. nigrum**

M_1 Grain smaller than 20 μm. . . . . **S. dulcamara**

K_1 No transverse furrow.

N_2 Pore more than 3 μm, edged by heavy costae. Grain *ca.* 50 μm. Columellae distinct . . . . **Nyssa**

N_1 Pore smaller.

O_2 *Equatorial limb semilobate*. Psilate . . **Rhamnus frangula**

O_1 *Equatorial limb semiangular*. Scabrate/microverrucate. *Cornus*. Cf. Table 7.2

P_2 Grain more than 50 μm. Tectum perforate . . . . **C. sanguinea**

P_1 Grain *ca.* 25 μm. No perforations . . . . **C. suecica t.**

H_1 Intercolpium convex.

Q_2 Pore edges distinctly protruding.

R_2 Endopore equatorially elongated. Grain less than 25 μm . . . . . .

S_3 Microechinate, vermiculate . . . . **Filipendula**

S_2 Psilate.

T_2 Tectum with minute perforations. *Endexine heavy* . . . . **Glaux**

T_1 Tectum not perforate. Incl. *Castanopsis* . . . . . . **Castanea t.**

S_1 Scabrate or indistinctly suprareticulate.

U_2 Operculate . . . . **Sanguisorba minor**

U_1 No operculum. Incl. *T. spadiceum, Medicago sativa, Astragalus arenarius, A. norvegicus* . . . . **Trifolium montanum t.**

R_1 Endopore circular. Furrow slit-shaped.

V_2 With vestibulum.

W_2 *Furrow very short, ca.* twice the diameter of the pore . . . . **Ludwigia**

W_1 Furrow longer, distinct . . . . **Hippophaë**

V_1 No vestibulum. Microechinate, columellae in vermiculoid pattern **Filipendula**

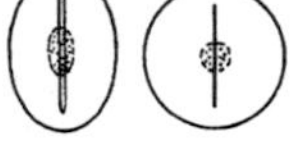

Q_1 Edges of pore not protruding, *endopore meridionally elongated, isodiametric* or indistinct . . . .

X_2 Grain perforate. *Furrow slit-shaped, endopore small, completely covered. Rumex* t. . . . .

Y_3 Grain smaller than 25(–28) μm . . . . **Oxyria t.**

Y_2 Grain 25–32 μm. Exine thicker . . . . **R. longifolius t.**

Y_1 Grain larger than 35 μm . . . . **R. aquaticus t.**

X_1 Not so.

Z_2 Grain 25–32 μm. Scabrate . . . . **Fagus**

Z_1 Grain less than 20 μm. Psilate . . . . **Crassula aquatica**

Table F11.2. Tricolporate. Echinate

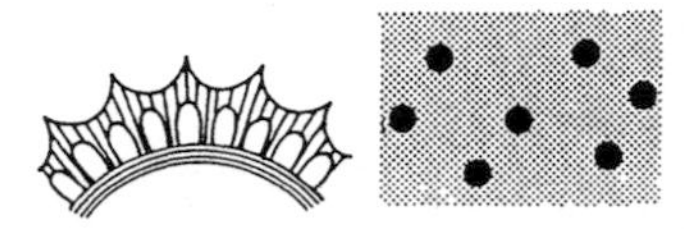

This table mainly comprises Compositae Asteroideae. Their exine is in typical cases multi-layered and consists of, in centrifugal direction: (1) foot layer; (2) inner columella layer, usually with scattered, coarse, often branched columellae; (3) paratectum formed by anastosmoses between the heads of the inner columellae; (4) outer columellae layer; columellae thinner and denser than in the inner layer; (5) (outer) tectum, which is very thin and densely perforated. It is a matter of convenience if one wants to consider the three outer layers (paratectum, outer columellae and tectum (s.s) as a tectum s.l.

A_2 *Columellae straight, not branched.* Spines with cylindrical base . . . .

- B_2 *Diameter of spines and columellae equal. Spines few, scattered*. . **Lonicera**
- B_1 *Spines much coarser than columellae, blunt,* frequently flaccid (bent over at base) . . . . **Rubus chamaemorus**

A_1 *Columellae branched, spines conical* from base. Asteroideae. . . . .

- C_2 Columellae indistinct in optical section. Incl. *Antennaria, Aster, Bellis, Bidens, Calendula, Erigeron, Filago, Galinsoga, Gnaphalium, Inula, Petasites, Senecio, Tussilago* . . . . **Solidago t.**
- C_1 Columellae prominent in optical section.
  - D_2 *Inner columella layer equal to or thicker than the outer*. . . . . . .
    - E_3 Spines vestigial.
      - F_2 Grains large, prolate with *longitudinal thickenings of intercolpia* (intersemiangulate) . . . . **Echinops**
      - F_1 Grains small, sphaerical–oblate . . . . **Artemisia t.**
    - E_2 *Spines blunt,* length equal to or smaller than basal diameter. Incl. *Arctium, Carlina, Carthamus, Onopordon* . . . . **Saussurea t.**
    - E_1 *Spines long, sharp.* Intercolpium in optical section distinctly crescentic. Incl. *Anthemis, Chrysanthemum, Matricaria, Serratula* . **Achillea t.**
  - D_1 *Inner columellae layer thinner than outer* . . . .
    - G_2 Spines vestigial . . . . **Centaurea scabiosa t.**
    - G_1 Spines distinct.

(H_2 Grains *ca.* 30 μm or smaller. Sphaerical. Colpi short. Cf. Table F7.3.)

H_1 Grains larger. Colpi long.

I_1 Oblate. Incl. *Carduus* . . . . **Cirsium t.**

I_2 Prolate. Incl. *C. alba, diffusa, jacea, phrygia* . . . . . **Centaurea nigra t.**

Table F11.3. Tricolporate. Striate–rugulate

As sculpturing elements, the striae are often difficult to observe and need a very careful focusing, especially in discordant grains. In the marginal figures for entries between I_2 and M_2 the outer furrow is indicated in black.

A_2 Operculate. See Special key . . . . **Rosaceae**

A_1 No operculum.

B_2 *Furrow transversal. Costae transversales present*. . . . . **Rhus**

B_1 Furrow not distinctly transversal. No costae transversales.

(C_2 Finely verrucate. *Furrow very short*, *ca.* twice the diameter of the pore **Ludwigia**

C_1 Striate–rugulate. Furrow much longer than the diameter of the pore.

D_2 Grain 25 μm or larger.

E_2 Pore indistinct.

F_2 Edge of furrow undulating. See Special key . . . . **Rosaceae**

F_1 Edge of furrow straight. Striae very distinct. Columellae fine. Menyanthaceae

G_2 Costae colpi distinct . . . . **Menyanthes**

G_1 Costae colpi absent or very thin. Columellae almost imperceptible . . . . **Fauria**

E_1 Pore distinct

H_2 *Polar area pointed.* Pore covered. Striae fine **Helianthemum**

H_1 Polar area rounded. Pore not or only partly covered. Gentianaceae

I_2 *All or almost all striae meridional* . . . .

J_2 Grain oblate or globular. P/E equal to or less than 1 **Lomatogonium**

J_1 Grain prolate, P/E more than 1.

K_2 Grain *ca.* 30 μm. **Gentiana**

K_1 Grain *ca.* 25 μm. **Centaurium**

I_1 Some striae not meridional.

L_2 *Intercolpium median with predominantly meridional striae* **Centaurium**

L_1 Intercolpium median rugulate.

M_2 Grain *ca.* 25 μm. *Intercolpium edges meridionally striated* **Blackstonia**

M_1 Grain equal to or smaller than 25 μm. *Intercolpium edges with groups of non-meridional striae* . . . . **Cicendia**

D_1 Grain smaller than 25 μm. Striae short.

N_2 *Furrow with coarse spines*. **Aesculus**

N_1 Furrow membrane without ektexine elements.

O_2 Sculpturing of intercolpium indistinct . . . . **Cicendia**

O_1 Sculpturing of intercolpium uniform, distinct.

P_2 Columellae distinctly visible. . . . . **Sempervivum**

P_1 Columellae indistinct–imperceptible . . . . **Sedum**

Table F11.4. Tricolporate. Suprareticulate–foveolate

A_2 *Heteropolar. Pores near the broad end* . . . . **Echium**

A_1 Isopolar.

B_2 Grain peroblate. Foveolate. Furrows very short. Costae colpi heavy. Exine more than 3 μm thick . . . . **Tilia**

B_1 Grain not peroblate. Furrow not very short.

C_2 Grain prolate. Exine more than 2 μm thick. Columellae very coarse.

D_2 *Grain oval.* Columellae uniform . . . .

E_2 Columellae simple. Furrows with broad costae. Margo heavy . . **Euphorbia**

E_1 Columellae branched. Margo indistinct. Grain dimorphic . . **Fagopyrum**

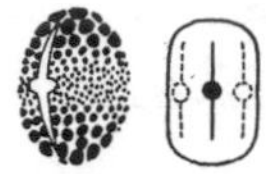

D_1 *Grain compressed oval. Columellae coarser* and more scattered at poles. . . **Polygonum bistorta t.**

C_1 Grains variable. Ektexine less than 2 μm thick. Columellae fine or indistinct.

F_2 *Equatorial limb subangular*. Grain oblate. Polar area very small. Costae transversales and pores distinct . . . . **Rhamnus catharticus**

F_1 Equatorial limb circular or *semiangular*. Grain subsphaerical or prolate. Polar area medium. Costae transversales present but not always very strong. Papilionaceae pp. (cf. Fægri, 1956).

G_2 Reticulum indistinct.

H_2 Grain prolate. Pore distinct, covered by a thin, structureless exine. Incl. *Vicia orobus, V. sepium, V. tenuifolium, Lathyrus montanus*. . . . . **Vicia cracca t.**

H_1 Grain subsphaerical. Pore covered, more or less distinct.

I_3 Grain more than 35 μm. Pore indistinct. Exine thin.

J_2 Furrow with *costae and equatorial bridge* . . . **Medicago sativa**

J_1 Not so . . . . **Medicago falcata**

I_2 Grain more than 30 μm. Exine thin. Incl. *Astragalus arenarius, A. norvegicus* . . . . **Astragalus spp.**

I_1 Grain less than 35 μm. Exine thick.

K_2 Grain 30–35 μm. Columellae distinct . . **Trifolium montanum**

K_3 Grain less than 25 μm. Columellae indistinct . . . . **Trifolium spadiceum**

G_2 Reticulum distinct, some *lumina blocked* . . . .

L_2 Polar area large . . . . **Coronilla emerus**

L_1 Polar area medium. Incl. *Oxytropis*. . . . . **Astragalus t.**

G_1 Reticulum distinct. No lumina blocked.

M_2 *More than 13–15 equatorial brochi* per intercolpium . . . .

N_2 Lumina with free columellae.

O_2 Muri broad. Pore diameter less than $\frac{1}{4}$ of polar axis length . **Astragalus frigidus**

O_1 Muri narrow. Pore diameter larger than $\frac{1}{4}$ of polar axis length **Astragalus penduliflorus**

N_1 No free columellae in lumina.

P_2 Grain more than 20 μm . . . . **Ononis t.**

P_1 Grain smaller than 20 μm . . . . **Astragalus alpinus**

M_1 *Less than 13–15 equatorial brochi* per intercolpium . . . .

Q_2 Meridional limb compressed oval. Furrow with heavy costae. Pore distinct, covered by a thin, almost structureless ektexine.

R_2 Grain *subsphaeroidal* . . . .

S_2 Lumina with perforations. Breadth of muri less than diameter of lumina . . . . **Lathyrus sativus**

S_1 No perforations in lumina. Muri very broad. . . **Lathyrus palustris**

R_1 *Grain prolate* . . . .

T_2 Grain more than 50 μm.

U_2 With distinct perforations in lumina . . . . **Pisum**

U_1 No perforations in lumina . . . . **Vicia faba**

T_1 Grain smaller than 50 μm.

V_2 Lumina with distinct perforations . . . . **Lathyrus maritimus**

V_1 No perforations in lumina. . . . . **Vicia t.**

Q_1 Meridional limb *circular* or more or less *rhomboidal*. Costae colpi variable. Pore covered by ektexine with distinct structure. *Trifolium* t.

W_2 Grain more than 40 μm. Incl. *Trifolium incarnatum, T. striatum.* . . **T. medium t.**

W_1 Grain smaller than 40 μm. Incl. *Medicago lupulina* . . . **Trifolium spp.**

Table F11.5. Tricolporate. Per-reticulate–frustillate

A_3 Reticulum indistinct or frustillate–perforate.

B_2 Columellae united only by their topmost parts. Furrow with equatorial bridge. Cf. Hottonia, Table F7.6. *Mercurialis.*

C_2 Exine more than 2 μm thick. Columellae grouped/irregularly distributed. Furrow membrane not with a row of granules . . . . . . **M. perennis**

C_1 Exine less than 1.5 μm thick. Columellae evenly distributed. Operculum formed by a (single) row of granules. . . . . **M. annua**

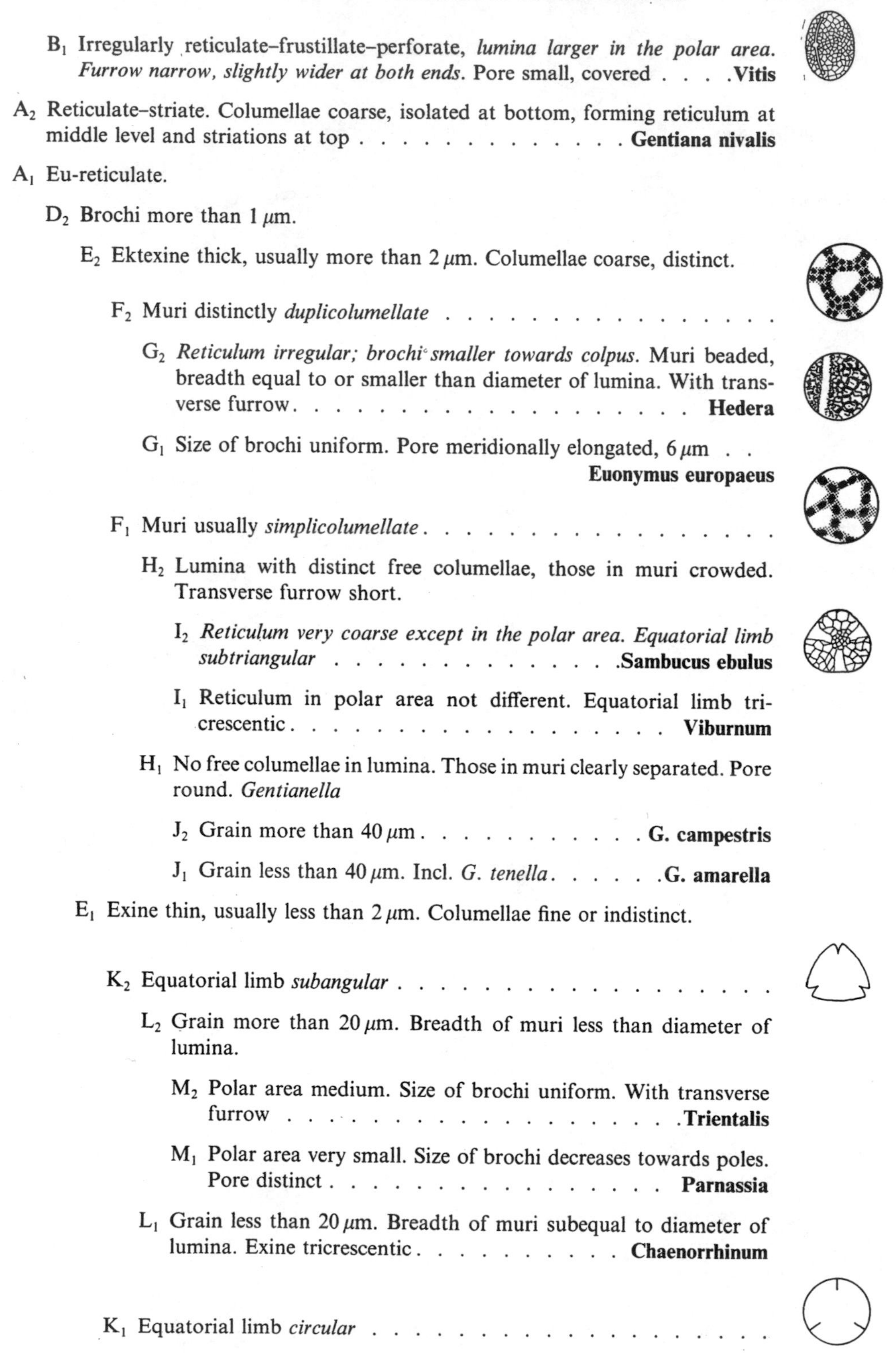

B_1 Irregularly reticulate–frustillate–perforate, *lumina larger in the polar area. Furrow narrow, slightly wider at both ends.* Pore small, covered . . . . **Vitis**

A_2 Reticulate–striate. Columellae coarse, isolated at bottom, forming reticulum at middle level and striations at top . . . . **Gentiana nivalis**

A_1 Eu-reticulate.

D_2 Brochi more than 1 μm.

E_2 Ektexine thick, usually more than 2 μm. Columellae coarse, distinct.

F_2 Muri distinctly *duplicolumellate* . . . .

G_2 *Reticulum irregular; brochi smaller towards colpus.* Muri beaded, breadth equal to or smaller than diameter of lumina. With transverse furrow. . . . . **Hedera**

G_1 Size of brochi uniform. Pore meridionally elongated, 6 μm . . **Euonymus europaeus**

F_1 Muri usually *simplicolumellate*. . . . .

H_2 Lumina with distinct free columellae, those in muri crowded. Transverse furrow short.

I_2 *Reticulum very coarse except in the polar area. Equatorial limb subtriangular* . . . . **Sambucus ebulus**

I_1 Reticulum in polar area not different. Equatorial limb tricrescentic. . . . . **Viburnum**

H_1 No free columellae in lumina. Those in muri clearly separated. Pore round. *Gentianella*

J_2 Grain more than 40 μm. . . . . **G. campestris**

J_1 Grain less than 40 μm. Incl. *G. tenella*. . . . . . **G. amarella**

E_1 Exine thin, usually less than 2 μm. Columellae fine or indistinct.

K_2 Equatorial limb *subangular* . . . .

L_2 Grain more than 20 μm. Breadth of muri less than diameter of lumina.

M_2 Polar area medium. Size of brochi uniform. With transverse furrow . . . . **Trientalis**

M_1 Polar area very small. Size of brochi decreases towards poles. Pore distinct . . . . **Parnassia**

L_1 Grain less than 20 μm. Breadth of muri subequal to diameter of lumina. Exine tricrescentic. . . . . **Chaenorrhinum**

K_1 Equatorial limb *circular* . . . .

N_2 Grain more than 20 μm. Transverse furrow/pore distinct.

O_2 *Transverse furrow long and bent*, almost equatorial . . . . . **Lysimachia nemorosa**

O_1 Not so.

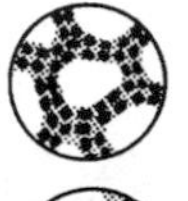

P_2 Distinctly *duplicolumellate. Lysimachia nummularia, L. vulgaris, Anagallis.* . . . . **Lysimachia t.**

P_1 *Simplicolumellate* or columellae indistinct . . . .

Q_2 Muri broad. With transverse furrow.

R_2 Columellae indiscernible. Margo heavy . . . . . **Lysimachia thyrsiflora**

R_1 Columellae distinct. Margo thin . . . . **Anagallis**

Q_1 Muri narrow. No transverse furrow. Incl. *Verbascum, Misopates* . . . . **Scrophularia t.**

N_1 Grain smaller than 20 μm. Transverse furrow indistinct. Incl. *Sambucus niger, S. racemosum* . . . . **Sambucus t.**

D_1 Micro-reticulate; brochi less than 1 μm.

S_2 Furrow slit-shaped. No margo. Pore covered. . . . . **Rumex sect. Acetosa**

S_1 Not so.

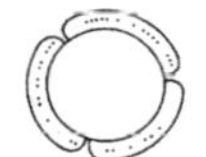

T_2 Endexine as thick as ektexine. Transverse furrow distinct, narrow **Glaux**

T_1 Endexine thin. No transverse furrow. . . . . **Hypericum**

Table F12. Stephanocolporate

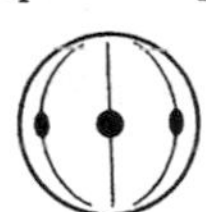

All grains are psilate–scabrate except where otherwise indicated. The transverse furrow in Boraginaceae is blunt-ended, pore-like. The number of furrows varies individually; numbers indicated represent averages.

A_2 More than 4 furrows. Cf. also Table F19.

B_2 Grain larger than 20 μm.

C_2 Transverse furrows present.

D_2 (Endo-)colpus aequatorialis present. No costae aequatoriales

E_2 More than 10 furrows. *Grain barrel-like* (depresed oval) *with an equatorial ridge*. . . . .

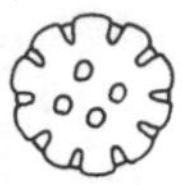

F_2 *Polar area with several isodiametric lacunae* . . . . **Polygala**

F_1 *Polar area coarsely vermiculate*, no round lacunae . . . . . **Utricularia**

E_1 *9–10 furrows*. Grain spherical. *No equatorial ridge* . . . .**Borago**

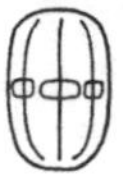

D_1 Mostly 8 meridional furrows. Grain compressed oval. *Transverse furrows distinctly separate, pore-like*. . . . .**Symphytum**

C_1 Transverse furrows absent.

G_2 5 furrows. Grain more than 40 μm, depressed oval to polyedric. Polar area with an intectate field of coarse columellae. Cf. M_2. . . . . . . **Viola arvensis**

G_1 6–7 furrows.

H_2 Reticulate. *Grain spherical*. . . . . **Pinguicula**

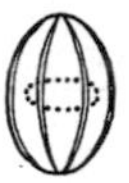

H_1 Psilate. *Grain prolate. Operculum very broad* . . . **Sanguisorba officinalis**

B_1 Grain smaller than 20 μm, *compressed oval* . . . .

I_2 Meridional furrows very short, hardly discernible. (Protruding poroid area with free granules ~ operculum) . . . . **Buglossoides arvensis**

I_1 Meridional furrows distinct.

J_2 Costae aequatoriales present. Sculpturing of equatorial belt almost invisible in LM. . . . .**Pentaglottis**

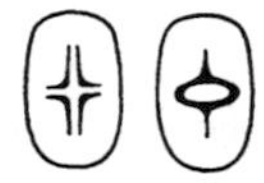

J_1 Costae transversales almost contiguous. Grain *sub-heterocolpate (three furrows with cross-shaped poroid area)* . . . . **Cynoglossum**

A_1 Four furrows.

K_2 Grain larger than 40 μm.

L_2 Transverse furrow distinct. *Grain oval* . . . . **Anchusa arvensis**

L_1 No transverse furrow.

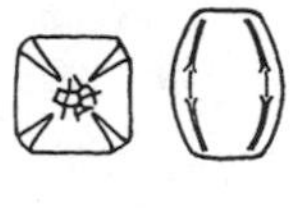

M_2 *Grain depressed oval* (barrel-shaped) *to polyedric*. Polar area with an *intectate field of coarse columellae* . . . . **Viola tricolor**

M_1 Grain oval. Exine structure uniform . . . . . . **Viola riviniana t.**

K_1 Grain smaller than 40 μm.

N_2 *Grain compressed oval.* Equatorial exine reticulate. Transverse furrow present

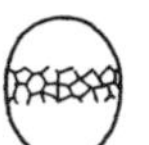

O_2 Exine thinner in polar area. *Equatorial reticulum coarse* . . . . **Nonea**

O_1 Exine thickness uniform.

P_2 Grain more than 20 μm.

Q_2 Meridional furrows with *margo of coarse columellae* **Pulmonaria**

Q_1 Meridional furrows not with coarse columellae . . . . **Anchusa officinalis**

P_1 Grain smaller than 20 μm. Sculpturing of equatorial belt nearly invisible in LM. Costae equatoriales present . . . . **Pentaglottis**

N_1 *Grain constricted oval.* No transverse furrows. *Pores* (but not constriction) *nearer one pole.* Incl. *Buglossoides purpureocoeruleum* . . . . **Lithospermum officinalis**

Table F13. Pericolporate

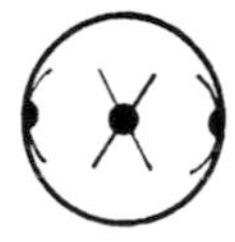

Within the area we know of no exclusively pericolporate pollen types. Tri- or stephanocolporate grains sometimes appear with more apertures than normal, and are then usually imperfectly isometric. The most important cases are those set out below.

A_2 Echinate, tectate, exine thick. Cf. Table 11.2. . . . . **Compositae asteroideae**

A_1 Psilate or reticulate.

B_2 Minutely microechinate. Perforations visible in LM. Transverse furrow sometimes present. Pore often missing, furrow membrane being ruptured **Polygonum oxyspermum t.**

B_1 Psilate with perforations and/or reticulate. Furrow slit-shaped. Pore covered. Cf. Table F11.1 . . . . **Oxyria, Rumex**

Table F14. Diporate

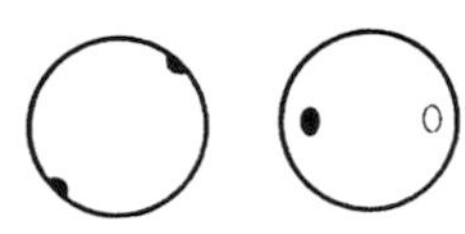

Colchicum is regularly, *Betula, Myriophyllum*, etc. are exceptionally diporate. Cf. also Table F6.

Table F15. Triporate

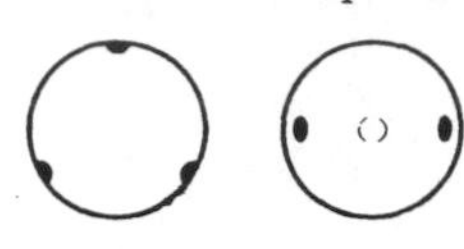

A_2 With *large, cylindrical vestibulum,* more than 10 μm deep. Onagraceae . . .

B_2 Grain 50 μm or larger. Incl. *Epilobium* . . . . **Chamaenerium t.**

B_1 Grain smaller than 50 μm . . . . **Circaea**

A_1 Vestibulum smaller, conical, or absent.

C_2 Distinctly echinate.

D_2 Grain more than 60 μm. Exine thick, more than 4 μm. Dipsacaceae.

E_2 Spines *short, broad-based, scattered.* Pore with *well developed anulus* **Dipsacus**

E_1 Spines *slender, dimorphic*, crowded. *Anulus indistinct* . . . **Knautia**

D_1 Grain smaller. Spines uniform, scattered: Campanulaceae.

F_2 Grain larger than 25 μm. Incl. *Phyteuma* . . . . . . **Campanula t.**

F_1 Grain smaller than 25 μm. . . . . **Jasione**

C_1 Psilate–scabrate.

G_2 Equatorial limb almost circular, without protruding pore edges.

H_3 Grain more than 20 μm. Micro-sculpturing faintly rugulate with minute spines. Columellae very distinct, scattered . . . . **Celtis**

H_2 Grain more than 20 μm. Micro-sculpturing apparently dotted **Cannabis**

H_1 Grain smaller than 20 μm. Psilate. Columellae delicate.

I_2 After a slight rise the tectum bends down, forming *a sunken pore* . **Humulus t.**

I_1 Tectum rises slightly at *pore, not sunken*. . . . .

J_2 With minute projections. Exine very thin. *Urtica.*

K_2 Spinules scattered . . . . **U. dioica**

K_1 Spinules closer together, more regularly distributed . . . . **U. urens**

J_1 Projections slightly coarser. Exine slightly thicker **Parietaria**

G_1 Equatorial limb semiangular or with protruding pore edges.

L_2 Micro-sculpturing and structure densely and *regularly dotted*. Juglandaceae, Myricaceae. . . . .

ca 5µm

M_2 Pores *not exactly in the equatorial plane*. Exine thick. Grain 30–50 µm **Carya**

M_1 Pores in the equatorial plane.

N_2 Grain more than 20 µm. Pore *broadly conical*, interior walls of pore chamber with irregular elements. . . . . **Myrica**

N_1 Grain smaller than 20 µm.

O_2 With 2 long pseudocolpi . . . . **Platycarya**

O_1 No pseudocolpi . . . . **Engelhardtia**

L_1 Micro-sculpturing and structure *more or less rugulate*, with scattered minute spinules on top of low vallae (ph!). Betulaceae . . . .

ca 5µm

P_2 Equatorial limb circular with protruding pores. Exdexine and tectum diverge in the pores, forming a vestibulum.

Q_2 *Vestibulum distinct. Tectum sharply thickened* at the pore. Cf. Table F11.1: V_2 *Ludwigia* . . . . **Betula**

Q_1 *Vestibulum not distinct*. Tectum not appreciably thickened at the pore. Incl. *Carpinus*. . . . . **Ostrya t.**

P_1 Equatorial limb semiangular, pore not protruding. Endexine and tectum do not diverge in the pore. *Depth of pore not much greater than thickness of the regular exine* . . . . **Corylus**

Table F16. Stephanoporate

A_3 Echinate. Spines distinct, regularly distributed. Campanulaceae.

B_2 Grain larger than 25 µm. Incl. *Phyteuma* . . . . **Campanula t.**

B_1 Grain smaller than 25 µm. . . . . **Jasione**

A_2 *Coarsely and softly rugulate* or suprareticulate. Peroblate . . . . **Ulmus**

A_1 Psilate, scabrate or very delicately rugulate.

C_2 Grain globular. Equatorial limb circular. Pores not distinctly protruding

D_2 Rugulate, spinulose. . . . . **Celtis**

D_1 Psilate. Columellae delicate.

E_2 Grain more than 20 μm . . . . **Cannabis**

E_1 Grain 20 μm or smaller.

F_2 *Pore sunken.* Cf. Table F15: I_2. . . . . **Humulus t.**

F_1 *Pore not sunken.*

G_2 Exine very thin, structure very delicate. Cf. Table F15 . . **Urtica**

G_1 Exine thicker, structure coarser. . . . . **Parietaria**

C_1 Grain oblate. Equatorial limb more or less angular and/or pore edges protruding.

H_2 Neighbouring *pores (anuli) connected with thickened bands (arci)*. Pore with thick anulus . . . . **Alnus**

H_1 No arci.

I_2 *Pores meridionally elongated* with endexinous anuli. *Myriophyllum* . . .

J_2 *Equatorial limb irregular* (basically ellipsoid). Grain triaxial. *Pores grouped, strongly protruding* . . . . **M. alterniflorum**

J_1 Equatorial limb regular. Pores equidistant.

K_2 Anulus distinct, *protruding* . . . . **M. spicatum t.**

K_1 Anulus externally *flat*, not protruding . . . . . **M. verticillatum**

I_1 Exine thickness uniform. Pores circular, without endexinous anuli.

L_2 Equatorial limb semiangular. Micro-sculpturing dense and regular. **Platycarya**

L_1 *Equatorial limb more or less circular.* Grain more than 40 μm. Micro-sculpturing faintly rugulate with minute spinules on vallae . **Carpinus**

Table F17. Periporate: Main key to subsections

Verrucate–echinate . . . . Table F17.1

Vermiculate–rugulate or reticulate . . . . Table F17.2

Psilate–scabrate . . . . Table F17.3

Table F17.1. Periporate. Verrucate–echinate

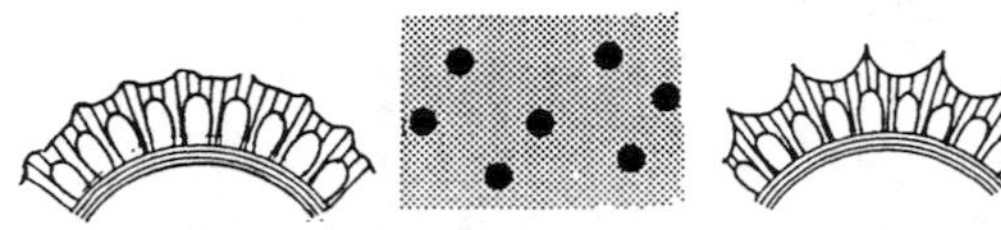

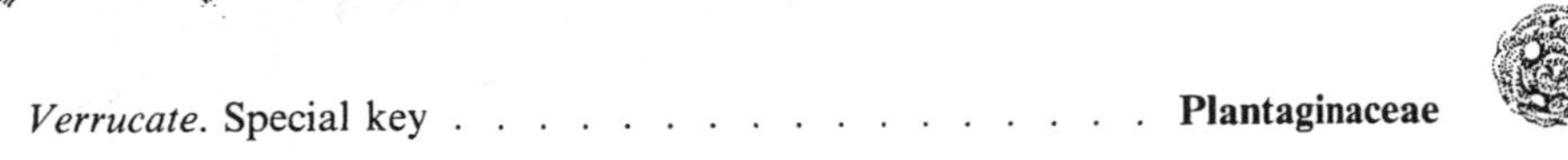

A_2 *Verrucate*. Special key . . . . **Plantaginaceae**

A_1 Echinate.

B_2 *Ca.* 100 pores. Grain large, spines strong . . . . **Malvaceae**

B_1 8–20 pores.

C_2 Pores without anuli.

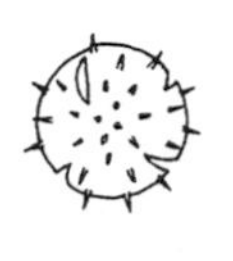

D_2 *Pores sharply delimited, elongated*, pore membrane without spines **Koenigia**

D_1 Pores less distinct, isodiametric; pore membrane with spines. Alismataceae.

E_2 *Spines with broad, compressed base* . . . . . . **Sagittaria**

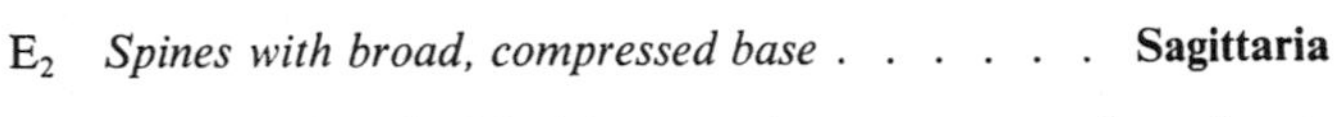

E_1 Spines with cylindrical base, on low verrucae .**Luronium t.**

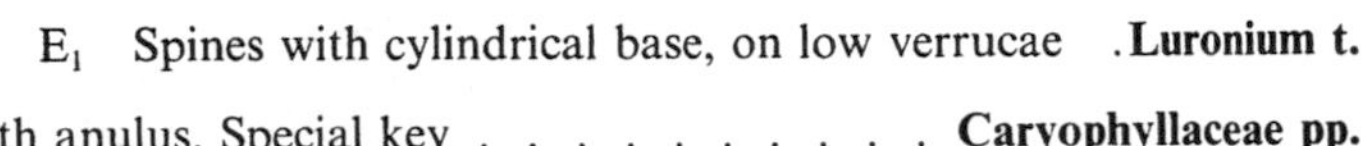

C_1 With anulus. Special key . . . . **Caryophyllaceae pp.**

Table 17.2. Periporate. Vermiculate–rugulate or reticulate

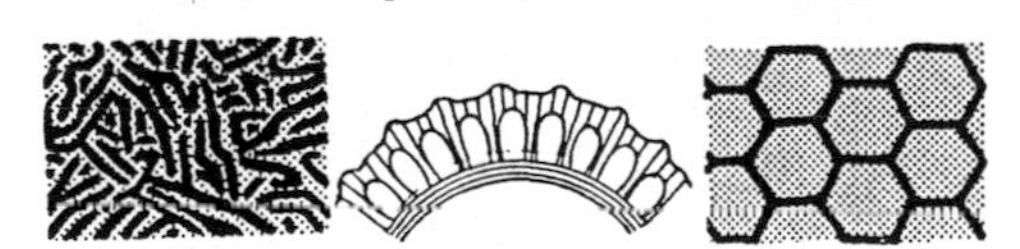

A_2 Vermiculate–rugulate.

B_2 Coarse columellae. More than 50 pores . . . . **Polemonium**

B_1 Columellae fine. Less than 50 pores . . . . **Buxus**

A_1 Reticulate.

C_2 Pores with anuli.

D_2 Columellae evenly distributed. Breadth of muri subequal to diameter of lumina . . . . **Liquidambar**

D_1 Not so. Special key . . . . **Caryophyllaceae pp.**

C_1 Anuli absent or indistinct.

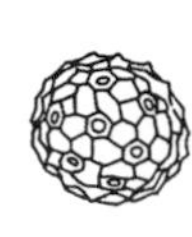

E_2 *Reticulum* regular, *very coarse. Bottom of lumina with a pore* and/or covered with granules. *Polygonum foliosum, P. hydropiper, P. lapathifolium, P. persicaria, P. tomentosum* . . **Polygonum sect. Persicaria**

E_1 Not so.

F_2 Pore indistinct, included in a brochus. Brochi with pore slightly larger than those without. Columellae elongate, club-shaped, fused at top, forming reticulum **Daphne**

F_1 Pore distinct, larger than brochi.

G_2 Edge of pore distinct. **Liquidambar**

G_1 Edge of pore not distinct . . . . **Buxus**

Table F17.3. Periporate. Psilate–scabrate

TECTUM SOLIDUM TECTUM PERFORATUM

A_2 *Grain irregular, usually pear-shaped. Pores indistinct.* Special key . . **Cyperaceae**

A_1 Grain regularly ellipsoidic.

B_2 Microechinate, dotted.

C_2 Columellae fine, evenly distributed. *Pores aggregated towards one pole* . **Juglans**

C_1 Columellae coarse and/or irregularly distributed. Pores evenly distributed.

D_2 Anulus distinct. Special key . . . . **Caryophyllaceae pp.**

D_1 *Anulus absent* or indistinct . . . .

E_2 *Surface undulating.* Special key . . . . **Plantaginaceae**

E_1 Surface even.

F_2 Grain more than 25 μm. Indistinctly intrareticulate.

G_2 Grain sub-polyhedric. *Exine between pores distinctly crescentic.* Pores round . . . . **.Alisma graminea**

G_1 Grain (sub-)globular. Exine not distinctly crescentic. Pores irregular, indistinct . . . . **Papaver argemone**

F_1 Grain 10–25 μm. No intrareticulum. . . . . . . **Thalictrum**

B_1 Not microechinate, dotted.

H_2 *Areas without tectum,* surrounding and connecting pores . . . . **Ribes**

H_1 Tectum not reduced.

I_2 Grain large (70–80 μm). Exine thick. . . . . **Calystegia**

I_1 Grain smaller.

J_2 Anulus present.

K_2 Anulus distinct, but often narrow and without structure.

L_2 Tectum without perforations. Number of pores usually higher than 50. Incl. Amaranthaceae . . . . **Chenopodiaceae**

L_1 Tectum perforate (indistinct in *Sagina*). Number of pores lower than 50.

M_2 6–12 large pores. *Tectum very thick. Anulus prominent.* Columellae very fine and short. . . . . **Fumaria**

M_1 Not so. Special key . . . . **Caryophyllaceae pp.**

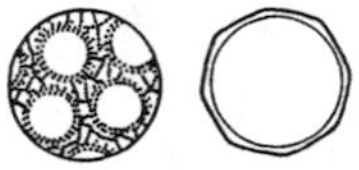

K_1 Anulus diffuse. *Columellae widely dispersed in interporia, dense on anuli. Grain usually polyhedric.* Intrareticulate. Alismataceae. . . . .

N_2 Grain less than 25 μm . . . . **Baldellia**

N_1 Grain more than 25 μm . . . . **Alisma plantago-aquatica**

J_1 No anulus.

O_2 Grain 10–25 μm. 4–12 pores.

P_2 Grain rounded. Distinct minute spines . . . . **Thalictrum**

P_1 Grain sub-polyhedric. Indistinctly dotted . . **Polycnemum/Paronychia t.**

O_1 Grain *ca.* 10 μm. No spines. 4 pores . . . . **.Herniaria**

Table F18. Syncolpate

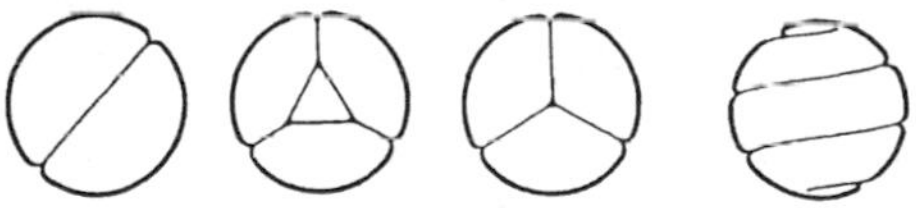

Colpi sometimes fuse at one pole also in grains not considered, here, syncolpate.

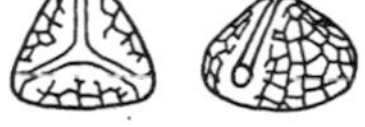

A_2 *Heteropolar*. Colpoid *furrows joining* at one pole. *Coarsely reticulate* **Thesium**

A_1 Isopolar.

B_3 Furrows meridional.

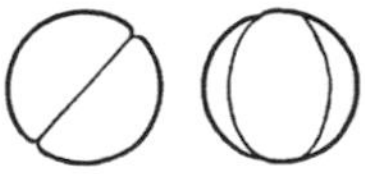

C_2 *Two furrows, fused to a ring*. Grain usually split, edges of colpus ragged. Psilate perforate–perfossulate. Cf. *Juniperus*, Table F2. . **Pedicularis**

C_1 More than two furrows.

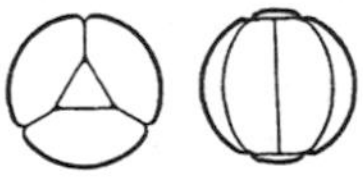

D_2 *Furrows bifurcate*, fused with neighbouring furrows, delimiting a polar field . . . .

E_2 Grain larger than 30 μm. Coarsely striated. Polar field large **Nymphoides**

E_1 Grain 10–25 μm, dimorphic. Minutely per-reticulate. *Primula*.

F_2 3–4 furrows. Size 10–15 μm . . . . **P. farinosa**

F_1 4–6 furrows. Size 20–25 μm . . . . **P. scandinavica t.**

D_1 *Furrows 3, not bifurcate* . . . .

G_2 Grain microverrucate . . . . **Loranthus**

G_1 Grain psilate. Short transverse furrow present. Incl. *Eucalyptus*. . . . **Myrtaceae**

B_2 *Furrows spirally oriented* . . . .

H_2 Grain with scattered spinules, not perforate . . . . **Eriocaulon**

H_1 Grain reticulate, heterobrochate. Exine usually splits . . . . . **Berberis**

B_1 Peri-syncolpate. *Exine divided into angular plates, usually falling apart* . . . .

I_2 Grain reticulate, heterobrochate . . . . **Berberis t.**

I_1 Grain pitted (perforated) . . . . **Mahonia t.**

Table F19. Heterocolpate

There are some grains in which there is a certain difference between furrows. However, we have here restricted ourselves to dealing with grains with colpoids or regular furrows without pores alternating with such that possess pores.

A_2 Tricolporate with meridional folds at both sides of the regular furrow. The folds may fuse at the pore(s). Cf. also Anthyllis, Table F11.11 . . . . . . **Verbena**

A_1 *Three furrows with pores alternate with three similar without* . . . .

B_2 Furrows approximately equally long. Grains dumbbell-shaped (constricted oval). Polar area index large, above 0.5. Boraginaceae.

C_2 Grains smaller than 10 μm. Incl. *Lappula* . . . . **Myosotis t.**

C_1 Grain more than 10 μm. . . . . **Mertensia**

B_1 Aporate furrows generally shorter than the porate ones. Polar area small. Grain not constricted. Lythraceae.

D_2 *Aporate furrows distinctly broader than the porate ones.* . . . . **Peplis**

D_1 Aporate furrows not broader than the porate ones. Grains dimorphic **Lythrum**

Table F20. Fenestrate

(A_3 *Non-meridional tectum-less areas* surrounding and connecting pores. . **Ribes**

A_2 Grains psilate.

B_2 *Polar area with small lacunae.* More than 8 furrows . . . . . **Polygala**

B_1 Polar area without lacunae. *Three meridional lacunae* alternate with porate furrows . . . . **Peplis**

A_1 *Grains* echinate, *lophate* . . . . **Compositae Cichorioideae**

Caryophyllaceae (periporate)

cf. Chanda, 1962; Fredskild, 1967
The section *Sperguleae* has colpate grains, cf. Table F7.1.

A_2 Grain *ca.* 10 µm. 1–8 pores. Sect. Paronychieae: *Herniaria, Illecebrum, Corrigiola* **Herniaria t.**

A_1 Grain larger than 10 µm.

B_3 Tectum with scattered distinct perforations similar in number to the columellae and spines, or per-reticulate. Sections Lychnidae, Diantheae; also *Honckenya*

C_2 Anulus broad. Usually 12 pores. Most species incl. *G. fastigiata, G. repens* . . . . **Gypsophila pp.**

C_1 Anulus less broad.

D_2 Less than 20 pores. Incl. *Dianthus, Tunica, Saponaria, Silene* pp. **Dianthus t.**

D_1 More than 20 pores. Incl. *Lychnis, Melandrium, Agrostemma, Viscaria, Silene* pp., *Honckenya* . . . . **Lychnis t.**

B_2 Tectum with indistinct perforations, distinctly scabrate. Pores small, more than 24. Incl. Chenopodiaceae. . . . . **Sagina**

B_1 Tectum with minute perforations that are much more numerous than columellae and spines. Section Sclerantheae, Alsineae, not *Honckenya*

E_2 Pores (+anuli) in depressions, surrounded by a zone without big columellae; the rest of the grain forming a hexagonal pattern of ridges. Pores 12 or about 20.

F_2 Ridges scabrate or with indistinct sculpturing . . . **Scleranthus**

F_1 Ridges echinate . . . . **Stellaria holostea**

E_1 Not so. Pores 6–*ca.*40. Incl. *Cerastium, Stellaria, Malachium, Arenaria* pp., ... **Cerastium t.**

Cyperaceae

Grains with perforate tectum with lacunae, i.e. areas where the tectum is broken up into small frustillae, usually forming an areolar sculpturing. The form and arrangement of lacunae is of great diagnostic value. They are most easily seen in strongly stained preparations. Columellae are sometimes clearly visible, each frustillum borne by several columellae. One or two of the lateral lacunae may be smaller than the others; the indices given below do not apply to these.

A_3 Lateral lagunae oblong, length twice as long as the breadth or more. Grain ovoid or pear-shaped.

B_2 Length of lateral lacunae more than half that of the grain . . . .

C_2 Grain more than 40 μm. Incl. *Heleocharis* spp., *Rhynchospora fusca* **Schoenus t.**

C_1 Grain less than 40 μm. Incl. *Heleocharis palustris* . **Schoenoplectus t.**

B_1 Length of lateral lacunae less than half of that of the grain. Grain more than 40 μm . . . . **Scirpus maritimus**

A_2 Lateral lacunae more or less circular or slightly elongated. Grain ovoid or pear-shaped.

D_2 2–3 small lacunae, length less than 20% that of the grain. Grain more than 40 μm, the narrow end usually projecting like a beak or a finger **Cladium mariscus**

D_1 4–6 lateral lacunae, their length normally more than 20% of that of the grain. Grain smaller than 40 μm, exceptionally (*Carex hirta*) 40–50 μm. The narrow end rounded: *Carex* type. . . . .

E_2 Length of lateral lacunae 20–35% of that of the grain. Incl. *Carex, Scirpus silvaticus* . . . . **Dulichium t.**

E_1 Length of lateral lacunae 30–50% of that of the grain. Incl. *Trichophorum* . . . . **Eriophorum t.**

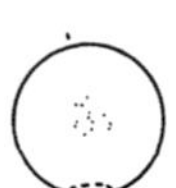

A_1 Lacunae irregular, rather large. Grain almost isodiametric, less than 30 μm. Structure rather coarse, columellae distinct . . . . . . **Rhynchospora alba t.**

Gramineae

Owing to the great importance of grasses, both in natural plant communities and in cultivated areas, and also the importance of certain taxa as cultivation indicators, it would be highly desirable to be able to identify grasses to species. However, the paucity of distinguishing characters sets very narrow limits. The Gramineae pollen type is easily recognized, but also very monotonous. The characters that have been most frequently used are quantitative: diameter of grain, of pore and of anulus. Like all such characters their value is limited, especially as size seems to change with time. By statistical methods it is possible to differentiate between the pollen of various taxa

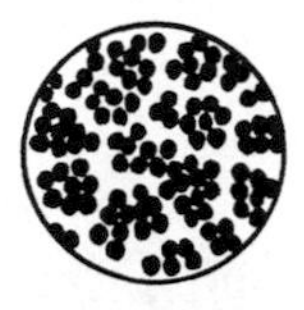

Fig. 12.5. Single- and double-grain structure in grass pollen.

if there is a sufficient number and no admixtures, but since the variation curves overlap, identification of individual grains remains impossible or doubtful.

When using data from the literature one must take into consideration that some refer to glycerol (jelly) preparations (e.g. Beug, 1961), others to silicon preparations (e.g. Andersen, 1978). For the same species, conversion factors vary between 1.1 and 1.3 for grain size and between 1.1 and 1.5 for pore data. Inasmuch as they have been used in the key below, Andersen's data have been modified by a factor of 1.2.

Grass pollen sometimes possess a soft, verrucate sculpturing, which is, however, difficult to observe in LM. It is not known if this is an artifact of preparation or a specific character, and it has so far not been used for diagnostic purposes.

In contrast, the structure shows differential patterns, which can be seen with adequate optical equipment, and which in northwest European grasses vary between two main types. In the one, *single-grain* structure, columellae are individually free, more or less crowded and more or less aggregated. In the *double-grain* structure some or all columellae are united into small frustillae, cf. Fig. 12.5.* As has been pointed out by Beug (1961) one can also distinguish between even distribution and aggregation of free columellae. In addition, columellae may form more or less distinct reticulae. In spite of a fairly extensive literature on the topic, a modern reconsideration of grass pollen morphology is overdue.

This key chiefly depends on the use of quantitative characters, the weakness of which has been mentioned. With few exceptions identification is subject to statistical uncertainties and even the comprehensive groups of the key are not too well defined. Measurements all refer to glycerol preparations.

A_3 Grain more than 60 μm . . . . **Zea mais**

A_2 Grain between 40 and 60 μm: Cereal type.

B_2 P/E ratio higher than 1.25. Pollen distinctly scabrate. Pore lateral . . . . **Secale cereale**

B_1 P/E ratio less than 1.25. Pore (sub-)apical.

C_2 Anulus diameter larger than 12 μm.

D_2 Exine with single-grain structure. . . . . **Triticum t.**

D_1 Exine with double-grain structure . . . . **Avena type.**

C_1 Anulus diameter between 10 and 12 μm: *Hordeum* type.

E_2 Grain more than 44 μm. Outer limit of anulus indistinct . . . . **Elymus s.l.**

E_1 Grain less than 47 μm.

* The terms *punctae, maculae* and *areolae*, which have been used in previous editions, are redundant and should be dropped now that improvement of optical equipment permits unequivocal recognition of the true nature of these structures. They are easily perceived, especially after staining and/or in swollen grains.

F_2 Outer limit of anulus sharp. Incl. *Hordeum, Agropyrum junceum* . . . . **Triticum t.**

F_1 Outer limit of anulus not sharp . . . . **Agropyrum t.**

A_1 Grain smaller than 40 μm. Anulus diameter less than 10 μm: wildgrass t.

G_2 Grain smaller than 26 μm. Columellae indistinct . . . . . . **Phragmites.**

G_1 Grain between 26 and 40 μm.

H_2 Double-grain structure. Columellae distinct. Incl. *Nardus, Cynosurus* **Dactylis t.**

H_1 Single-grain structure, some aggregation opposite the pore. Incl. most wild-grasses . . . . **Festuca t.**

Plantaginaceae

Cf. Bassett and Crompton, 1967; S. T. Andersen, 1970

A_2 Pore with distinct anulus.

B_2 Pore operculate.

C_2 More than 8 pores; grain distinctly micro-echinate . . **P. lanceolata**

C_1 Less than 8 pores; not distinctly micro-echinate, though minute punctae may be seen in Ph . . . . **P. coronopus**

B_1 Pore rarely operculate, but with isolated granules. 5–9 pores. Distinctly micro-echinate . . . . **P. maritima, P. alpina**

A_1 Pores without distinct anulus, with granules, no operculum

D_2 Grain normally more than 30 μm. Pores numerous, sharply delimited . **Littorella uniflora**

D_1 Grain less than 30 μm.

E_2 Pore sharply delimited, 5–9. Distinctly micro-echinate. . **P. maritima**

E_1 Pore not sharply delimited.

F_2 Punctae indistinct or invisible . . . . **P. major**

F_1 Punctae distinct (Ph).

G_2 Verrucae rather indistinct or very small. Pores numerous **P. tenuiflora**

G_1 Verrucae distinct.

H_2 Verrucae very coarse. . . . . **P. media**

H_1 Verrucae less coarse . . . . **P. montana**

Rosaceae

Cf. Eide (1981).

A_2 With tectate operculum.

B_2 Psilate-scabrate-echinate, oblate-globular. *Sanguisorba.*

C_2 Operculum narrow, furrow very short . . . . **S. minor**

C_1 Operculum as broad as the intercolpium, furrow long . . **S. officinalis**

B_1 Striate.

D_2 Vallae coarse. Perforations not visible.

E_2 Vallae rarely parallel. Operculum extremely narrow, pore always distinct. . . . .**Sibbaldia procumbens**

E_1 Vallae often parallel. Operculum narrow to broad, pore not always distinct.

F_2 Vallae broad, diffuse.

G_2 Prolate. Colpi constricted, ends rounded . **Comarum palustre**

G_1 Not prolate. Colpi not constricted, ends acute.

H_2 Polar area flattened. . . . . **Fragaria vesca**

H_1 Polar area not flattened . . . .**Potentilla spp.**

F_1 Vallae narrow, distinct.

I_2 Colpus ends rounded . . . . **Fragaria viridis**

I_1 Colpus ends acute . . . .**Potentilla spp.**

D_1 Vallae fine. Tectum perforate.

J_2 Vallae transversal, extremely fine. Operculum extremely narrow. . . . **Agrimonia**

J_1 Vallae submeridional to subtransversal. Operculum short. . . **Rosa**

A_1 No tectate operculum, but colpus membrane occasionally with loose columellae.

K_4 Clavate–baculate–echinate . . . . **Rubus chamaemorus**

K_3 Microechinate–echinate. No perforations . . . . **Filipendula**

K_2 Rugulate–microstriate/psilate

L_3 Rugulate–microstriate. No costae colpi or granules on furrow membrane **Mespilus germanica**

L_2 Rugulate to psilate, costae colpi and granules usually present . . . . . **Cotoneaster**

L_1 Psilate. Polar projection usually subtriangular. . . . . . . **Alchemilla**

K_1 Striate.

M_2 Vallae coarse. Colpus with distinct bridge . . . . **Geum**

M_1 Vallae fine.

N_2 *Vallae paired.* Striation weak . . . . **Crataegus**

N_1 Vallae simple.

O_2 Vallae indistinct. Perforations large. . . . . **Rubus**

O_1 Vallae distinct.

P_2 Vallae high, frequently short . . . . **Rosa**

P_1 Vallae low and narrow.

Q_2 Vallae straight, moderately branched.

R_2 Perforations distinct. Vallae without dense structure . **Sorbus**

R_1 Perforations may be unobservable. Vallae with dense structure **Prunus**

Q_1 Vallae curved, branched, anastosmosing.

S_2 Vallae short, strongly curved . . . . **Prunus padus**

S_1 Vallae moderately curved.

T_2 With costae colpi, size *ca.* 27 μm . . . . **Malus**

T_1 No costae colpi. Size *ca.* 22 μm . . . . **Dryas**

13. Morphological glossary

Definitions given below are mainly those that have been used in this book and make no pretence of completeness, nor do the etymological derivations. For a more general discussion reference is made to Kremp (1965). Plural nominative is indicated after the solidus. Note especially the plurals of the fourth declination: arcus/arcus. Adjectival forms and derivatives: *tri*-colpate, *semi*-lobate, *a*-tectate, rug*ula*, bacul*oid* are usually included under the main word (noun).

Even though modern microscopes permit an experienced worker to distinguish much clearer today than when this book was first published, we have kept the limit between distinctly visible (discernible) and indistinct at *ca.* 1 μm.

The terms given below are partly general terms used in a specific way, partly technical terms coined exclusively for palynomorphology.

above. Further away from the centre of the grain.

anulus (anus = ring). Thicker or thinner part of the exine, surrounding a pore. Annulus (annus = year) is a wrong derivation.

aperture/rate (apertus = open). Openings or thinner parts of the exine usually placed in a regular geometric pattern. Provide exits for the pollen tube on its emergence and make possible changes of the size of grains with changing water content. The term *omniaperturate* is logically impossible, cf. p. 211. Like the old nursery rhyme: unless there is dough around it, there is no hole for the doughnut. *Inaperturate*: without visible apertures.

apiculate. With protracted polar areas (cf. Fig. 11.3).

apocolpium. The polar area beyond the ends of colpi (and margines, if any) (cf. Fig. 11.13).

arcus/us (= bow). Thickened bands connecting the anuli in some porate grains.

aerolate (dim. of area). Sometimes used as a synonym for frustillate (q.v.). In this book used in a general sense. Cf. key Gr.

baculum/-a (= rod). Radially elongated projection (height ⩾ 2 × diameter) with blunt head. In this book not used for sculpturing elements; see columella.

beaded. Ridges or muri the tops of which are formed by discrete heads of columellae.

below, beneath. Nearer the centre of the grain.

blocked (lumina). Some lumina (q.v.) are apparently filled up with exinous (tectal) material.

bordered. Apertures surrounded by exine of a special character, including anulate and marginate apertures.

brochus/-i, brochate (Gr. = noose). The unit of a reticulum measured from the middle of adjacent muri (cf. Fig. 12.3).

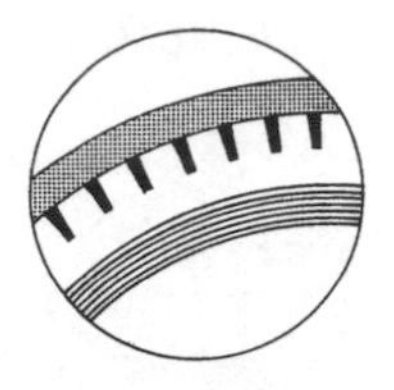

Fig. 13.1. Cavate exine. Endexine: concentric lines; columellae: black; tectum: shaded.

calymmate (Gr. = covered). Tetrad with a common exine of all grains. In *acalymmate* tetrads the individual grains have developed exines.
cavate (cavum/-a = cave). A hollow area between tectum and foot layer/endexine not traversed by columellae, probably formed by the rupturing of columellae and splitting apart of the layers of the exine. Broken columellae may adhere to the one or the other side (cf. Fig. 13.1; cf. *vesiculate*).
clava/-ae; clavate (=rod). Baculate (q.v.) elements constricted at base.
columella/ae, columellate. (dim. of columna = column). Rod-like structural elements between foot-layer/endexine and tectum. In intectate grains equal to baculum which is the corresponding sculptural element. Usage of the two words is confused both between authors and within the same author. Columella is the general term, baculum should be restricted to the sculptural element (cf. Fig. 11.4).
colpoid (colpus-like, q.v.). Used in a general sense, but also for a colpus-like exine feature, a fold, that does not penetrate the exine (cf. also *pseudocolpus*).
colporate. Aperture with different outer and inner geometry, usually poroid inside, colpoid outside. Usually only one pore per furrow.
colpus/-i, colpate. An aperture (q.v.), the length of which is greater than the width and ends usually acute. The term, which has become a standard, is unfortunate as its linguistic derivation is completely wrong. It was introduced by Erdtman (1943a) in feminine form (colpa/ae) which does not seem to exist anywhere. It has later been given masculine form, referring to Greek: kolpos. Unfortunately, this word means something that bulges out: womb, bosom, which might indicate the intine/cytoplasm bulging out of the aperture when a living grain is wetted, but evidently this is not what Erdtman (1943) had in mind. A Latin stem *colp-* does not exist. Wodehouse used the adjectival term, which is very apt for the wetted grain (cf. Wodehouse, 1935, PI. IX). For the aperture he consistently used the term *(germinal) furrow*, which is too narrow in relation to the potential functional of the aperture.
concordant pattern. In a tectate grain: the pattern formed by columellae is identical with the sculpturing (cf. Fig. 11.8).
costa/ae (=rib). Thickened edges accompanying apertures: *c. pori* = anulus. *c. transversales, c. aequatorialis*, edging the corresponding colpi.
crescentic. See *tricrescentic.*
crest (crista = cockscomb, etc.). An elongated sculpturing element, usually with a sharp upper edge, frequently echinate, height/width $\geqslant$ 1.
cristate. See crest.
discordant pattern. In a tectate grain: the pattern formed by the columellae differs from that of the sculpturing (cf. Fig. 11.8).
distal. The part of the grain (supposed to have been) on the outer side of the tetrahedral tetrad.
duplicolumellate. With columellae standing in pairs, especially two rows in muri (cf. Fig. 11.9).
dyad, see tetrad.
echinate (echinus = sea urchin). With conical, acute, sharp, sculptural elements. The term *echinus* for the individual element is linguistically indefensible: *spine.*
ectine. Obsolete for ektexine q.v.
ektexine. Outer part of the exine (cf. Fig. 11.4).
ekto-, see *endo-.*
endexine. Inner part of the exine (cf. Fig. 11.4).
endine. Obsolete for endexine, q.v.
endo-. Features of the inner part of the exine, e.g. *endopore*, pore in the endexine conform with a pore or a colpus in the ektexine. *Endo-crack*: rupture(?) in the endexine.
equatorial. See *transverse. Equatorial bridge*: a strip of tectum material crossing the colpus in the equator.
exine. Outer layer of the pollen grain, enclosing intine and cytoplasm (cf. Fig. 11.4). *Exine grenue*: exine in which the protosporopollenine (q.v.) appears as granules, not on lamellae.
exitus (=departure). Any place on the surface of the grain supposed to be preformed for the penetration of the wall by the emerging pollen tube: the function of a pore, a colpus, a thin area, a papilla.
exospore. Corresponds functionally, in cryptograms, to exine in phanerograms.
exterior. On the outer side of the exine.
fenestrate. See Table F.
fimbriae, /ate. (=fringe). With long, hair-like appendages.
fold. Sculpturing element formed by the folding up of the outermost layer of the grain (spore) coat. Upper edge blunt. Usually with interior cavity.
foot layer. Inner layer of ektexine (cf. Fig. 11.4).
fossula/-ae, fossulate (fossa = ditch). See *frustillate.*
foveola, /-ate (fovea = pit). Badly defined thinner round parts of the exine, exitus or not; never with anulus.
frustillum/-a, frustillate (frustum = broken-off piece). Small area of tectum covering two or more columellae, separated by groves (fossulae) cutting through the tectum (cf. Fig. 11.9).

furrow. *Colpus*, q.v.

gemma/ae, gemmate (= bud). Globular sculpturing element with narrow base.

granule. In this book used in a general sense only. With some authors = *columella*.

harmomegathos. Mechanism permitting size changes of the pollen grain.

heterobrochate. Reticular sculpturing (q.v.) with small and large brochi mixed irregularly.

heterocolpate. Some colpi regular, with (endo-)pore, other pseudocolpi without pore. See Table F.

intectate, see *tectum*.

intercolpium. The meridional segment between two colpi (see Fig. 13.2).

interior. Inside the surface coat of the grain.

interporium (rarely used). In tri- and stephanoporate grains the segment defined by meridians touching two adjacent pores or their anuli. The definition in Iversen and Troels Smith (1950: 34) is not quite clear and has been misunderstood (see Fig. 13.2).

intersemiangular. See Fig. 11.3.

intexine. Obsolete for endexine (q.v.).

intine. The (non-fossilized) intermediate stratum in a pollen grain.

intra-. Pattern in the structure below a tectum, e.g. *intrareticulate* formed by columellae in a reticulate pattern below a tectum which may be psilate, scabrate or whatever.

isometric = *peri-* (q.v.).

Körnchen. Columellae (obsolete).

lacuna. More or less circular, indefinite opening in the ektexine, not a pore (cf. keys F and Cy).

laesura/-ae (= scar: laedo = damage). Closed slits in pteridophyte spores, functioning like, but not homologous with, colpi.

latitude. Meridional distance from the equator of the grain.

lobate (Gr. lobos = (ear-)lobe, etc.). See Fig. 11.3.

longitude. Distance as measured along the equator of the grain.

lophate (Gr. lophos = crest, esp. cockscomb). With a very coarse reticulum (two to four brochi per quadrant), formed by crests (q.v.).

lumen/-ina (= light). Openings in the reticulum, measured between muri (cf. Fig. 12.13).

macula. Obsolete term (cf. key Gr.).

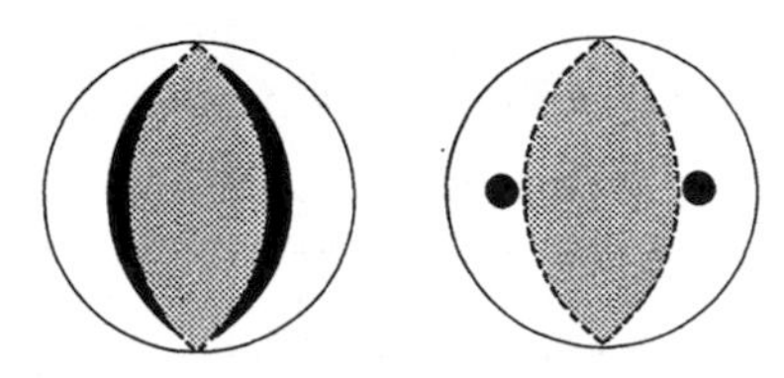

Fig. 13.2. Intercolpium (left) and interporium. Apertures indicated in black.

margo/-ines (= edge). Ektexine edging a colpus and differing in structure or sculpturing from the rest of the grain.

meridional (meridies = south). The direction on the surface of a rotational grain from pore to pole.

micro-. The prefix usually indicates that the element is smaller than 1 μm (cf. introduction to this chapter). *Microtubuli*: (sub)microscopic channels in the exine.

monad. See tetrad.

multicolumellate. Elements comprising a group of (more than three) columellae.

murus/-i (= *wall*). The walls defining a reticulum (*cf. Fig. 12.3*).

nexine. Inner part of exine (cf. Fig. 11.4).

oblate. In general terms any body the quatorial diameter of which is longer than the polar axis. Erdtman has divided the shape into sub-oblate, oblate and peroblate. In that restricted sense the middle term must be eu-oblate.

oncus. A non-sporopollenin body filling up some pore openings.

operculum/-a, operculate (= cover). Small tecta or isolated tectum elements on the aperture membrane (see Fig. 11.12).

oval. See Fig. 11.3.

pan- = *peri-* (q.v.).

pattern. Geometrical configuration of structural elements.

per-. A feature penetrating through the tectum down to the footlayer/endexine.

perforation. See *tubuli*.

peri- (-colpate, -porate). Elements (apertures) evenly distributed over the whole surface of the grain, not restricted to polar or equatorial positions.

perine (from Gr. peri = around and exine q.v.). An outermost, usually thin and evanescent, envelope of pteridophyte spores.

pit. Negative sculptural element, more or less isodiametric.

polar area = *apocolpium* (q.v.). *Polar area index*: the relation between the diameter of the apocolpium and the equatorial diameter. In tricolpate grains the former is often taken as the distance between the ends of two colpi (margines) (cf. Fig. 11.13).

polar axis. Theoretically, the axis between the proximal and the distal pole of the grain. Practically, the axis of maximum symmetry.

pollenkitt. Amorphic masses covering and/or accumulating in intercolumellar space, especially in grains of entomophilous flowers.

polyad. See *tetrad*.

polyplicate. See Table F.

pore/-i, porate (Gr. poros = opening). Aperture with length/breadth ratio less than 2, ends rounded. *Dupli-, tripliporate*: with two or three, pores, respectively, per colpus. *Pororate*: with pores both in ekt- and endexine.

prolate. Any body the polar axis of which is longer than the equatorial diagrameter (cf. *oblate*).
pro(to)- (-columellae, -sporopollenin). Precursors in the early stages of development of a pollen grain.
proximal. Opposite of *distal* (q.v.).
pseudocolpi. In heterocolpate grains: the colpi that do not carry pores.
psilate (psilos = nude). Sculpturing, the elements of which are smaller than 1 μm.
punctae (obsolete). Cf. p. 285.
reticulum (dim. of rete = net). A network-like pattern formed by ridges or by corresponding arrangement of columellae or both, the columellae often united in their upper parts.
rugula/-ae, *rugulate* (dim. of ruga = fold). Elongated (length ⩾ 2 × breadth) elements irregularly distributed; cf. *striate*. NB: ruga *sensu* Erdtman = colpus.
scabrate (scaber = rough). Sculptural elements < 1 μm. The noun *scabra/-ae* is linquistically meaningless, but useful.
sculpturing (sculpo = carve). The geometry of the outside of the grain, i.e. those elements that are shown by a SEM photo. In a tectate grain sculpturing is supra-tectal only, in in- and semitectate grains sculpturing comprises more or less the whole depth of the exine. Positive sculpturing elements project beyond the general surface of the grain, negative ones go under that surface.
saccate = vesiculate *(q.v.)*.
semiangular. Cf. Fig. 11.3.
sexine. Outer part of the exine (cf. Fig. 11.4).
simplicolumellate. Columellae dispersed one by one, especially in a simple row in muri (cf. Fig. 11.9).
spine. Acute sculpturing element with more or less isodametric cross-section, ⩾ 1 μm (cf. *echinate*, *spinule*).
spinule, *-/ose*. Micro-spine, usually as a secondary sculpturing superimposed on a coarser pattern (cf. Fig. 13.3).
spir-. With apertures (colpi) in a spiral pattern (see Table F).
stephano- (Gr. stephanos = wreath). Several (more than three) equal features set at the same – equatorial – level on the surface of the grain, usually more than three apertures. *Stephanocolp(or)ate* furrows are meridional. Non-meridional furrows belong in the periporate pattern.

Fig. 13.3. Isolated columellae (intectate) of *Linum catharticum*, diameter *ca.* 0.5 μm, with spinules on top.

stria/-ae, *striate*. (= groove). Subparallel elongated elements, usually a sculpturing feature (cf. Fig. 11.7).
structure ('texture'). Construction of the exine below the tectum.
sub-angular. See Fig. 11.3.
supra-. A feature exclusively belonging to the sculpturing of a tectate grain not reflected in the structure (cf. Fig. 11.7).
syn- (-colpate). See Table F. The opposite (for a dichotomic key) would be choricolpate; not used.
tectum, */-ate* (= roof). The outermost, more or less complete covering layer of the grain (cf. discussion, Chapter 11.5). *Intectate*: grain without tectum; *atectate*, primitively tectum-less; *etectate*, tectumless by reduction; *semitectate*, grains with large openings in the tectum, covering less than ± 75% of the surface (cf. Fig. 11.4).
tetrad. Pollen grains which have not separated after meiosis, and are dispersed as one unit. The interior walls are reduced with a thin simple exine or none (see *calymmate*). *Dyad*: the corresponding unit consisting of two grains (*monads*). *Polyad*: the same, unit consisting of more than four monads.
texture = *structure* (q.v.).
theca. The pollen-bearing part of the stamen.
transverse. Running parallel with, usually in the equatorial plane. *Transverse colpi* are discrete in contrast to *equatorial* colpus, which runs uninterrupted around the entire equator.
trichotomocolpate. See Table F.
tricrescentic. In polar view of tricolpate grains: exine very thick in the median of the intercolpium, gradually thinning towards the colpus.
tryphine = *pollenkitt* (q.v.).
tubuli (*micro-tubuli*). Channels through the exine. Straight on (if visible) seen as *perforations*.
Ubisch bodies. Small, usually unstructured sporopollenin bodies formed in the thecae.
undulate (unda = wave). With subparallel, densely placed, slightly elongated sculptural elements with a rounded profile, breadth/height > 1.
vallum /-a. Ridges in striate and regulate sculpturing, verging towards tangentially elongated verrucae (q.v.).
vermiculate. Rugulate (q.v.) with contorted ridges (vallae).
verrucae/-ae, *verrucate* (= wart). A sculpturing element ⩾ 1 μm, ideally with a cylindrical base and a rounded head, diameter subequal to height. In practice intermediate between a conical spine and a round gemma (q.v.).
vesiculate. Corresponding to *cavate* (q.v.); in non-

columellate pollen, partly filled by a three-dimensional sporopollenin network.

vestibulum. Cavity forming the pore, usually separated from the interior of the grain by a low rise of, or by separation of the layers of the exine, the outer ones protruding.

viscin: Thread-like spropollenin bodies connecting pollen grains in some families.

visibility. Cf. introduction to this list.

Zwischenkörper (obs.) = *oncus* (q.v.).

References

Aaby, B. 1975 Cykliske klimavariationer de sidste 7500 år påvist ved undersøgelse af højmoser og marine transgressions-faser.—Dan. geol. unders. Årb. 1974: 91–104.

Aaby, B. 1986 Trees as anthropogenic indicators in regional pollen diagrams from eastern Denmark—Pp. 73–94 in Behre, 1986.

Aaby, B., and H. Tauber 1974 Rates of peat formation in relation to degrees of humification and local environment, as shown by studies of a raised bog in Denmark.—Boreas 4: 1–17.

Aario, L. 1940 Waldgrenzen und subrezente Pollenspektren in Petsamo Lappland.—Ann. acd. sci. fenn. A 54, 8, 120 pp.

Aario, L. 1944 Über die pollenanalytischen Methoden zur Untersuchung von Waldgrenzen.—Geol. fören. Stockh. förh. 66: 337–354.

Adam, D. P. 1967 Nomogram for Mosimann's Beta-compound multinomial correlation coefficient.—Am. j. sci. 265: 907–908.

Adam, D. P., C. W. Ferguson, and V. C. Lamarck Jr 1967 Enclosed bark as a pollen trap.—Science 157: 1067 only.

Afzelius, B. M. 1956 Electron-microscope investigations into exine stratification.—Grana Palynol. 1, 2: 22–37.

Afzelius, B. M., G. Erdtman and F. S. Sjöstrand 1954 On the fine structure of the outer part of the spore wall of Lycopodium clavatum as revealed by the electron microscope.—Sven. bot. tidskr. 48: 155–161.

Aitchison, J. 1981 A new approach to null correlations to proportions.—Math. geol. 13: 175–189.

Ambach, W., S. Bortenschlager, and H. Eisner 1966 Pollen-analysis investigations of a 20 m firn pit on the Kesselwandferner (Ötzal alps).—J. glaciol. 6: 233–236.

Andersen, A. 1954 Two standard pollen diagrams from South Jutland.—Dan. geol. unders. 2 rk. 80: 188–209.

Andersen, S. T. 1960 Silcone oil as a mounting medium for pollen grains.—Dan. geol. unders. 4 rk. 4, 1. 24 pp.

Andersen, S. T. 1961 Vegetation and its environment in Denmark in the Early Weichselian Glacial (Last Glacial).—Dan. geol. unders.—2 rk. 75. 175 pp.

Andersen, S. T. 1967 Tree-pollen rain in a mixed deciduous forest in south Jutland (Denmark).—Rev. palaeobot. palynol. 3: 267–275.

Andersen, S. T. 1970 The relative pollen productivity and pollen representation of north European trees, and correction factors for tree pollen spectra.—Dan. geol. unders. 2 rk. 96. 99 pp.

Andersen, S. T. 1974 Wind conditions and pollen deposition in a mixed deciduous forest. I–II.—Grana 4: 57–77.

Andersen, S. T., and T. Bertelsen 1972 Scanning electron microscope studies of pollen of cerals and grasses.—Grana 12: 79–86.

Anderson, E. 1955 Pollenspridning och avståndsisolering av skogsfröplantager.—Norlands Skogsvårdsförb. tidskr. 1955. 33.

Andersson, G. 1902 Hasseln i Sverige fordom och nu. —Sver. geol. unders. Ser. Ca 3. 168 pp.

Argus, G. W. 1974 An experimental study of hybridization and pollination in Salix.—Can. j.bot. 52: 1613–1619.

Assarson, G., and E. Granlund 1924 En metod för pollenanalys av minerogena jordarter.—Geol. fören. Stockh. förh. 46: 76–82.

Audran, J.-C., and E. Masure 1976 Précisions sur l'infrastructure de l'exine chez les Cycadales.—Pollen spores 18: 5–26.

Auer, V. 1956–70 The Pleistocene of Fuego-Patagonia I–V.—Ann. acad. sci. fenn. A 3; 45, 50, 60, 80, 100. 1300 pp.

Avetisyan, E.M. Аветисян, Е.М. 1950 Упрощенный ацетолизный метод обработки пыльцы. — Бот. журн. 35: 385–386.

Bahnson, H. 1968 Kolorimetriske bestemmelser af humificeringstal i højmosetorv fra Fuglsø mose på Djursland.—Medd. dan. geol. for. 18: 55–63.

Bailey, J. W. 1960 Some useful techniques in the study and interpretation of pollen morphology.—J. Arnold arb. 41: 141–148.

Bakels, C. C. 1988 Pollen from plaggen soils in the Province of North Brabant, the Netherlands.—In

Groenman-van Wateringe, W. and M. Robinson (eds) Man-made soils. BAR Rep. Internat. ser. In press.
Bartlein, P. J., I. C. Prentice, and T. Webb III 1986 Climatic response surfaces from pollen data for some eastern North American taxa.—J. biogeogr. 13: 35–57.
Bassett, I. J., and C. W. Crompton 1967 Pollen morphology and chromosome numbers of the family Plantaginaceae in North America.—Can. j. bot. 46: 349–361.
Bassett, I. J., and J. Terasmäe 1962 Ragweeds, Ambrosia species, in Canada and their history in postglacial time. —Can. j. bot. 40: 141–150.
Bastin, B. 1982 Premier bilan de l'analyse pollinique de stalagmites holocènes en provenance des grottes belges.—Rev. belg. geogr. 106: 88–97.
Bates, C. D., P. Coxon, and P. L. Gibbard 1978 A new method for preparation of clay-rich sediment samples for palynological investigations.—New phytol. 81: 459–463.
Battarbee, R. W., and M. J. Kheen 1982 The use of electronically counted microspheres in absolute diatom analysis.—Limnol. oceanogr. 27: 184–188.
Baudisch, O., and E. v. Euler 1935 Über den Gehalt einiger Moorerdarten an Carotinoiden.—Ark. kemi miner. geol. 11A, 21.
Behre, K.-E. 1976 Beginn und Form der Plaggenwirtschaft in Nordwestdeutschland nach pollenanalytischen Untersuchungen in Ostfriesland.—Neue Ausgrab. Forsch. Nierdersachsen 10: 197–224.
Behre, K.-H. 1981 The interpretation of anthropogenic indicators in pollen diagrams.—Pollen spores 23: 225–245.
Behre, K.-E. 1986 Anthropogenic indicators in pollen diagrams.—Balkema (Rotterdam) 246 pp.
Behre, K. E., and D. Kucăn 1986 Die Reflektion archäologisch bekannter Siedlungen in Pollendiagrammen verschiedener Entfernung. Beispiele aus der Siedlungskammer Flögeln, Nordwestdeutschland.—Pp. 94–114 in Behre, 1986.
Beijerinck, W. 1934 Humusortstein und Bleichsand als Bildungen entgegensetzter Klimate.—Proc. k. akad. wet. Amst. 37: 93–98.
Benninghof, W. S. 1962 Calculation of pollen and spores density in sediments by addition of exotic pollen in known quantities.—Pollen spores 4: 332–333.
Benninghof, W. S. 1965 Atmospheric particulate matter of plant origin.—Pp. 133–144 in Tsuchiya, H. H., and A. A. Brown (eds) Proc. Atmospheric biol. conf. 1965.
Berglund, B. E. 1973 Pollen distribution and deposition in an area of south-eastern Sweden—some preliminary results.—Pp. 117–129 in Birks and West, 1973.
Berglund, B. E. (ed.) 1986 Handbook of Holocene palaeoecology and palaeohydrology.—Wiley-Interscience (New York). 869 pp.
Berglund, B. E., and M. Ralska-Jasiewiczowa, M. 1986 Pollen analysis and pollen diagrams.—Pp. 455–484 in Berglund, 1986.
Berglund, B. E., U. Emanuelson, S. Persson, and T. Persson 1986 Pollen/vegetation relationships in grazed and mowed plant communities of South Sweden.—Pp. 37–52 in Behre, 1986.
Berglund, B., G. Erdtman, and J. Praglowski 1959 Några ord om betydelsen av inbäddningsmediets brytningsindex vid palynologiska undersökningar.—Sven. bot. tidskr. 52: 462–468.
Beug, H.-J. 1961 Leitfaden der Pollenbestimmung. Lief. 1.—Fischer (Lieberst.-Stuttgart). 63 pp.
Bierly, E. W., and E. W. Hewson 1962 Some restrictive meteorological conditions to be considered in the design of stacks.—J. appl. meteorol. 1: 383–390.
Birks, H. H. 1985 Comparing 'isophyte' and 'isopoll' maps.—J. biogeogr. 12: 90–91.
Birks, H. J. B. 1970 Inwashed pollen spectra at Loch Fada, Isle of Skye.—New phytol. 69: 807–820.
Birks, H. J. B. 1973 Past and present vegetation of the Isle of Skye—a paleoecological study.—Cambridge University Press. 415 pp.
Birks, H. J. B. 1974 Numerical zonations of Flandrian pollen data.—New phytol. 73: 351–358.
Birks, H. J. B. 1978 Geographic variation of Picea abies (L.) Karsten pollen in Europe.—Grana 17: 149–160.
Birks, H. J. B. 1980 Quaternary vegetational history of West Scotland.—5. internat. palynol. conf. Excursion guide C 8.
Birks, H. J. B. 1981 The use of pollen analysis in the reconstruction of past climates: a review.—Pp. 111–138 in Wigley, T. M. L., J. J. Ingram, and G. Farmer (eds) Climate and history. Cambridge University Press.
Birks, H. J. B. 1985 A pollen-mapping project in Norden for 0–13 000 BP.—Univ. Bergen Bot. inst. Rapp. 38. 41 pp.
Birks, H. J. B. 1986a Lat-Quaternary biotic changes with particular reference to north-west Europe.—Pp. 3–65 in Berglund, 1986.
Birks, H. J. B. 1986b Numerical zonation, comparison and correlation of Quaternary pollen-stratigraphical data.—Pp. 743–774 in Berglund, 1986.
Birks, H. J. B. and H. H. Birks 1980 Quaternary paleoecology.—E. Arnold (London). 289 pp.
Birks, H. J. B. and A. D. Gordon 1985 Numerical methods in pollen analysis.—Academic Press (London, etc.). 317 pp.
Birks, H. J. B. and S. M. Peglar 1980 Identification of Picea pollen of Late Quaternary age in North America: a numerical approach.—Can. j. bot. 58: 2043–2058.
Birks, H. J. B. and R. G. West (eds.) 1973 Quaternary plant ecology.—Blackwells (London, etc.). 326 pp.
Blackmore, S., and S. H. Burnes 1986 Freeze fracture and cytoplasmatic maceration of pollen grains.—Grana 25: 41–45.
Blackmore, S., and P. R. Crane 1988 The ontogeny and evolution of apertures in pollen and spores.—Bot. rev. In press.

Blackmore, S., and H. G. Dickinson 1981 A simple technique for sectioning pollen grains.—Pollen spores 23: 281–286.

Blackmore, S., and I. N. Ferguson 1986 Pollen and spores. Form and function.—Linn. soc. symposium ser. 12. 443 pp.

Blytt, A. 1876 Essay on the immigration of the Norwegian flora during alternate rainy and dry periods. —Cammermeyer (Christiana), 89 pp.

Bonny, A. P. 1972 A method for determining absolute pollen frequencies in lake sediments.—New phytol. 71: 391–403.

Bonny, A. P. 1976 Recruitment of pollen to the seston and sediment of some Lake District lakes.—J. ecol. 64: 859–887.

Bonny, A. P. 1978 The effect of pollen recruitment processes on pollen distribution over the sediment surface of a small lake in Cumbria.—J. ecol. 66. 385–416.

Bonny, A. P., and P. Allen 1983 A comparison of pollen data from Tauber traps paired in the field with simple cylindrical collectors.—Grana 23: 51–58.

Booberg, G. 1930 Gisselåsmyren.—Norrländskt handbibliotek 12. 329 pp.

Borchgrevink, A.-B. 1974 Torv 1.—Kulturhistorisk leksikon for nordisk middelalder. 18: 520–523. Gyldendal (Oslo).

Boros, A., and M. Járai-Komlódi 1975 An atlas of recent European moss spores.—Akad. kiado (Budapest). 466 pp.

Borowik, H. 1963, 1966 The trapping of scots pine and oak pollen in Bialowieza National Park (Pol.-Engl.) I–II.—Acta soc. bot. polon. 32: 655, 35: 159.

Borse, C. 1939 Über die Frage der Pollenproduktion, Pollenzerstörung und Pollenverbreitung in ostpreussischen Waldgebieten.—Schr. phys.-ökon. Ges. Königsb. (Pr.) 71: 127–144.

Bortenschlager, S. 1969 Pollenanalyse des Gletschereises —Grundlegende Fragen der Pollenanalyse überhaupt. —Ber. dtsch. bot. Ges. 81 (1968): 491–497.

Bowman, P. W. 1931 Study of a peat bog near the Matamuk River, Quebec, Canada, by the method of pollen analysis.—Ecology 12: 694–708.

Brattegard, O. 1951–52 Edvard Edvardsen. Bergen.—Bergens hist. for. skr. 55, 58. 526 + 377 pp.

Brinkmann, P. 1934 Zur Geschichte der Moore, Marschen und Wälder Nordwest-deutschlands. III. Das Gebiet der Jade.—Bot. jahrb. 64: 369–442.

Brookes D., and K. W. Thomas 1968 The distribution of pollen grains in microscope slides. Part I. The nonrandomness of the distribution.—Pollen spores 9 (1967): 621–630.

Brooks, J., and W. C. Elsik 1975 Chemical oxidation (using ozone) of the spore wall of Lycopodium clavatum.—Grana 14 (1974): 85–91.

Brooks, J., and G. Shaw 1968a The post-tetrad ontogeny of the pollen wall and the chemical structure of the sporopollenin in Lilium henryi.—Grana 8: 227–234.

Brooks, J., and G. Snaw 1968b Chemical structure of the exine of pollen walls and a new function for carotinoids in nature.—Nature 219: 522–623.

Brooks, J., and G. Shaw 1968c Identity of sporopollenin with older kerogen and new evidence for the possible biological source of chemicals in sedimentary rocks. —Nature 220: 678–679.

Brooks, J., and G. Shaw 1969 Evidence for extraterrestrial life: Identity of sporopollenin with the insoluble organic matter present in the Orgeuil and Murray meteorites and also with some terrestrial microfossils. —Nature 223: 756 only.

Brooks, J., and G. Shaw 1973a The role of sporopollenin in palynology.—Pp. 80–90 in M. I. Neustadt (ed.): Problems of palynology. Proc. III internat. palynol. conf.

Brooks, J., and G. Shaw 1973b Recent developments in the chemistry, biochemistry and post-tetrad ontogeny of sporopollenin derived from pollen and spores.—Pp. 98–114 in Heslop-Harrison, 1973.

Brooks, J., P. R. Grant, M. D. Muir, P. van Gijzel, and G. Shaw (eds) 1971 Sporopollenin.—Academic Press (London), 718 pp.

Brown, C. A. 1960 Palynological technique.—Baton Rouge. 188 pp.

Brummitt, R. K., J. K. Ferguson, and J. W. Poole 1980 A unique and extraordinary pollen type in the genus Crossandra (Acanthaceae).—Pollen spores 22: 11–16.

Buchmann, S. L., C. E. Jones, and L. J. Colin 1985 Vibratile pollination of Solanum douglasii and S. xanti (Solanaceae) in southern California.—Waksmann j. biol. 35: 1–25.

Burden, E. T., J. H. McAndrews, and G. Norris 1986 Palynology of Indian and European forest clearance and farming in lake sediment cores from Awanda provincial park, Ontario.—Can. J. earth sci. 23: 43–54.

Burrows, F. M. 1975a Calculation of the primary trajectories of dust seeds, spores and pollen in unsteady winds.—New phytol. 75: 389–403.

Burrows, F. M. 1975b Wind-borne seed and fruit movement.—New phytol. 75: 405–418.

Cain, S. A., and D. E. Cain 1948 Size-frequency characteristics of Pinus echinata.—Bot. gaz. 110: 325–330).

Carniel, K. 1972 Electronenmikroskopische Analyse der Pollenentwicklung bei Heleocharis palustris.—Oesterr. bot. Z. 120: 223–234.

Casparie, W. G. 1969 Bult- und Schlenkenbildung im Hochmoortorf.—Vegetatio 19: 146–180.

Cerceau, M.-T. 1959 Clé de détermination d'Ombellifères de France et d'Afrique du Nord d'après leurs grains de pollen.—Pollen spores 1: 145–190.

Cerceau, M.-T., M. Hideaux, L. Marceau, and F. Roland 1970 Cassure du pollen par les ultrasons pour l'étude structurale de l'exine au microscope électroni-

que à balayage.—C. r. hebd. acad. sci. Paris D 270: 66–69.
Chaloner, W. G. 1986 Electrostatic forces in insect pollination and their significance in exine ornament.—Pp. 103–108 in Blackmore and Ferguson, 1986.
Chamberlain, A. C. 1967 Deposition of particles to natural surfaces.—Pp. 138–164 in Gregory and Monteith, 1967.
Chamberlain, A. C., and R. C. Chadwick 1972 Deposition of spores and other particles on vegetation and soil.—Ann. appl. biol. 71: 141–158.
Chanda, S. 1962 On the pollen morphology of some Scandinavian Caryophyllaceae.—Grana palynol. 3, 3: 67–89.
Christensen, B. Brorson 1946 Measurement as a means of identifying fossil pollen.—Dan. geol. unders, 4 rk. 3, 2. 20 pp.
Christensen, B. Brorson 1949 Om mikrotomsnit af pollenexiner.—Medd. dan. geol. foren. 11: 441–444.
Christensen, B. Brorson 1954 New mounting media for pollen grains.—Dan. geol. unders. 2 rk. 80: 7–11.
Christensen, P. B. 1986 Pollen morphological studies in the Malvaceae.—Grana 25: 94–117.
Christie, A. D., and J. C. Ritchie 1969 On the use of isentropic trajectories in the study of pollen transport. —Nat. can. 96: 531–549.
Claugher, D., and J. R. Claugher 1987 Betula pollen grain substructure revealed by fast atom etching.—Pollen spores 29: 5–20.
Clausen, K. E. 1960 A survey of variation in pollen size within individual plants and catkins of three taxa of Betula.—Pollen spores 2: 299–304.
Cleve-Euler, A. 1951–5 Die Diatomeen von Schweden und Finland I–V.—K. sven. vetensk. akad. handl. 4. ser. 2, 1; 3, 3; 4, 1: 4, 5; 5, 4. 961 pp.
Comanor, P. L. 1967 Forest vegetation and the pollen spectra: an examination of the usefulness of the R values.—Diss. abstr. B 28: 1804 B only.
Corbet, S. A., J. Beament and D. Eisikowitch 1982 Are electrostatic forces involved in pollen transfer?—Plant, cell envir. 5: 125–129.
Correns, C. 1922 Geschlechtsbestimmung und Zahlenverhältnis der Geschlechter beim Sauerampfer (Rumex acetosa).—Biol. Zentralbl. 42: 465–480.
Cour, P. 1974 Nouvelles techniques de détection des flux et des retombées polliniques: étude de la sédimentation des pollens et des spores à la surface du sol.—Pollen spores 16: 103–142.
Cour, P. and M. van Campo 1980 Prévisions des récoltes à partir de l'analyse pollinique de l'atmosphère.—C.r. hebd. acad. sci. Paris 290 D: 1043–1046.
Couteaux, M. 1977 A propos de l'interprétation des analyses polliniques de sédiments mineraux, principalement archéologiques.—Pp. 259–267 in Laville, H., and J. Renault-Miskovski (eds) Approche écologique de l'homme fossile.—Suppl. Bull. ass. fran. ét. Quat. 47.
Craig, A. J. 1972 Pollen influx to laminated sediments: a pollen diagram from northeastern Minnesota.—Ecology 53: 46–57.
Crane, P. E. 1986 Form and function in wind dispersed pollen.—Pp. 197–202 in Blackmore and Ferguson, 1986.
Cuellar, H. S. 1967 Description of a pollen release mechanism in the flower of the Mexican hackberry tree, Celtis laevigata.—Southw. nat. 12: 471–474.
Cundill, P. R. 1984 The use of mosses in modern pollen studies at Worton Lochs, Fife.—Trans. bot. soc. Edinb. 44: 375–383.
Cundill, P. R. 1986 A new design of pollen trap for modern pollen studies.—J. bigeogr. 13: 83–89.
Currier, P. J., and P. O. Kapp 1974 Local and regional pollen rain components at Davis Lake, Montcalm county, Michigan.—Mich. acad. 7: 211–225.
Cushing, E. J. 1961 Size increases in pollen grains mounted in thin slides.—Pollen spores 3: 265–274.
Cushing, E. J. 1967 Evidence for differential pollen preservation in late Quaternary sediments in Minnesota.—Rev. palaeobot. palynol. 4: 87–101.
Cwynar, L. C., E. Burden, and J. H. McAndrews 1979 An inexpensive method for concentrating pollen and spores from fine grained sediments.—Can. j. earth sci. 16: 1115–1120.
Dahl, A. O. 1969 Wall structure and composition of pollen and spores.—Pp. 35–48 in Tschudy and Scott, 1969.
Dahl, E. 1951 On the relation between summer temperature and the distribution of alpine vascular plants in the lowlands of Fennoscandia.—Oikos 3: 22–52.
Danielsen, A. 1970 Pollen-analytical late Quaternary studies in the Ra district of Østfold, Southeast Norway. —Univ. Bergen. Årbok 1969. Mat.-natv. ser. 14. 146 pp.
Danielsen, R. 1986 Hydrosere og myrutvikling i ei oceanisk myr på Fedje, Hordaland.—Thesis, cand. scient. Univ. Bergen. 58 pp.
Davis, M. B. 1961 The problem of rebedded pollen in late-glacial sediments at Taunton, Massachusetts.—Am. j. sci. 259: 211–222.
Davis, M. B. 1963 On the theory of pollen analysis.—Am. j. sci. 261: 897–912.
Davis, M. B. 1966 Determination of absolute pollen frequency.—Ecology 47: 310–311.
Davis, M. B. 1967a Pollen deposition in lakes as measured by sediment traps.—Geol. soc. am. bull. 78: 849–858.
Davis, M. B. 1967b Pollen accumulation rates at Rogers Lake, Connecticut, during late- and postglacial time. —Rev. palaeobot. palynol. 2: 219–230.
Davis, M. B. 1968 Pollen grains in lake sediments: re-

deposition caused by seasonal water circulation.—Science 162: 796–799.

Davis, M. B. 1969 Climatic changes in southern Connecticut recorded by pollen deposition at Rogers Lake.—Ecology 50: 409–422.

Davis, M. B. 1973 Pollen evidence of changing land use around the shores of Lake Washington.—Northw. sci. 47: 133–148.

Davis, M. B. 1973 Redeposition of pollen grains in lake sediments.—Limnol. oceanogr. 18: 44–52.

Davis, M. B. (ed.) 1986 Vegetation–climate equilibrium.—Vegetatio 67: 65–141.

Davis, M. B., and L. Brubaker 1973 Differential sedimentation of pollen grains in lakes.—Limnol. oceanogr. 18: 635–646.

Davis, M. B., and E. S. Deevey 1964 Pollen accumulation rates: estimates from late-glacial sediment of Rogers Lake.—(Mich. univ. Great Lakes res. div. Contr. 38.) Science 145: 1293–1295.

Davis, M. B., L. B. Brubaker, and J. M. Beiswenger 1971 Pollen grains in lake sediments: Pollen percentages in surface sediments in southern Michigan.—Quat. res. 1: 450–467.

Davis, R. B. 1967 Pollen studies of near-surface sediments in Maine Lakes.—Proc. 7 congr. INQUA, 7: 143–173.

Davis, R. B., L. A. Brewster, and J. Sutherland 1969 Variation in pollen spectra within lakes.—Pollen spores 11: 557–571.

Degerbøl, M., and B. Fredskild 1970 The urus (Bos primigenius Bojanus) and Neolithic domesticated cattle (Bos taurus Linné) in Denmark.—Biol. skr. dan. vidensk. selsk. 17, 1. 177 pp.

Degerbøl, M., and H. Krog 1951 Den europœiske sumpskilpadde (Emys orbicularis) i Danmark.—Dan. geol. unders. 2 rk. 78.

Delcourt, P. A., H. R. Delcourt, and T. Webb III 1984 Atlas of mapped contributions of dominance and modern pollen percentage for important tree taxa in eastern North America.—AASP Contrib. Ser. 14. 131 pp.

Demianowicz, Z. 1961 Pollenkoeffizienten als Grundlage der quantitativen Pollenanalyse des Honigs.—Pszczeln. zesz. nauk. 5: 95–105.

Dickinson, H. G. 1976 The deposition of acetolysis-resistant polymers during the formation of pollen.—Pollen spores 18: 321–334.

Digerfeldt, G. 1966 A new type of large-capacity sampler.—Geol. fören. Stockh. förh. 87 (1965): 425–430.

Digerfeldt, G. 1976 A pre-Boreal water-level change in Lake Lyngsjö, Central Halland.—Geol. fören. Stockh. förh. 98: 329–336.

Digerfeldt, G., and U. Lettevall 1969 A new type of sediment sampler.—Geol. fören. Stockh. förh. 91: 388–406.

Digerfeldt, G., R. W. Battarbee, and L. Bengtsson 1975 Report on annually laminated sediment in Lake Järlasjön, Nacka, Stockholm.—Geol. fören. Stockh. förh. 97: 29–40.

Di-Giovanni, F., P. M. Beckett, and J. R. Flenley 1988 Modelling of dispersion and deposition of tree pollen within a forest canopy.—Grana. In Press.

Dimbleby, G.W. 1957 Pollen analysis of terrestrial soils.—New Phytol. 56: 12–28.

Dricot, E. 1961 Analyse pollinique d'un profil de sable à Averbode (Campine belge).—Agricultura 2. ser. 9: 651–655.

Ducker, S. C., J. M. Pettitt, and B. Knox 1978 Biology of Australian seagrasses: Pollen development and submarine pollination in Amphibolis aquatica and Thalassiodendron ciliatum (Cymodoceaceae).—Austr. j. bot. 26: 265–285.

Dumait, P., L. Marceau, C. Devin, and M. van Campo 1963 Nouvelle méthode de concentration des pollens dans les sédiments pauvres par microflottation.—C. r. hebd. seanc. acad. sci. Paris 256: 231–233.

Dunbar, A., and S.-O. Strandhede 1973 Pollen development in the Eleocharis palustris group (Cyperaceae) I-II.—Bot. not. 126: 197–265.

Dupont, L.M. 1985 Temperature and rainfall variation in a raised bog ecosystem.—Thesis, Amsterdam. 62 pp. Also in Rev. palaeobot. palynol. 48: 71–159.

Durham, O. C. 1946 The volumetric incidence of atmospheric allergens. III. Rate of fall of pollen grains in still air.—J. allergy. 17: 70–86.

Du Rietz, G. E. 1954 Die Mineralbodenwasserzeigergrenze als Grundlage der natürlichen Zweigliederung nord— und mitteleuropäischer Moore.—Vegetatio 5–6: 511–586.

Dyakowska, J. 1937 Researches on the rapidity of the falling down of pollen of some trees.—Bull int. acad. pol. sci. lettr. Cl. sci. math. nat. Sér. B 1. Bot. 1936: 155–68.

Dyakowska, J. 1964 The variability of the pollen grain of Picea excelsaLink.—Acta bot. soc. polon. 33: 727–748.

Dyakowska, J., and J. Zurzycki 1959 Gravimetric studies on pollen.—Bull. int. acad. pol. sci. lettr. Cl. sci. math. nat. 7, Sér. B 1. Bot. 11–16.

Echlin, P. 1973 The role of the tapetum during microsporogenesis of angiosperms.—Pp. 41–61 in Heslop-Harrison, 1973.

Eckblad, F. -E. 1975 Tilletia sphagni, Helotium schimperi or what?—Pollen spores 17: 423–428.

Ehrenberg, C. G. 1838 Beobachtungen über neue Lagen fossiler Infusorien und das Vorkommen von Fichtenblüten neben deutlichem Fichenholz, Haifischzahnen, Echinoiden und Infusorien in Volhynischen Feuersteine der Kreide.—Monatsber. Berliner Akad. 1838: 104.

Eide, F. G., and Aa. Paus 1982 Vegetasjonshistoriske undersøkelser på Kårstø, Tysvær kommune, Rogaland.—Univ. Bergen. Bot. inst. Rapp. 23. 45 pp.

Eisenhut, G. 1961 Untersuchungen über die Morphologie und ökologie der Pollenkörner heimischer und fremdländischer Waldbäume.—Forstwiss. Forsch. 15.

Elsik, W. C. 1966 Biological degradation of fossil pollen grains and spores.—Micropalaeontology 12: 515–518.

Eneroth, O. 1951 Undersökning rörande möjligheterna att i fossilt material urskilja de olika Betula-arternas pollen.—Geol. fören. Stockh. förh. 73: 343–390.

Erdtman, O. G. E. 1921 Pollenanalytische Untersuchungen von Torfmooren und marinen Sedimenten in Südwest-Schweden.—Ark. f. botanik. 17, 10. 173 pp.

Erdtman, G. 1927–45 Literature on pollen-statistics (and related topics) published before 1927/published 1945. —Geol. fören. Stockholm förh. 49–67.

Erdtman, G. 1931 Worpswede-Wabamum. Ein pollenstatistisches Menetekel.—Abh. nat. Ver. Bremen 28 (1931): 11–17.

Erdtman, G. 1934 Über die Verwendung von Essigsäureanhydrid bei Pollen-untersuchungen.—Sven. bot. tidskr. 28: 354–358.

Erdtman, G. 1935 Pollen statistics.—Pp. 110–125 in Wodehouse 1935.

Erdtman, G. 1937 Pollen grains recorded from the atmosphere over the Atlantic.—Meddel. Göteb. bot. trädg. 12: 186–196.

Erdtman, G. 1938 Pollenanalys och pollenmorfologi.—Sven. bot. tidskr. 32: 130–132.

Erdtman, G. 1943a An introduction to pollen analysis. —Chronica botanica (Waltham, Mass.). 238 pp.

Erdtman, G. 1943b Pollenspektra från svenska växtsamhällen. Jämte pollen-analytiska markstudier i södra Lappland.—Geol. fören. Stockh. förh. 65: 37–66.

Erdtman, G. 1945a Pollen morphology and plant taxonomy III. Morina L. With an addition on pollenmorphological terminology.—Sven. bot. tidskr. 39: 187–191.

Erdtman, G. 1949 Palynological aspects of the pioneer phase in the immigration of the Swedish flora II.—Sven. bot. tidskr. 43: 46–55.

Erdtman, G. 1952 Pollen morphology and plant taxonomy. Angiosperms. (An introduction to palynology I). —Almqvist and Wiksell (Stockholm). 539 pp.

Erdtman, G. 1957, 1965 Pollen and spore morphology/ Plant taxonomy. Gymnospermae, Pteridophyta, Bryophyta II (Illustrations). III (Text).—Almqvist and Wiksell (Stockholm). 151 pp.

Erdtman, G. 1962 Palynologiska perspektiv.—Sven. naturvetensk. 15: 219–227.

Erdtman, G. 1969 Handbook of palynology.—Munksgaard (København). 486 pp.

Erdtman, O. G. E. and H. Erdtman 1933 The improvement of pollen analysis technique.—Sven. bot. tidskr. 27: 347–357.

Erdtman, G. and P. Sorsa 1971 Pollen and spore morphology/Plant taxonomy IV. Pteridophyta (Text and additional illustrations).—Almqvist and Wiksell (Stockholm). 302 pp.

Ertman, G., B. Berglund, L. Praglowski, S. Nilsson 1961, 1963 An introduction to a Scandinavian pollen flora. I–II.—Almqvist and Wiksell (Stockholm). 92 + 84 pp.

Ernst, O. 1934 Zur Geschichte der Moore, Marschen und Wälder Nordwest-Deutschlands IV. Untersuchungen in Nordfriesland.—Schr. naturwiss. Ver. Schleswig-Holstein 20: 209–329.

Evans, G., S. Limbrey, and H. Cleeve (eds) 1975 The effect of man on the landscape: the highland zone.—Counc. Brit. archaeol. res. rep. 11. 129 pp.

Fægri, K. 1936 Einige Worte über die Färbung der für die Pollenanalyse hergestellten Präparate.—Geol. fören. Stockh. förh. 58: 439–443.

Fægri, K. 1939, K. 1939 Single-grain pollen preparations.—Geol. fören. Stockh. förh. 61: 513–514.

Fægri, K. 1940 Quartärgeologische Untersuchungen im westlichen Norwegen. II. Zur spätquartären Geschichte Jærens.—Bergens mus. årb. 1939–40. Naturvitensk rk. 7. 201 pp.

Fægri, K. 1944a Studies on the Pleistocene of Western Norway. III. Bømlo.—Bergens Mus. årb. 1943 Naturvitensk. rk. 8. 100 pp.

Fægri, K. 1944b On the introduction of agriculture in Western Norway.—Geol. fören. Stockh. förh. 66: 449–462.

Fægri, K. 1945 A pollen diagram from the sub-alpine region of central South Norway.—Nor. geol. tidsskr. 25: 99–126.

Fægri, K. 1950a On the value of palaeoclimatological evidence.—Centen. proc. R. meteorol. soc. 1950: 188–195.

Fægri, K. 1950b Studies on the Pleistocene of western Norway. IV. On the immigration of Picea abies (L.) Karst.—Univ. Bergen. årb. Naturv. rk. 1. 1949. 52 pp.

Fægri, K. 1951 An unrecognised source of error in pollen analysis.—Geol. fören. Stockh. förh. 73: 51–56.

Fægri, K. 1954a On age and origin of the beech forest (Fagus silvatica L.) at Lygrefjorden, near Bergen (Norway).—Dan. geol. unders. 2 rk. 80: 230–249.

Fægri, K. 1954b Some reflections on the trophic system in limnology.—Nytt mag. bot. 3: 43–49.

Fægri, K. 1956 Palynological studies in NW European Papilionaceae.—Univ. Bergen Bot. Mus. (Mimeographed). 13 pp.

Faegri, K. 1961 Palynology of a bumble-bee nest.—Veröff. geobot. Inst. ETH, Stift. Rübel, Zürich 37: 60–67.

Fægri, K. 1968 A note on the maritime forest limit in south-east Alaska.—Årb. Univ. Bergen. Mat.—naturvitensk. rk. 1968. 5. 20 pp.

Fægri, K. 1970 A pollen diagram from Voss, W. Norway. —Colloq. geogr. 12: 125–131.

Fægri, K. 1971 The preservation of sporopollenin membranes under natural conditions.—Pp. 256–270 in Brooks *et al*, 1971.

Fægri, K. 1973 In memoriam O. Gunnar E. Erdtman 1887–1973.—Pollen spores 15: 5–12.

Fægri, K. 1978 What is the polar axis?—Grana 17: 15–16.

Fægri, K. 1979 The problem of polyphyletic origin with special reference to angiosperms.—Taxon 29: 312–315.

Fægri, K. 1981 Some pages of the history of pollen analysis.—Striae 14: 42–47.

Fægri, K. 1985 The importance of palynology for the understanding of the archaeological environment in northern Europe.—[Pp. 333–346 in Renault Miskovsky *et al.* (eds.) Palynologie et archéologie]. CNRS Centr. Rech. archéol. Notes monogr. techn. 17.

Fægri, K. 1989. Strandkjeks. Naduren. In print.

Fægri, K., and P. Deuse 1960 Size variations in pollen grains with different treatments.—Pollen spores 2: 293–298.

Fægri, K., and H. Gams 1937 Entwicklung and Vereinheitlichung der Signaturen für Sediment- und Torfarten.—Geol. fören. Stockh. förh. 59: 273–284.

Fægri, K., and P. Ottestad 1948 Statistical problems in pollen analysis.—Univ. Bergen. årb. 1948. Naturvitensk. rk. 3. 27 pp.

Fægri, K., and L. van der Pijl 1979 The principles of pollination ecology. 3rd ed.—Pergamon (Oxford, etc.) 244 pp.

Fagerlind, F. 1952 The real significance of pollen diagrams.—Bot. not. 1952: 185–244.

Fedorova, R.V. Федорова, Р.В. 1952a Количественные закономерности распространения пыльцы древесных пород воздушным путем. – Ак. наук. СССР. Тр. геогр. 52: 91–103

Fedorova, R.V. 1952b Распространение пыьцы и спор текучими водам. – Ак. наук. СССР. 52: 45–72.

Fedorova, R.V. 1959a Распространение пыльцы березы воздушным путем. – Ак. наук. СССР. Тр. инст. геогр. 77: 140–144.

Fedorova, R.V. 1959b Рассеивание воздушным путем пыцьцы злаков. – Ак. наук. СССР. Тр. инст. геогр. 77: 145–156.

Feuer, S., and J. Kujit 1980 Fine structure of mistletoe pollen. III. Large-flowered neotropical Loranthaceae and their Australian relatives.—Am. j. bot. 67: 34–50.

Firbas, F. 1934 Über die Bestimmung der Walddichte und der Vegetation waldloser Gebeite mit Hilfe der Pollenanalyse.—Planta 22: 109–145.

Firbas, F. 1935a Über die Wirksamkeit der natürlichen Verbreitungsmittel der Waldbäume.—Natur u. Heimat 6 (3): 65–73.

Firbas, F. 1935b Die Vegetationsentwicklung des mitteleuropäischen Spätglazials—Bibl. bot. 112. 68 pp.

Firbas, F. 1937 Der pollenanalytische Nachweis des Getreidebaus.—Zeitschr.Bot. 31: 447–478.

Firbas, F. 1939 Vegetationsentwicklung und Klimawandel in der mitteleuropäischen Spät—und Nacheiszeit.'—Naturwissensch. 27: 81–108.

Firbas, F. 1949–52 Waldgeschichte Mitteleuropas I–II.—Fischer (Jena). 480 + 256 pp.

Firbas, F., and F. Broihan 1936 Das Alter der Trockentorfschichten im Hils.—Planta 26: 291–302.

Firbas, F., and H. Losert 1949 Untersuchungen über die Entstehung der heutigen Waldstufen in den Sudeten.—Planta 36: 478–506.

Firbas, F., and H. Sagromsky 1947 Untersuchungen über die Grösse des jährlichen Pollenniederschlages vom Gesichtspunkt der Stoffproducktion.—Biol. Zentralbl. 66: 129–140.

Fischer, H. 1890 Beiträge zur vergleichenden Morphologie der Pollenkörner.—Breslau (Thesis) 72 pp.

Flenley, J. R. 1971 Measurements of the specific gravity of the pollen exines.—Pollen spores 13: 170–186.

Flenley, J. R. 1973 The use of modern pollen samples in the study of vegetational studies of tropical regions.—Pp. 131–143 in Birks and West, 1973.

Florin, M. B. 1945 Skärgårdstall och 'strandskog' i västra Södermanlands pollendiagram.—Geol. fören Stockh. förh. 67: 511–533.

Flynn, J. J., and J. R. Rowley 1970 Wandmikrotubuli in Pollenkörnern.—Zeiss Inf. 18: 40–45.

Fokkema, N. J. 1971 The effect of pollen in the phyllosphere on colonization by saprophytic fungi and on infection by Helminthosporium sativum and other leaf pathogens.—Netherl. j. plant pathol. 77 Suppl. 1. 60 pp.

Ford, J. 1972 Monad pollen development; the initiation of cytoplasmic polarity—Cytobios 6: 81–87.

Fredskild, B. 1967 Palaeobotanical investigations at Sermermiut, Jakobshavn, West Greenland.—Medd. Grønland 178, 4. 54 pp.

Fredskild, B. 1973 Studies in the vegetational history of Greenland.—Medd. Grønland 198, 4. 245 pp.

Frenzel, B. 1969 Floren—und Vegetationsgeschichte seit dem Ende des Tertiärs (Historische Geobotanik).—Fortschr. Bot. 31: 309–319.

Frey, D. G. 1960 The ecological significance of cladoceran remains in lake sediments.—Ecology 41: 684–699.

Fries, M., and U. Hafsten 1965 Asbjørnsen's peat sampler—the prototype of the Hiller sampler.—Geol. fören. Stockh. förh. 87: 307–313.

Fritzche, C. J. 1837 Über den Pollen.—Mém. sav. étrang. acad. St. Petersburg 3: 649–769.

Fröman, I. 1944 De senkvartära strandförskjutningarna som växtgeografisk faktor i belysning av murgrönans geografi i Skandinavien-Baltikum.—Geol. fören. Stockh. förh. 66: 655–681.

Funkhouser, J. W., and W. R. Evitt 1959 Preparation techniques for acid-insoluble microfossils.—Micropaleontology 5: 369–375.

Gams, H. 1938 Vorschläge zur Vereinheitlichung der Zeichen für Mikrofossildiagramme, Waldgeschichtliche Karten und Moorprofile.—Chron. bot. 4: 121–123.
Geinitz, F. E. 1887 Geologische Notizen aus der Lüneburger Heide.—Jahresh. Naturwiss. Ver. Fürstentum Lüneburg 10: 36–42.
Geologklubben vid Stockholms högskole 1947 Kvartärgeologiskt möte den 5–9 november 1945.—Geol. fören. Stockh. förh. 69: 205–247.
Germeraad, J. H., and J. Muller 1971 A computer-based numerical coding system for the description of pollen grains and spores.—Rijksmus. Geol. mineral. Leiden. I–II.
Gifford, F. A. 1976 Turbulent diffusion—typing schemes: a review.—Pp. 68–86 in Shroud, R. L. (ed.) Consequences of effluent release. Nuclear safety 17.
Gimingham, C. H. 1972 Ecology of heathlands.—Chapman & Hall (London) 266 pp.
Gluzbar, E.A. 1968 Building a fossil spore library, methods of isolating single spores and pollen grains. – Paleontol. j. 2: 126. iii (Transl. from Палеонтол. журн. 2: 122–126).
Godwin, H. 1938 Data for the study of post-glacial history.—New phytol. 37: 329–332.
Godwin, H. 1954 Recurrence-surfaces.—Dan. geol. unders. 2 rk. 80: 22–30.
Godwin, H. 1958 Pollen-analysis in mineral soil. An interpretation of a podzol pollen-analysis by Dr. G. W. Dimbleby.—Flora 146: 321–327.
Godwin, H. 1968 The origin of the exine.—New phytol. 67–676.
Godwin, H., P. Echlin, and B. Chapman 1967 The development of the pollen-grain wall in Ipomoea purpurea (L.) Roth.—Rev. palaeobot. palynol. 3: 181–195.
Göppert, H. R. 1836 De floribus in statu fossilis commentatio.—Nov. acta acad. Leopold.—Carol. natur. cur. 18, I: 547–572.
Göransson, H. 1982 Neolitikums begynnelse i Østergötland, Sverige, enligt pollenanalytiske data.—Pp. 99–124 in Sjøvold, T. (ed.) Introduksjon av jordbruk i Norden. Universitetsforlaget (Oslo, etc.)
Göransson, H. 1984 Pollen analytical investigations in the Sligo area.—Pp. 154–193 in Burenholt, G. (ed.) The archaeology of Carrowmore. Theses and papers in North-European archaeology. Stockholm.
Goldstein, S. 1960 Degradation of pollen by phycomycetes.—Ecology 41: 543–545.
Gordon, A. D., and H. J. B. Birks 1972 Numerical analysis methods in Quaternary palaeoecology. I. Zonation of pollen diagrams.—New phytol. 71: 961–979.
Goss, J. A. 1968 Development, physiology and biochemistry of corn and wheat pollen.—Bot. rev. 34: 333–358.
Graham, A. 1962 The role of fungal spores in palynology.—J. paleontol. 36: 60–68.
Grangeon, W., C. H. Greber, M. Locquin, and J. Roger 1962 Utilisation d'une machine taxinomique dans une branche des sciences naturelles: la palynologie.—Bull. bur. rech. géol. min. I: 1–15.
Granlund, E. 1932 De svenska högmossarnas geologi. —Sver. geol. unders. Ser. C 373. 193 pp.
Gregory, P. H. 1945 The dispersion of air-borne spores. —Trans. Br. mycol. soc. 28: 26–72.
Gregory, P. H. 1951 Deposition of air-borne Lycopodium spores on cylinders.—Ann. appl. biol. 38: 357–376.
Gregory, P. H., and J. L. Monteith (eds) 1967 Airborne microbes.—17th symp. soc. gen. microbiol. Imp. coll. Lond. April 1967. Cambridge University Press.
Greig, J. 1982 Interpretation of pollen spectra from urban deposits.—Pp. 47–65 in Hall and Kenward, 1982.
Grichuk, M. P. and V. P. Grichuk Гричук, М.П. и В.П. Гричук 1960 О приледниковой растительности на территории СССР. – Перигляциальные явления на территории СССР, 66–100. Изв. Моск. унив.
Groner, U. 1985 Palynologie der Karsthöhlensedimente im Hölloch, Zentralschweiz.—Thesis. Zürich. 172 pp.
Gross, H. 1935 Der Döhlauer Wald in Ostpreussen.—Beih. Bot. Zentralbl. 53B: 414–431.
Grosse-Brauckmann, G. 1961 Zur Terminologie organogener Sedimente.—Geol. Jahrb. 79: 117–144.
Grosse-Brauckmann, G., and E. Stix 1979 Beziehungen swischen Pollenkonzentrationen in der Luft und Pollenniederschlag.—Flora 168: 53–84.
Guillet, B. 1971 Etude palynologique des podzols III. —Pollen spores 12: 421–446.
Guinet, P. 1987 Geographic patterns of the main characters in genus Acacia (Leguminosae) with particular reference to subgenus Phyllodinae.—Pp. 293–311 in Blackmore and Ferguson, 1986.
Guinet, P., and A. L. Thomas 1973 Interprétation de la répartition dissymmetrique des couches de l'exine dans les pollens composés. Consequences relatives a la notion de l'aperture.—C.r. hebd. seances. acad. sci. Paris 276: 1545–1548.
Hafsten, U. 1951 A pollen investigation of two peat deposits from Tristan da Cunha.—Res. Nor. sci. exp. Tristan da Cunha 1937–38, 22. 42 pp.
Hafsten, U. 1956 Pollen-analytic investigation on the late Quaternary development of the inner Oslo fjord area. —Bergens Mus. årb. 1956 Naturvitensk. rk. 8. 161 pp.
Hafsten, U. 1959 Bleaching + HF + Acetolysis: a hazardous preparation process.—Pollen spores 1: 77–79.
Hafsten, U. 1960a Pleistocene development of vegetation and climate in Tristan da Cunha and Gough Island. —Årb. univ. Bergen Mat. naturvitensk. rk. 1960, 2. 48 pp.
Hafsten, U. 1960b Pollen-analytic investigations in

South Norway.—Pp. 434–462 in Holtedahl, O. (ed.) Geology of Norway. Nor. geol. unders. 208.

Hafsten, U. 1961 Pleistocene development of vegetation and climate in the southern High Plains, as evidenced by pollen analysis.—Pp. 59–91 in Wendorf, F. (ed.) Paleoecology of Llano Estacado. Fort Burgwin res. center. Publ. 1.

Hafsten, U. 1985 The immigation and spread of spruce forest in Norway, traced by biostratigraphical studies and radiocarbon datings. A preliminary report.—Nor. geogr. tidsskr. 39: 99–108.

Halden, B. E. 1917 Om torvmossar och marina sediment inom norra Hälsinglands litorinaområde.—Sver. geol. unders. Ser. 227 pp.

Hall, A. R., and H. K. Kenward 1982 Environment archaeology in the urban context.—CBA res. rep. 43.

Hamilton, A. C., and R. A. Perrott 1980 Modern pollen deposition in a tropical African mountain.—Pollen spores 22: 437–468.

Hansen, H. P. 1949 Pollen contents of moss polsters in relation to forest composition.—Am. Midl. nat. 42: 473–479.

Hardy, K. R., and H. Ottersten 1968 Two scales of convection in the clear atmosphere.—Proc. int. conf. cloud phys. Toronto: 534–538 (Am. meteorol. soc.).

Harris, T. M. 1971 [Discussion].—P. 271 only in Brooks *et al.*, 1971.

Havinga, A. J. 1963 A palynological investigation of soil profiles developed in cover sand.—Meded. landbouwhogesch. Wageningen 63: 1–92.

Havinga, A. J. 1971 An experimental investigation into the decay of pollen and spores in various soil types. —Pp. 446–478 in Brooks *et al.*, 1971.

Hebda, R. J., and J. N. A. Lott 1974 Effects of different temperatures and humidities during growth on pollen morphology: an SEM study.—Pollen spores 15: 563–572.

Hecht, A. (ed.) 1985 Paleoclimate analysis and modeling. —Wiley-Interscience (New York, etc.)

Hedberg, H. D. (ed.) 1972 An international guide to stratigraphic classification, terminology, and usage. —Lethaia 5: 283–295.

Heide, K. 1984 Holocene pollen stratigraphy from a lake and a small pond in north-central Wisconsin, USA. —Palynology 8: 3–20.

Heide, K., and R. H. W. Bradshaw 1982 The pollen–tree relationship within forests of Wisconsin and Upper Michigan, USA.—Rev. Palaeobot. Palynol. 36: 1–23.

Heim, J. 1970 Les relations entre les spectres polliniques récents et la végétation actuelle en Europe occidentale. —Thèse Louvain. 181 pp.

Heim, J. 1971 Etude statistique sur la validité des spectres polliniques provenant d'échantillons de mousses.—Lejeunea n.s. 58. 33 pp.

Heslop-Harrison, J. 1962 Origin of exine.—Nature 195: 1069–1071.

Heslop-Harrison, J. 1968a Tapetal origin of pollen-coat substances in Lilium.—New phytol. 67: 779.

Heslop-Harrison, J. 1968b Pollen wall development.— Science 161: 236–237.

Heslop-Harrison, J. 1969 An acetolysis-resistant membrane investing tapetum and sporogenous tissue in the anthers of certain compositae.—Can. j. bot. 47: 541–542.

Heslop-Harrison, J. (ed.) 1973 Pollen: development and physiology.—Butterworth (London). 378 pp.

Heslop-Harrison, J., and Y. Heslop-Harrison 1980 Cytochemistry and function of the Zwischenkoerper in grass pollens.—Pollen spores 22: 5–10.

Hesmer, H. 1933. Die natürliche Bestockung und die Waldentwicklung auf verschiedenartigen märkischen Standorten.—Z. Forst- Jagdwesen 65: 505.

Hesse, M. 1980 Entwicklungsgeschichte und Ultrastruktur von Pollenkitt und Exine bei nahe verwandten entomophilen und anemophilen Angiospermensippen der Alismataceae, Lemnaceae, Juncaceae, Cyperaceae, Poaceae und Araceae.—Plant syst. evol. 134: 229–267.

Hesse, M. 1981 Viscinfäden bei Angiospermen—homologe oder analoge Gebilde?—Microskopie 38: 85–89.

Hesse, M. 1984 An exine model for viscin threads.— Grana 29: 69–75.

Hesse, M., and K. Kubitzki 1983 The sporoderm ultrastructure in Persea, Nectandra, Hernandia, Gomortegea and some other lauralean genera.—Plant syst. evol. 141: 299–311.

Hesse, M., and M. Waha 1983 The fine structure of the pollen wall in Strelitzia reginae (Musaceae).—Plant syst. evol. 144: 285–293.

Hesse, M., and R. Zetter 1984 Viscin threads of a miocene species of the Onagraceae.—Pollen spores 26: 95–100.

Hesselman, H. 1919a Iakttagelser över skogsträdpollens spridningsförmåga.—Meddel. statens skogsförsöksanst. 16: 27–53.

Hesselman, H. 1919b Om pollenregn på hafvet och fjärrtransport af barrträdspollen.—Geol. för. Stockh. förh. 16: 89–99, 105, 107.

Heusser, C. J. 1978 Modern pollen spectra from Oregon. —Bull. Torrey bot. club 105: 14–17.

Heusser, C. J., and W. L. Balsam 1977 Pollen distribution in the Northeast Pacific Ocean.—Quat. res. 7: 45–62.

Hicks, S. 1986 Modern pollen deposition records from Kuusamo, Finland, II.—Grana 25: 183–204.

Hicks, S., and Hyyvärinen, V.—P. 1986 Sampling modern pollen deposition by means of 'Tauber traps'; some considerations.—Pollen spores 28: 219–242.

Hirst, J. M., and G. W. Hurst 1967 Long-distance spore transport.—Pp. 307–344 in Gregory and Monteith, 1967.

Ho, R. H. and J. H. Owens 1973 Microstrobili of lodgepole pine.—Can. j. forest. res. 8: 453–456. (Quoted from Biol. Abs. 57: 54524.)

Høeg, O. A. 1924 Pollen on humble-bees from Novaya Zemlya.—Rep. sci. res. Nor. exped. N. Zemlya 1921. 27. 18 pp.

Holling, R., and F. Overbeck 1960 Über die Grösse der Stoffverluste bei der Genese von Sphagnumtorfen.—Flora 150: 192–205.

Holmsen, G. 1919 Litt om grangrænsen i Fæmundstrakten.—Tidsskr. skogbr. 27: 39–48.

Holst, N. O. 1909 Postglaciala tidsbestämningar.—Sver. geol. unders. Ser. C 216. 74 pp.

Hoogenraad, H. R. 1935 Studien über die sphagnicolen Rhizopoden der niederländischen Fauna.—Arch. Protistenkd. 84: 1–100.

Hopkins, J. 1950 Different flotation of and deposition of conifer and deciduous tree pollen.—Ecology 31: 633–641.

Huntley B., and H. J. B. Birks 1983 An atlas of past and present pollen maps for Europe, 0–13 000 years ago. —Cambridge University Press (Cambridge).667 pp.

Hustedt, F. 1930 Bacillariophyta (Diatomeae).—In Die Süsswasserflora Mittel-Europas herausg. v. A. Pascher 10. 466 pp. Now replaced by Krammer and Lange-Bertalot, 1986, 1988.

Hustedt, F. 1930-62 Die Kieselalgen.—In Rabenhorst, L. Kryptogamenflora von Deutschland, Österreich and der Schweiz. 7, 1-2. 920, 845 pp.

Hyde, H. A. 1955 Oncus, a new term in pollen morphology.—New phytol. 54: 255–256.

Hyde, H. A. 1972 Atmospheric pollen and spores in relation to allergy. I.—Clin. aller. 2: 153–179.

Hyde, H. A., and D. A. Williams 1945a Pollen of lime (Tilia sp.).—Nature 155: 457 only.

Hyde, H. A., and D. A. Williams 1945b Studies in atmospheric pollen. II. Diurnal variation in the incidence of grass pollen.—New phytol. 44: 83–94.

Hyde, H. A., and D. A. Williams 1946 Studies in atmospheric pollen. III. Pollen production and pollen incidence in ribwort plantain (Plantago lanceolata L.). —New phytol. 45: 271–277.

Ichikura, M., and Y. Iwanami 1981 Studies on fall-velocities of pollen grains.—Jap. j. palynol. 27: 5–13.

Ikuse, M. 1956 Pollen grains of Japan.—Hirokawa (Tokyo) 303 pp.—Jap. J. palynol. 27: 5–13.

Ikuse, M. 1965 On the number of pollen grains contained in anthers of some plants.—Quart. rev. 4: 144–149. (Quoted from Exc. bot. 11: 171).

Istomina, E.S. Истомина, Е.С., М. Коренева и С.Н. Тюремнов 1938: Атлас растиельных остатков, встречаемых Б торфе. – Ленинград (Акад. наук. СССР)

Iversen, J. 1936 Sekundäres Pollen als Fehlerquelle.—Dan. geol. unders. 4 rk. 2, 15. 24 pp.

Iversen, J. 1941 Landnam i Danmarks Stenalder.—Dan. geol. unders. 2 rk. 66. 68 pp.

Iversen, J. 1944 Viscum. Hedera, and Ilex as climate indicators.—Geol fören. Stockh. förh. 66: 463–483.

Iversen, J. 1947a [Discussion interventions].—In Geologklubben 1947.

Iversen, J. 1947b Plantevækst, dyreliv og klima i det senglaciale Danmanrk.—Geol. fören. Stockh. förh. 69: 67–78.

Iversen, J. 1949 The influence of prehistoric man on vegetation.—Dan. geol. unders. 4 rk. 3, 6. 25 pp.

Iversen, J. 1954 The late-glacial flora of Denmark and its relation to climate and soil.—Dan. geol. unders. 2 rk. 80: 87–119.

Iversen, J. 1956 Forest clearance in the Stone Age.—Scientific American 194: 36–40.

Iversen, J. 1958 The bearing of glacial and interglacial epochs on the formation and extinction of plant taxa. —Uppsala univ. årssk. 6: 210–215.

Iversen, J. 1960 Problems of the early Post-Glacial forest development in Denmark.—Dan. geol. unders. 4 rk. 4, 3. 32 pp.

Iversen, J. 1964 Retrogressive vegetational succession in the Post-Glacial.—J. ecol. 52 Suppl: 59–70.

Iversen, J. 1969 Retrogressive development of a forest ecosystem demonstrated by pollen diagrams from fossil mor.—Oikos Suppl. 12: 35–49.

Iversen, J. 1973 The development of Denmark's nature since the last glacial.—Dan. geol. unders. 5.rk. 7. 126 pp.

Iversen, J., and J. Troels-Smith 1950 Pollenmorfologiske Definitioner og Typer.—Dan. geol. unders. 4 rk. 3, 8. 54 pp.

Jacobs, B. F. 1986 Identification of pine pollen from the southwestern United States.—Pp. 155–168 in Jacobs *et al.*, 1986.

Jacobs, B. F., P. L. Fall, and O. K. Davis 1986 Late Quaternary vegetation and climates of the American southwest.—Contrib. Am. ass. stratigr. palynol. 16. 190 pp.

Janssen, C. R. 1966 Pollen spectra from the deciduous and coniferous forests of northeastern Minnesota.—Ecology 47: 804–835.

Janssen, C. R. 1980 Some remarks on facts and interpretation in Quaternary palynostratigraphy.—Bull. ass. franc. étude Quat. 1980: 171–176.

Janzon, L. Å. 1981 Airborne pollen grains under winter conditions.—Grana 20: 183–185.

Jaworka, Z. 1971 The results of palynological research of icemarginal lake deposits in Jelenia Gora.—Kwart. geol. 15: 947–954. (Quoted from Exc. bot. 22: 95).

Jentys-Szaferowa, J. 1928 La structure des membranes du pollen de Corylus, de Myrica et des espèces européenes de Betula et leur détermination à l'état fossile. —Bull. internat. acad. polon. sci. et lettr. Cl. sci. math. nat. Sér. B 1. Bot. 1928: 75–125.

Jentys-Szaferowa, J. 1959 A graphical method of comparing the shapes of plants.—Rev. Polish acad. sci. 4: 9–38.

Jessen, K. 1920 Moseundersøgelser i det nordøstilge Sjælland.—Dan. geol. unders. 2 rk. 34. 268 pp.

Jessen, K. 1935 Archaeological dating in the history of North Jutland's vegetation.—Acta Archaeol. 5: 185–214.

Jessen, K. 1938 Some west baltic pollen diagrams.—Quartär 1: 124–139.

Jessen, K. 1949 Studies in the late Quaternary deposits and flora-history of Ireland.—Proc. R. Ir. acad. 52, B: 85–290.

Jørgensen, S. 1967 A method for absolute pollen counting.—New phytol. 66: 489–493.

Johnsen, J. 1978 Outwash of terrestric soils into Lake Saksunarvatn, Faroe Islands.—Dan. geol. unders. AÅrb. 1977: 31–37.

Jones, D. J. 1958 Displacement in microfossils.—Sedim. petrol. 28: 453–467.

Jorde, W., and H. F. Linskens 1974 Zur Persorbtion von Pollen und Sporen durch die intakte Darmschleimhaut.—Acta allergol. 29: 165–175.

Juvigne, E. 1973 Densité des exines de quelques espèces de pollens et spores fossiles.—Ann. soc. géol. belg. 96: 363–372. (Quoted from Biol.Abstr. 58: 21195).

Käpylä A., and A. Penttionen 1981 An evaluation of the microscopical counting method of the tapes in Hirst-Burkard pollen and spore traps.—Grana 20: 131–142.

Kaland, P. E. 1976 Adsorbsjon av humussyrer i trekull. —Arkeo 1976: 9–11.

Kaland, P. E. 1986 The origin and management of Norwegian coastal heaths as reflected by pollen analysis. —Pp. 19–38 in Behre, 1986.

Kaland, P. E., K. Krzywinski, and B. Stabell 1983 Radio-carbon dating of transitions between marine and lacustrine sediments and their relation to the development of lakes.—Boreas 13: 243–258.

Kaland, P. E., and Stabell, B. 1981 Method for absolute diatom-frequency analysis and combined diatom and pollen analysis of sediments.—Nord. J. bot. 1: 679–700.

Kapp, R. O. 1969 How to know pollen and spores.—Brown (Dubuque). 249 pp.

Kappen, L., and Straka, H. 1988 Pollen and spores transport into the Antarctic.—Polar biol. 8: 73–180.

Katts, N.Y. Кацц, Н.Я. и Кацц, В. 1933 Атлас растиельных остатков в торфе.–Ленингад (Сельхошз)

Kidson, E. J., and G. L. Williams 1969 Concentration of palynomorphs by use of sieves.—Oklahoma geol. notes. 29: 117–119.

Kidson, E. J. and G. L. Williams 1971 A device for the manipulation of micro-fossils.—Pollen spores 13: 359–364.

Kiese, O. 1972 Bestandesmeteorologische Untersuchungen zur Bestimmung des Wärmehaushalts eines Buchenwaldes.—Ber. Inst. Met. Klimatol. Techn. Univ. Hannover. 6. 132 pp.

Kirchheimer, F. 1940 Hundert Jahre Pollenforschung im Dienste der Paläobotanik.—Planta 31: 414–417.

Kjellman, W., T. Kallstenius, and O. Wagner 1950 Soil sampler with metal foils: device for taking undisturbed samples of very great length.—R. Swed. geotechn. inst. Proc. 1. 75 pp.

Klaus, W. 1953 Zur Einzelpräparation fossiler Sporomorphen.—Mikroskopie 8: 1–14.

Knoll, F. 1930 Über Pollenkitt und Bestäubungsart.—Z. Bot. 23: 609–675.

Knoll, F. 1932 Über die Fernverbreitung des Blütenstaubes durch den Wind.—Forsch. Fortschr. 8: 301–302.

Knoll, F. 1936 Eine Streuvorrichtung zuer Untersuchung der Pollenverkittung.—Oesterr. bot. Z. 85: 161–182.

Knox, R., and J. Heslop-Harrison 1969 Cytochemical localization of enzymes in the wall of the pollen grain. —Nature 233: 92–94.

Köhler, E., and P. Brückner 1982 Die Pollenkörner der afrikanischen Buxus—und Notobuxus-Arten (Buxaceae) und ihre systematische Bedeutung.—Grana 21: 71–82.

Koski, V. 1970 A study of pollen dispersal as a mechanism of gene flow in conifers.—Metsätutkimuslaistoksu julkasuja (Comment. inst. forst. fenn.) 70, 4. 78 pp.

Kozumplik, V., and B. R. Christie 1972 Dissemination of orchard-grass pollen.—Can. j. plant sci. 52: 997–1002 (Quoted from Biol. abstr. 55: 52544).

Krammer, K., and H. Lange-Bertalot 1986, 1988. Bacillariophyta. Vol. 2, 1/2 in Ettly, H. *et al.* (eds) Die Süsswasserflora von Mitteleuropa.

Kremp, G. O. W. 1965 Morphologic encyclopedia of palynology.—University of Arizona Press (Tucson). 185 pp.

Kress, W. J. 1986 The use of ethanolamine in the study of pollen wall stratification.—Grana 25: 31–40.

Kress, W. J., D. E. Stone, and S. C. Sellers 1978 Ultrastructure of exineless pollen in Heliconia (Heliconiaceae).—Am. j. bot. 65: 1064–1076.

Krzywinski, K. 1976 En registrering av resent pollenregn målt i forskjellig vegetasjon på Milde, Bergen.—Thesis Univ. Bergen. 215 pp.

Krzywinski, K. 1977 Different pollen deposition mechanisms in forest: a simple model.—Grana 16: 199–202.

Krzywinski, K. 1979 Preliminær undersøkelse av planterester i latrine.—Arkeo 1: 31–33.

Krzywinski, K., and K. Fægri 1974 Etnobotanisk bidrag til funksjonsanalyse.—Arkeo 1: 33–40.

Krzywinski, K., S. Fjelldal, and E. C. Soltvedt 1983 Recent palaeoethnobotanical work at the Mediaeval excavations at Bryggen, Bergen, Norway.—Pp. 145–169 in Proudfoot, B. (ed.) Site, environment, and economy. BAR internat. ser. 173.

Krzywinski, K., and E. C. Soltvedt 1988 A mediaeval brewery (1258–1450) in Bryggen, Bergen.—The Bryggen Papers. Suppl. ser. 2. In press.

Kummel, B., and D. Raup 1966 Handbook of paleontological technique.—Freeman (San Francisco). 852 pp.

Kupias, K. R., A. Koivikko, and Y. Mäkinen 1981 Liberation of Taraxacum and Leucanthemum pollen in the air through mechanical agitation.—Grana 20: 199–205.

Kuprianova, L.A. Л. А. Куприянова 1957 Анализ пыльцы растительных остатков из желудка березовского мамонта (К Вопросу о характере растительности эпохи березовского мамоита) Pp. 331–358 in Сворник памяти Африкана николаевича криштофовича

Kuyl, O. S., U. Muller, and H. T. Waterbolk 1955 The application of palynology to geology with reference to western Venezuela.—Geol. Mijnb. N. S. 17: 49–76.

Kvamme, M. 1988 Lokale pollendiagram og bosetningsforhold.—36 pp. I Näsman, U. and Lund, J. (eds) Folkevandringstiden i Norden. Univ. forlag. Århus. In press.

Låg, J. 1949 Forstyrret lagrekke i torv ved vatnet Spålen i Norderhov.—Nor. geol. t. 28: 33–39.

Lanner, R. 1966 Needed: A new approach to the study of pollen dispersal.—Silvae genet. 15: 50–52.

Larsen, E., F. Eide, O. Longva, and J. Mangerud 1984 Allerød-Younger Dryas climatic inferences from cirque glaciers and vegetational development in the Nordfjord area, western Norway.—Arctic alpine res. 16: 137–160.

Lie, S. E., B. Stabell, and J. Mangerud 1983 Diatom stratigraphy related to Late Weichselian sea-level changes in Sunnmøre, western Norway.—Nor. geol. unders. 380: 203–219.

Linskens, W. F. (ed.) 1960 Pollen physiology and fertilization—North Holland (Amsterdam).

Liu, K.-B., and Colinvaux, P. W. 1988 A 5200-year history of Amazon rain forest.—J. biogeogr. 15: 231–248.

Livingstone, D. A. 1955 A light-weight piston sampler for lake deposits.—Ecology 36: 137–139.

Løvlie, R., K. Krzywinski, and J. Kjøde 1979 Impregnation of organic lake sediments for palaeomagnetic measurements.—Phys. earth planet int. 20: 22–24.

Louveaux, J. 1955 Introduction à l'étude de la recolte du pollen par les abeilles (Apis mellifica L.).—Physiol. comp. oecol. 4: 1–54.

Louveaux, J. 1970 Atlas photographique d'analyse pollinique des miels.—Minist. agric. Serv. repress. fraudes controle qual. 3. 52 pl.

Louveaux, J., A. Maurizio, G. Verwohl 1970 Methods of melissopalynology.—Bee world 51: 125–138.

Ludlam. F. H. 1967 The circulation of air, water, and particles in the troposphere.—Pp. 1–17 in Gregory and Monteith, 1967.

Lüdi, W. 1937 Die Pollensedimentation im Davoser Hochtale.—Ber. geobot. Inst. Rübel Zürich 1936: 107–127.

Lüdi, W. 1947 Der Pollengehalt von Oberflächenproben am Katzensee bei Zürich.—Ber. geobot. Inst. Rübel Zürich 1946: 82–92.

Lüdi, W., and V. Vareschi 1936 Die Verbreitung und der Pollenniederschlag der Heufieberpflanzen im Hochtale von Davos.—Ber. geobot. Inst. Rübel Zürich 1935: 47–111.

Lundqvist, G. 1924 Utvecklingshistoriska insjöstudier.—Sver. geol. unders. Ser. C 330. 129 pp.

Lundqvist, G. 1938 Sjösediment från Bergslagen (Kolbäcksåns vattenområde).—Sver. geol. unders. Ser. C 420. 186 pp.

Lundqvist, J., and K. Bengtsson 1970 The red snow. A meteorological and pollen-analytical study of long-transported material from snowfall in Sweden.—Geol. fören. Stockh. för. 92: 288–307.

McAndrews, J. H. 1967: Pollen analysis and vegetational history of the Itaska region, Minnesota.—Pp. 219–236 in Cushing, E. J. and H. E. Wright, (eds) Quaternary palaeoecology. Yale University Press (New Haven).

McAndrews, J. H. 1984 Pollen analysis of the 1973 ice core from Devon Island glacier, Canada.—Quat. res. 22: 68–76.

McAndrews, J. H., A. A. Berti, and G. Norris 1973 Key to the Quaternary pollen and spores of the Great Lakes region.—R. Ont. mus. Life sc. Misc. publ. 61 pp.

McAndrews, J. H., and D. M. Power 1973 Palynology of the Great Lakes: the surface sediments of Lake Ontario.—Can. j. earth sci. 10: 777–792.

McAndrews, J. H., and G. Sansom 1977 Analyse pollinique et implications archéologique et géomophologiques, Lac de la Hutte Sauvage (Mushuau-Nipi), Nouveau-Québec.—Geogr. phys. Quat. 31: 177–183.

McAndrews, J. H., and A. R. Swanson 1967 The pore number of periporate pollen with special reference to Chenopodium.—Rev. palaeobot. palynol. 3: 105–117.

McAndrews, J. H., and H. E. Wright Jr 1969 Modern pollen rain across the Wyoming basin and the northern Great Lakes (USA).—Rev. palaeobot. palynol. 9: 17–43.

MacDonald, G. M., and Ritchie, J. C. 1986 Modern pollen spectra from the western interior of Canada and the interpretation of Late Quaternary vegetation development.—New phytol. 103: 245–268.

McDonald, J. E. 1962 Collection and washout of airborne pollens and spores by raindrops.—Science 135: 435–437.

Maher, L. J. 1964 Ephedra pollen in the sediments of the Great Lakes region.—Ecology 45: 391–395.

Maher, L. J. 1972 Nomograms for computing 0.95 confidence limits of pollen data.—Rev. palaeobot. palynol. 13: 85–93.

Malgina, E.A. Малгина, Е.А. 1966 Об интерпретации ресултатов споровопылцево анализа четвертицных отлошений средней Азии. – In Нейштадт, М. И. (ред.): Значение палинологичецкого анализа для стратиграфии и палеофлористики: 256–262.

Malmström, C. 1923 Degerö stormyr.—Medd. Statens skogsförsöksanst. 20, 1. 176 pp.

Mamakova, K. 1968 Lille Bukken and Lerøy—two pollen diagrams from western Norway.—Årb. univ. Bergen. Mat.-naturvitensk. rk. 1968, 4. pp.

Mandrioli, P., M. G. Negrini, G. Cesari, and G. Morgan 1984 Evidence for long-range transport of biological and anthropogenic aerosol particles in the atmosphere. —Grana 23: 43–53.

Mangerud, J., S. T. Andersen, B. E. Berglund, and J. J. Donner 1974 Quaternary stratigraphy of Norden, a proposal for terminology and classification.—Boreas 3: 109–128.

Manten, A. A. 1969 Bibliography of palaeopalynology 1836–1966.—Rev. palaeobot. palynol. 8: 1–570.

Markgraf, V. 1980 Pollen dispersal in a mountain area. —Grana 9: 127–146.

Martens, P., and L. Waterkeyn 1962 Structure du pollen 'ailé' chez les conifères.—La Cellule 62: 171–222.

Martin, P. S., and J. E. Mosimann 1965 Geochronology of pluvial Lake Cochise, southern Arizona. III. Pollen statistics and Pleistocene metastability.—Am. j. sci. 263: 313–358.

Martin, P. S., B. E. Sabels, and D. Shutler Jr 1961 Rampant cave coprolite and ecology of the Shasta Ground sloth. Amer. j. sci. 259: 102–127.

Martin, P. S., and F. W. Sharrock 1964 Pollen analysis of prehistoric human faeces: A new approach to ethnoboany. Amer. antiq. 30: 168–180.

Maurizio, A. 1949 Pollenanalytische Untersuchungen an Honig und Pollen-Höschen.—Beih. schweiz. Bienen-Ztg. 2: 320–455.

Maurizio, A. 1953 Weitere Untersuchungen an Pollen-Höschen.—Beih. schweiz. Bienen-Ztg. 2: 487–556.

Maurizio, A., and J. Louveaux 1967 Les méthodes et le terminologie en melissopalynologie.—Rev. palaeobot. palynol. 3: 291–295.

May, K. R. 1967 Physical aspects of sampling airborne microbes.—Pp. 60–80 in Gregory and Monteith, 1967.

Mepham, R. H., and A. R. Lane 1968 Exine and the role of the tapetum in pollen development.—Nature 219: 961–962.

Meriläinen, Y. 1967 The diatom flora and hydrogen ion concentration.—Ann. bot. fenn. 4: 51–58.

Meyer-Berthaud, B. 1986 Melissotheca: a new pteridosperm pollen organ from the Lower Carboniferous of Scotland.—Bot. j. Linn. Soc. 93: 277–290.

Middeldorp, A. A. 1982 Pollen concentration as a basis for indirect dating and quantifying net organic and fungal production in a peat bog ecosystem.—Rev. palaeobot. palynol. 37: 225–282.

Mikkelsen, V. 1949 Has temperature any influence on pollen size?—Physiol. plant. 2: 323–324.

Mirams, R. R. 1953 A comparison of the gravity and impact methods of collecting spores and pollen grains from the atmosphere at Wellington, N.Z.—N.Z. j. sci. technol. B 34: 378–383.

Miyoshi, N. 1969 Light- and electron microscope studies of spores in the musci. I.—Hikobia 5: 172–177.

Moe, D. 1970 The post-Glacial immigration of Picea abies into Fennoscandia.—Bot. not. 123: 61–66.

Moe, D. 1974a Studies in the post-Glacial vegetation development on Hardangervidda, Southern Norway. I. The occurrence and origin of pollen of plants favoured by man's activity.—Nor. arch. rev. 6 (1973): 67–73.

Moe, D. 1974b Identification key for trilete microspores of Fennoscandian pteridophytes.—Grana 14: 132–142.

Monoszon, M.X. Моносзон, М. Х. 1959a Рассеивание воздушным путем пылы маревых. – Работ. спор.-пыльц анализ./Мат. геоморф. палео-геогр. СССР. 21: 157–165.

Monoszon, M.X. 1959b Описание пыльцы представителей семейства Ulmaceae, произрастающих на территории СССР./для целей пыльцевого анализа/Ibid. 187–198.

Monoszon, M.X. 1961 О барициях морфологиеских признаков пылци некоторых видов дуба. – Доки. ак. наук. СССР. 140: 1456.

Monoszon, M.X. 1971 О видовых определьниях пыльцов ныльцов некотопых видоб родов Alnus и Alnaster. – Виол наук 14: 65–74. (Quoted from Biol. Abstr. 54.44736).

Moore, D. G. 1961 The free-corer: sediment sampling without wire and winch.—J. Sedim. petrol 31: 627–631.

Moore, P. D. and Webb, J. A. 1978 An illustrated guide to pollen analysis.—Hodder and Stoughton (London, etc.) 133 pp.

Mosimann, J. E. 1962 On the compound multinomial distribution, the multivariate β-distribution and correlation among proportions.—Biometrika 49: 65–82.

Mosimann, J. E. 1963 On the compound negative multinomial distribution and correlation among inversely sampled pollen counts.—Biometrika 50: 47–54.

Müller, I. 1947 Der pollenanalytische Nachweis der menschlichen Besiedlung im Federsee- und Bodenseegebiet.—Planta 35: 70–87.

Müller, P. 1937 Das Hochmoor von Etzelwill.—Ber. geobot. inst. Rübel Zürich 1938: 85–106.

Muller, J. 1959 Palynology of recent Orinoco delta and shelf sediments.—Micropaleontology 5: 1–32.

Muller, J. 1970 Pollen morphology of the genus Lepisanthes (Sapindaceae) in relation to taxonomy. —Blumea 18: 507–561.

Murkaite, R. 1970 (Pollen dispersal in Picea abies).—Tr. litov. nauchnoissled. inst. lesn. khoz. 11: 123–134. (Quoted from Biol. Abstr. 54: 56459)

Nakagava, S., and M. Katsuda 1975 On the pollinosis caused by Dalmatian chrysanthemum.—Jap. j. palynol. 15: 45–56.

Nejschtadt, M.I. Нейштадт, М.И. 1952 Споровопыльцевой метод в СССР. – Москва (Акад. Наук.) 221 pp.

Nejschtadt, M.I. 1957 История лесов и палеогеография в голоцене. – Москва (Акад. Наук.) 403 pp.

Nejschtadt, M.I. 1960 Палинология в СССР./1952–1957гг/. – Москва (Акад. Наук.) 271 pp.

Neustadt, M. J. 1961 Zur Geschichte der Seen im Holozän.—Verh. Int. Ver. Limnol. 14: 279–284.

Newmark, F. M., and A. M. Salley 1972 Gravity slide study at McMurdo Station, Antarctica.—Ann. allergy 30: 67–68.

Nichols, H. 1967a The suitability of certain categories of lake sediments for pollen analysis.—Pollen spores 9: 615–620.

Nichols, H. 1967b The disturbance of arctic lake sediments by 'bottom ice'.—Arctic 20: 213 only.

Nicklas, K. J. 1981 Simulated wind pollination and airflow around ovules of some early seed plants.—Science 211: 275–277.

Niklas, K. J., and E. J. Buchman 1986 Wind pollination in two sympatric species of Ephedra.—Am. j. bot. 73: 674 only. (Abstr.).

Øvstedal, D. O. 1985 The vegetation of Lindås and Austrheim, western Norway.—Phytocoenologica 13: 323–449.

Ogden, E. C., G. S. Raynor, and J. V. Hayes 1971 Travels of airborne pollen.—Progr. rep. 11. N.Y. State mus. sci. serv. Mimeographed. 25 pp.

Ogden, E. C., G. S. Raynor, J. V. Hayes, D. M. Lewis, and J. H. Haines 1974 Manual for sampling airborne pollen.—Hafner (New York). 182 pp.

Ohnstad, F. E., and J. G. Jones 1982 The Jenkins surface-mud sampler. Users manual.—Freshw. biol ass. Occ. pap. 15. 45 pp.

Olausson, E. 1957 Das Moor Roshultsmyren.—Lunds univ. årsskr. NF 2, 53, 12. 72 pp.

Oldfield, F. 1959 The pollen morphology of some of the West European Ericales. Preliminary descriptions and a tentative key to their identification.—Pollen spores 1: 19–48.

Olsson, I. U. 1986 Radiometric dating.—Pp. 273–312 in Berglund, 1986.

Ording, A. 1934 Om nye metoder og hjelpemidler ved pollenanalytiske undersøkelser.—Meddel. (17) nor. skogforsøksves. 5: 159–196.

O'Rourke, M. K., and M. D. Lebowitz 1984 A comparison of regional atmospheric pollen with pollen collected at and near homes.—Grana 23: 55–64.

O'Rourke, M. K., and J. I. Mead 1986 Late Pleistocene and Holocene pollen records from two caves in the Grand Canyon of Arizona, USA.—Pp. 169–185 in Jacobs *et al.*, 1986.

Osvald, H. 1923 Die Vegetation des Hochmoores Komosse.—Svenska växtsociol. sällsk. handl. 1. 436 pp.

Overbeck, F. 1934 Zur kenntnis der Pollen mittel-und nordeuropäischer Ericales.—Beih. Bot. Centralbl. 51. II: 566–583.

Overbeck, F. 1947 Studien zur Hochmoorentwicklung in Niedersachsen und die Bestimmung der Humifizierung bei stratigraphisch-pollenanalytischen Mooruntersuchungen.—Planta 35: 1–56.

Overbeck, F. 1958 Ein Ergänzungsgerät zum Dachnowsky-Moorbohrer.—Veröff. geobot. Inst. Rübel Zürich 34: 119–120.

Overbeck, F. 1975 Botanisch-geologische Moorkunde. —Wacholtz (Neumünster). 719 pp.

Overbeck, F., and H. Schmitz 1931 Zur Geschichte der Moore, Marschen und Wälder Nordwestdeutschlands. I. Das Gebiet von der Niederweser bis zur unteren Ems.—Mitt. Prov.-St. Naturdenkmalpflege Hannover 3: 1–179.

Overbeck, F., K. Münnich, L. Aletsee, and F. R. Averdieck 1957 Das Alter der "Grenzhorizonts' norddeutscher Hochmore nach Radiocarbon-Datierungen.—Flora 145: 37–71.

Pande, G. K., R. Pakrash, and M. A. Hassam 1972 Floral biology of barley (Hordeum vulgare L.)—Ind. j. agric. sci. 48: 697–703. (Quoted fr. Biol. Abstr. 56: 293 10.)

Parker, R. L. 1926 The collection and utilization of pollen by honeybees.—Cornell univ. agric. exp. st. Mem. 98. 54 pp.

Paus, Aa. 1982 Paleo-økologiske undersøkelser på Frøya, Sør-Trøndelag.—Thesis (cand.real.) Univ. Trondheim.

Paus, Aa. 1988 Late Weichselian vegetation, climate, and floral migration at Sandvikvatn, North Rogaland, southwestern Norway.—Boreas 17: 133–139.

Pennington, W. 1947 Studies of the Post-Glacial history of British vegetation. VII. Lake sediments: Pollen diagrams from the bottom deposits of the North Basin of Windermere.—Philos. trans. r. soc. Edinb. 233: 137–175.

Pennington, W. 1975 The effect of Neolithic man on the environment in north-west England: the use of absolute pollen diagrams.—Pp. 74–86 in Evans *et al.*, 1975.

Pennington, W., E. Y. Haworth, A. P. Bonny, and J. P. Lishman 1972 Lake sediments in northern Scotland. —Philos. trans. r. soc. London B 264: 191–294.

Pettitt, J. M. 1966 Exine structure in some fossil and recent spores as revealed by light and electron microscopy.—Bull. Brit. mus. nat. hist. Geol. 13: 221–257.

Pettitt, J. M., and A. C. Jermy 1975 Pollen in hydrophilous angiosperms.—Micron 5: 377–405.

Pohl, F. 1937 Die Pollenerzeugung der Windblütler.—Bei. bot. Centralb. 56 A: 365–470.

Pokrovskaya, E.M. Покорвская И.М./ред./1950 Пыльцевой аналив. – Москва. – Cf. French translation 1958: Ann. serv. d'inf. géol. BRGM.

Pokrovskaya, E.M. 1966 Паиеопалиногия. I. – III. – тр. всесоюз. научн. – исслеоговательского геол. инст. (ьсегеи) Нов. сер. 141. 1158 pp.

Potonié, R. 1934 Zur Mikrobotanik der Kohlen und ihrer Verwandten. I. Zur Morphologie der fossilen Pollen und Sporen.—Arb. Inst. Paläobot. Petrogr. Brennsteine 4: 5–125.

Potter, L. D. 1967 Differential pollen accumulation in water-tank sediments and adjacent soils.—Ecology 48: 1041–1043.

Potter, L. D., and J. Rowley 1960 Pollen rain and vegetation, San Augustin Plains, New Mexico.—Bot. Gaz. 122, 1.

Potzger, J. E. 1975 A borer for sampling in permafrost.—Ecology 36: 161 only.

Potzger, J. E., A. Courtemanche, M. Sylvio, and Fr. M. Hueber 1957 Pollen from moss polsters on the mat of Lac Shaw bog, Quebec, correlated with a forest survey.—Butler univ. bot. stud. 13 (1956): 24–35.

Praglowski, J. R. 1966 On pollen size variations and the occurrence of Betula nana in different layers of a bog.—Grana palynol. 6: 528–543.

Praglowski, J. R. 1970 The effects of pre-treatment and the embedding media on shape of pollen grains.—Rev. palaeobot. palynol. 10: 203–208.

Prentice, H. C., O. Mastenbroeck, W. Berendsen, and P. Hogeweeg 1984 Geographic variation in the pollen of Silene latifolia (S. alba, S. pratensis): a quantitative morphological analysis of population data.—Can. j. bot. 62: 1259–1267.

Prentice, I. C. 1986 Forest-composition calibration of pollen data.—Pp. 799–816 in Berglund, 1986.

Punt, W. 1962 Pollen morphology of the Euphorbiaceae with special reference to taxonomy.—Wentia 7: 1–116.

Puttock, C. F., and C. J. Quinn 1980 Perispore morphology and the taxonomy of the Australian Aspleniaceae.—Austr. j. bot. 28: 305–322.

Raynor, G. S. 1971 Wind and temperature structure in a coniferous forest and a contiguous field.—Forest sci. 17: 351–363.

Raynor, G. S., L. Cohen, J. V. Hayes, and E. C. Ogden 1966 Dyed pollen grains and spores as tracers in dispersion and deposition studies.—J. appl. microbiol. 5: 728–729.

Raynor, G. S., E. Maynard, A. Irving, and E. C. Ogden 1961 Pollen sampling and dispersion studies at Brookhaven national laboratory.—APCA Jounr. 11: 557–562, 584.

Raynor, G. S., E. C. Ogden, and J. V. Hayes 1972a Dispersion and deposition of corn pollen from experimental sources.—Agron. j. 64: 420–427.

Raynor, G. S., E. C. Ogden, and J. V. Hayes 1972b Dispersion and deposition of timothy pollen from experimental sources.—Agric. meteorol. 9 (1971–1972): 347–366.

Raynor, G. S., E. C. Ogden, and J. V. Hayes 1974 Enhancement of particulate concentrations downwind of vegetative barriers.—Agric. meteorol. 13: 181–188.

Reader, R. J. D., and J. M. Stewart 1972 The relationship between net primary production and accumulation for a peatland in southeastern Manitoba.—Ecology 53: 1024–1037.

Reddi, C. S., and N. S. Reddi 1986 Pollen production in some anemophiles.—Grana 25: 55–61.

Reissinger, A. 1936 Methode der Bohrungen in Seen zur Untersuchung von Sedimentschichten.—Internat. Rev. ges. Hydrobiol. Hydrogr. 33: 1–24.

Reitsma, T. 1966 Pollen morphology of some European Rosaceae.—Acta bot. neerl. 15: 290–307.

Reitsma, T. 1969 Size modification of recent pollen grains under different treatments.—Rev. palaeobot. palynol. 9: 175–252.

Rempe, H. 1937 Untersuchungen über die Verbreitung des Blütenstaubes durch die Luftströmungen.—Planta 227: 93–147.

Renault-Miskovski, J., and M. Girard 1978 Analyse pollinique du remplissage pleistocène inferieur et moyen de la grotte du Vallonet (Rocquebrune-Cap-Martin, Alpes-Maritimes).—Geol méditerr. 5: 385–402.

Renault-Miskovski, J., Bui-Thi-M., and Girard, M. 1985 Palynologie archèol.—CNRS Centre rec. archéol. Notes monogr. techn. 17. 502 pp.

Reyment, R. A., R. E. Blackith, and N. A. Campbell 1984 Multivariate morphometrics. 2nd edn.—Academic Press (London, etc.) 233 pp.

Risch, C. 1940 Die Pollenkörner der in Deutschland vorkommenden Labiaten.—Verh. Bot. Ver. Prov. Brandenburg 80, 21.

Ritchie, J. C. 1974 Modern pollen assemblages near the arctic tree line, Mackenzie delta region, Northwest Territories.—Can. j. bot. 52: 381–396.

Ritchie, J. C., and S. Lichti-Federovich 1963 Contemporary pollen spectra in central Canada. I. Atmospheric samples at Winnipeg, Manitoba.—Pollen spores 5: 95–114.

Ritchie, J. C., and S. Lichti-Federovich 1967 Pollen dispersal phenomena in arctic–subarctic Canada.—Rev. palaeobot. palynol. 3: 255–266.

Robertsson, A.-M. 1973 Late-glacial and pre-Boreal pollen and diatom diagrams from Skurup, Southern Scania.—Sver. geol. unders. Ser. C 679. 75 pp.

Roland, F. 1966, 1968 Étude de l'ultrastucture des apertures: I. Pollen à pores. II. Pollen à sillons.—Pollen spores 8: 409–419, 10: 479–520.

Roland-Heydacker, P. 1979 Aspects ultrastructuraux de

l'ontogenèse des pollens et tapis chez Mahonia aquifolia Hutt. Berberidaceae.—Pollen spores 21: 259–278.
Rowell, T. K., and J. Turner 1985 Litho-, humic and pollen stratigraphy at Quick Moss, Northumberland. —J. ecol. 73: 11–25.
Rowley, J. R. 1964 Formation of the pore in pollen of Poa annua.—Pp. 59–69 in Linskens, 1964.
Rowley, J. R. 1967 Fibrils, microtubules, and lamellae in pollen grains.—Rev. palaeobot. palynol. 3: 213–226.
Rowley, J. R., and A. O. Dahl 1956 Modifications in design and use of the Livingstone piston sampler.—Ecology 37: 849–851.
Rowley, J. R., and A. Dunbar 1967 Sources of membranes for exine formation.—Sven. bot. tidskr. 61: 49–64.
Rowley, J. R., and J. J. Flynn 1969 Membranes, pressure, and exine forms: a reinterpretation of exine formation on aborted pollen grains.—Pollen spores 11: 169–180.
Rowley, J. R., and J. Rowley 1956 Vertical migration of spherical and aspherical pollen in a Sphagnum bog. —Proc. Minn. acad. sci. 24: 29.
Rowley, J. R., and D. Southworth 1967 Deposition of sporopollenin on lamellae of unit membrane dimension.—Nature 213: 703–704.
Rowley, J. R., and K. M. Walch 1972 Recovery of introduced pollen from a mountain glacier stream.—Grana 12: 146–152.
Rowley, J. R., A. O. Dahl, and J. S. Rowley, 1981 Substructure in exines of Artemisia vulgaris (Asteraceae).—Rev. palaeobot. palynol. 35: 1–38.
Rudolph, K., and F. Firbas 1927 Die Moore des Riesengebierges.—Beih. Bot. Centralbl. 43 II: 69–144.
Rybníčková, E., and K. Rybníček 1971 The determination and elimination of local elements in pollen spectra from different sediments.—Rev. palaeobot. palynol. 11: 165–176.
Rybníčková, E., and K. Rybníček 1972 Erste Ergebnisse paläogeobotanischer Untersuchungen des Moores bei Vracov, Südmähren.—Fol. geobot. phytotax. 7: 285–308.
Rybníčková, E., and K. Rybníček 1985 Paleogeobotanical evaluation of the Holocene profile from the Režbinec Fish-pond.—Fol. geobot. phytotax. 20: 418–438.
Saad-Liman, S. B., and M. A. Nabli 1984 Ultrastructure of the exine of Borago officinalis (Boraginaceae).—Grana 27: 1–10.
Samuelsson, G. 1915 Über den Rückgang der Haselgrenze und anderer pflanzengeographischer Grenzlinien in Skandinavien.—Bull. geol. inst. Uppsala 13: 93–114.
Sandegren, R. 1916 Hornborgasjön.—Sver. geol. unders. Ser. Ca 14. 94 pp.
Sangster, A. G., and H. M. Dale 1961 A preliminary study of differential pollen grain preservation.—Can. j. bot. 39: 35–43.
Sangster, A. G., and H. M. Dale 1964 Pollen grain preservation of under-represented species in fossil spectra.—Can. j. bot. 42: 437–449.
Sarvas, R. 1955 Ein Beitrag zur Fernverbreitung des Blütenstaubes einiger Waldbäume.—Z. Forstgenet. Forstpflanzenzücht. 4: 137–142.
Scamoni, A. 1949 Beobachtungen über den Pollenflug der Kiefer und Fichte.—Fortswissensch. Zentralb. 68: 735–751.
Scamoni, A. 1955 Über den gegenwärtigen Stand unseres Wissens vom Pollenflug der Waldbäume.—Z. Forstgenet. Forstpflanzenzücht. 4: 145–149.
Schnetter, R. 1972 Estudio de la sedimentacion actual del polen en Santa Marta, Colombia. S.A.—Caldasia 11: 93–98.
Schoch-Bodner, H. 1940 The influence of nutrition upon pollen grain in Lythrum salicaria.—J. genet. 40: 393–402.
Scott, H. G., and C. J. Stojanovich 1963 Digestion of juniper pollen by collembola.—Florida entomol. 46: 189–191. (Quoted from Biol. Abstr. 45: 39875).
Scott, R. K. 1970 The effect of weather on the concentration of pollen within sugar beet crops.—An. appl. biol. 66: 119–127.
Seiwald, A. 1980 Beiträge zur Vorgeschichte Tirols IV: Natzer Plateau—Villanderer Alm.—Ber. nat. med. Ver. Innsbruck 67: 31–72.
Semerikov, L. F., and N. V. Glotov 1971 Evaluation of the isolation in populations of durmast oak (Quercus petraea Liebl.).—Genetika 7: 65–71 (Quoted from Biol. abstr. 52: 88138).
Shapiro, J. 1958 The core-freezer—a new sampler for lake sediments.—Ecology 39: 758.
Shellhorn, S. H., H. H. Hull, and P. S. Martin 1964 Detection of fresh and fossil pollen with fluorochrome. —Nature 202: 315–316.
Simmons, I. G. 1975 The ecological setting of Mesolithic man in the Highland zone.—Pp. 57–63 in Evans *et al.*, 1975.
Simonsen, S. 1980 Vertikale variasjoner i holocen pollensedimentasjon i Ulvik, Hardanger.—A. m. S. Varia 8. 68 pp.
Sitte, P. 1960 Die optische Anisotropie von Sporodermen.—Grana palynol. 2, 2: 16–40.
Skre, O. 1982 The amounts and properties of transported air at two Norwegian stations as functions of wind direction and weather type.—Grana 20: 169–178.
Skvarla, J. J., and D. E. Anderegg 1972 Infestation of cedar pollen by Rhizopidium (Chytridiomyces).—Grana 12: 47–51.
Skvarla, J. J., and D. A. Larson 1966 An electron microscope study of pollen morphology in the Compositae with special reference to the Ambrosiineae.—Grana palynol. 6: 210–269.
Skvarla, J. J., and C. C. Pyle 1968 Techniques of pollen

and spore electron microscopy. II. Ultramicrotomy and associated techniques.—Grana palynol. 8: 255–270.

Slade, D. H. 1968 Meteorology and atomic energy 1968. —U.S. Atomic energy comm. Div. techn. inf. 445 pp.

Smirnov, N. N. 1964 On the quantity of allochthonous pollen and spores received by the Rybinsk reservoir. —Hydrobiologia 24: 421–429.

Smit, E. M., and Janssen, C. R. 1983 Late Holocene vegetational diversity of the Dommel River, the Netherlands.—Ber. Rijksd. Oudheidk. bodemonderz. 33: 95–106.

Smith, A. G. 1970 The influence of Mesolithic and Neolithic man on British vegetation.—Pp. 81–96 in Walker, D. and West, 1970.

Smith, A. G., and J. R. Pilcher 1973 Radiocarbon dates and vegetational history of the British Isles.—New phytol. 72: 903–914.

Smith, A. J. 1959 Description of the Mackereth portable core sampler.—J. sediment. petrol. 29: 246–250.

Sorsa, P. 1964 Studies on the spore morphology of Fennoscandian fern species.—Ann. bot. fenn. 1: 179–201.

Southworth, D. 1983 Exine development in Gerbera jamiesonii (Asteraceae: Mutisiae).—Am. j. bot. 70: 1038–1047.

Spieksma, F. T. M., and J. F. den Tonkelaar 1986 Four-hourly fluctuations in grass pollen concentrations in relation to wet versus dry weather, and to short versus long over-land advection.—Int. j. biometerol. 30: 351–358.

Stanley, E. A. 1965 Use of reworked pollen and spores for determining Pleistocene-Recent and intra-Pleistocene boundaries.—Nature 206: 289–291.

Stanley, E. A. 1966 The problem of reworked pollen and spores in marine sediments.—Mar. geol. 4: 397–408.

Stanley, E. A. 1967 Palynology of six oceanbottom cores from the southwestern Atlantic ocean.—Rev. palaeobot. palynol. 2: 195–203.

Stanley, E. A. 1969 Marine palynology.—Oceanogr. mar. bio. ann. rev.: 277–292.

Steinberg, K. 1944 Zur spät- und nacheiszeitlichen Vegetationsgeschichte des Undereichsfeldes.—Hercynia 3: 529–587.

Stephen, J. C., and J. R. Quinsby 1934 Anthesis, pollination and fertilisation in Sorghum.—J. agric. res. 49: 123–136.

Steusloff, U. 1905 Torf- und Wiesenkalk-Ablagerungen im Rederang- und Moorseel Becken.—(Diss. Rostock) Arch. Ver. Fr. Naturgesch. Mecklenburg 59. 64 pp.

Stevenson, A. L. 1968 A new technique for obtaining uniform volume sediment samples for pollen analysis. —Pollen spores 10: 463–463.

Stix, E., and G. Grosse-Brauckmann 1970 Der Pollen- und Sporengehalt der Luft und seine tages- und jahreszeitlichen Schwankungen unter mitteleuropäischen Verhältnissen.—Flora 159: 1–37.

Stockmarr, J. 1972 Tablets with spores used in absolute pollen analysis.—Pollen spores 13: 615–621.

Straka, H. 1985 L'histoire tardi-et postglaciaire de la végétation de l'Eifel volcanique (Allemagne de l'ouest) —Ecol. méditerr 11: 99–105, cf. 201.

Straka, H. 1986 English terminology for the figures and lettering in the pollen formulas in Palynologia madagassica et mascarenica.—Pp. 147–158 in Lienau, K., H. Straka, and B. Friedrich (eds) 1986. Palynologia madagassica et mascarenica, families 167 to 181. Tropische und subtropische Pflanzenwelt 55. Ak. Wiss. Lit. Mainz.

Strasburger, E. 1928 Das botanische Praktikum. 7. Aufl. N.M. Koernicke.—Fischer (Jena) 883 pp.

Strøm, K. M. 1934 A new sampler lead.—Nor. geol. tidsskr. 14: 162–166.

Sundelin, U. 1917 Fornsjöstudier inom Stångåns och Svartåns vattenområden.—Sver. geol. unders. Ser. Ca 16. 290 pp.

Swain, A. M. 1980 Landscape pattern and forest history in the Boundary Waters Canoe area, Minnesota: a pollen study from Hug Lake.—Ecology 61: 747–754,

Szafer, W. 1935a The significance of isopollen lines for the investigation of geographical distribution of trees in the post-glacial period.—Bull. Internat. acad. polon. scient. et lettr. Cl. sci. math. nat. Sér. B 1. Bot. 1935: 235–239.

Szafer, W. 1935b The method of isopollens applied to the investigation of the history of trees by means of pollen analysis.—Zesd. int. bot. congr. Amsterdam Proc. vol. II: 100–102.

Tatzreiter, S. 1985 Präparation von Pollen und Sporen für das Rasterelektron-Mikroskop und Lichtmikroskop unter Verwendung von Rhodaniden.—Grana 24: 33–44.

Tauber, H. 1965 Differential pollen dispersion and the interpretation of pollen diagrams.—Dan. geol. unders. 2 rk. 89. 69 pp.

Tauber, H. 1967 Investigations of the mode of transfer of pollen in forested areas.—Rev. palaeobot. palynol. 3: 277–286.

Tauber, H. 1974 A static non-overload pollen collector. —New phytol. 73: 359–369.

Tauber, H. 1977 Investigations of aerial pollen transport in a forested area.—Dan. bot. ark. 32. 1. 121 pp.

Terasmäe, J. 1951 On the pollen morphology of Betula nana.—Sven. bot. tidskr. 45: 358–361.

Terasmäe, J. R., and R. J. Mott 1964 Pollen deposition in lakes and bogs near Ottawa, Canada.—Can. j. bot. 42: 1355–1363.

Thanikaimoni, G. 1968 Pollen morphological terms. Proposed definitions. I.–IV int. palynol. conf. Lucknow (1966–67) 1: 228–239.

Thanikaimoni, G. 1970 Les palmiers: palynologie et systematique.—Inst. franc. Pondichery. Tra. sect.

scient. techni. II. (1970) 286 pp.
Thanikaimoni, G. 1972 Index bibliographique sur la morphologie des pollens d'angiospermes.—Inst. fr. Pondichéry. Trav. sect. sc. tec. XII, 1. 337 pp./Supplement. 2. 163 pp.
Thomas, A. L., and B. Lugardon 1972 Sur la structure fine des tetrades de deux Annonacées (Asteranthe asterias et Hexalobus monopetalus).—C. r. hebd. acad. sci. Paris 275 D: 1749-1752.
Thomas, K. W. 1964 A new design for a peat sampler. —New phytol. 63: 422.
Thompson, R. S. 1986 Palynology and Neotoma middens.—Pp. 89–112 in Jacobs *et al.*, 1986.
Thorarinsson, S. 1944 Tefrokronologiska studier på Island.—Geogr. ann. 1944: 1-398.
Ting, W. S. 1966 Determination of Pinus pollen species by pollen statistics.—Univ. Calif. publ. geol. sc. 58. 168 pp.
Tolonen, K. 1967 Soiden Kehityshistorian tutkimusmentelmistä II. Turvekairoista.—Suo 1967, 6. (Reprint?) 7 pp.
Tolonen, K. 1986 Charred particle analysis.—Pp. 485–495 in Berglund, 1986.
Traverse, A. 1988 Palaeopalynology.—Unwin Hyman (Boston, etc.). 600 pp.
Traverse, A., and R. H. Ginsburg 1966 Pollen and associated microfossils in marine surface sediments of the Great Bahama Bank.—Rev. palaeobot. palynol. 3: 243–254.
Treibs, A. 1934 Chlorophyll und Häminderivate in bituminösen Gesteinen, Erdölen, Erdwachsen und Asphalten.—Leibigs Ann. Chem. 510: 42.
Troels-Smith, J. 1954 Ertebøllekultur-Bondekultur. Resultater av de siste 10 aars undersøgelser i Aamosen. —Aarb. nord. oldk. hist. 1953: 1–62.
Troels-Smith, J. 1955 Karakterisering af jordarter.—Dan. geol. unders. 4 rk. 3, 10 pp.
Troels-Smith, J. 1956 Neolithic period in Switzerland and Denmark.—Science 124: 876–881.
Troll, W. 1928 Über Antherenbau, Pollen und Pollination von Galanthus.—Flora N.F. 23: 321.
Tschudy, R. H., and R. A. Scott 1969 Aspects of palynology.—Wiley-Interscience (New York, etc.). 510 pp.
Tsukada, M. 1958 Untersuchungen über das Verhältnis zwischen dem Pollen-gehalt der Oberflächenproben und der Vegetation des Hochlandes Shiga.—J. inst. polytech. Osaka City Univ. Ser. D 9: 217–234.
Tsukada, M., S. Sugita, and Y. Tsukada 1986 Oldest primitive agriculture and vegetational environment in Japan.—Nature 322: 632–634.
Tutin, W. 1969 The usefulness of pollen analysis in interpretation of stratigraphic horizons, both Late-glacial and Post-glacial.—Mitt. Int. Ver. Limnol. 17: 154–164.
Tyldesley, J. B. 1973 Long-range transmission of tree pollen to Shetland. I–III.—New phytol. 72: 175–190, 691–697.
Ueno, J. 1960 Studies in pollen grains of gymnospermae. —J. inst. polytechn. Osaka city univ. Ser. D. II: 109–136.
Ueno, J. 1973–74 The fine structure of pollen surface IV, V Gymnospermae (A), (B)—Rep. fac. sci. Shizuoka univ 8: 101–115, 9: 79–94.
Usinger, H. 1980 Une relation entre le taille des pollens et le climat.—Mém. mus. nat. hist. nat. N.s. 27: 51–55.
van Campo, M. 1961 Méchanique aperturale.—Grana palynol. 2, 2: 93–97.
van Campo, M. 1971 Précisions nouvelles sur les structures comparées des pollens de gymnospermes et angiospermes.—C.r. hebd. acad. sci. Paris 272: 2071–2074.
van Campo, M. 1975 Pollen analyses in the Sahara.—Pp. 46–64 in Wendorf. F. and P. Martin (eds) Problems in prehistory. North Africa and the Levant. Southern Methodist University Press (Dallas).
van Campo, M. 1978 La face interne de l'exine.—Rev. palaeobot. palynol. 26: 301–311.
van Campo, M., and P. Guinet 1961 Les pollens composés. L'exemple des mimosacées.—Pollen spores 3: 201–218.
van Campo, M., and N. Hallé 1959 Palynologie africaine III. Les grains de pollen des Hippocratéacées d'Afrique de l'Ouest.—Bull. inst. fr. d'Afr. noire 21 Sér. A 3: 807–899.
van Campo, M., and B. Lugardon 1973 Structure grenue infratectale de l'ectexine des pollens de quelques Gymnospermes et Angiospermes.—Pollen spores 15: 171–188.
van Campo, M., and J. Sivak 1972 Structure alvéolaire de l'extexine des pollens à ballonets des Abietacées. —Pollen spores 14: 115–141.
van Campo-Duplan, M. 1950 Recherches sur la phylogénie des Abietinés d'après leurs grains du pollen. —Trav. lab. for. Toulouse 2, 4: 1–183.
van Geel, B. 1976 A paleoecological study of Holocene peat bog sections, based on the analysis of pollen, spores and macro and microscopic remains of fungi, algae, cormophytes and animals.—Thesis Amsterdam. 75 pp.
van Geel, B., and T. van der Hammen 1978 Zygnemataceae in Quaternary Columbian sediments.—Rev. palaeobot. palynol. 25: 377–392.
van Geel, B., S. J. P. Bohnke, and H. Dee 1981 A palaeoecological study of an Upper Glacial and Holocene sequence from "De Borcher", the Netherlands. —Rev. palaeobot. palynol. 31: 367–448.
van Geel, B., D. P. Hallewas, and J. P. Pals 1983 A late Holocene deposit under the Westfriese zeedijk near Enkhuizen (Prov. of Nord-Holland, the Netherlands): palaeoecological and archaeological aspects.—Rev. palaeobot. palynol. 38: 299–335.
van Gijzel, P. 1967 Palynology and fluorescence microscopy.—Rev. palaeobot. palynol. 2. 49–79.
Vareschi, V. 1942 Die pollenanalytische Untersuchung

der Gletscherbewegung.—Veröff. geobot. Inst. Rübel Zurich 19. 144 pp.
Vasanthy, G., and S. A. J. Pocock 1986 Radial through rotate symmetry of striate pollen of the Acanthaceae. —Can. journ. bot. 64: 3050–3058.
Vicari, H. 1936 Untersuchungen über die Membranen rezenter und fossiler Sporen und Pollen.—Thesis. Bern. 51 pp.
von Post, L. 1916 Om skogsträdpollen i sydsvenska torfmosslagerföljder.—Geol. fören. Stockh. förh. 38: 384–390.
von Post, L. 1918 Skogsträdpollen i sydsvenska torvmosslagerföljder.—Forh. 16. skand. naturforskermøte 1916: 433–465. (Transl. by M. B. Davis and K. Fægri in Pollen et spores 9: 376–401.)
von Post, L. 1924 Ur de sydsvenska skogernas regionala historia under post-artisk tid.—Geol. fören. Stockh. förh. 46: 83–128.
von Post, L. 1929 Die Zeichenschrift der Pollenstatistik. —Geol. fören. Stockh. förh. 41: 543–565.
von Post, L. 1947 [Interventions]—In Geologklubben 1947.
von Post, L. 1950 [Review]—Geol. fören. Stockh. förh. 72: 363–364.
von Post, L., and E. Granlund 1926 Södra Sveriges torvtillgångar. I.—Sver. geol. unders. Ser. C 335. 127 pp.
von Post, L., and R. Sernander 1910 Pflanzenphysiognomische Studien auf Torfmooren in Närke.—Geologkongr. Guide excursion A 7, 48 pp.
Vorren, K-D. 1979 Anthropogenic influence on the natural vegetation in coastal North Norway during the Holocene.—Nord. arkeol. rev. 12: 1–21.
Vuorela, I. 1973 Relative pollen rain around cultivated fields.—Acta bot. fenn. 102. 27 pp.
Vuorela, I. 1985 The peat knife, an improved tool for collecting peat samples for pollen analysis.—Mem. soc. faun. flor. fenn. 61: 103–105.
Wagenitz, G. 1955 Über die Änderung der Pollengrösse von Getreiden durch verschiedene Ernährungsbedingungen. Ber. dtsch. Bot. Ges. 58: 297–302.
Waha, M. 1984 Zur Ultrastrucktur und Funktion pollenverbindender Fäden bei Ericaceae und anderen Angiospermenfamilien.--Plant syst. evol. 147: 189–203.
Walch, K. H., J. R. Rowley, and N. J. Morton 1970 Displacement of pollen grains by earthworms.—Pollen spores 12: 39–44.
Walker, D. 1970 Direction and rate in some British post-glacial hydroseres.—Pp. 117–139 in Walker and West, 1970.
Walker, D., and P. M. Walker 1961 Stratigraphic evidence of regeneration in some Irish bogs.—J. ecol. 49: 169–185.
Walker, D., and R. G. West (eds) 1970 Studies in the vegetational history of the British Isles.—Cambridge University Press. 266 pp.
Walker, J. W. 1974 Evolution of exine structure in the pollen of primitive angiosperms.—Am. j. bot. 61: 891–902.
Walker, J. W. 1976 Evolutionary significance of the exine in the pollen of primitive angiosperms.—Pp. 207–308 in Ferguson, J. K., and J. Muller (eds) The evolutionary significance of the exine.—Linn. soc. sympos. ser. 1.
Walker, J. W., and J. A. Doyle 1975 The bases of angiosperm phylogeny: palynology.—Ann. Miss. bot. gdn. 62: 664–723.
Walker, J. W., and J. J. Skvarla 1975 Primitive columellaless pollen: a new concept in the evolutionary morphology of Angiosperms.—Science 187: 445–447.
Webb, T. III 1985 Holocene palynology and climate.—Pp. 163–195 in Hecht, 1985.
Webb, T. III 1986 Is vegetation in equilibrium with climate? How to interpret late-Quaternary pollen data. —Vegetatio 67: 75–91.
Webb, T. III and R. A. Bryson 1972 Late- and post-Glacial climatic change in the northern Midwest, USA: quantitative estimates derived from fossil pollen spectra by multivariate statistical analysis.—Quatern. res. 2: 70–115.
Weber, C. A. 1893 Über die diluviale Flora von Fahrenbrug in Holstein.—Bot. Jahrb. 18, Beibl. 43: 1–13.
Weber, H. A. 1918 Über spät- und postglaziale lakustrine und fluviatile Ablagerungen in der Wyhraniederung bei Lobstadt und Borna und die Chronologie der Postglazialzeit Mitteleuropas.—Abh. nat. Ver. Bremen 19: 187–267.
Welinder, S. 1985 Comments on early agriculture in Scandinavia.—No. Arch. rev. 18: 94–96.
Welten, M. 1944 Pollenanalytische und stratigraphische Untersuchungen in der prähistorische Höhle des 'Chilchli' im Simmental.—Ber. geobot. Inst. Rübel Zürich 1943: 90–100.
Welten, M. 1952 Über die spät- und postglaziale Vegetationsgeschichte des Simmentals sowie die frühgeschichtliche und historische Wald- und Weiderodung auf Grund pollenanalytischer Untersuchungen.—Veröff. geobot. Inst. Rübel Zürich 26. 135 pp.
Welten, M. 1957 Über das glaziale und spätzglaziale Vorkommen von Ephedra am nordwestlichen Alpenrand.—Ber. d. schweiz. Bot. Ges. 67: 33–54.
Welten, M. 1958 Pollenanalytische Untersuchung alpiner Bodenprofile: historische Entwicklung des Bodens und säkulare Sukzession der örtlichen Pflanzengesellschaften.—Veröff. geobot. Inst. Rübel Zürich 33: 253–274.
Wendelbo, P. 1961 Studies in Primulaceae III. On the genera related to Primula with special reference to their pollen morphology.—Univ. i Bergen Årb. Mat-naturv. rk. 1961, 19. 31 pp.
Westenberg, J. 1947 Mathematics of pollen diagrams.

I–II.—Proc. k. ned. akad. wet. 50, 5–6. 22 pp.
Westenberg, J. 1964 Nomograms for testing significance in a 2 × 2 contingency table for sample size of 10 to 500.—Proc. k. ned. akad. wet. Ser. A 67: 441–466.
Westenberg, J. 1967 Testing significance of difference in a pair of relative frequencies in pollen analysis.—Rev. palaeobot. palynol. 3: 395–369.
Whitehead, D. R., and M. C. Sheehan 1971 Measurement as a means of identifying fossil maize pollen. II. The effect of slide thickness.—Bull. Torrey bot. club 98: 268–272.
Więçkowski, S., and K. Szszepanek 1963 Assimilatory pigments from subfossil fir needles (Abies alba Mill.). —Acta soc. bot. polon. 32: 101–
Wille, N. 1879 Ferskvandsalger fra Novaja Semlja samlede av Dr. F. Kjellman på Nordenskiölds expedition 1875.—Öfvers. kgl. vetensk.-akad. förhandl. 36, 5: 13–74.
Wilson, I. T. and J. E. Potzger 1943 Pollen records from lakes in Anoka County Minnesota: A study on methods of sampling.—Ecology 24: 382–392.
Witte, H. 1905 Stratiotes aloides L. funnen i Sveriges postglaciala aflagringar.—Geol. fören. Stockh. förh. 27: 432–451.
Witte, H. J. L., and B. van Geel 1985 Vegetational and environmental succession and net production between 4500 and 800 B.P. reconstructed from a peat deposit in the western Dutch coastal area (Assendelver Polder). —Rev. palaeobot. palynol. 45: 239–300.
Wodehouse, R. P. 1928 The phylogenetic value of pollen-grain characters.—Bull. Torrey bot. club 42: 891–934.
Wodehouse, R. P. 1935 Pollen grains.—McGraw-Hill (New York). 574 pp.
Wodehouse, R. P. 1945 Hayfever plants.—Chronica botanica (Waltham, Mass.) N.S. 15. 245 pp.
Woodhead, N., and L. M. Hodgson 1935 A preliminary study of some Snowdonian peats.—New phytol. 43: 263–282.
Wright, H. E. Jr, D. A. Livingstone, and E. J. Cushing 1965 Coring devices for lake sediments.—Pp. 494–520 in Kummel, B. and D. Raup 1965 Handbook of paleontological techniques. Freeman (San Francisco, etc.).
Yamanoi, T. 1969 Studies on the grain size of recent and fossil pollens with special reference to their chemical treatment.—Jap. j. palynol. 4: 11–18.
Zaklinskaya, E.D. Заклинская, Е.Д., С.Н. Наумов, А.Н. Сладков 1960 Таксономия и номенклат ископаемых пыльцы и спор. – Междчнародн. геол. конгр. XXIсесс. Докл . советских геол.: 167.
Zander, E. 1935–49 Blütengestaltung und Herkunftbestimmung bei Blütenhonig. I–IV.—Berlin, Leipzig, München. (Various publishers.)
Zetsche, F. 1929 Die chemischen Grundlagen der Pollenanalyse.—Mitt. naturforsch. Ges. Bern. 1928:
Zoller, H. 1960a Pollenanalytische Untersuchungen zur Vegetationsgeschichte der insubrischen Schweiz.—Denkschr. Schweiz. naturforsch. Ges. 83: 46–156.
Zoller, H. 1964 Zur postglazialen Ausbreitungsgeschichte der Weisstanne (Abies alba Mill.) in der Schweiz.—Schw. Zeitschr. Forstwesen 1964: 681–700.
Züllig, H. 1956a Sedimente als Ausdrücke des Zustandes eines Gewässers.—Schweiz. Z. Hydrol. 18: 7–142.
Züllig, H. 1956b Das kombinerte Ramm-Kolben-Lot, ein leichtes Bohrgerät zur vereinfachten Gewinnung mehrerer meterlanger, ungestörter Sedimentprofile. —Schweiz. Z. Hydrol. 18: 208–124.

Index to identification keys

The number in front of the colon sign indicates the table, those after the sign indicate the places in the table. Numbers in parentheses refer to incidental mentioning. The abbreviations Cy, Gr, Pl and Ro refer to the special indexes for Cyperaceae, Gramineae, Plantaginaceae and Rosaceae.

Index of plant names

Plant names in illustrations and captions are not always included. Names occurring in the identification keys are found in the 'Index to identification keys' which starts on p. 315

General index

Including index of chapter sections (§)